Equilibria and Kinetics of Biological Macromolecules

Equilibria and Kinetics of Biological Macromolecules

Jan Hermans
Barry Lentz

WILEY

Published by John Wiley & Sons, Inc., Hoboken, New Jersey.
Published simultaneously in Canada.

Library of Congress Cataloging-in-Publication Data:

Hermans, Jan.
Equilibria and kinetics of biological macromolecules / by Jan Hermans and Barry Lentz.
p. ; cm.
Includes index.
ISBN 978-1-118-47970-4 (cloth)
I. Lentz, Barry. II. Title.
[DNLM: 1. Biocompatible Materials–pharmacokinetics. 2. Biophysical Processes. 3. Macromolecular Substances–pharmacokinetics. 4. Molecular Conformation. QT 37]
RM301.5
615.7–dc23

2013013996

10 9 8 7 6 5 4 3 2 1

For Harold Scheraga

Contents

Preface

It is only by attempting to explain our science to each other that we find out what we really know.
—J. M. Ziman, Nature 252: 318–324 (1969)

This book has grown out of circa 12 years of collaborative teaching of a 6-credit biophysics course that forms the core of the didactic teaching for the Molecular and Cellular Biophysics Program at UNC CH. Thus the book is directed at an audience of first year graduate students. However, the book has grown well beyond the content of those courses, also thanks to input and suggestions from colleagues who have shared our teaching the course (see Acknowledgments), and it is our hope that it will prove useful to working biochemists who seek a deeper understanding of modern biophysics.

The book is not meant to be a complete text in biophysics, as it focuses on the input of physics and physical chemistry to experimental studies and theoretical models of equilibria and kinetics of biological macromolecules (largely, proteins). A chapter is devoted to methods of molecular simulations; applications of molecular dynamics are included in several chapters. On the other hand, we limited the size of this book by devoting no space to spectroscopy and structure determination.

The book assumes some knowledge of physics and/or physical chemistry, but in Part 1, the chapters on thermodynamics, simple quantum mechanics and molecular structure and intra- and intermolecular forces shore up what may be shaky backgrounds of some students, and provide references for later chapters. Part 1 concludes with a chapter on water and the hydrophobic effect.

Two chapters in Part 2 introduce various ensembles of statistical mechanics, and these are followed by the aforementioned chapter on molecular simulations.

Next, in Part 3, we discuss equilibria of binding of "ligands" to macromolecules from different standpoints: chemical equilibrium theory, thermodynamics, and statistical mechanics. These are followed by a discussion of linked equilibria, and a chapter that focuses on hemoglobin as an example of allosteric control of function.

Part 3 concludes with a chapter on charge–charge interactions of macromolecules in solution.

In Part 4, we deal with folding equilibria. A brief overview of the physics of polymer solutions is followed by a chapter on the theory of helix-coil transitions of polypeptides and its many applications, and it ends with a section on helix-coil equilibria of double-stranded nucleic acids. This is followed by a long chapter on equilibria of protein folding. Part 4 concludes with a chapter on elasticity with elastin and tenascin as examples of two different mechanisms.

The final part of the book is devoted to kinetics. The first chapter describes kinetic measurement methods and a variety of kinetic models, ranging from simple rate equations to transition state theory. This is followed by a chapter on experiments and theory of kinetics of protein folding. Part 5 concludes with a chapter on stochastic processes and theories from the Langevin equation to Kramers' theory of reaction rates.

Finally, in a series of Appendices we have covered technical (mostly mathematical) details which we had skipped earlier to make the main content of this book easier to follow.

The authors will maintain a web page devoted to corrections and discussion of this book. Please consult the authors' personal web pages at the University of North Carolina.

Acknowledgments

This book's inception was in the form of lecture notes for the introductory class in molecular biophysics given at UNC each fall semester. An enormous help has been the feedback we received from students taking the class.

We have received help from many colleagues. We are grateful to professors Papoian (now at the University of Maryland) and Dokholyan, who have each taught part of the course, for letting us base important sections of the book on new presentations given by them in their lectures. Individual chapters have had input from Gary Ackers at Washington University, from Gary Pielak in the UNC Chemistry Department, from Gary Felsenfeld at the NIH, from Andy McCammon at UCSD, from Weitao Yang at Duke, from Austen Riggs at the University of Texas, from Robert Baldwin at Stanford and from Hao Hu at the University of Hong Kong.

We thank Dr. M. Hanrath, University of Cologne for the computer drawings of hydrogen atom wave functions shown in Chapter 2, and Dr. Chad Petit for microcalorimeter results discussed in Chapter 8. Some figures of molecular structures were prepared with the vmd graphics program.* We acknowledge many answers to questions involving basic Physics, found by consulting Wikipedia.

JH and BRL

September 2012

*Humphrey, W., Dalke, A., Schulten, K. VMD - visual molecular dynamics. *J. Mol. Graphics Modell.* **14**: 33–38 (1996).

Part 1

Basic Principles

In our treatment of the biophysics of macromolecules, we must assume some knowledge of basic physics and physical chemistry. Part 1 is a compendium ("review" if you like) of those aspects of these subjects that the reader will be expected to have mastered. This is not meant to take the place of a thorough textbook dealing with these topics (some recommended texts are listed), but rather as a summary of key concepts and information.

We start with thermodynamics, which is unique as it expressly assumes no models of molecular structure and intermolecular interactions, while this is otherwise the case for all other topics treated in this book. As thermodynamics remains an essential tool of modern molecular physics, one simply must know thermodynamics, so we start with it.

We then attend to three basic motions of massive particles such as nuclei or even whole molecules: free translation, free rotation, and movement in a potential. For simplicity, we limit ourselves to the quantum mechanical treatment of each, although the reader will surely recall classical treatments that occupied parts of basic physics courses, which describe the high temperature limits of these motions. The motion of electrons, however, can be described only via quantum mechanics. So, the fourth essential model we review is that of the Hydrogen Atom. This masterpiece of late nineteenth and early twentieth century physics provides the basis for all of chemistry and for what we think we know about molecules and their interactions, which is the topic of the third chapter of this section. After reviewing

Equilibria and Kinetics of Biological Macromolecules, First Edition. Jan Hermans and Barry Lentz.

how quantum methods are used to calculate molecular structure and energies, and the difficulties of doing these calculations on the grand scale required by studies of macromolecules, we introduce the approximation of molecular mechanics.

Finally, in Chapter 4 this basic material is applied to the structure of liquid water and the thermodynamics of the hydrophobic effect.

1

Thermodynamics

One does not understand thermodynamics, one can only know it
—Jan Hermans

As a biophysicist, you must know thermodynamics
—Barry Lentz

1.1 INTRODUCTION

Thermodynamics describes the relation between different forms of energy, their interconversion, and the exchange of energy between physical systems. Thermodynamics is applicable to energy management in all situations. It was developed in the context of the industrial revolution, with an important goal being the design of more efficient versions of newly invented machines, first the steam engine, later such devices as the internal combustion engine and the refrigerator. Thermodynamics also describes how the *total energy* of a system is partitioned between *useful energy* (available to do work) and *wasted energy* (that associated with the randomness of a system), and establishes conditions that must be met for a system to not undergo spontaneous change, that is, to be at equilibrium. The branch of thermodynamics that concerns us most deals with the energetics of chemical systems and systems containing interacting molecules. However, thermodynamics does not formally assume a molecular nature of matter, but is simply a formal description of the relationship between work, heat, and energy. Three laws, which are based on "everyday" observations, form the foundation of thermodynamics. The surprisingly profound conclusions that follow from these laws have been verified extensively.

Thermodynamics strikes many as a boring formalism, seemingly devoid of the interesting intellectual content of quantum and statistical mechanics. Indeed, one

Equilibria and Kinetics of Biological Macromolecules, First Edition. Jan Hermans and Barry Lentz.

can think of thermodynamics as a bookkeeping tool that tracks otherwise obscure relations between different forms of energy storage, and in doing so keeps the biophysicist from many an egregious error. At the same time, the very fact that a complex framework of relations can be built on a few fundamental laws should be a source of marvel, as is the insight of the scientists who developed thermodynamics in the nineteenth century. The development of thermodynamics on the basis of a few laws resembles the development of mathematics from a small number of axioms. However, the axioms of mathematics can be chosen by the mathematician, while the laws of thermodynamics are based on observations of our physical world, and these laws could be changed only on the basis of radically new experimental findings.

This chapter is not a textbook on thermodynamics; it is presumed that students using this book have had an introductory physical chemistry course that treated chemical thermodynamics in some detail. It is also presumed that many who have had such a course do not remember it very well. Thus, our goal is to review briefly the fundamental *concepts* of thermodynamics and then to give them a context in terms of solutions of macromolecules and their interactions with other molecules.

1.2 THE FUNDAMENTAL POSTULATES OR LAWS OF THERMODYNAMICS

1.2.1 Systems

A system is a part of the universe in which we have interest for a particular problem. In biology, it is often some collection of molecules. It is separated by some boundary from the rest of the universe (its *surroundings*; Fig. 1.1).

Open systems exchange energy and matter with their surroundings.

Closed systems exchange energy but not matter with their surroundings.

Isolated systems exchange neither energy nor matter with their surroundings.

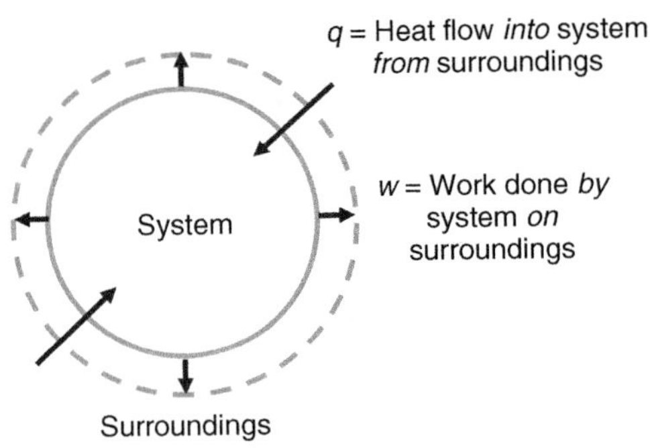

FIGURE 1.1 A *closed system* exchanges energy in the form of heat and work but not matter with its surroundings. If no heat is exchanged ($q = 0$), the process is *adiabatic*. An open system can also exchange matter with the surroundings.

1.2.2 States and State Functions

The *state* of a closed system can be changed by the exchange of energy with the outside (surroundings), and can also change spontaneously. *Thermodynamics is concerned with the* equilibrium *states that are the outcome of spontaneous change*, and with the processes *by which change from one equilibrium state to another occurs*. Many equilibrium states are *metastable*; for example, a mixture of oxygen and hydrogen gases is stable, but can be ignited to explode (spontaneous change) and form water vapor. A state is defined in terms of characteristic properties, such as temperature, density, pressure, and chemical composition. The energy of a state is one of its fundamental characteristics, and is therefore called a *state function*. By definition, a state function depends on certain properties of a system such as the number of molecules composing it (N), the volume (V), perhaps the pressure on the system, and a very interesting property called temperature (T).

Observation tells us that not all these properties are independent; that is, if we set the values of some, then others are fixed by these assignments. Aside from the extensive (how big the system is) property N, the thermodynamics of a closed system are defined by two additional properties, which are referred to as *independent* variables of the system. All other properties of the system, including its state functions, are *dependent* properties of the system. There is nothing holy or sacrosanct about an independent variable, these are defined by the experimental conditions we use to observe the system and are those *properties over which we exercise control*. However, once we choose these independent properties or variables, the values of the state functions for the system are defined and can be obtained by the laws of thermodynamics. Thermodynamic state functions depend only on the values of these independent properties and not on how the system reached this state.

1.2.3 The First Law and Forms of Energy—Energy a State Function

Classical mechanics introduces three forms of energy: kinetic energy, potential energy, and work. *Kinetic energy* is evident in an object's motion. The *potential energy* of an object is latent energy that allows the object to do work or to acquire kinetic energy. *Work* has associated with it a *force* and a *path*; force acting along the path changes the potential energy and/or the kinetic energy of an object. *Thermodynamics* considers an additional category of energy, *heat*, and is concerned solely with the relationships between and interconversion of heat, work, and energy. We stress that thermodynamics does not distinguish between kinetic and potential energy, nor does it bother itself with motion—these issues are totally the venue of mechanics. These two independent areas of physics came together only in the latter half of the nineteenth century through the collaboration of the Scottish mathematician James Scott Maxwell (kinetic theory of gases) and the Austrian physicist Ludwig Eduard Boltzmann (the Boltzmann distribution) to develop a statistical description of the average speed of molecules in a gas. This is the Maxwell–Boltzmann distribution, which forms the basis of *statistical mechanics* (also called *statistical thermodynamics*; see Chapter 5).

The First Law of thermodynamics states that the energy of a system and its surroundings is conserved. The "everyday" experience of doing *work* to move a mass up a hill against the force of gravity leads to the concept that the work done is converted into *potential energy*, which remains hidden, until the object is released and rolls back down. In the absence of friction, energy is conserved during the rollback down the hill, and the object acquires a new form of energy, *kinetic energy*. There are many familiar examples of converting energy into work or heat or capturing work as energy. Several are illustrated in Fig. 1.2. The invention of the steam engine stimulated the development of thermodynamics. In it and its modern-day replacement, the internal combustion energy, the energy released in the form of heat produced when hydrocarbons react with oxygen to form CO_2 and H_2O, causes this gas mixture (or water vapor in the steam engine) to expand and this produces pressure–volume work (PV work) on a piston that is captured as the work needed to increase the kinetic energy of a vehicle. Thus, by virtue of the First Law, *heat* also must be considered a form of energy. In this example, a chemical reaction liberates energy in the form of heat. By virtue of the First Law, the *chemical* (or *internal*) *energy* of the reactants must decrease by a like amount. Similarly, a charged battery possesses potential energy that is released when electrons are allowed to flow through a wire to drive an electric motor that

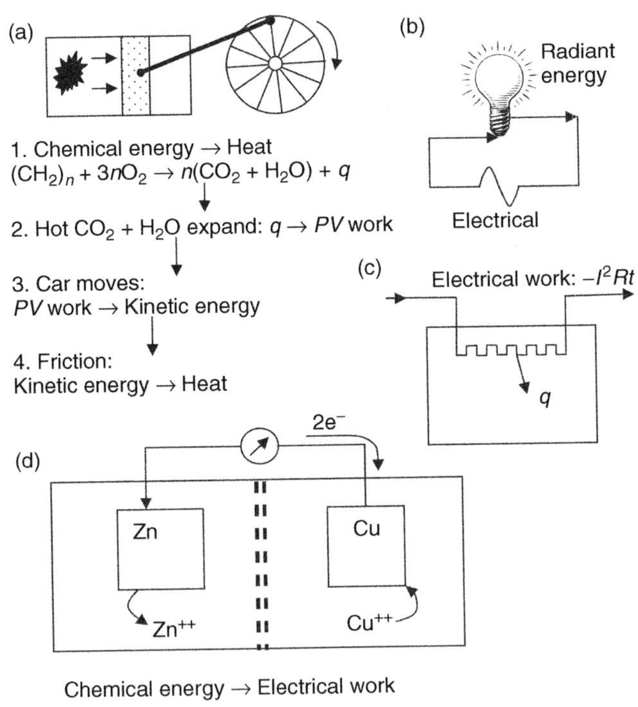

FIGURE 1.2 Examples of interconversion of different forms of energy. (a) Internal combustion engine, (b) light bulb, (c) electric water heater, and (d) flashlight battery.

performs work. This process can also be used to produce heat by running a current I through a resistor R.

Heat has been traditionally defined in terms of the amount needed to change the temperature of 1 g of water by 1 °C, the calorie. In modern usage, heat is treated as energy and expressed in appropriate standard international units of energy. Thus, the calorie is now defined by the equation 1 cal = 4.184 J. Physical scientists preferentially use the standard units, while nutritionists have adhered to the calorie.

To make a more formal definition of the First Law, we note that the total energy of a closed system can be changed by two means: by *work* (work done by the system) or by the transfer of *heat* into the system. The internal energy *decreases* as a result of work w done *by* the system, and *increases* as a result of heat q transferred *into* the system. Thus, the First Law states that the change of the internal energy is[1]

$$\Delta U_{a\rightarrow b} = U_b - U_a = q - w. \tag{1.1}$$

The First Law requires *the internal energy, U to be a state function*, that is, to depend only on the internal state of a system, as determined by its characteristics such as temperature, volume, and composition

$$\begin{aligned} \Delta U_{a\rightarrow b} &= -\Delta U_{b\rightarrow a} \\ q - w &= -(q' - w'). \end{aligned} \tag{1.2}$$

(the primed quantities are for a process that produces state a from state b). Were the First Law not to hold, it would be possible to build a perpetual motion machine ("of the first kind"), a device that indefinitely continues to produce energy, a situation that all our experience tells us is impossible.

We then have for any small change in the system that

$$dU = \partial q - \partial w. \tag{1.3}$$

Work and heat are definitely *not* state functions, as one can raise a system's temperature by transferring into it heat from a bath, but also by performing work, for instance, electrical energy applied to an electrical heating element (Fig. 1.2c), or mechanical energy applied by stirring; by writing ∂q and ∂w (rather than dq and dw), we indicate that q and w are not state functions.

We now understand heat as kinetic and potential energy that is distributed *randomly* over the atoms making up any chemical or physical system, as described in detail in chapters 5 and 6 on Statistical Thermodynamics. Note however, that this insight was unavailable when thermodynamics was first developed as a science.

1.2.4 Temperature and the Ideal Gas or Kelvin Scale

Simple "everyday" experience tells us that two systems in contact through a wall that allows the flow of heat will change until they reach *thermal equilibrium*, and if

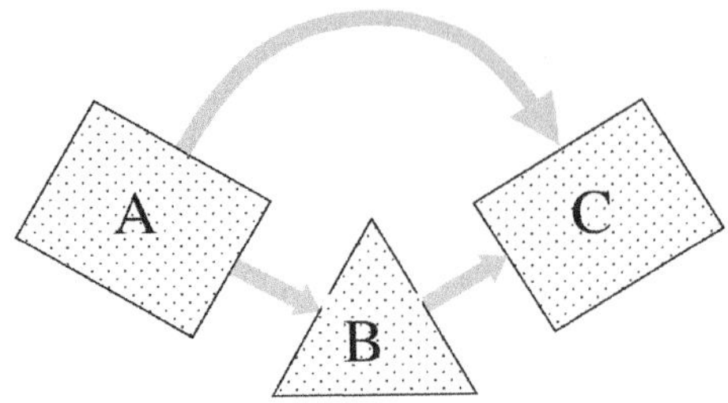

FIGURE 1.3 If heat flows from A to B and from B to C, then (i) heat will also flow from A to C, and (ii) A is said to be hottest and have the highest temperature, and C is coolest and has the lowest temperature. When heat flow ceases, the systems are said to be in thermal equilibrium with each other, are equally "hot" and have the same temperature.

two systems are in thermal equilibrium with a third, they are in thermal equilibrium with each other (Fig. 1.3). We say the two systems have a common property called *temperature*. If two systems are not in thermal equilibrium, they are at different temperatures. Heat flows from high to low temperature, and temperature orders "hotness." (This is sometimes called the *Zeroth Law of Thermodynamics*.) A rise of a system's energy content in the form of heat corresponds to a rise in the temperature.

Early scales of temperature (such as Celsius' scale) depended on two sharply defined experimental points (0 °C as the melting temperature of ice, 100 °C as the boiling temperature of water at 1 atm pressure) and interpolation assuming linear expansion of liquid volume (e.g., mercury).

The Kelvin temperature scale is set by relating temperature to physical properties of an ideal gas, as follows.

Because, in the gaseous state, molecules interact only slightly, the gaseous state is a natural starting point for theories of matter. The *ideal or perfect gas* is a *hypothetical state in which the molecules do not interact at all.* One approximates an ideal gas by diluting a real gas, that is, by increasing its volume and thus lowering its pressure. Thermal motions in an ideal gas consist of internal vibrations within each molecule, and of rotational and translational motions of the gas molecules. As the molecules do not interact (except by rare collisions), the thermal energy does not depend on the volume occupied by a sample.

It is known *from observation* (Boyle's Law) that dilute gases, which we expect to be close to ideal, follow a simple relationship between pressure and volume

$$PV = \text{constant}, \tag{1.4}$$

and that this product increases with temperature. By now setting this proportional to the absolute temperature, that is,

$$PV = Nk_{\rm B}T, \tag{1.5}$$

where N is the number of molecules, one obtains the Kelvin scale; the value of the proportionality constant, k_B, called *Boltzmann's constant*,[2] is fixed by retaining the 100° interval of the Celsius scale.

1.2.5 The Second Law: Real and Reversible Processes

Three simple examples suffice to show that work is not a state function. A viscous liquid can absorb energy by transfer of heat from warmer surroundings, or by work in the form of agitation; in either case the end result is a resting, but warmer liquid. A gas can expand against a piston and in so doing perform work on the environment, or it can expand into a vacuum, and perform no work (Fig. 1.4b). Heat generated by combustion of fuel can be used to drive machinery (steam or internal combustion engine, steam turbine), but the heat can also be used to warm the environment directly, without any work being generated.

In the first example, *work is turned into heat and "lost,"* and everyday experience shows many such instances of friction. In the second and third examples, the *ability to perform a certain amount of work is lost.*

The Second Law reflects this experience by stating that in any real process some ability to perform work is lost. The magnitude of work performed by a system in a real process is less than the maximum possible, and, if a fixed amount of work is performed on the system, then the system's ability to perform work is increased by a smaller amount (and, perhaps, not increased at all).

The Second Law states that it is not possible to create a machine that, for example, captures heat to do an equivalent amount of work (e.g., PV work due to expansion), and then applies this work to generate a high energy state (e.g., an electrical potential) that can then be used to generate an equivalent amount of heat (I^2R heating) that can be used to do an equivalent amount of work, etc. As energy is conserved according to the First Law, if each of these processes were completely efficient, we would have a perpetual motion machine ("of the second kind"), which, by our experience, is not possible.

A quantitative statement of this law requires that we distinguish between *reversible* and *irreversible* processes.[3] In brief, if a *reversible process*, say A → B

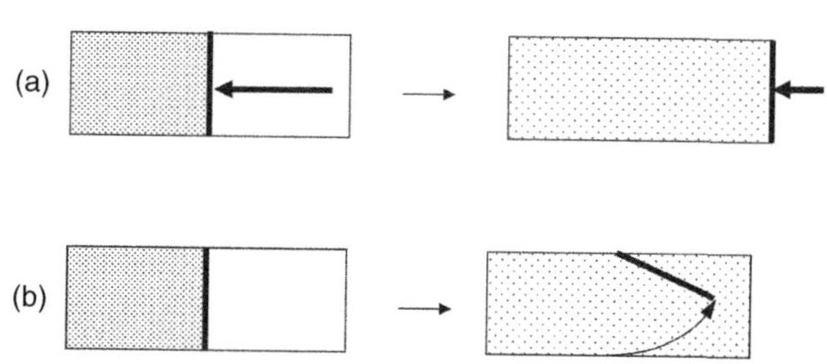

FIGURE 1.4 (a) When a gas expands against a force (indicated by the arrows) acting on a piston, the gas performs so-called PV work. (b) When the gas is allowed to expand into a vacuum by removal of a partition, no work is performed.

is repeated in reverse following exactly the same path, and the complete process is A → B → A, then there is no *net* exchange of work or heat with the environment, that is, any heat or work expended in running the system through the first leg is recovered during the second leg, and vice versa. However, reversible change is an idealization that we can never achieve in real processes. We approximate a reversible process by carrying out the change very slowly, that is, in infinitesimal steps. While a reversible A → B → A process (e.g., a swinging pendulum) can continue indefinitely, such a process can only be imagined and never be achieved in a real system. Thus, real processes are *irreversible* and behave according to the Second Law.

1.2.6 New State Functions: Free Energy and Entropy

We have seen that the energy of a system is defined by its current state (pressure, volume, temperature, and contents), and not by its history; on the other hand, work and heat are definitely *not* state functions. However, we *can* now ask *how much work might be performed by a system in an optimally chosen process, and thereby define a new state function, the free energy, A, the process to take place at constant temperature.*

Because total energy is conserved, a change in A cannot exceed the concomitant change of the internal energy, U of the system. The function $U-A$, is of course also a state function, which is written as TS, where S is our *second new state function, the entropy*

$$A = U - TS. \tag{1.6}$$

Naturally, this will serve to determine A only if we can determine S, and this is where the (famous) Carnot cycle comes into play.[4]

We will not describe the Carnot cycle here as it involves steps that are adiabatic, making the description obscure. Instead, we describe a simple scheme that employs compression and expansion of an ideal gas in a heat pump (a machine that exchanges work and heat flow) as shown in Fig. 1.5. As mentioned, the (N) molecules in an ideal gas do not interact, the internal energy is independent of P or V, which, themselves, are related by the gas law (Eq. 1.5).

We now pick a temperature, T_1 and a volume V_1 and compress the gas from that starting point to a smaller volume V_2, by moving a piston. When the moving piston collides with the gas molecules, it increases their velocities, and thus the internal energy increases and the temperature rises. (The increase of the internal energy exactly equals the amount of work done by pushing the piston.) However, if the system is placed in a heat reservoir[5] at constant temperature T_1, then the excess heat, q_1 will flow from the gas into the heat bath, and the energy of the gas does not change. Because the internal energy of the gas does not change, the work done on the gas is equal to the amount of heat transferred. Because, according to our definition of the free energy, the work done in this process can be equated with

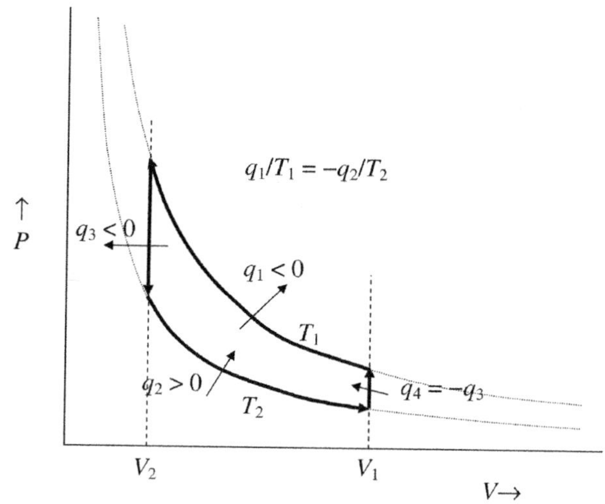

FIGURE 1.5 Four-step cycle (not a Carnot cycle) of isothermal compression, cooling by contact with a bath, isothermal expansion and heating by contact with a bath, applied to an ideal gas. Arrows indicate heat flow into or out of the system in each step.

the free energy change, we have

$$\Delta A_1 = -\int PdV = -Nk_\mathrm{B}T_1 \ln(V_2/V_1) = -q_1$$

$$\Delta A_1 = \Delta U - q_1 = \Delta U - T_1\frac{q_1}{T_1}, \tag{1.7}$$

ΔU being zero.

We can compress the gas from V_2 to V_1 at another temperature, T_2, which gives the same equation relating ΔA_2, T_2, V_1/V_2, q_2, and ΔS_2 and thus

$$\Delta S_2 = -\Delta S_1.$$

We can then construct the cyclic process of Fig. 1.5 by compressing from V_1 to V_2 at T_1, then cooling to T_2, expanding back to V_1 at T_2, and heating back up to T_1. Because the internal energy is independent of P and V, the heating and cooling steps produce/require exactly opposite changes of energy, $q_3 = -q_4$. In a cyclic process, the net change of any state function is zero. The change in energy for the entire cycle is obviously zero.

$$\Delta U_1 + \Delta U_2 + \Delta U_3 + \Delta U_4 = 0 + 0 + q_3 + q_4 = 0. \tag{1.8}$$

Now, by *defining the entropy as the integral of the heat exchanged divided by the temperature in a reversible (or quasi-static) process,*

$$dS = \frac{dq}{T}$$

$$S = \int \frac{dq}{T}, \tag{1.9}$$

the entropy changes in the heating and cooling steps are also equal and opposite,[6] so that for this cycle

$$\Delta S_{\text{cycle}} = \Delta S_1 + \Delta S_2 + \Delta S_3 + \Delta S_4 = 0. \tag{1.10}$$

It is easily shown that this equation holds for any cycle that combines two such cycles that share a part of their circumference, and as any closed cycle can be decomposed in smaller cycles of the form of Fig. 1.5, it follows that the *entropy as defined by* Eqs. 1.6 and 1.9 *is a state function for ideal gas systems*. This can then be generalized to any other system by an argument that invokes the Second Law for this system thermodynamically coupled to an ideal gas.[4]

The *work* done to run the cycle $\left(-\int PdV\right)$, which equals the area inside the closed curve in Fig. 1.5), is not zero, and this energy ends up as the difference between the *heat* given off in the compression leg and that taken up in the expansion leg of the cycle. When run in the indicated direction, each cycle transfers heat from the cooler heat reservoir at T_2 to the warmer heat reservoir at T_1 in the amount $q_2 = -\Delta A_2$, and thus acts as a (completely impractical) heat pump.

1.2.7 Entropy Tends to Increase

The transfer of heat between two systems at different temperatures occurs in one direction and is an irreversible process. The entropy of the colder system increases by dq/T_1, and that of the warmer decreases by dq/T_2; the net change is positive, and in the absence of performance of work

$$dS = dq(1/T_1 - 1/T_2) > 0.$$

Many processes produce a rise or fall in the temperature of the system. In order to maintain the system at a constant temperature, the necessary heat is exchanged with a heat reservoir, and one sees that this cannot be done reversibly without an increase of the entropy (of system plus reservoir) unless the temperature of the heat reservoir is at all times exactly the same as that of the system (which would require an infinitely slow process).

In a system kept at constant temperature by contact with a heat reservoir (an isothermal system), a real, irreversible process is one in which the work done by the system is less than the maximum possible, that is, is less than the decrease in free energy,

$$-dA \geq \partial w. \tag{1.11}$$

Given the relation between A, U, and S, we then have

$$-(dU - TdS) \geq \partial w$$

$$dS \geq \frac{dU + \partial w}{T} = \frac{\partial q}{T}, \tag{1.12}$$

and one sees that the entropy increases by *more* than the amount corresponding to the heat transferred into or out of the system.

The general conclusion is that in all real processes,[7] because of the Second Law, *the entropy of the universe (the system plus its reservoir) increases* ($dS_U > 0$). This is a familiar statement of the Second Law. It is also the least transparent statement, although it follows completely from the more common-sense statement that heat does not flow from a lower to a higher temperature system.[8]

1.2.8 The Second Law and Equilibrium

A fundamental concept of thermodynamics is that of *equilibrium, a state from which spontaneous change* (i.e., not induced by an exchange of work or heat) *is not possible.* The majority of applications of thermodynamics in molecular biophysics consider conditions needed to establish equilibrium and the deviations from equilibrium if these conditions are not met.

For an isolated system (no w or q), the Second Law requires that entropy of a system increases (irreversible or spontaneous change) or remains the same (reversible change). Thus, at equilibrium, the entropy of an isolated system tends to become larger, until it reaches a maximum, which can be expressed as

$$dS \geq 0 \text{ in the absence of exchange of heat or work.}$$

We can now see that the entropy plays a critical role in thermodynamics, as it allows us to *express the equilibrium condition as the maximum of a state function. The rest is algebra.*

If the system is able to exchange heat or work with a reservoir, we have

$$\partial w = \partial q - dU \leq TdS - dU$$

$$dU \leq TdS - PdV - \partial w', \tag{1.13}$$

where w has been separated into work for expansion (PV work) and other work, w'.[9] In this expression the differential of the state function U is expressed as a function of the independent variables S and V, and is the state function that is minimized when entropy and volume are constant ($dS = dV = 0$; and no other work is done). This expresses the *combined First and Second Laws.* The equal sign holds for reversible processes. However, a condition of constant entropy is not easily realized experimentally, and its meaning is difficult to grasp. In the next section, we discuss which state functions are minimized when temperature, rather than entropy, is constant, and when work is done (pressure, not volume, being constant).

1.3 OTHER USEFUL QUANTITIES AND CONCEPTS

1.3.1 Gibbs and Helmholtz Free Energies and Enthalpy

We now have the basis for the normal treatment of thermodynamics. In this treatment, it is convenient to define two new functions, in addition to U, S, and A. These functions are the enthalpy, H and the Gibbs free energy, G; the previously introduced state function A is distinguished as the Helmholtz free energy. (Older literature tends to use the symbol F for A, while some still use F for G.) We then have

$$\begin{aligned} H &= U + PV \\ A &= U - TS \\ G &= A + PV = H - TS\,. \end{aligned} \tag{1.14}$$

By combining Eqs. 1 and 1.14 we get the following expressions for dH, dA, and dG,[10]

$$\begin{aligned} dH &\leq TdS + VdP - \partial w' \\ dA &\leq -SdT - PdV - \partial w' \\ dG &\leq -SdT + VdP - \partial w'. \end{aligned} \tag{1.15}$$

This allows us to identify state functions that are minimized under three different sets of equilibrium conditions:

1. At equilibrium, the Helmholtz free energy, A, is a minimum at constant T and V.
2. At equilibrium, the enthalpy, H, is a minimum at constant S and P.
3. At equilibrium, the Gibbs free energy, G, is a minimum at constant T and P.

As a corollary,

1. A is the state function defined by independent variables T, V, and composition N_i.
2. H has the same form as U except that its independent variable is P instead of V.
3. G is the state function defined by independent variables T, P, and composition N_i.

Within narrow margins, biological systems operate at constant temperature. Not surprisingly, applications of thermodynamics in molecular biophysics rely on state functions A and G whose minima define the equilibrium condition at constant temperature. Specifically, spontaneous processes at constant T result in decreases in free energy until they reach equilibrium, at which point the free energy change is zero:

$$\delta A \leq 0, \quad \text{in a system held at constant volume and temperature,} \tag{1.16}$$

$$\delta G \leq 0, \quad \text{in a system held at constant pressure and temperature;} \tag{1.17}$$

for each, the equal sign holds at equilibrium.

At equilibrium, A or G is at a minimum, and any perturbation of the conditions of the system causes an increase of A or G,

$$dA \geq 0 \text{ for perturbations from equilibrium at constant } V, \tag{1.18}$$

$$dG \geq 0 \text{ for perturbations from equilibrium at constant } P. \tag{1.19}$$

As biophysical systems are studied at constant pressure, the Gibbs free energy is normally the more useful. Note however that, in most experiments with solutions, the changes in PV are so small that a distinction between A and G has no noticeable effect.[11]

1.3.2 Chemical Potential

A fundamental concept of thermodynamics, and certainly one of the most useful in chemistry, is that of *chemical potential*, which is crucial to a description of the thermodynamics of mixtures (and hence of solutions). It is defined as the partial derivative of the Gibbs free energy with respect to the amount, N_i of component i,[12] while T, P, and the amounts, $N_{j \neq i}$ of all other components are taken as defined (i.e., constant or fixed), so that

$$\mu_i = \left(\frac{dG}{dN_i}\right)_{T,P,N_{j \neq i}} = \left(\frac{dA}{dN_i}\right)_{T,V,N_{j \neq i}}. \tag{1.20}$$

The chemical potential describes the intrinsic or *intensive free energy* that a substance has in a mixture (or in a pure state).

If we are describing an open system, we must extend the total differentials of U, A, H, and G (Eqs. 1.13 and 1.15), which reflect the combined First and Second Laws, in order to take into account the dependence on n_i, and this gives

$$dU \leq TdS - PdV - \partial w' + \sum \mu_i dN_i$$

$$dH \leq TdS + VdP - \partial w' + \sum \mu_i dN_i$$

$$dA \leq -SdT - PdV - \partial w' + \sum \mu_i dN_i$$

$$dG \leq -SdT + VdP - \partial w' + \sum \mu_i dN_i, \tag{1.21}$$

where ∂w is any additional work done by the system.

1.3.3 Fundamental Relationships Between State Functions

Our statement of the combined First and Second Laws (Eq. 1.13) expresses dU as an exact differential with respect to the independent variables V and S. Several properties of exact differentials are quite useful in thermodynamics. (See Section A9.7, "Useful relations between partial differential quotients.")

Single derivatives. The first is that the total differential of any function is given by the sum of the partial derivatives of that function with respect to each independent variable of the function times the differentials of the variable (Eqs. A9.16 and A9.17).

Using the definition of a total differential, we get the following expressions for T, P, and μ_i in terms of partial derivatives of U

$$T = \left(\frac{\partial U}{\partial S}\right)_{V,N_i}; \quad P = -\left(\frac{\partial U}{\partial V}\right)_{S,N_i}; \quad \mu_i = \left(\frac{\partial U}{\partial N_i}\right)_{S,V,N_j}. \tag{1.22}$$

Application to the total differentials of the other state functions we have defined (G, H, A; see Eq. 1.15) gives the following additional relations between state variables and state functions

$$S = -\left(\frac{\partial G}{\partial T}\right)_{P,N_i}; \quad V = -\left(\frac{\partial G}{\partial P}\right)_{T,N_i}; \quad \mu_i = \left(\frac{\partial G}{\partial N_i}\right)_{T,P,N_j}$$

$$T = \left(\frac{\partial H}{\partial S}\right)_{P,N_i}; \quad V = \left(\frac{\partial H}{\partial P}\right)_{S,N_i}; \quad \mu_i = \left(\frac{\partial H}{\partial N_i}\right)_{S,V,N_j}$$

$$S = -\left(\frac{\partial A}{\partial T}\right)_{V,N_i}; \quad P = -\left(\frac{\partial A}{\partial V}\right)_{T,N_i}; \quad \mu_i = \left(\frac{\partial A}{\partial N_i}\right)_{T,V,N_j}. \tag{1.23}$$

Double derivatives. The second property that we can exploit is that the order of partial differentiation can be switched according to the Euler chain rule (Eq. A9.13). For example,

$$\left(\frac{\partial}{\partial V}\left(\frac{\partial A}{\partial T}\right)_V\right)_T = \left(\frac{\partial}{\partial T}\left(\frac{\partial A}{\partial V}\right)_T\right)_V$$

$$\text{i.e., } \left(\frac{\partial S}{\partial V}\right)_T = -\left(\frac{\partial P}{\partial T}\right)_V. \tag{1.24}$$

This is a so-called Maxwell relation. The following Maxwell relations (in which S does not appear) are particularly useful,

$$\left(\frac{\partial V}{\partial N_i}\right)_{P,V} = -\left(\frac{\partial \mu_i}{\partial P}\right)_{V,T}$$
$$\left(\frac{\partial P}{\partial N_i}\right)_{V,T} = -\left(\frac{\partial \mu_i}{\partial V}\right)_{P,T}$$
$$\left(\frac{\partial \mu_i}{\partial N_j}\right)_{N_i,T} = \left(\frac{\partial \mu_j}{\partial N_i}\right)_{N_j,T}. \tag{1.25}$$

Multiple variables. As noted, what we call "independent variables" are not really independent but are what we choose to control, and control does not always mean that we keep the independent variable constant. For example, P and V might both vary, while we control how V varies, and we must then consider how state functions vary when both P and V vary. This is again accomplished using additional rules about differentials.

As an example of application of Eq. A9.15, consider the internal energy, U, which is a function of both V and P. For example, we may ask how U depends on P when V also varies, which is formally expressed by

$$\left(\frac{\partial U}{\partial P}\right)_T = \left(\frac{\partial U}{\partial P}\right)_V + \left(\frac{\partial U}{\partial V}\right)_P \left(\frac{\partial V}{\partial P}\right)_T. \tag{1.26}$$

The first term on the right is zero because no PV work is done at constant V. Substituting the expression for $(\partial U/\partial V)_P$ derived earlier (Eq. 1.22), we get

$$\left(\frac{\partial U}{\partial P}\right)_S = P\left(\frac{\partial V}{\partial P}\right)_T = P\kappa, \tag{1.27}$$

where κ is the isothermal compressibility.

Temperature dependence of energy and free energy. The temperature dependence of both the energy and of the enthalpy is called a *specific heat*

$$\left(\frac{\partial U}{\partial T}\right)_{V,N_i} = C_V \quad \text{and} \quad \left(\frac{\partial H}{\partial T}\right)_{P,N_i} = C_P; \tag{1.28}$$

C_V is the specific heat at constant volume, and C_P the specific heat at constant pressure. The temperature dependence of the free energy can be expressed as

$$\left(\frac{\partial A}{\partial T}\right)_V = -S \quad \text{and} \quad \left(\frac{\partial G}{\partial T}\right)_P = -S, \tag{1.29}$$

or instead as[13]

$$\frac{\partial(A/T)}{\partial(1/T)} = U \quad \text{and} \quad \frac{\partial(G/T)}{\partial(1/T)} = H. \tag{1.30}$$

Thus, a complete knowledge of A or G implies a knowledge of all thermodynamic functions.

1.3.4 The Gibbs–Duhem Equation and Equilibrium

At fixed intensive variables T, P, and μ_i,[14] we can integrate the exact differential for dG

$$dG = -SdT + VdP + \sum \mu_i dN_i \tag{1.31}$$

to obtain

$$G = \sum N_i \mu_i . \tag{1.32}$$

Note that this relationship between the free energy and the chemical potential of all species applies only for the Gibbs free energy at fixed T and P. If we take the total differential of this, we obtain

$$dG = \sum N_i d\mu_i + \sum \mu_i dN_i . \tag{1.33}$$

If we equate the total differential of G from this equation with the total differential form of dG given in Eq. 1.21 we obtain the *Gibbs–Duhem equation*[15]

$$\sum N_i d\mu_i = VdP - SdT, \tag{1.34}$$

which, at constant T and P, becomes

$$\sum N_i d\mu_i = 0. \tag{1.35}$$

The usefulness of this expression will become clear as we apply it in a variety of situations. We shall see that Gibbs–Duhem relates the chemical potential of solvent to that of solutes (Eq. 1.49), defines the condition of chemical equilibrium and leads to the definition of the equilibrium constant (Eq. 1.67), and is again used in a derivation of linkage relations in Chapter 9.

1.3.5 Relation Between Heat Capacity and Other Functions

We said earlier that if A is completely known, then other thermodynamic functions are also known. But we can just as well base a knowledge of A, U, and S on a knowledge of the specific heat, C_V. The reason for doing this is that specific heat of very many systems can be measured accurately with a calorimeter.

To begin with, the energy U and the entropy S are integrals of the specific heat (heat is transferred slowly, at constant volume)

$$\Delta U = \int dq = \int dT \frac{dq}{dT} = \int dT\, C_V$$
$$\Delta S = \int \frac{dq}{T} = \int \frac{dT}{T} \frac{dq}{dT} = \int dT \frac{C_V}{T} . \tag{1.36}$$

The entropy S of all systems is set equal to zero when the absolute temperature, T, is zero. (This is the so-called *Third Law*.) Classical chemical thermodynamics by convention sets the energy of each pure chemical element to zero when T is zero; current practice sets the energy at $T = 0$ to the quantum-mechanical ground state energy, ε_0.

The free energy is then given by these two equations

$$\Delta A = -\int S\, dT$$

$$\frac{\Delta A}{T} = \int U\, d(1/T), \tag{1.37}$$

where $\Delta A = A(T) - \varepsilon_0$. If the calorimetric measurements are done at constant pressure, then equivalent expressions relate C_P, H, and G.

1.4 THERMODYNAMICS OF THE IDEAL GAS

Pressure and volume of an ideal gas are related by Boyle's ideal gas law, $PV = Nk_BT$ (Eq. 1.5). Boyle's law is an empirical relation, which we now understand to apply only when each gas molecule behaves *independently* of all the others.

Accordingly, the free energy depends on the volume according to

$$A = -\int PdV = -\int \frac{Nk_BT}{V} dV = \mathrm{C} - Nk_BT \ln V = Nk_BT \ln(V/V^\bullet)$$

$$\mathrm{C} = -Nk_BT \ln V^\bullet, \tag{1.38}$$

where the integration constant C is set by choosing a fixed reference volume $V^\bullet$. The volume $V^\bullet$ represents a *standard state*, that is, a state relative to which we can define the chemical potential of the gas at any other experimentally defined volume, V (or pressure P, related to V by the ideal gas law). In principle, the choice of standard state is arbitrary, but, in practice, convention sets the pressure of the standard state at 1 bar, and $V^\bullet$ then depends on T according to the ideal gas law.[16] The equation for A states that if we change either P or V away from *standard conditions*, A varies as the natural logarithm of the ratio of the new volume to the volume at standard conditions.

By differentiating the free energy A of the ideal gas in Eq. 1.38 with respect to N at constant T and V, one obtains according to Eq. 1.23 an expression for the chemical potential

$$\mu = k_BT \ln V^\bullet - k_BT \ln V$$

$$\mu = k_BT \ln\left(\frac{V^\bullet}{N}\right) + k_BT \ln\frac{N}{V} = \mu^\bullet + k_BT \ln\frac{N}{V}, \tag{1.39}$$

where the constant term $\mu^\bullet$ is the chemical potential of the gas at a standard state, which is here taken as $N/V = 1$; $\mu^\bullet$ is still a function of T. One can also express

μ as a function of T and $P = Nk_BT/V$,

$$\mu = \mu^{\bullet} + k_BT\ln(N/V) = \mu^{\circ} + k_BT\ln P. \tag{1.40}$$

The constant terms $\mu^{\bullet}$ and μ° both represent the chemical potential of the gas at the conventional standard state of a gas at 1 bar, but their values differ according to the different choice of either volume or pressure as state function.

The terms indicated with $\mu^{\bullet}$ and μ° are termed the *unitary* or *standard chemical potential* and are independent of concentration; however, their values depend on the choice of standard state. The terms in $\ln(N/V)$ and $\ln P$ (the *cratic terms*) contain the concentration dependence and are related to the entropy of the gas, which becomes greater the more the gas is dilute.

The ideal gas law applies also when the molecules in the gas are not all of the same kind, and the gas can contain a mixture of different components. We say that in an ideal gas mixture each component contributes to the total pressure as if it were the only gas occupying the volume, that is, $P = P_1 + P_2 + P_3 + \cdots$, where P_j is the partial pressure of gas j, which is X_jP, where X_j is the mole fraction of gas j in the mixture $\left(X_j = N_j/\sum_i N_i\right)$.

It is then easy to show that the equivalent expression for the chemical potential for a component of a gas mixture is

$$\begin{aligned}\mu_i &= \mu_i^{\bullet} + k_BT\ln(N_i/V)\\ \mu_i &= \mu_i^{\circ} + k_BT\ln P_i.\end{aligned} \tag{1.41}$$

Alternatively, if we make such a mixture at constant pressure and temperature, then the volume of the mixture will be the sum of the volumes of all the gases that we mix $\left(V = \sum_i V_i = \sum_i k_BTN_i/P_i\right)$ and the total number of molecules will be $N = \sum_i N_i$. The free energy change for creating this mixture is the sum of the free energies for expanding each of the component gases from volume V_i to volume V.

$$\Delta A = -k_BT\sum_i N_i\ln(V/V_i) = k_BT\sum_i N_i\ln X_i \tag{1.42}$$

This is the *free energy of mixing* the gases at constant T and P. We note here that this same expression describes the free energy of mixing of an ideal mixture. This is a mixture in which all molecular species interact in an identical manner; thus, in a two-component system 1–1, 2–2, and 1–2 interactions are equivalent in this model system.

1.5 THERMODYNAMICS OF SOLUTIONS

Most biophysical experiments are done in solution. Fortunately, the thermodynamics of dilute solutions are relatively easy to describe on the basis of experiments that allow one to relate these to the thermodynamics of gases.

1.5.1 "Ideal" Dilute Solutions

The thermodynamics of solutions can be related to those of dilute gases by experiments that consist of equilibrating the solution with its vapor and measuring the concentration of the solute in both phases. At low concentration of solute in the liquid, the ratio between the concentration of a particular solute,[17] component 1, in solution $c_{1,\mathrm{s}}$ and the concentration in the vapor $c_{1,\mathrm{v}}$ is a constant whose value is found to be specific for that solute and that solvent:

$$\frac{c_{1,\mathrm{s}}}{c_{1,\mathrm{v}}} = \frac{(N_1/V)_\mathrm{s}}{(N_1/V)_\mathrm{v}} = K_{\mathrm{v}\to\mathrm{s}}. \tag{1.43}$$

Here, $K_{\mathrm{v}\to\mathrm{s}}$ is the *partition coefficient* or equilibrium constant for transferring solute from vapor to solution phase. This relation (called *Henry's Law*) holds only in the limit as the concentration of solute approaches zero, but in practice it holds over a sufficient range of concentration that accurate values of transfer equilibrium constants can be determined. In an ideal gas, the molecules are assumed not to interact with each other. In a dilute solution, we assume that solute molecules interact only with the surrounding solvent molecules but not with other solute molecules. This model is called the *ideal solution.*

We now make use of the fact that G is a minimum for the equilibrated system; consequently, transfer of a *small* amount of solute from solution to vapor or vice versa causes balancing changes in G of the gas and solution according to

$$dG = dN_1(\mu_{1,\mathrm{s}} - \mu_{1,\mathrm{v}}) = 0, \tag{1.44}$$

so that the two chemical potentials are equal,

$$\mu_{1,\mathrm{s}} = \mu_{1,\mathrm{v}}. \tag{1.45}$$

(This is true for all components in all phase equilibria.) First, substituting the expression for $\mu_{1,\mathrm{v}}$ of the ideal gas, Eq. 1.39, and second using the proportionality of $c_{1,\mathrm{s}}$ and $c_{1,\mathrm{v}}$, Eq. 1.43, one obtains the following expression for the chemical potential of the solute in an ideal solution,

$$\begin{aligned} \mu_{1,\mathrm{s}} &= \mu_{1,\mathrm{v}}^{\bullet} + k_\mathrm{B}T \ln c_{1,\mathrm{v}} \\ \mu_{1,\mathrm{s}} &= \mu_{1,\mathrm{v}}^{\bullet} + k_\mathrm{B}T \ln(c_{1,\mathrm{s}}/K_{\mathrm{v}\to\mathrm{s}}) = \mu_{1,\mathrm{s}}^{\circ} + k_\mathrm{B}T \ln c_{1,\mathrm{s}}. \end{aligned} \tag{1.46}$$

The standard or unitary chemical potential in the vapor and solution differ; $\mu_{1,\mathrm{v}}^{\circ}$ is the chemical potential of pure component 1 in the vapor at a concentration (particle density) $c_{1,\mathrm{v}} = (N_1/V)_\mathrm{v} = 1$, and $\mu_{1,\mathrm{s}}^{\circ}$ is the chemical potential of component 1 in solution at a concentration $c_{1,\mathrm{s}} = (N_1/V)_\mathrm{s} = 1$.

The standard chemical potential of solute in an ideal solution is the standard chemical potential of solute in the ideal gas phase plus the free energy of transferring a molecule of solute from the vapor to the solution phase,

$$\mu_{1,\mathrm{s}}^{\circ} = \mu_{1,\mathrm{v}}^{\bullet} - k_\mathrm{B}T \ln(K_{\mathrm{v}\to\mathrm{s}}). \tag{1.47}$$

The *value* of the standard chemical potential depends not only (as expected) on the state for which it is defined but also on the units in which the concentration is expressed.[16] Unless it is explicitly stated otherwise, one should assume that concentration units are in moles per liter, and, if we use any concentration scale other than the molarity scale, then the value of $\mu_{i,s}^{\circ}$ must be altered by subtracting $k_B T$ times the natural log of the factor that converts molarity to the new concentration unit, and, if we wish to compare the standard states of a solute in two different solutions, we must use the same units of concentration for both.

Using units of mole fraction, the chemical potential of an ideal solution is

$$\mu_{1,s} = \mu_{1,s}^{*} + k_B T \ln X_{1,s}, \tag{1.48}$$

where the standard chemical potential, $\mu_{i,s}^{*}$, is the chemical potential of pure component $i (X_i = 1)$ surrounded by solvent. The observant reader will note that this seems nonsensical, as for $X_i = 1$ no solvent is present. Indeed $\mu_{i,s}^{*}$ is the value required to obtain $\mu_{i,s}$ equal to $\mu_{1,s}^{*} + k_B T \ln X_{1,s}$ for proper dilute solutions, that is, for $X_{1,s} \ll 1$; $\mu_{i,s}^{*}$ represents the (imaginary) state of pure compound 1 interacting only with the solvent.

The last step is to derive an equation for the *chemical potential of the solvent* in an ideal solution. We start with the Gibbs–Duhem equation (Eq. 1.35), which becomes for just two components

$$N_1 d\mu_1 + N_0 d\mu_0 = 0. \tag{1.49}$$

If we wish to focus on how the chemical potentials change with N_1, we can divide both sides of Eq. 1.49 by dN_1,

$$N_1 \frac{d\mu_1}{dN_1} + N_0 \frac{d\mu_0}{dN_1} = 0$$

$$\frac{d\mu_0}{dN_1} = -\frac{N_1}{N_0}\frac{d\mu_1}{dN_1}. \tag{1.50}$$

With this expression for the chemical potential of the dilute solute ($N_1 \ll N_0$), Eq. 1.46, one has with the mole fraction of solute, $X_1 = N_1/(N_1 + N_0)$

$$\frac{d\mu_0}{dN_1} = -\frac{N_1}{N_0}\frac{d(-k_B T \ln X_1)}{dN_1} = -\frac{k_B T}{N_0}\left(1 - \frac{N_1}{N_1 + N_0}\right) \approx -k_B T \frac{1}{N_0}, \tag{1.51}$$

and integration gives

$$\mu_{0,s} = \text{constant} - k_B T \frac{N_1}{N_0} \approx \mu_{0,s}^{\circ} - k_B T\, X_1 = \mu_{0,s}^{\circ} - k_B T\, (1 - X_0). \tag{1.52}$$

At low total solute concentrations, the expression for μ_0 can also be written as

$$\mu_{0,s} = \mu_{0,s}^{\circ} + k_B T \ln(1 - X_1) = \mu_{0,s}^{\circ} + k_B T \ln X_0. \tag{1.53}$$

If one equates $\mu_{0,\mathrm{s}}$ with the chemical potential of an ideal gas of Eq. 1.39, one sees that the (partial) pressure of solvent in the vapor is proportional to the mole fraction of solvent in dilute solution, always assuming that both vapor and solution behave ideally. This is the classical form of *Raoult's Law.*

In summary, the chemical potentials of components of a dilute solution are

$$\mu_0 = \mu_{0,\mathrm{s}}^{\circ} - k_{\mathrm{B}}T\sum_{i>0} X_i$$

$$\mu_{i>0} = \mu_{i,\mathrm{s}}^{*} + k_{\mathrm{B}}T \ln X_i = \mu_{i,\mathrm{s}}^{\circ} + k_{\mathrm{B}}T \ln c_i, \tag{1.54}$$

where $\mu_{0,\mathrm{s}}^{\circ}$ is the chemical potential of pure solvent at T, P, but $\mu_{i,\mathrm{s}}^{*}$ of the dilute solute is not the chemical potential of pure solute, but rather the value needed to give the *actual* value of $\mu_{i,\mathrm{s}}$ of a dilute solution ($X_i \ll 1$), when substituted in Eq. 1.46, and similarly for $\mu_{i,\mathrm{s}}^{\circ}$.

1.5.2 Nonideal Solutions

If the proportionality between $c_{1,\mathrm{v}}$ and $c_{1,\mathrm{s}}$ does not hold in Eq. 1.46, the solution is said to be "*nonideal*" and a correction is needed. This deviation from ideal behavior is due to interactions between solute molecules. However, at equilibrium the chemical potential of solute in solution and vapor are equal whether or not either phase is ideal. Thus, we can develop a description of a nonideal solution just as we did for an ideal solution and, in this way, hold on to the simplified functional forms for the concentration dependence of ideal solutions by replacing concentration by *activity*, a,

$$\mu_i = \mu_{i,\mathrm{s}}^{\circ} + k_{\mathrm{B}}T \ln a_i = \mu_{i,\mathrm{s}}^{\circ} + k_{\mathrm{B}}T \ln \gamma_i c_i, \tag{1.55}$$

where γ_i is the activity coefficient of component i in the real solution. As can be seen from this discussion, deviations from $\gamma_i = 1$ imply the existence of significant interactions between solutes in a mixture, and deviations are most severe for strongly interacting solutes at high concentrations. Of course, this is all easy to write down, and even easier to say, but it is rarely easy to measure the activities of solutes in a multicomponent solution. The reality is, then, that we usually end up using the thermodynamic expressions for ideal solutions, even though these probably do not hold in a cell. We shall use activities wherever this is appropriate (e.g., we shall express equilibrium constants as ratios of activities, not concentrations), but in practical applications ideal behavior is assumed and activities are nearly always equated with the corresponding concentrations.

Only quite small molecules are sufficiently volatile to allow one to establish the ideality of their solutions by measurements of vapor pressure. However, the concentration dependence of the chemical potential of macromolecules can be determined via the technique of light scattering (see Section 6.7), and these measurements give us confidence, for example, that dilute protein solutions follow close to ideal thermodynamics. Molecules of nucleic acids are highly charged and interact strongly in water; however, the interaction is greatly reduced in solutions of reasonable ionic strength.

The activity of many inorganic ions can be measured with potentiometric methods (which are otherwise not discussed in this book). A very important application of potentiometric measurement is the pH meter. Thus, pH is defined as

$$\mathrm{pH} = -{}^{10}\log a_{\mathrm{H}^+}. \tag{1.56}$$

The pH of 0.1 N HCl is 0.11, not 0.10, which indicates that this moderately concentrated acid solution is significantly nonideal.

1.5.3 Osmotic Effects

Equation 1.51 shows that solutes change the activity of water in a manner that is roughly proportional to their concentration. For relatively inert solutes (i.e., those that interact weakly with other solutes and even with the solvent), this expression can hold up to significant concentrations. Some such solutes, called *osmolytes*, occur in cells; by lowering the activity of cellular water, they serve to maintain cellular water content under drought conditions. As biological macromolecules interact strongly with water, a change in the activity of water brought about by osmolytes such as glycerol or sucrose can have significant effects on, for example, equilibria between conformation states or equilibria for forming macromolecular complexes. This will be taken up in Chapter 9 and again in Chapter 16.

Osmotic pressure reflects a tendency of solutions to become more dilute. We can think of the free energy as a potential with a corresponding "force" that drives the system to lower values of the potential by making the solution more dilute, similar to the pressure that tends to expand the volume of a gas. This thermodynamic "force" is due to entropy and manifests itself clearly in a situation where compartments can be separated by a wall containing passages through which the solvent can pass, but the solute cannot; such a wall is called a *semipermeable wall* or membrane. Dialysis membranes pass solvent and solutes of low molecular weight, but not macromolecules; membranes used in desalination pass water, but not NaCl; lipid bilayer membranes are impermeable to many compounds, while others can pass across the membrane by dissolving in the membrane, by taking advantage of fluctuations in membrane structure, by forming hydrogen-bonded chains (in the case of water), or, in the case of cell membranes, by diffusing through specific protein channels.

If a solution and a pure solvent are equilibrated in two compartments (A containing solution; B containing solvent; Fig. 1.6) separated by a semipermeable membrane, the solvent passes from compartment B into A in order to maximize the entropy of the system. Equilibrium is reached when the pressure in compartment A is higher than that in compartment B by a certain amount, namely, when the work required to move the solvent from B to A against this pressure is equal to the decrease in free energy (increase in entropy) associated with solvent moving

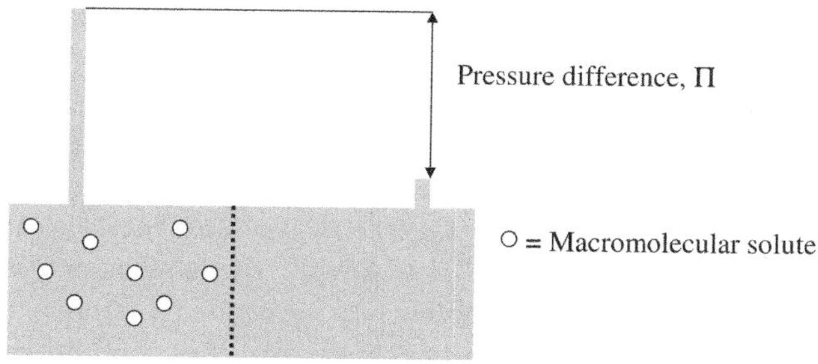

FIGURE 1.6 Osmotic pressure resulting from equilibration of solvent across a semipermeable membrane.

into compartment A. The free energy to transfer a solvent molecule from pure solvent into a solution can be written as a sum of three terms, the first for removing the solvent molecule from the solvent at constant pressure, the second for moving it to a system with a larger pressure, and the third for adding it to the solution at constant pressure,

$$\Delta G = (-\mu_0^\circ) + \Pi V_0 + \mu_0, \tag{1.57}$$

where V_0 is the volume per solvent molecule, and the difference in pressure between the two compartments is called the *osmotic pressure*, denoted with the symbol Π ($\Pi = P_A - P_B$).

In an ideal solution, the chemical potential of the solvent is given by Eq. 1.54. Setting ΔG in this equation to zero gives for the osmotic pressure that

$$-\mu_0^\circ + \Pi V_0 + \mu_0^\circ - k_B T X_1 = 0$$

$$\Pi_{\text{ideal}} = k_B T \frac{X_1}{V_0} \approx \frac{k_B T N_1}{V} = k_B T \frac{c_1}{M_1}, \tag{1.58}$$

where N_1 is the number of moles of solute and c_1 the concentration in units of mass/volume, M_1 being the solute's molecular mass. Osmotic pressure reflects solution entropy; it is one of several *colligative properties*,[18] which all depend on the number of particles in a solution (and hence offer opportunities for measurement of molecular weight). If the solute is not a single particle, we must multiply c by the number of moles of particles per mole of solute (e.g., 2 for NaCl).

1.6 PHASE EQUILIBRIA

1.6.1 Equality of Chemical Potential

Melting, boiling, and sublimation are familiar transitions between different phases of the same pure material, brought about by changes in temperature and pressure. Less familiar are the transitions between different solid phases; for example, the relatively open structure of ice I (ice-cube ice) collapses to more compact forms of ice at high pressures.

When two phases of a pure compound are in equilibrium, the chemical potential of the compound is the same in each phase; one can prove this easily by considering the transfer of a small amount, dN, of compound from one phase (B) to the other phase (A), for which the free energy change is $dG = dN(\mu_A - \mu_B)$; this must be zero at equilibrium according to Eq. 1.19. Equality of chemical potential in the two phases holds only at specific combinations of pressure and temperature.

If the system contains more than one compound, the equilibrium condition at constant P and T

$$dG = -dG_A + dG_B = -\left(\sum \mu_{i,A} dN_i\right) + \left(\sum \mu_{i,B} dN_i\right) = 0, \qquad (1.59)$$

for *any* choice of the dN_i, which gives as the *essential condition of phase equilibrium* that the chemical potential of each component is the same in every phase (or state)

$$\mu_{i,A} = \mu_{i,B}. \qquad (1.60)$$

1.6.2 Transfer Free Energy

The chemical potential of a compound in solution can be related experimentally to the chemical potential of the pure compound, if one is able to establish equilibrium with a phase containing the same compound in pure form (crystalline solid or dilute gas) for which the chemical potential is known by standard methods described in any thermodynamics text. One then has

$$\mu_{1,c} = \mu_{1,s} = \mu^{\circ}_{1,s} + k_B T \ln a_{1,s} \text{ (for a crystalline phase)}$$

or

$$\mu_{1,v} = \mu^{\circ}_{1,v} + k_B T \ln c_{1,v} = \mu^{\circ}_{1,v} + k_B T \ln a_{1,s} \text{ (for a vapor phase)}. \qquad (1.61)$$

If the pure phase is a vapor, one may determine the standard chemical potential in the solution, $\mu^{\circ}_{1,s}$, by extrapolating to low concentration, where activity equals concentration. If the pure phase is a crystal, and the solution is not dilute, then the extrapolation to dilute solution can be done by measuring the chemical potential of the solvent (from equilibrium with its vapor) and applying the Gibbs–Duhem relation to relate its change upon dilution to the corresponding change in the chemical potential of the solute.

Such a difference between standard chemical potentials is called a *transfer free energy*. Measurement presupposes measurable solubility in the chosen solvent and either a measurable volatility or existence of a crystalline phase. While the *concept* of a transfer free energy applies also to solutions of a macromolecule, its measurement is usually impossible because neither the gaseous form nor a pure crystalline form of the macromolecule can be realized experimentally. Transfer free energies of small molecules related to the structure of proteins have been measured and play an important role in understanding the effect of the solvent on the stability of native conformations of proteins and nucleic acids. These include values based on solubility of amino acids and peptides, and on transfer equilibrium to the dilute gas phase of a series of more volatile molecules related to amino acid side chains. This will be discussed in detail in Chapter 16.

1.6.3 Phase Diagram of a Pure Compound

For a system containing only a single component, the chemical potential of each phase is a different function of T and P; consequently, two phases can coexist in equilibrium only at certain combinations of T and P, at which the chemical potentials happen to have the same value. If, say, T is changed, two phases no longer coexist and the system exists as a single-phase region of its phase diagram. Indeed, for most combinations of T and P, only one phase is stable. However, P can be adjusted at the new value of T to reestablish phase equilibrium. Thus, for each T, there is a unique P for which two phases coexist. These T and P pairs define *coexistence curves* separating regions of stability of single phases. A coexistence curve establishes a *relation* between T and P, that is, P is a function of T or *vice versa*. A *triple point* is where two coexistence curves intersect, that is, where three phases (A, B, and C, which can be a crystalline form, liquid, and vapor) are at equilibrium. At this triple point, $\mu_A(T, P) = \mu_B(T, P) = \mu_C(T, P)$, and we have two equilibrium constraints.

The results can be represented in the form of a *phase diagram*. A simple example illustrates how such a diagram is interpreted (Fig. 1.7). The slope of the solid–liquid coexistence curve, AB, is negative because liquid water has higher density than ice. (This is atypical; most solids have higher density than their liquid forms.) The triple point occurs when ice and liquid water are in equilibrium at near-zero atmospheric pressure; the pressure of the water vapor in equilibrium with these two phases is only 0.006 atm (4.6 mmHg). The phase diagram also displays a so-called *critical point*: distinction between liquid and vapor phases usually becomes less marked at high temperature and pressure, as the vapor becomes denser as a result of the increase in pressure and the liquid expands because of the increase in temperature; finally, liquid and vapor phases become indistinguishable and the coexistence curve between these two phases ends.

1.6.4 The Gibbs Phase Rule

More complicated phase diagrams result in systems containing more than one component. These can be constructed using a relation known as the *Gibbs phase rule*. In general, we have seen that the number of independent intensive variables (those not dependent on N, the number of molecules), required to define the thermodynamic state of a system is 2 (e.g., T and P or T and V/N). This is called the *number of degrees of freedom* (n_F) of the system. Gibbs recognized that, if more than one phase was present, each equality of chemical potentials across all phases (Eq. 1.60) imposed an additional constraint on the system that reduced n_F, while

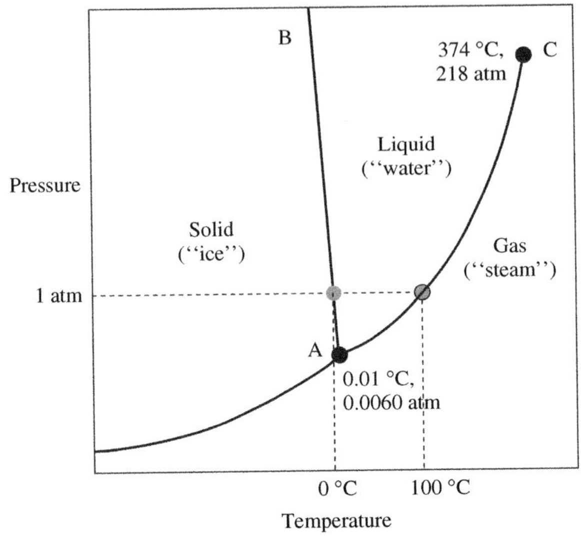

FIGURE 1.7 Schematic phase diagram separating solid, liquid, and vapor phases of water, showing triple point (A), critical point (C), and melting and boiling points at atmospheric pressure.

conversely the variable composition of each phase produced additional degrees of freedom. The phase rule states that n_F is two plus the number of components present beyond one ($n_C - 1$), and reduced by the number of constraints, which is equal to the number of phases present, less one ($n_P - 1$),

$$n_F = 2 + (n_C - 1) - (n_P - 1) = 2 + n_C - n_P.$$

We will not go into the phase rule or the process of constructing phase diagrams here in detail (any text on Physical Chemistry will treat this).

Typically, in biophysics, the phases contain *mixtures of components*. If two phases are in equilibrium, the chemical potentials of each component, μ_j, must be the same in each phase, and, of course, these now also depend on the concentrations of all components,

$$\mu_{j,A}(T, P, \{N_{i,A}\}) = \mu_{j,B}(T, P, \{N_{i,B}\}) \tag{1.62}$$

For a two-component system, three independent variables define the state of the system (T, P, $X_1 = 1{-}X_2$). If we fix P, the condition of phase equilibrium (Eq. 1.62) establishes a relationship between T, $X_{1,A}$, and $X_{1,B}$ that defines coexistence lines of a *temperature-composition phase diagram.*

The most commonly encountered phases in biophysics are vapor and dilute aqueous solution; with the exception of ice crystals, pure solid phases are uncommon. Of great importance is the phospholipid lamellar (bilayer) phase that provides the essential structure of biological membranes that delimit and compartmentalize the cell. This phase self-assembles in water because water is largely excluded from the lipid, and the lipid is very poorly soluble in water. However, solutes of medium polarity can and do partition between the aqueous and lipid lamellar phases; the ability to equilibrate between bilayers and aqueous phases is thought to be the basis of activity of some anesthetics. Equilibration of immiscible liquids produces two liquid phases, each of which consists predominantly of one of the components, with a small admixture of the other (e.g., ether and water). Typical for macromolecular solutes is coexistence of a *solvent-rich phase* with a *solute-rich phase.* Examples include precipitation of proteins by addition of a poor solvent component and formation of protein crystals from supersaturated mother liquor.

1.7 CHEMICAL EQUILIBRIA

When a chemical reaction takes place, the amounts of the reactants decrease and the amounts of the products increase. For a reaction having reactants with *stoichiometries* ν_i, a small change in the extent of reaction (represented by δs) leads to changes in the amounts of each of the components present in the reaction mixture

$$\delta N_i = \nu_i \delta \mathrm{s} \tag{1.63}$$

For example,

$$\mathrm{A} + 2\mathrm{B} \leftrightarrow 2\mathrm{C}$$

$$\nu_{\mathrm{A}} = -1,\ \nu_{\mathrm{B}} = -2,\ \nu_{\mathrm{C}} = +2. \tag{1.64}$$

Whenever a reaction has proceeded to equilibrium (at constant T and P), a further backward or forward reaction must produce an increase in the Gibbs free energy, that is, G is at a minimum with respect to s.

$$\frac{\partial G}{\partial s} = \frac{\partial \sum N_i \mu_i}{\partial s} = 0$$

$$\Rightarrow \sum \nu_i \mu_i = 0 \tag{1.65}$$

This relationship can also be obtained from the Gibbs–Duhem equation.

Recalling that the chemical potential of any substance in solution is related to the activity via Eq. 1.55, we have

$$\sum \nu_i \mu_i = \sum \nu_i \mu_i^\circ + k_{\mathrm{B}} T \sum (\nu_i \ln a_i) = 0 \tag{1.66}$$

This condition for equilibrium can be rewritten as

$$\prod_i {a_i}^{\nu_i} = K_P$$

$$-k_{\mathrm{B}} T \ln K_P = \sum_i \nu_i \mu_i^\circ. \tag{1.67}$$

This defines K_P, the *equilibrium constant* at constant P in terms of the standard chemical potentials of the reactants and the stoichiometry of the reaction. In the case of the above-mentioned simple example this becomes

$$K_P = \exp[-(2\mu_C^\circ - \mu_A^\circ - 2\mu_B^\circ)/k_B T] = \left(\frac{a_C{}^2}{a_A a_B{}^2}\right)_{eq}. \tag{1.68}$$

For an ideal solution, the activities can be replaced by concentrations.

It is common to write

$$-k_B T \ln K_P = \sum_i \nu_i \mu_i^\circ = \Delta G^\circ, \tag{1.69}$$

where ΔG° is the *standard free energy change for the reaction*. Formally, this represents the free energy change when each of the reactants is taken from, and each of the products is placed in a solution in which its activity is equal to 1. For an ideal solution, these solutions have concentration equal to 1. As noted previously, the standard state represented by μ_i^0 is not pure reactant i, but is an imaginary state in which each component is surrounded only by the solvent (see discussion of Henry's Law). In many biophysics publications, standard free energy changes are (alas) reported as ΔG, with omission of the superscript.[19]

When a chemical equilibrium is written as in Eq. 1.63, there is no clear distinction between reactants and products. By virtue of the Second Law, a chemical equilibrium will favor that side of the equation having lower free energy, and in a chemical reaction this would be the product side, with the other side, of higher free energy, then containing the reactants. Nevertheless, many important metabolic reactions seemingly proceed from reactants to products of *higher* free energy. However, upon closer study this is found to never be the case, as such an unfavorable reaction is always *coupled* to another reaction that is very favorable. Say, we have a synthesis of a compound AB, from reactants A and B,

$$\mathrm{A + B \rightleftharpoons AB} \tag{1.70}$$

that is unfavorable (as indicated by the longer arrow pointing to the left). Then, either this reaction can proceed in two (or more) favorable steps, in which first one (or more) reactants is converted to a different form

$$\begin{aligned} &\mathrm{A + XY \rightleftharpoons AX + Y} \\ &\mathrm{AX + B \rightleftharpoons AB + X,} \end{aligned} \tag{1.71}$$

or, the favorable and unfavorable reactions may be directly coupled in a single reaction, as in

$$\mathrm{A + XY + B \rightleftharpoons AB + X + Y.} \tag{1.72}$$

In both cases, the breakdown of one compound (XY) "produces" the free energy (and more) needed for the biosynthesis of the other (AB). Compounds such as

XY are called *high energy compounds*; a much encountered example is adenosine triphosphate (breaking down to adenosine diphosphate and phosphate ion, or to adenosine monophosphate and pyrophosphate ion).

1.8 TEMPERATURE DEPENDENCE OF CHEMICAL EQUILIBRIA: THE VAN'T HOFF EQUATION

As the standard free energy of a reaction can be found from the equilibrium constant, the enthalpy, which is related to the temperature dependence of the free energy by Eq. 1.30, can be found from the temperature dependence of the equilibrium constant as

$$\Delta H^{o} = \frac{\partial[\Delta G^{o}/T]}{\partial[1/T]} = -k_{B}\frac{\partial \ln K_{P}}{\partial[1/T]} = k_{B}T^{2}\frac{\partial \ln K_{P}}{\partial T} \tag{1.73}$$

This is the van 't Hoff equation. It states quantitatively the common observation that "an endothermic equilibrium ($\Delta H^{\circ} > 0$) shifts to the right" when the temperature is raised.

1.9 MICROCALORIMETRY

A classical approach to measuring thermodynamic functions is to use a calorimeter (a "heat meter"). It is easy to accurately supply energy in the form of heat input to a sample by passing an electric current through a heating element embedded in the sample, and thereby measure the specific heat, C_V, by monitoring the resultant change in temperature. Two types of sensitive microcalorimeters are available, a scanning type, in which the temperature is varied over an interval by continuous heating, and a mixing type, in which successive aliquots of reactant are added to the sample. Scanning microcalorimetric studies of proteins were pioneered by J. Sturtevant; subsequently, much work has been done with instruments developed by J. Brandts in the United States. and by P. Privalov in the (then) Soviet Union; elements of both these instruments have been incorporated in commercially available instruments. We describe here both scanning and mixing type instruments, according to one particular design strategy.

A *differential scanning calorimeter* is one in which a sample (e.g., protein plus buffer) and a control (only buffer) are isolated as much as possible from the environment and from each other. The instrument is surrounded by a water bath that tracks as closely as possible the temperature of the samples, in order to suppress heat flow between reservoir and sample ("pseudo-adiabatic"). The reservoir is gradually heated and the temperature differences between the sample (ΔT_1) and the buffer (ΔT_2) ampoules and the reservoir are measured (Fig. 1.8). A feedback circuit causes electrical heaters in contact with each ampoule to provide enough heat so as to keep $\Delta T_1 \approx \Delta T_2 \approx 0$. The block is designed so that heat cannot flow

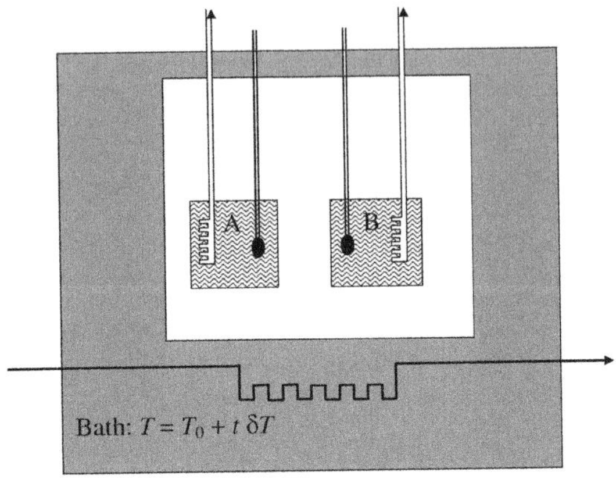

FIGURE 1.8 Diagram of a pseudo-adiabatic differential scanning calorimeter. The bath is heated at a steady rate. The cells are isolated from the bath; the temperature in each cell is sensed separately, and feedback mechanisms supply current to maintain each cell at the temperature of the bath. The difference in current supplied to the two cell heaters is recorded.

between the sample and reference, so that any heat that is produced (exothermic change) or absorbed (endothermic change) by the sample relative to the buffer reference is recorded as a difference in electrical heat energy supplied to the sample and reference ampoules ($[I^2Rt]_1 - [I^2Rt]_2$), as measured with a sensitive ammeter and recorded as a function of temperature. A temperature-induced transition or change in the sample ampoule is thus recorded as a difference in heat flow into the sample versus the reference. Sophisticated modern calorimeters can measure very small differences in heat flow (~10 ncal/s).

By measuring the heat required to take a sample relative to buffer from (nearly) 0 K to higher temperature by heating one can obtain values of C_V or C_P as a function of temperature. This can be integrated to give both internal energy and entropy changes; see Eqs. 1.36 and 1.37.

If the volume of the system is held constant during the heating process so that no PV work is done on or by the system, the procedure yields C_V, S, U, and A. If the pressure is held constant, and the volume changes during heating, the system will do work ($P\Delta V$) on its surroundings, thus reducing the change in system internal energy as a result of input of the same amount of heat. In this case, C_P, S, H, and G are the state functions obtained from the calorimetric measurement.

The enthalpy of a temperature-induced transition can be measured with such an instrument, provided that the enthalpy of the transition, and therefore the temperature dependence of the equilibrium constant (Eq. 1.73), is sufficiently large. This will be discussed in Chapter 16 where we describe how microcalorimetry is applied to study the folding–unfolding transitions of proteins.

A calorimeter can also be set up as a mixing calorimeter to measure reaction enthalpies. Figure 1.9 shows an *isothermal titration calorimeter* used for this

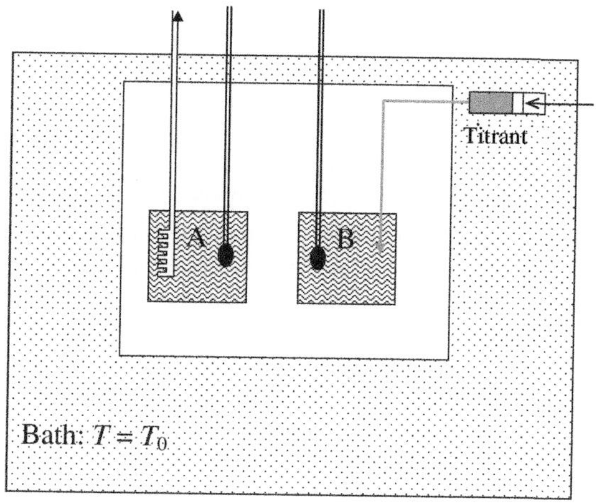

FIGURE 1.9 Diagram of an isothermal titration calorimeter (design principle similar to Fig. 1.8). At the start, sample (B) and reference (A) have identical contents. Titrant is added in small aliquots to the sample cell, and current is supplied to the reference to maintain the same temperatures in the two cells. (If the reaction is endothermic, heating is applied to cell B.) If necessary, the temperature of the bath can also be adjusted.

purpose. A sample solution is placed in an ampoule and stirred continually with a very precise stirring device so that the heat of stirring can be carefully recorded. Another solution (titrant) is added from a chamber at the same temperature as the sample. With each addition of titrant, one adds I^2Rt heat to the reference ampoule (also containing sample solution) to keep the temperatures of the sample and reference identical (if the reaction is exothermic. In the case of an endothermic reaction, the compensating heat is supplied to the sample). An example of data from such an experiment is discussed in Section 8.3 (Fig. 8.7).

Note that calorimetric enthalpy and heat capacity are intrinsically determined per unit of sample mass (e.g., gram), while enthalpy and heat capacity derived from the temperature dependence of equilibrium constants via the van't Hoff equation are determined per mole.

NOTES

1. As a convention in this book, a change of a thermodynamic function, such as $U_b - U_a$, is denoted either as $\Delta U_{a\to b}$ or as ΔU_{ab}. The corresponding notation for equilibrium constants is $K_{ab} = K_{a\to b} = c_b/c_a$, and similarly for rate constants k_{ab} or $k_{a\to b}$ represents the forward rate constant in the reaction $a \to b$.
2. Boltzmann's constant, k_B is expressed in *units* of energy/(temperature × amount of material). The *value* of k_B depends on the choice of units for any of these three quantities. If the amount of material, N, is expressed in number of molecules versus number of moles, the values of k_B differ by a factor of Avogadro's number, N_A. At one time, we distinguished between these two cases by assigning the name Boltzmann's constant (k_B)

to the value per molecule, and the name gas constant ($R = N_A k_B$) to the value per mol. In modern physics, the convention is to refer only to k_B, with the units being understood from the context in which it is used.

3. Also called, respectively, quasi-static and real processes.
4. Complete discussions of the Carnot cycle and how it leads to the conclusion that S is a state function can be found in advanced texts on thermodynamics.
5. A "heat reservoir" is a body at constant volume that does no work but simply provides or absorbs heat. The heat taken up or given off is thus equal to its change in internal energy (ΔU) and is thus a state function of the reservoir.
6. Because the dependence of U on T is the same at both pressures.
7. A (real) adiabatic process can be accomplished as an isothermal real process followed by exchange of heat with a bath at a different temperature. The entropy of the universe rises in both steps.
8. If an ideal gas expands without performing work (when the volume is suddenly increased), the ability to perform work decreases, so that A decreases. But no work is done, so the energy U is constant. In this *irreversible* process the entropy has decreased but no heat has been exchanged.
9. Other types of work include electrical work ($e\Delta E_{el}$), volume change of a gel ($\Pi\Delta V$), and mechanical work including elastic extension ($F\Delta L$), where e = charge, E_{el} = electrostatic potential, Π = osmotic pressure, F = force, L = distance or length.
10. For example, to derive the expression for dA, the definition $A = U - TS$ gives $dA = dU - TdS - SdT$. Then, substitution for dU according to the fundamental statement of the first and second laws, Eq. (1.13) gives $dA = dU - TdS - SdT \leq TdS - PdV - \partial w - TdS - SdT$, or $dA \leq -SdT - PdV - \partial w$.
11. However, for an ideal gas $G = A + PV = A + N k_B T$.
12. In this chapter, the symbol N represents number of molecules, but can equally well be taken to represent number of mols.
13. This is derived as $\partial(A/T)/\partial(1/T) = A + (1/T)\partial A/\partial(1/T) = A - T\partial A/\partial T = A + TS = U$, where we have substituted $\partial A/\partial T$ from Eq. (1.23).
14. The *proportions* of different components and P and T are held constant, so that the μ_i are constant during the integration.
15. J. Willard Gibbs was a great American scientist, professor at Yale, whose name is immortalized in the term Gibbs free energy, G.
16. Unless specifically stated otherwise, the choice of standard state is the pure gaseous, liquid, or crystalline solid state of a compound at $T = 25°\text{C}$ and pressure $P = 1$ bar $= 100\,\text{kPa} = 10^5\,\text{N M}^{-2} \approx 1.013\,\text{atm}$.
17. It is usual to number the components of a mixture $1, 2, \ldots, i$; if one of the components is in great excess, that is, a solvent, then it is given index 0.
18. Colligative properties include depression of freezing point and elevation of boiling point.
19. Note that $\ln K_P$ is properly defined only if K_P is unitless, which is the special case in which the number of molecules on each side of the reaction is equal (i.e., if $\sum \nu_i = 0$), or, in the general case, if the activities are expressed in mole fraction units. When an equilibrium constant is given in concentration units other than mole fraction, and $\sum \nu_i \neq 0$, it is ideally first converted to mole fraction units before one calculates ΔG° (something that admittedly is not always done).

2

Four Basic Quantum Mechanical Models of Nuclear and Electronic Motion: A Synopsis

Each theory is a model of reality
—Eisenberg and Crothers

2.1 INTRODUCTION

Newton's mechanics (or classical mechanics) provide an adequate description of large bodies, but small, rapidly moving particles such as electrons require a quantum mechanical treatment. The need for quantum mechanics was revealed by development of accurate spectroscopy in the late nineteenth and early twentieth centuries. Quantum mechanics has become indispensable as a basis for understanding the laws of chemistry: chemical bonding, chemical reactivity, interaction of molecules with radiation, and molecular energetics. Quantum mechanics teaches primarily that energy and mass are irrevocably linked. This fundamental postulate has the corollaries (i) that the energy of a system cannot be varied continuously, but is "quantized", (ii) that very low mass particles have some of the properties of waves, and (iii) and that energy previously thought to be represented by waves has some of the properties of particles.

Exact quantum mechanical descriptions of molecules require a great deal of mathematical sophistication. Indeed, the quantum mechanical description of biological macromolecules in solutions is so difficult, that it is more often bypassed

Equilibria and Kinetics of Biological Macromolecules, First Edition. Jan Hermans and Barry Lentz.

in favor of approximations; frequently, these are expressed in terms of classical (Newtonian) mechanics. Thus, as biophysicists, we usually can get by with an understanding of the basic principles of the quantum view of matter in order to understand the nature of the chemical bond or to interpret spectroscopic observations on macromolecules. Here we review these principles, emphasizing aspects that are particularly relevant to the main topics of this text.

2.2 FUNDAMENTAL HYPOTHESES OF QUANTUM THEORY

2.2.1 The Energy of Electromagnetic Radiation is "Quantized": the "Wave-Particle Duality"

This hypothesis was suggested to Planck and Einstein by measurements of the energy of electrons emitted by a metal surface struck by light of different frequencies (*photoelectric effect*). At the time, physicists found it very surprising that electrons were not emitted until the frequency of the light reached a threshold, and, above this frequency, the energy of emitted electrons increased linearly with frequency, with proportionality constant $h = 6.626 \times 10^{-11}$ J s. Bohr had already suggested that electrons move in orbits of discrete energies, and thus Planck's interpretation of the photoelectric effect (Fig. 2.1) was that light energy is also "discretized", and transmitted in small increments, "wave packets", E_ν whose magnitude is given by the product of Planck's constant, h, and the frequency of the radiation, ν

$$E_\nu = h\nu. \tag{2.1}$$

(The threshold frequency of the photoelectric effect corresponds to the energy needed to release an electron from the metal surface, with the electron's energy corresponding to the difference between E_ν and the release energy.) Einstein took this to mean that light possessed a particle property, and termed Planck's wave packets *photons*, or particles of light.

Dilute gases emit and absorb light at sharp maxima at discrete wavelengths, rather than in a continuous spectrum. The idea that light energy is transmitted in

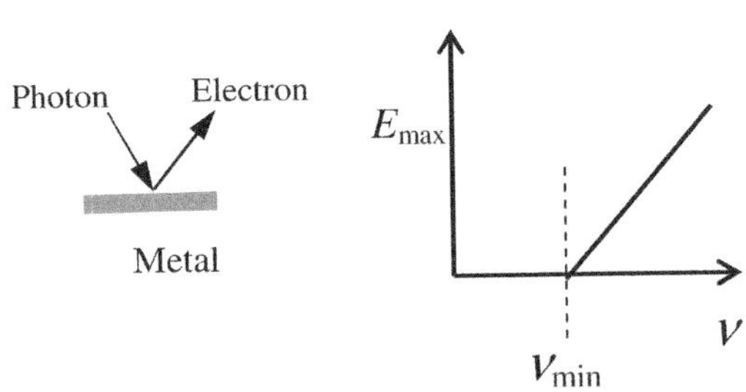

FIGURE 2.1 The photoelectric effect. Electrons are emitted only when $\nu > \nu_{max}$. Maximum energy of electron, E_{max} varies linearly with ν when $\nu > \nu_{max}$.

discrete packets then leads to the conclusion that atomic and molecular energies are restricted to specific energy levels, and that the spectra accurately reflect the energy differences between these, via Eq. 2.1.

2.2.2 Small Particles Have a Wave-Like Nature

De Broglie took this a logical step further and hypothesized that all particles had discrete energies defined by their wave-like nature. He proposed this relationship in his PhD thesis as an explanation of the observation that electron beams are diffracted like X-rays similar to electromagnetic radiation. As the energy of a particle moving in the absence of a potential is just kinetic ($p^2/2m$), he proposed that the wavelength associated with a moving particle with momentum $p = mv$ (mass m, velocity v) is

$$\lambda = \frac{h}{p} = \frac{h}{mv}. \tag{2.2}$$

2.2.3 Heisenberg's Uncertainty Principle

This corollary of de Broglie's proposal states that the exact position, x, and momentum, p, of a particle cannot be specified simultaneously:

$$\Delta x \Delta p \approx \Delta E \Delta t \approx h. \tag{2.3}$$

As noted, a similar relation exists between the uncertainty in the energy of a particle, ΔE, and the length of time, Δt, over which the particle can be observed or detected.

2.2.4 The Schrödinger Equation

A corollary of the Planck/Einstein and de Broglie hypotheses is Schrödinger's equation that describes the behavior of particles in terms of the following "wave" equation

$$\left[-\frac{h^2}{8\pi^2 m}\frac{\partial^2}{\partial x^2} + V(x)\right]\psi(x) = \widehat{H}\psi = E\psi(x)$$

$$-\frac{h^2}{8\pi^2 m}\nabla^2\psi(x) \equiv -\frac{h^2}{8\pi^2 m}\frac{\partial^2}{\partial x^2}\psi(x) = [E - V(x)]\psi(x). \tag{2.4}$$

∇^2 is the Laplace operator, which is here expressed in one linear dimension (see Appendix I). Schrödinger was a mathematician who studied wave motion and wrote this equation as a means to test de Broglie's proposal. The term in brackets is the *Hamiltonian operator* (or *energy operator*, $\widehat{H}$), and E is the expectation value of the energy operator obtained by $\widehat{H}$ operating on the *wave function* ψ. The Hamiltonian operator has two parts. The second part, which is the potential energy,

V, simply multiplies ψ. By carrying it to the right hand side, one sees that the remaining first part of the operator, when operating on ψ produces the product of ψ and the difference between total and potential energy, that is, the kinetic energy. Therefore, the first term in $\widehat{H}$ is called the *kinetic energy operator*, which takes the same form for any particle. Note that this operator determines the curvature of the wave function; consequently, more gently curved wave functions are associated with lower kinetic energy. The dependence of V on the coordinates, x, varies for different models and is responsible for the different behaviors of particles in different physical situations.

The form in which the Schrödinger equation is given here is for stationary states, that is, time-invariant states. A more general formulation describes both the time and position variation of the wave function, $\Psi(x,t)$.

In general, the Schrödinger equation has an infinite set of stationary solutions, ψ_i, called the *eigenfunctions* of the Hamiltonian operator, each one with its corresponding expectation value or *eigenvalue* of the Hamiltonian, as defined by the Schrödinger equation. An eigenvalue represents the total energy of the system, $E_{i,}$ in a particular *state* described by a wave function or eigenfunction. While the position of a particle in the ith quantum state is not defined by the wave function, Max Born proposed that the square of the value of the wave function, $\Psi_i(x,t)\Psi_i^*(x,t)$, is the probability that it can be found at position x at time t with energy $E_i = (p_i^2/2m)$, where $\Psi_i^*(x,t)$ is the complex conjugate of $\Psi_i(x,t)$. If $\Psi_i(x,t)$ is real, as in the models discussed next, $\Psi_i^*(x,t) = \Psi_i(x,t)$. This provided a link between what many (e.g., Born, Heisenberg, Bohr) saw as merely theoretical conjecture and the experiments that allow us to observe what is "real". While this is the most common interpretation of $\Psi_i(x,t)$, debate over its meaning and use continues even today. Application to the wave function for a stationary state (here identified with lower-case ψ) gives the probability distribution for the electron as

$$\psi_i(x)\psi_i^*(x)dx = P_i(x)dx = \rho_i(x)dx. \tag{2.5}$$

This is also the electron density, ρ, a quantity that is accessible via accurate X-ray diffraction measurements.

2.3 THREE SIMPLE MODELS OF NUCLEAR MOTION

2.3.1 Introduction

The Schrödinger equation has simple solutions for three basic kinds of motion that play a central role in classical mechanics: translation, vibration, and rotation. The quantum mechanics descriptions of these motions are needed when the particle masses or the particle energies are very small; for macroscopic particles and larger energies (which in molecular systems mean at higher temperatures), the quantum and classical mechanics descriptions become indistinguishable. As we shall see, these simple models are a good starting point for describing molecular motions (global translation, global and internal rotation, and internal vibrations).

2.3.2 Particle in a Box

A particle confined to a box has potential energy equal to zero inside the box, but when it reaches the extremities of the box or is outside the box, it experiences an infinite potential (Fig. 2.2). As the particle cannot be outside the box, the wave function there is zero, and because the wave function is continuous, it is zero at the edge of the box. The particle's kinetic energy is not zero, and the Schrödinger equation is written as shown below. Its stationary (i.e., time-independent) solutions (inside the box, i.e., $0 < x < a$) are simple sinusoidal functions in x (Fig. 2.2), the linear position in the box:

$$-\frac{h^2}{8\pi^2 m}\frac{\partial^2\psi}{\partial x^2} = E\psi$$
$$\psi_n = (2/a)^{1/2}\sin(n\pi x/a)$$
$$E_n = \frac{n^2h^2}{8ma^2}, \quad n = 1, 2, 3, \ldots, \tag{2.6}$$

where a is the length of the box and m the mass of the particle. The factor $(2/a)^{1/2}$ is determined by the fact that the particle must be somewhere in the box for any n:

$$\int \psi_n\psi_n^* dx = 1, \tag{2.7}$$

where ψ_n^* is the complex conjugate of ψ. Note that the wave functions are "orthogonal", which means that

$$\int \psi_i\psi_{j\neq i}^* dx = 0. \tag{2.8}$$

Orthogonality is a general property of any wave function that satisfies the Schrödinger equation. We can equate the energy, E_n with the mean kinetic energy of a particle in the nth state:

$$E_n = \frac{n^2h^2}{8ma^2} = \left\langle\frac{mv^2}{2}\right\rangle_n$$
$$\langle v^2\rangle_n = \left(\frac{nh}{2ma}\right)^2. \tag{2.9}$$

From this, one sees that the (root-mean-square) velocities of the particle in successive states are evenly spaced, although this is not so for the energies, which are proportional to the mean-square velocities. We note that E_n (mean kinetic energy of state n) is smaller for larger boxes and thus for more gently curved wave functions. This general principle applies to all quantum mechanical systems, from the particle in a box to molecular orbitals to the "strings" inherent in the quantum mechanical description of gravity.

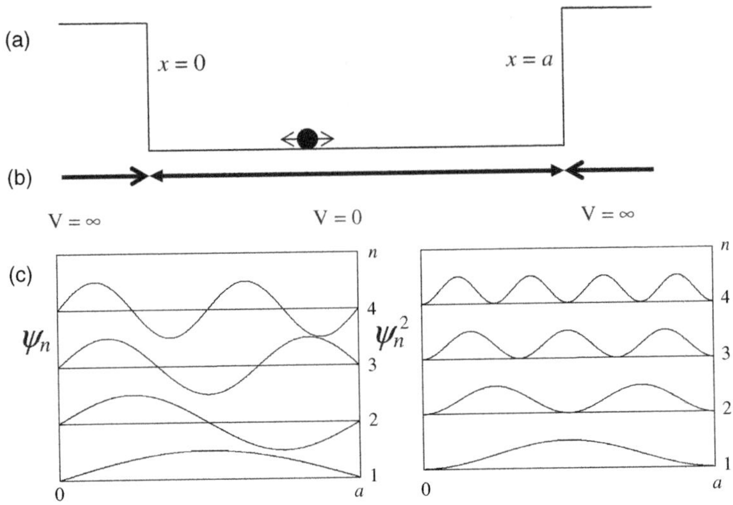

FIGURE 2.2 (a) Particle in a box is free to move between two walls, at $x = 0$ and at $x = a$. (b) The potential energy of the particle is zero inside the box, infinite outside it. (c) (left) The wave functions for quantum numbers 1, 2, 3, and 4; (right) the corresponding probability distributions.

There are three *uniquely quantum mechanical results* that cannot be predicted classically:

1. There are *nodes* in the wave function, or locations in the box where the particle has zero probability of being found (Eq. 2.6 and Fig. 2.2).
2. There is no state for which the particle is fixed in space ($n = 0$) and has zero energy. This reflects the fact that a particle whose position is defined within certain limits cannot be standing still because of the Uncertainty Principle (Eq. 2.6).
3. *Tunneling*: The particle is found with zero probability outside the box only if the potential energy outside the box is infinite or $V \gg E_n$. If the potential at the wall is finite, there is still a finite probability that the particle will occur where $V > E$, that is, at the wall and outside the wall. This is termed *tunneling*. In general, electrons can tunnel between low energy states as long as the potential barrier between these is not too large relative to the energy of the electron. Tunneling is not evident in our analysis of the particle in the box because this analysis assumes that V is infinite outside the box; it *is* evident in the behavior of the quantum mechanical harmonic oscillator as we shall see in the following section.

2.3.3 Harmonic Oscillator

This model extends our discussion of motion in a single linear dimension to the situation where the particle has both potential and kinetic energy. The potential

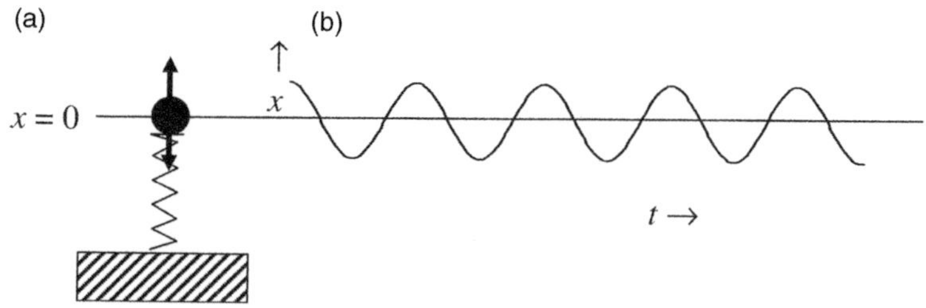

FIGURE 2.3 (a) A harmonic oscillator: Mass on a spring is free to oscillate up (positive x direction) and down (negative x) about an equilibrium position ($x = 0$). (b) The position of a macroscopic oscillating mass changes with time, t according to a cosine function, $x = \cos(2\pi\nu t)$, where ν is the frequency.

energy rises as a quadratic function of the position of the particle as it deviates from its equilibrium position at $x = 0$ (Fig. 2.3):

$$V = \frac{1}{2}kx^2. \tag{2.10}$$

According to classical mechanics, the particle oscillates about its equilibrium position with a frequency, ν

$$\nu = \frac{1}{2\pi}\sqrt{\frac{k}{m}}, \tag{2.11}$$

and the energy can have any value dictated by the maximum initial displacement of the particle from its equilibrium position.

In order to treat the problem quantum mechanically, one substitutes the potential energy of the harmonic oscillator in the Schrödinger equation, which gives

$$\left[-\frac{h^2}{8\pi^2 m}\frac{d^2}{dx^2} + \left(\frac{1}{2}kx^2\right)\right]\psi = E\psi. \tag{2.12}$$

The solution to this form of the Schrödinger equation is complicated by the existence of the potential energy term, but is still a set of oscillating functions given as $H_n(\xi) \times \exp(-\beta x^2/2)$, with $H_n(\xi)$ being the set of Hermite polynomials, $H_n(\xi)$, with $\xi = \beta^{1/2}x$ and $\beta = \frac{2\pi}{h}(mk)^{1/2}$]. The first two Hermite polynomials are given in Table 2.1. Higher order Hermite polynomials are found recursively with

$$H_{n+1}(\xi) = 2\xi H_n(\xi) - 2(n-1)H_{n-1}(\xi). \tag{2.13}$$

The probability distribution for finding the particle at any point, x, in its one-dimensional space is again found by taking the square of the wave function (see Fig. 2.4). Unlike the particle in a box, the particle on a spring is not confined to a particular region of space but can explore larger regions as it occupies higher energy states.

TABLE 2.1 The first two Hermite polynomials with corresponding harmonic oscillator wave functions and energy levels

n	E_n	$H_n(\xi)$	ψ_n
0	$\frac{1}{2}h\nu$	1	$\exp(-\beta x^2/2)$
1	$\frac{3}{2}h\nu$	2ξ	$(2\beta)^{\frac{1}{2}} x \exp(-\beta x^2/2)$

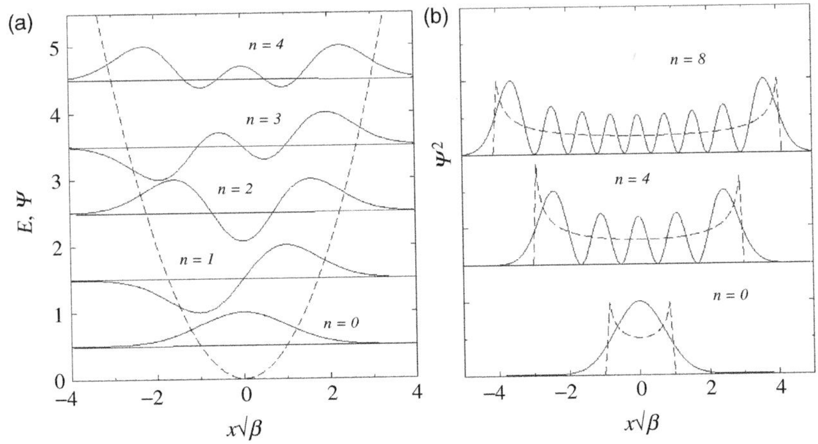

FIGURE 2.4 The quantum mechanical harmonic oscillator. (a) The curves are wave functions, ψ, for the first five energy levels, $h\nu/2$, $3h\nu/2$, $5h\nu/2$, $\ldots$, $9h\nu/2$ (which are represented by the horizontal lines). The dashed line follows the potential energy of the oscillator, $(1/2)Kx^2$. Each wave function extends slightly beyond this curve; this is called *tunneling*. (b) The probability distributions, ψ^2, for three energy levels are shown. The dashed curves show the corresponding energy and position distributions for classical oscillators with the same energy. From Kauzmann, W. *Quantum Chemistry*. New York: Academic Press, 1957.

The energy can assume any of the following discrete values ($n \geq 0$):

$$E = \left(n + \frac{1}{2}\right) h\nu. \qquad (2.14)$$

Note that the energy levels are evenly spaced.

As for the particle in a box, we find that the wave function has nodes. However, the lowest energy level is not zero, but equal to $1/2h\nu$, the *ground state energy*.

2.3.4 Rigid Rotator

The rotation of any polyatomic molecule *in a particular conformation* can be treated in terms of rotations about the three principal axes of a *rigid* asymmetric object ("top") having three unequal principal moments of inertia[1]: I_1, I_2, I_3. The largest possible moment of inertia of a rigid rotator is I_1, the smallest is I_3, and I_2 is

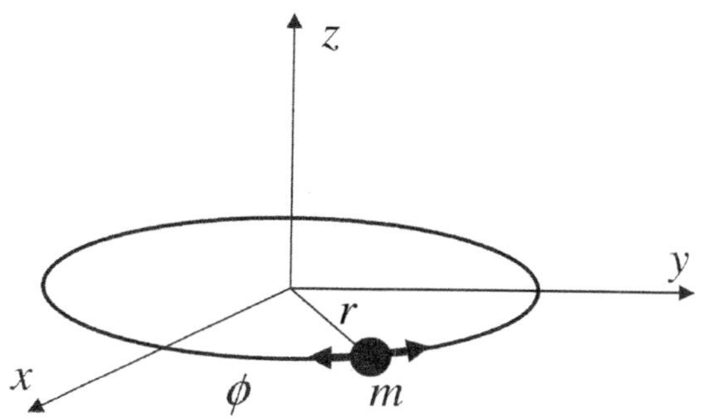

FIGURE 2.5 The simplest one-dimensional rigid rotator.

the moment about the axis perpendicular to the axes of the other two principal moments.

The general quantum mechanical treatment of rotation of an asymmetric top is complex. Even the treatment of a symmetric top (two equal moments) is complex. Fortunately, important features of the quantum mechanical treatment of rotation are contained in the treatment of *one-dimensional rotation*, that is, rotation about *a single, fixed axis*. The model rotator consists of a mass of magnitude m moving along a circle with radius r, sweeping out an angle ϕ (Fig. 2.5). Note that this figure shows motion in two linear dimensions, although it can be described in terms of one circular coordinate. The Schrödinger equation for this one-dimensional rotator can be most conveniently written in terms of the single spherical coordinate ϕ (see Appendix I). As is true for the particle in a box, the potential energy, V is zero for the rigid rotator.

Accordingly, the Schrödinger equation reduces to

$$r^2\nabla^2\psi = \frac{\partial^2\psi}{\partial\phi^2} = \frac{-2EI}{\hbar^2}\psi. \tag{2.15}$$

This is very similar to the Schrödinger equation for a particle in a one-(linear)-dimensional box, which has an oscillatory wave function. However, the fact that φ is an angle means that the wave function must be periodic, that is, satisfy the condition that $\psi(\varphi) = \psi(\varphi + 2\pi)$. This has two consequences. First, it sets a boundary condition. Second, in order for the wave function to be a real oscillatory function in x and y (sin or cos), it must be a complex oscillatory function in $\varphi(e^{\pm in\varphi})$. This leads to the following energy levels and associated normalized wave functions:

$$\psi_n = \left(\frac{1}{2\pi}\right)^{1/2} \exp(in\phi)$$

$$E_n = \frac{n^2\hbar^2}{2I}, \quad n = 0, \pm 1, \pm 2, \ldots . \tag{2.16}$$

Because these wave functions are complex, the product of ψ and its complex conjugate, ψ^* describes the probability of finding the particle at any point in its circular orbit in one of the stationary states. Unlike for the particle in a box or the harmonic oscillator, this function has no nodes, that is, there is an equal probability of finding the particle at any point (φ) on the circle, $1/(2\pi)$. One sees that for the rigid rotator, as for the harmonic oscillator, the level with $n = 0$ is allowed, but $E_n = 0$ in this state, a result different than seen for the harmonic oscillator. Recall that $n = 0$ was not allowed for the particle in a box because of the Uncertainty Principle.

A *new quantum mechanical property* is illustrated by the rigid rotator: the energy levels of the rotator are *degenerate*, with there being two wave functions (corresponding to clockwise and counterclockwise rotation) for each energy level.

For most moderately large and heavy molecules (certainly for all macromolecules but probably not for H_2), the moments of inertia are large, and the energy difference between successive levels is small, that is, $\Delta E = E_{i+1} - E_i = (2n+1)\hbar^2/(2I) \ll k_B T$. This can simplify a statistical description of the rotational contribution of particles to the thermodynamic properties of a system. For this reason, it is often convenient to compare the rotational properties of a particle to the experimental temperature. For this, we define the rotational temperature as $\Theta_{rot} = \hbar^2/2Ik_B$. Then, $E_n = n^2 k_B \Theta_{rot}$. If $\Theta_{rot} \ll T$, then the rotational energy levels are very closely spaced, and we say that the particle behaves classically with respect to rotation.

If the Schrödinger equation for the rotator is written, as of course it should be, in three dimensions, the result is similar, except that the nth energy level is $\omega_n = (2n+1)$ fold degenerate, which corresponds to a *discrete* number of directions that the axis of rotation can assume in space. Although the *energy levels* are not evenly spaced, the number of *states* for a given energy interval (density of states) approaches a constant value as n becomes large.

2.4 HYDROGEN ATOMIC ORBITALS: A SIMPLE MODEL OF ELECTRONIC MOTION IN ATOMS

Until now, we have been dealing with the motions of nuclei. We now turn our attention to the motion of electrons around nuclei. Of course, these motions contribute along with nuclear motions to the total energy of a macromolecule. Because the mass of the electron is so small, *electronic motion and energy in chemical systems can be described only with quantum mechanics.*

The hydrogen atom, with one nucleus and one electron, is the most important *simple* quantum mechanical system in chemistry. It is the basis of most quantum mechanical interpretations of visible and UV spectroscopic experiments and provides the framework for our understanding of atomic structure, UV, visible and IR spectroscopy, and, ultimately, of chemical bonding.

The hydrogen atom contains a nucleus and one electron, the nucleus and electron having, respectively, charge $+e$ and $-e$, and the potential energy depends on the distance between nucleus and electron, r according to Coulomb's law (see Chapter 3):

$$V(r) = \frac{-e^2}{4\pi\varepsilon_0 r}. \tag{2.17}$$

This problem combines the complications of the three-dimensional rotator with the complications of adding a potential energy function to the rotational motion. Because the problem has spherical symmetry, and because the problem deals with only two interacting particles, an analytical solution to the Schrödinger equation is possible. Because there are only two particles, we can simplify to examining the motion of a single "reduced mass" $(m_p m_e)/(m_p + m_e)$ round a fixed center. Because the proton is 1837 times more massive than the electron, we can treat the problem essentially as the motion of an electron around a fixed proton. The approximation of considering nuclei to be fixed in order to consider the motion of the much less massive electrons is termed the *Born–Oppenheimer approximation*, and is indispensable in treating problems more complex than the H-atom. The Schrödinger equation for this problem is

$$\widehat{H}\psi = -\frac{h^2}{8\pi^2 m}\nabla^2\psi + V\psi = -\frac{h^2}{8\pi^2 m}\nabla^2\psi - \frac{e^2}{4\pi\varepsilon_0 r}\psi = E\psi. \tag{2.18}$$

The motion occurs in three dimensions, and this equation is most conveniently solved by expressing the Laplace operator in spherical polar coordinates (see Appendix I). The set of wave functions that satisfy the Schrödinger equation are considerably more complex than any we have thus far encountered (Table 2.2). Four quantum numbers (only three of these are shown in this table) define the wave functions, but the energy of the system is determined only by the *principal quantum number*, n:

$$E_n = -\frac{1}{n^2}\frac{e^2}{2a_0}, \quad n = 1, 2, 3, \ldots$$

$$a_0 = \frac{h^2}{4\pi m e^2} = 0.0529\ \text{nm} = 0.529\ \text{Å}, \tag{2.19}$$

where a_0 is called the *Bohr radius*. Except for the ground state, that is, for $n > 1$, the energy levels are *degenerate*, with the degenerate states being described by two additional quantum numbers, the *azimuthal quantum number*, l (for the value of the angular momentum) and the *magnetic quantum number*, m (for its direction). The value of l can range from 0 to $n - 1$, the value of m from $-l$ to $+l$. The azimuthal quantum numbers are usually indicated with the familiar designations $s(l = 0)$, $p(l = 1)$, $d(l = 2)$, $f(l = 3)$.

All wave functions having quantum number $l = 0$ (s orbitals) are spherically symmetrical, and *no* wave functions for quantum number $l \neq 0$ have spherical symmetry. As is true for the particle in a box and harmonic oscillator, the H-atom probability functions have nodes, either planar (one plane for p, two for d, three for f), conical (two for d, three for f), or spherical. The squares of the wave functions are the probability distribution functions of electrons around the nucleus that define

TABLE 2.2 The first few hydrogen atom wave functions

Shell	n	l	m	Orbital	Wave function $\psi(n, l, m)$
K	1	0	0	1s	$(1/\sqrt{\pi})(Z/a_0)^{3/2}e^{-\sigma}$
L	2	0	0	2s	$(1/4\sqrt{2\pi})(Z/a_0)^{3/2}(2-\sigma)e^{-\sigma/2}$
	2	1	0	$2p_z$	$(1/4\sqrt{2\pi})(Z/a_0)^{3/2}\sigma e^{-\sigma/2}\cos\theta$
	2	1	± 1	$2p_x$	$(1/4\sqrt{2\pi})(Z/a_0)^{3/2}\sigma e^{-\sigma/2}\sin\theta\cos\phi$
				$2p_y$	$(1/4\sqrt{2\pi})(Z/a_0)^{3/2}\sigma e^{-\sigma/2}\sin\theta\sin\phi$
M	3	0	0	3s	$[1/(81\sqrt{3\pi})](Z/a_0)^{3/2}(27-18\sigma+2\sigma^2)e^{-\sigma/3}$
...					

Z is the nuclear charge, a_0 is the Bohr radius, and $\sigma = Zr/a_0$.

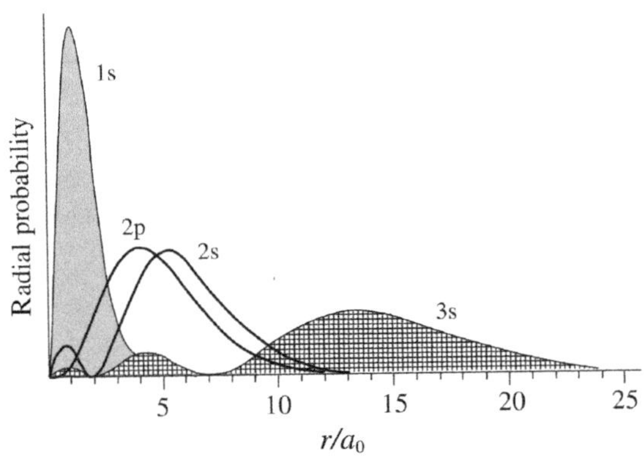

FIGURE 2.6 Electron probability, $\psi\psi^*$ of hydrogen atom wave functions versus distance from the nucleus, r. The 1s and 3s wave functions are highlighted. The probabilities shown are integrals over the surface of a sphere with radius r.

the atomic "orbitals".[2] The wave functions are antisymmetric[3] across their nodal planes, although the probability distributions they define are symmetric.

As shown in Fig. 2.6, the electron density, averaged over all orientations in space, for each wave function extends farther in space the higher the principal quantum number. The first few asymmetric hydrogen atom wave functions are illustrated with surfaces of constant fixed amplitude, $|\psi|$ in Fig. 2.7.

The equations for the wave functions for degenerate energy levels (i.e., all states above the ground state) as given in Table 2.2 are arbitrary. This is because two or more wave functions for the same energy level (same quantum number n) can be used in a linear combination, for example,

$$\psi_{n,h} = a_1\psi_{n,1} + a_2\psi_{n,2}, \tag{2.20}$$

to give a wave function that also satisfies the Schrödinger equation for that same energy level. Such a new wave function $\psi_{n,h}$ is called the *hybrid wave function.* (A resulting set of hybrid wave functions must be normalized and mutually

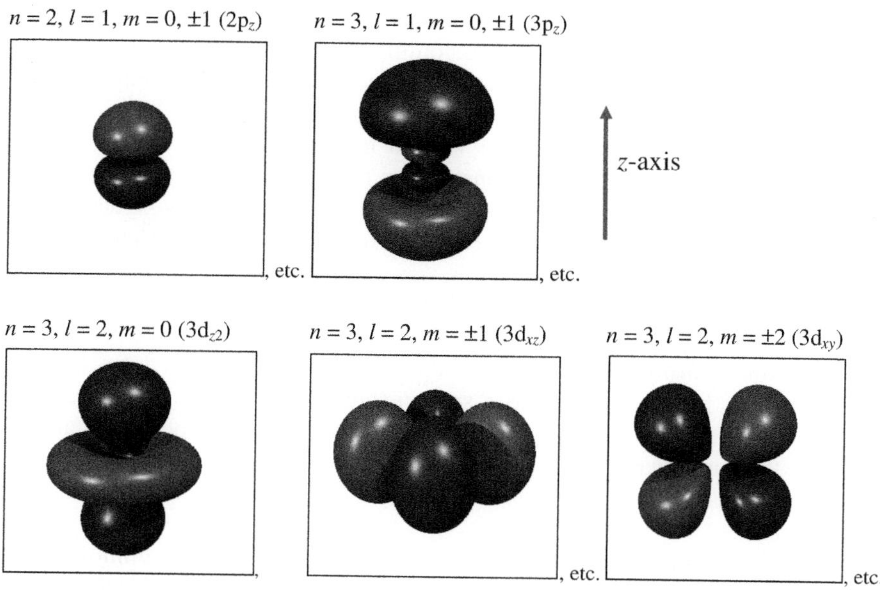

FIGURE 2.7 Asymmetry of some hydrogen atom wave functions. The direction of the z-axis is "up" tilted toward the viewer. Not shown are wave functions that are obtained from these by exchange of x, y, and z axes, indicated with "etc." Colors indicate positive (red) and negative (blue) values of ψ. From M. Hanrath, University of Cologne, http://www.tc.uni-koeln.de/people/hanrath/. (*See insert for color representation of the figure.*)

orthogonal.) Hybrid wave functions are useful in descriptions of chemical bonding, as will be discussed briefly in the following chapter. The wave functions of Table 2.2 are the basis for understanding the electronic ground states of atoms with more than one electron, as discussed in the next section.

In addition, every state (including the ground state) is twofold degenerate due to the fourth quantum number that is not shown in the functions shown in Table 2.2. This is the *electron spin quantum number*, m_s, which can have the values $+1/2$ and $-1/2$, called *spin up* and *spin down* to indicate the opposite directions of the spin magnetic moment of an electron. This quantum number is not given in Table 2.2 because it is not defined by the nonrelativistic Schrödinger equation given in Eq. 2.21 but requires a relativistic treatment due to Dirac (not discussed here). The degeneracy corresponding to m and m_s is broken in a magnetic field, which is the basis for nuclear magnetic resonance (NMR) spectroscopy.

2.5 MANY ELECTRON ATOMS

2.5.1 Analytical Approximations, Separation of Wave Functions, Pauli Exclusion, and Variation Principles

While the two-body hydrogen atom allows for a simple solution, once we add another electron (or another nucleus), the problem becomes mathematically intractable; the hydrogen atom is the only atom for which the Schrödinger equation has an analytical solution. Accordingly, a number of different methods have been developed to approximate the electronic wave functions of many electron atoms and molecules with analytical expressions.

An important approximation is to write a many electron wave function as a product of one electron wave functions. Doing this, one ignores the *correlation* of the electrons' positions; the electron–electron interactions can now be described in terms of the Coulomb energy of the *charge distribution*, defined in terms of each electron's wave function, or "orbital".

Whatever analytical form is chosen for the wave function, this is subject to the following two important principles.

The *Pauli exclusion principle* states that two electrons cannot have the same wave function; because the electron spin quantum number can have two values ($+1/2$ and $-1/2$), two, and only two, electrons can have the same orbital and still not have the same wave function. A wave function for n electrons will thus be approximated as a product of n electronic wave functions, which can be identical only in pairs.

The *variation principle* states that the expectation value of the Hamiltonian (i.e., the system energy) in the true atomic or molecular ground state wave function is the lower bound of the energy of *any* atomic or molecular wave function, and this principle can be applied to improve any approximate wave function by systematically adjusting its parameters.

2.5.2 Aufbau and Hartree-Fock Methods

An approximation that has proved quite successful is to describe the electron distribution of other atoms in terms of hydrogen wave functions that are *scaled* to reflect both the larger nuclear charge and the presence of additional electrons. This way, the He atom's ground state is described in terms of two electrons of opposite spin occupying the same modified 1s orbital, which is a simple exponentially decaying function of the electrons' distance from the nucleus (Table 2.2). This exponential function contains the nuclear charge Z; this is chosen as an adjustable parameter, $Z < 2$ in order to account for shielding of the nuclear charge by the other electron's charge distribution. The value of Z is set by using the variation principle by minimizing the total energy (electrons' kinetic energy plus the Coulomb energy for interactions of the electrons with the nucleus and of the two electron distributions).

For atoms with more than two electrons, additional orbitals are needed, and these will have to be based on hydrogen wave functions of higher principal quantum number and higher energy. A simple approximation is to assume that all interior electrons provide charge "shielding" for outer electrons. The outer electrons can then be treated successively by the single electron wave functions of the hydrogen atom, with a varying net "nuclear" charge, Z_{eff}, that is calculated using the wave functions for all interior electrons. This is the *aufbau* ("build-up") *method* for electron energy levels. The more effectively the other electrons shield the nuclear charge, the closer to one will be the value of Z_{eff}, and the less strongly the electron is bound. The electrons in inner shells provide excellent shielding for electrons in outer shells, while electrons in the same shell provide the weakest shielding. Consequently, Z_{eff} for the one electron of Na that is placed in the M shell is small, while the opposite is true for one of the seven electrons in the M shell of Cl; this renders it easy to move one electron from Na to Cl to form a Na^+ and a Cl^- ion, that are further stabilized by interactions with water, or with other ions in a NaCl crystal.

In order to take a more accurate account of electron interactions, one can follow an iterative procedure in which the wave functions of all but one electron are fixed and the wave function of the variable electron is adjusted to minimize the energy. Then, that electron's wave function is fixed and another is freed, and the wave function of the freed electron is obtained. This procedure continues until the energy of each electron converges. In this way, one can describe all the electronic states and energies of even the largest atom. The procedure is called the *Hartree-Fock Self-Consistent Field Method* and is widely used also in studies of molecular structure. We shall return to this in more detail in our discussion of electronic structure of molecules in Chapter 3.

The point of our discussion of atomic structure is twofold: first, to show that the H-atom is the most important simple model in chemistry and is the basis of all we know about atomic structure, and, second, to set up our discussion of molecular structure.

NOTES

1. The moment of inertia is a measure of the amount of energy needed to set an object spinning about an axis. For an object consisting of several point masses, one has the following simple formulation $I = \sum_i m_i r_i^2$ where the summation is over all point masses, with m_i mass and m_i distance from the chosen axis of rotation. A body can spin freely with constant angular velocity only about a principal axis. All three principal axes of inertia pass through the object's center of mass. If the object has symmetry, one may use one or both of the following rules to find the principal axes: (i) a plane of symmetry is perpendicular to a principal axis and (ii) an axis of symmetry is a principal axis.
2. By comparison with an earlier description of atomic structure given by Bohr, these are called *orbitals*. This is an unfortunate nomenclature, as it suggests that electrons, like planets, move in predictable orbits around nuclei, which, of course, as quantum mechanical particles, they do not.
3. A symmetric function has the same value on opposite sides of the symmetry element (inversion center, mirror plane, or rotation axis). An antisymmetric function has the same magnitude, but opposite sign at positions related by the symmetry operation.

SUGGESTED READING

Eisenberg, D.S., Crothers, D. *Physical chemistry: With Applications to the Life Sciences*. Menlo Park, CA: Benjamin/Cummings, 1979.

Atkins, P., de Paula, J. *Physical Chemistry*, 7th ed, New York: Freeman, 2001.

Kauzmann, W. *Quantum Chemistry*. New York: Academic Press, 1957

3

Molecular Structure and Interactions

3.1 INTRODUCTION

In this chapter we extend the methods of quantum mechanics (QM) to describe molecular systems. We see that all cases, starting with the simple hydrogen molecule ion, with just two nuclei and one electron, require approximations and can only be solved numerically, at considerable computational cost if high accuracy is a goal. We then show how, with the quantum mechanical model as basis, it is possible to develop the molecular mechanics (MM) model, with a continuous energy function that approximates the quantum mechanical description, and allows one to perform tens of thousands of calculations in the time required for a single accurate *ab initio* evaluation of structure and energy.

3.2 CHEMICAL BONDING: ELECTRONIC STRUCTURE OF MOLECULES

3.2.1 Principle

As noted in the previous chapter, once the complexity rises above that of the hydrogen atom, no analytical solutions of the Schrödinger equation exist. While we have seen how to solve this problem fairly simply for multielectron atoms by approximating these in terms of the spherically symmetric H-atom, the same approach does not apply to molecules, which cannot be approximately treated as

Equilibria and Kinetics of Biological Macromolecules, First Edition. Jan Hermans and Barry Lentz.

spherically symmetric even for the simplest molecule, the hydrogen molecule ion, H_2^+, containing just two nuclei and a single electron.

In treating this and other molecules, it is always assumed that the motion of the nuclei can be ignored when treating the motion of electrons, as nuclei are 1837 times more massive than electrons. (This is the Born–Oppenheimer approximation.) This assumption has the effect of ignoring coupling between nuclear and electronic motions. Given fixed nuclear positions, an approximation to the electronic wave function must be found, so that the energy can be calculated; the nuclear positions can then be optimized by seeking the structure of lowest energy. This leaves two problems to be solved, namely, how to construct an approximate wave function and how to optimize it. The answers are conceptually simple:

1. Approximate wave functions can be constructed as linear combinations of simpler functions, with sets of atomic wave functions being ideally suitable functions (although not the only sets of functions used for this purpose). Using the historic name "orbital" for wave function, this consists of constructing a *molecular orbital* (MO) as a *linear combination of atomic orbitals* (LCAO) of the isolated atoms.
2. The MO wave function is optimized by adjusting the coefficients multiplying the component wave functions in order to lower the energy, using the variation principle. This has already been mentioned in our discussion of the structure of many electron atoms in Section 2.5.

3.2.2 LCAO–MO Method and Variation Principle

The LCAO–MO method uses a special property of the Hamiltonian operator, namely, that it is a *Hermitian* operator. Hermitian operators have several useful properties. One is that their eigenfunctions are *orthogonal* and form a *complete set of basis functions* (see textbooks on Quantum Chemistry, e.g., Bohm, for a complete discussion). Orthogonality has been introduced in the previous chapter (Eq. 2.8). "Completeness" means that any wave function that satisfies the Schrödinger equation for a particular Hamiltonian can be represented as a sum of contributions from the members of an infinite complete set of wave functions of another Hamiltonian having the same form. Thus, if $\{\psi_i\}$ is the set of wave functions of the hydrogen atom Hamiltonian, then any wave function for any molecular arrangements of nuclei and electrons (same basic Hamiltonian, but more complex geometry) can be written as a linear combination (here indicated with Ψ_j) of the "basis" set

$$\Psi_j = \sum a_{ij}\psi_i, \tag{3.1}$$

where a_{ij} is a coefficient that gives the contribution of ψ_i to Ψ_j. This means that any MO (wave function for a collection of atoms) can be represented as an LCAO of these atoms.

The appropriate linear combination is found using the variation principle (see Section 2.5). For a true wave function that satisfies the Schrödinger equation, the ratio $(\widehat{H}\psi)/\psi$, which is equated with the energy, is invariant, that is, independent of the coordinates. However, this is not the case for a function that is not a true wave function, in which case one defines the energy as an average (the expectation value) over the wave function. The expectation value (or observed value) of any operator in a quantum state is written as the integral over all space of that operator operating on the wave function, the result multiplied by the wave function of that quantum state; this is then divided by the integral of the square of the wave function. The energy in quantum state j is thus written as

$$\bar{E}_j = \frac{\int \Psi_j \widehat{H} \Psi_j^* d\mathbf{x}}{\int \Psi_j \Psi_j^* d\mathbf{x}}, \tag{3.2}$$

where the horizontal bar indicates the expectation value and **x** *represents the multi-dimensional vector of all spatial coordinates.*

Accordingly, the jth molecular (or atomic) orbital wave function (Ψ_j) can be obtained by application of the variation principle, by finding the coefficients a_{ij} that minimize the energy $\bar{E}_j$ of that molecular (or atomic) quantum state.

3.2.3 Examples: Hydrogen Molecule Ion and Hydrogen Molecule

As a first example, we show how this procedure leads to a set of MO wave functions for the H^+-ion molecule. If the two nuclei, A and B, are far apart, the single electron can reside close to one or the other. If this is treated as two independent systems, one a hydrogen atom and the other a hydrogen nucleus, the "molecule" is described with wave function 1*s* centered on one of the two nuclei. (We consider the electronic ground state.) If the electron is free to be on either nucleus, the wave function must reflect this equal probability, plus the fact that the electron density near each nucleus is just half what it would be in the absence of the other nucleus. The wave function is then

$$\Psi_\pm = (\psi_{1s_A} \pm \psi_{1s_B}). \tag{3.3}$$

It is easily verified that Ψ_+ and Ψ_- are orthogonal, while Ψ_+ and Ψ_- are not normalized. Using the wave function $\Psi_\pm$ also when the distance is not large, one finds that the single electron distributes between the two nuclei as shown in Fig. 3.1. Addition of the two component wave functions gives additional electron density in the space between the two nuclei, which lowers the energy. In contrast, by subtracting the two wave functions, the space between the two nuclei becomes more devoid of electrons than if there had been no interaction, which is unfavorable.

This leads to the basic picture of the *chemical bond*: a linear combination of valence atomic orbitals (AOs) of a pair of atoms is used to describe a redistribution of electrons. In the *bonding orbital*, the electrons are distributed over larger distances than in the AOs, and this lowers the *kinetic* energy.[1] (Fig. 3.1). Recall that the kinetic energy operator contains a factor $\nabla^2\psi$ which represents the curvature of the wave function). The increased probability of the electron being located between

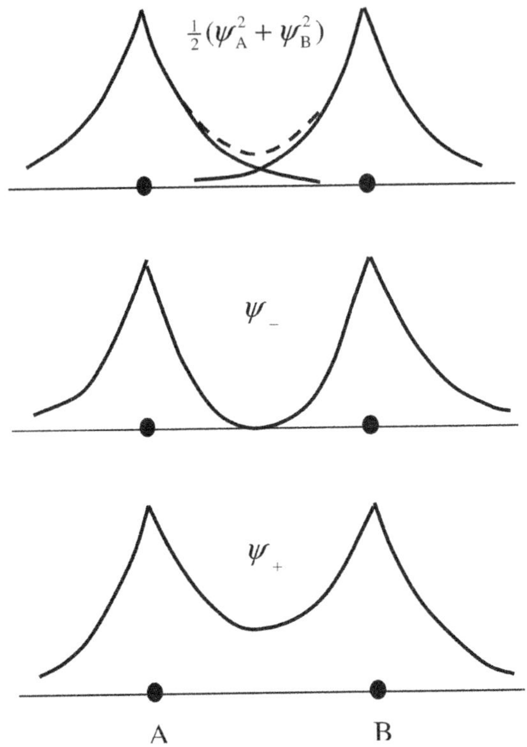

FIGURE 3.1 Schematic electronic probability distributions for the hydrogen molecule ion, with nuclei at A and B. Top: In the absence of interaction, the electron has equal probability of occupying the 1s orbitals of either atom. Middle: The antibonding molecular orbital is obtained by subtracting the wave functions ψ_A and ψ_B; bottom: The bonding orbital obtained when ψ_A and ψ_B are added. In this last case, the electron has highest probability in the space between the two nuclei.

adjacent atomic nuclei in bonding orbitals leads to a lower *potential* energy (electrons closer to the positively charged nuclei), while the closer proximity of the nuclei increases the potential energy. A bond is formed if the sum of these energy terms has a (negative) minimum value.

An *antibonding orbital* is an LCAO that satisfies symmetry considerations and the Pauli principle but has a higher energy than either the AOs or the bonding orbital. Antibonding orbitals (Fig. 3.1) have more kinetic energy (ground state energy of a particle in a box increases when the box size decreases) and a somewhat higher potential energy due to the lower probability of finding electrons between the nuclei.

The wave function Ψ_+ of Eq. 3.3 can be improved by including terms in higher energy wave functions for the two hydrogen nuclei A and B, with the coefficients found by application of the variation principle to minimize the expectation value of the energy, $\bar{E}_+$. In practice, the number of terms will be limited by computer resources. There being only a single electron in this special case, only the kinetic energy and the electrostatic interaction between the electron distribution and the nuclei need be considered. Optimization of the internuclear distance, r can be achieved rapidly by standard minimization methods if one simultaneously calculates $\bar{E}_+$, $\partial \bar{E}_+/\partial r$, and $\partial^2 \bar{E}_+/dr^2$.

The energy and bond length of the *hydrogen molecule* are approximated by the same procedure, with both the electrons assigned to the Ψ_+ wave function, with opposite spin. As described for isolated atoms in Section 2.5, the energy of interaction between the two electrons is usually calculated as the Coulomb interaction between two electronic charge distributions (also called the exchange energy), ignoring the *correlation energy*, a correction that accounts for the fact that the electron distribution of electron number 2 depends on the position of electron number 1. The problem of computing the correlation energy has not yet been completely solved; an approximation can be included at significant added computational cost.

3.2.4 Hybrid Wave Functions and Chemical Bonds

The so-called hybrid orbitals provide a special example of effective use of linear combinations of wave functions. We illustrate that here for atoms with from three to nine electrons, that is, the series from Li to F. We see that in each of these the 1s orbital is filled with two electrons, while the remainder has available the 2s and 2p orbitals. The 2s wave function has spherical symmetry, so that $\psi(\mathbf{r}) = \psi(-\mathbf{r})$, and the 2p wave functions, although possessing a direction in space, still are antisymmetric, for which $\psi(\mathbf{r}) = -\psi(-\mathbf{r})$. However, an asymmetric wave function extending more in one direction than the opposite direction can interact more favorably with a second nucleus than either. Because of the different symmetries, a linear combination of 2s and 2p wave functions will be asymmetric about the origin (but still possess rotational symmetry about the long axis).

Four asymmetric $2sp^3$ wave functions can be constructed as linear combinations of the 2s and 2p wave functions, according to

$$(2sp^3)_1 = \frac{1}{\sqrt{4}}(2s + 2p_x + 2p_y + 2p_z); \quad (2sp^3)_2 = \frac{1}{\sqrt{4}}(2s - 2p_x - 2p_y + 2p_z);$$
$$(2sp^3)_3 = \frac{1}{\sqrt{4}}(2s + 2p_x - 2p_y - 2p_z); \quad (2sp^3)_4 = \frac{1}{\sqrt{4}}(2s - 2p_x + 2p_y - 2p_z); \tag{3.4}$$

where 2s stands for ψ_{2s}, etc. These wave functions have the same principal quantum number ($n = 2$) and the same energy as the 2s and 2p wave functions. The four sp^3 wave functions are asymmetrically elongated in directions enclosing angles of 109.47...° and form a tetrahedral arrangement. For an atom, such as carbon, that is known to form tetrahedral molecules, this set of wave functions would be a rational first approximation for combining with the 1s wave function of a hydrogen atom located as indicated in Fig. 3.2 to form MOs of methane.

The same hybrid orbitals can be used when the central atom is oxygen. In that case, four electrons are paired to occupy two of the hybrid orbitals, and the other

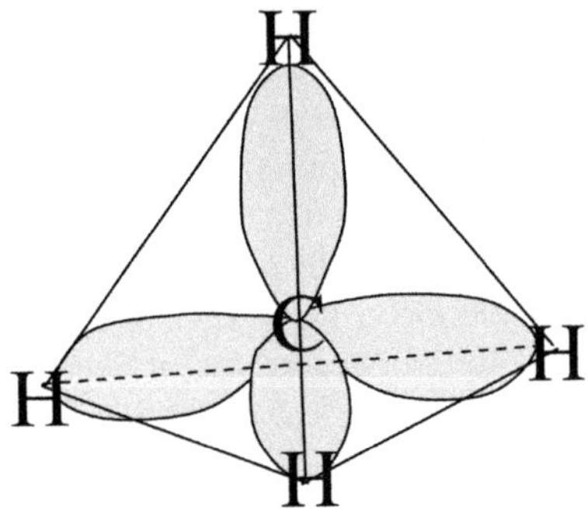

FIGURE 3.2 Hybrid sp^3 wave functions of a carbon atom; these are available to form bonding orbitals with the 1s wave functions of hydrogen atoms placed in the indicated positions. (A more precise rendering would show that each hybrid wave function does not have a node at the location of the nucleus and extends significantly opposite the main lobe.)

two hybrid orbitals contain unpaired electrons and can be used to form chemical bonds, for example, with hydrogen atoms. Repulsion between the pairs of electrons in the first two orbitals distorts the geometry, resulting in a smaller H—O—H bond angle of circa 105°.

3.2.5 *Ab Initio* Calculations

For any molecular geometry, in principle, a set of wave functions (MOs) and energies can be found by taking linear combinations of atomic wave functions of all the atoms in the molecule, and electrons can be assigned to these by procedures similar to those used for assigning electrons to atomic wave functions. This method is rigorous and exact within its *assumption* (*independence of electrons*) as long as we consider all wave functions of all contributing atoms as part of the basis functions for MOs. In practice, it is not possible to use an infinite set of basis functions for each atom of a molecule, and it is impractical even to use a very large set. However, if one uses a reasonably large set of wave functions for all atoms involved in a small molecule, and calculates full MOs, this is termed an *ab initio* approach, which yields a very good approximation to the actual energy, wave function, and electronic distribution; these calculations are done with the Hartree–Fock self-consistent field method. *Ab initio* methods have been extended to include electron correlation, that is, interactions between individual electrons, but at a significant increase in computational cost.

As an alternative to atomic wave functions, other basis sets can be used. Gaussian functions have found favor because the exchange integrals are more readily evaluated. The much used Gaussian program (www.gaussian.com), originally developed by John Pople, employs these. However, many other excellent programs are in use.

We mention briefly an alternative method, called *density functional theory* (DFT), which focuses on the electron density (the square of the electronic wave function) rather than on the wave function itself. DFT methods give mathematically more tractable forms of the electron density and produce approximate results comparable to, often better than, those obtained with Hartree–Fock *ab initio* approaches, and at much lower computational cost.

From the viewpoint of biophysics, we have emphasized the *ab initio* description for several reasons. First, it is needed to understand how we establish the electronic probability distribution function and energy of a molecule. Second, it establishes the basis for understanding the forces between nuclei in a molecule. Third, as computers have become more powerful, it has become increasingly commonplace to calculate molecular energies using QM, and one should have some idea of what is involved in such calculations. One needs it also to discuss the spectroscopy of macromolecules, but that material is not considered in this book.

3.2.6 Approximate Methods

As the assumption (independence of individual electrons) that was made in order to develop the treatment based on MOs is fairly severe, it is often not worth trying to be too rigorous using this approach. This led chemists to a number of approximate approaches. A number of so-called *semiempirical* methods have been developed, in which the calculation is greatly simplified by shortcuts that restrict the solution of the quantum mechanical problem. In some cases, experimental information is used to restrict that part of the calculation that is based on QM, and in other cases this information is based on results of *ab initio* calculations.

A serious problem of currently available methods for approximating the electronic structure of molecules is that these computations scale *at least* as the third power of the number of electrons. For studies of macromolecules it is only practical to use methods that scale linearly with the size of the system, that is, with the number of atoms (or number of electrons). Recently, methods have been worked out in which a macromolecular system is subdivided into (overlapping) parts, the QM calculation is performed for each part, and the results for all parts reconciled in an additional step. For a protein molecule one might choose these parts to be individual amino acid residues, each with its immediate surroundings consisting of all atoms within a set distance of any of the residues atoms. Chemical bonds that have to be "cut" in order to isolate the atoms of a subsystem create open valences that must be reconciled after all the local calculations have been performed. An empirical applied electrostatic field approximates longer range electrostatic forces. This is the *divide-and-conquer* approach.[2]

A difficulty even with this approach applied to macromolecules is the very long calculation times required (at least on present-day computers). A common solution is found by realizing that in many instances such high accuracy is needed only at a particular site. Thus, in simulations of enzyme catalysis in which chemical bonds are broken and remade, the substrate and the residues that participate directly in catalysis should be treated at a high level, but surrounding atoms can be represented more approximately. For example, in studies of bond breaking and bond making in enzyme–substrate complexes, the computational system typically treats a core of substrate and nearby side chains with an accurate *ab initio* QM calculation, while the remainder of the protein, including a solvent layer, is treated with MM with explicit representation of all atoms.[3] An example of application of such a *QM/MM method* is discussed in Section 18.15.2.

3.3 EMPIRICAL CLASSICAL ENERGY EXPRESSIONS

A macromolecular system will contain a large number of atoms and chemical bonds, but in most problems of interest very few of these bonds will be broken and remade, and often none will be. In many studies based on an ability to calculate the energy and forces of a macromolecular system, most or all of the system remains close to equilibrium, and the energy can be expressed with good accuracy in terms of the *deviations* from equilibrium. In the so-called MM method, the inter- and intramolecular energy is expressed entirely in terms that are reasonable approximations only under near-equilibrium conditions. Experience shows that this adequately describes the behavior of liquids and solutions of macromolecules at equilibrium in a "normal" temperature range (0–100 °C). Evaluation of the MM energy expression is extremely fast: millions of successive evaluations of energy and forces can be done in the time required for a single *ab initio* calculation. However, when interpreting results of simulations with an MM force field, one must always keep in mind the inherent approximations and how these might affect the outcome. One of the primary approximations is the *empirical potential energy function* required to express the energy of a system in terms of the coordinates of all atoms in the system. In this section, we describe the empirical potential energy functions used to estimate the internal energy of a macromolecule. In the remainder of this chapter, we describe the empirical potential energy functions used to approximate the noncovalent forces between molecules.

3.3.1 Bond Stretch Energy

The internal energy of a diatomic molecule can be expressed in the following series expansion

$$l = l_0 + \delta l$$

$$E_{\text{bond}}(l) = \frac{K_l}{2}(\delta l)^2 + \frac{K_{3,l}}{6}(\delta l)^3 + \cdots, \tag{3.5}$$

where l is the bond length, l_0 its equilibrium value, and E_{bond} the energy needed to change the length from its equilibrium value. If the factors $K_{3,l}$ and higher are zero, this is tantamount to treating the bond as a harmonic spring with force constant K_l. If not, the spring is anharmonic (not having the potential energy function of a harmonic oscillator, $\frac{1}{2}kx^2$).

This is a simple mechanical expression for the energy, which can be differentiated with respect to the atomic coordinates in order to give the atomic forces corresponding to the bond's being stretched or compressed to a nonequilibrium

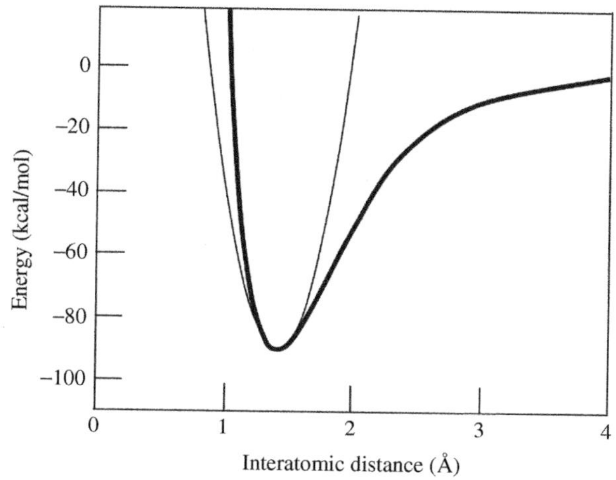

FIGURE 3.3 Comparison of Morse potential for HBr (heavy line) and quadratic potential (thin line).

length. We still need QM to determine the appropriate value of the force constant(s); these are set by fitting an expression of the form of Eq. 3.5 to energies of an A–B molecule calculated using *ab initio* methods for a series of different bond lengths. The actual potential energy function obtained by this procedure is anharmonic and is similar to that shown in Fig. 3.3.

A more accurate empirical form of this potential is shown in Fig. 3.3 and is termed the *Morse potential*, which is given by

$$E_{\text{morse}}(l) = D_e\left[1 - e^{-a(l-l_0)}\right]^2 - D_e, \tag{3.6}$$

where D_e is the dissociation energy, l_0 the equilibrium bond length, and $2a^2D_e$ equals the second derivative d^2E/dl^2 at the energy minimum, that is, is equal to K_l, if K_l is the bond-stretching force constant of Eq. 3.5. The functional form given in Eq. 3.6 has the fully dissociated state, at infinite separation as the reference state of zero energy, while the series expansion of Eq. 3.5 has zero energy at the equilibrium bond distance, for $l = l_0$.

It is easy to see that chemical bonds treated quantum mechanically behave as *anharmonic* springs because at large separation of the two atoms, A and B, the energy levels off, as is evident from the fact that covalent bonds do break if sufficiently excited by thermal energy, while for zero internuclear distance the energy is infinitely large. However, note that the Morse potential approaches the harmonic potential for l close to l_0. The anharmonic nature of the interatomic potential is experimentally shown by the fact that infrared (IR) spectra are temperature dependent, with vibrational transitions occurring at lower frequencies at higher temperatures. We will see in later chapters that increased temperature reflects occupancy of higher quantum states of a system. Therefore, the energy level spacings of the actual interatomic potential decrease with increasing energy, which indicates a broadening of the potential well at higher energies, as for the Morse potential. IR spectroscopy provides an experimental tool for obtaining parameters for the bond-stretching potential.[4]

Which of these functions is used is purely a matter of computational expediency as the parameters of all three forms can be adjusted to match either experimental properties or accurate quantum mechanical calculation. The parameters of any form of the bond-stretching potential need to be determined only once, as experience has shown that the values of equilibrium bond length and force constant(s) are quite insensitive to the environment in which the A–B molecule is placed (in a vacuum, in an organic solvent, in aqueous solution, or bound to a protein, etc.). These constants are even insensitive to changes in chemical bonding several atoms removed, so that a single set of parameters accurately represents the deformation of C—C bonds in saturated aliphatic hydrocarbons and in hydrocarbon chains attached to other moieties, including the side chains of amino acids such as leucine and valine.

3.3.2 Bend and Twist Energy

In an entirely analogous manner, the internal energetics of a triatomic molecule (A – B – C) can be expressed as a sum of two contributions from bond length stretching plus a term for deforming the bond angle, θ

$$\theta = \theta_0 + \delta\theta$$

$$E_{\text{bond angle}}(\theta) = \frac{K_\theta}{2}(\delta\theta)^2 + \frac{K_{3,\theta}}{6}(\delta\theta)^3 + \cdots. \tag{3.7}$$

Cross terms involving two different bond lengths, or bond angles, or a bond length and a bond angle, etc. may be included to improve the fit of the empirical force field to the underlying quantum mechanical energetics.

A molecule with more than three atoms can take up multiple conformations in which atoms are fixed at different coordinates by making rotations about chemical bonds (while not stretching any chemical bonds). In the case of proteins, the peptide backbone has three chemical bonds about which torsions can occur: N—C^α, C^α—C, and C—N. One of these (C—N) has considerable double bond character (~40% as estimated by Linus Pauling), meaning that C^α—C—N—C^α together with the peptide group's O and H atoms are restrained to a plane. Thus, only rotations of the angles φ and ψ, as illustrated in Fig. 3.4, can lead to distinctly different backbone conformations, that is, define what generally is meant by the term *protein conformation*. There are of course additional bonds in the amino acid side chains about which rotation can occur and lead to additional conformations of the whole protein. Thus, terms for dihedral angle rotations must be included in empirical potential functions. The energy associated with torsional deformation from the lowest energy conformations can also be expressed in terms of Taylor series in the dihedral angles, ρ. However, it is more convenient to use a cosine form.

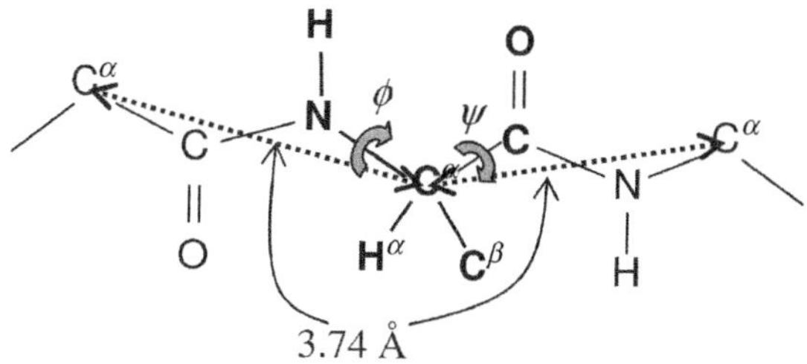

FIGURE 3.4 Alanine residue (bold) in a polypeptide chain. Hydrogen atoms bonded to C^{β} and substituents of terminal C^{α} atoms are not shown. The peptide groups are relatively rigid and the distance between successive C^{α} atoms is nearly constant at 3.74 Å, whereas rotation about the N—C^{α} and C^{α}—C bonds provides the principal internal degrees of freedom.

In the general case, one uses several cosine terms, that is, a *Fourier expansion in* ρ:

$$\rho = \rho_0 + \delta\rho$$

$$E_{\text{dihedral}}(\rho) = \sum_n \frac{K_{\rho,n}}{2}\{1 - \cos(n \cdot \delta\rho)\}. \tag{3.8}$$

For example, for groups of four atoms that exist in two approximately coplanar conformations, one needs at least two cosine terms. One of these ($n = 2$) corresponds to the existence of two energy minima (at $\rho = 0$ and $\rho = \pi$, the *cis* and *trans* conformations), while the other ($n = 1$) corresponds to a single energy minimum and produces an energy difference between the former two minima. If only a single term is used, the coefficient K_ρ corresponds to the energy barrier separating the minima.

It has been found that coefficients for these expressions depend on the type of bond, but do not vary much from molecule to molecule (as long as the chemical environment is similar), and depend only weakly on the physical environment.

Tables 3.1 and 3.2 give values of some of these constants, used in the charmm and amber/opls force fields. These illustrate the fact that different groups of researchers have obtained slightly different "best" values of force field parameters, depending on their choice of model compounds, on the level of the quantum

TABLE 3.1 Molecular mechanics parameters for bond lengths and bond angles of aliphatic hydrocarbons (Eqs. 3.5 and 3.7; charmm force field)

Atoms	Eqm. Value	K
C—C	1.53	445
C—H	1.11	644
C—C—C	114	107
C—C—H	110	106
H—C—H	109	71

Units are Å, degrees, kcal/(mol Å^2), and kcal/(mol rad^2).

TABLE 3.2 Some molecular mechanics parameters for the peptide group (Eqs. 3.5, 3.7 and 3.8; opls/amber force field)

Atoms	Eqm. Value	K_l	Atoms	Eqm. Value	K_θ	Atoms	Eqm. Value	K_ρ	n
C^α—C^β	1.53 Å	536	N—C^α—C	110°	126	C^α—C—N—C^α	180°	6.1	2
C—H	1.09	680	C^α—C—N	117°	140	C^α—C—N—C^α	180°	2.3	1
C^α—C	1.52	634	C—N—C^α	122°	100	O—C—N—C^α	0°	6.1	2
N—C	1.34	980	O—C—C^α	120°	160	O—C—N—H	180°	4.9	2
N—C^α	1.45	674	O—C—N	123°	160	C^α—C—N—H	0°	4.9	2
N—H	1.01	868	C—C^α—C^β	111°	126	H—N(—C)—C^α	180°	4.0	1
C=O	1.23	1140				O—C(—N)—C^α	180°	42.0	1

Units are Å, degrees, K_l: kcal/(mol Å^2), K_θ and K_ρ: kcal/(mol rad^2). In the rightmost four columns, the first five lines refer to torsion about the peptide bond between successive residues, and the last two refer to deformation of planar groups consisting of three atoms bonded to a common fourth (N and C, respectively).

mechanical calculation used, and on whether experimental data (from x-ray crystallography and IR and Raman spectra of small molecules) were used. The most widely used force fields today are amber (originated by Peter Kollman at UCSF)[5], charmm (originated by Martin Karplus at Harvard)[6], and gromos (originated by Wilfred van Gunsteren at the University of Groningen, later the ETH, Zürich)[7]. The opls parameters (originated by Bill Jorgensen at Yale)[8] describe intermolecular interactions and have been combined with intramolecular parameters from the amber force field.

3.4 NONCOVALENT FORCES BETWEEN ATOMS AND MOLECULES

We have seen in Section 3.2 that interatomic forces can be attributed to two effects: the electron distribution that determines the electrons' kinetic energy, and the Coulomb energy of the interactions between electronic and nuclear charges. If the contacts are intimate (as when atoms form a chemical bond), both contributions are large. However, for noncovalent interactions between molecules, especially at larger separations, the Coulomb contribution makes the major contribution. As a first approximation, intermolecular forces can be treated as the sum of Coulomb interactions between fixed charge distributions of interacting molecules (which may depend on the conformation of each molecule individually). In a more accurate approximation, a molecule's charge distribution responds to the electric field generated by the surrounding molecules (*polarization*), and the energy is calculated as the sum of Coulomb and polarization energies.

These interactions are called *polar interactions*, and can be evaluated in terms of a distribution of partial point charges; an alternative is to treat these interactions in terms of multipole expansions of the electric fields of the molecules. Both the point charge distributions and the multipole expansions are approximations of the actual electronic distribution of the molecule. Obviously, higher accuracy is attained using the electronic and nuclear distribution as the basis for calculating polar interactions, but this requires QM.

A pair of noble gas atoms attract, in spite of the fact that neither atom has a charge and that both have filled valence s orbitals, which give them approximately spherical charge distributions. One concludes that these so-called *nonpolar interactions* require a quantum mechanical treatment. However, we shall see that such interactions can be represented with reasonable accuracy in terms of a simple function, the Lennard-Jones potential. In order to describe these, we must first deal with polar interactions.

We deal successively with these approximations in this section.

3.4.1 Polar Interactions

Point charge approximation. Polar molecules can often be approximated as a collection of fixed partial point charges (Fig. 3.5). And Coulomb's law (see subsequent text) can be used to evaluate the electrostatic interatomic forces. This is a decent approximation if (i) the interatomic distances exceed the extent of the electronic wave functions that determine the electronic distribution and (ii) the approach of two molecules does not significantly perturb the electronic distribution of either molecule.

The magnitude of the Coulomb force between *point charges* is strictly distance dependent and not orientation dependent. However, the force between two collections of point charges will depend on all intercharge distances and unit vectors, and, as such, will depend on the orientation between these fixed charge collections. With access to computers, it is convenient to use the point charge model in order to compute the dependence of energy and force on distance and orientation of complex particles. This is the approach commonly taken in MM, with atoms being assigned partial charges to reflect net molecular charge and the distribution of electrons. These partial charges can be estimated from electronic wave functions calculated on the basis of QM. See, for example, Table 3.3.

This approach is useful as long as the electronic distributions remain fixed in space as the particles interact. Of course, electrons do not remain unperturbed when subject to another electric field, so, in general, the molecules become *polarized* as

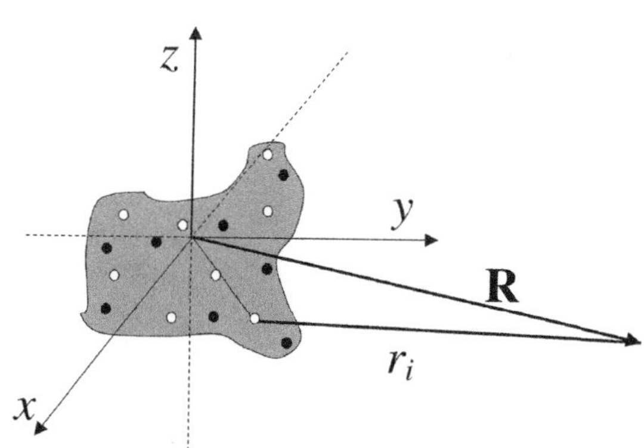

FIGURE 3.5 Macromolecule as a collection of negative and positive point charges.

TABLE 3.3 Partial charges of the alanine residue in a peptide (opls/amber force field)

Atom	Partial charge	Atom	Partial charge
N	−0.50	C^{β}	−0.18
H_N	0.30	H^{β}	0.06
C^{α}	0.14	C	0.50
H^{α}	0.06	O	−0.50

The unit of charge is the charge on an electron (1.60×10^{-19} Coulomb).

they approach. Accounting for polarization, which can be quite strong for ion binding to proteins, is a major challenge in MM. The partial charges in Table 3.3 have been empirically adjusted in order to account for the average polarization in a condensed polar medium and concomitant formation of hydrogen bonds, such as occurs in a folded protein and in aqueous solution. In a hydrogen bond, a polar hydrogen atom, such as the H_N atom of alanine and the H of H_2O, will be attracted to electronegative atoms, such as O of the peptide backbone or water, and the resulting electric field will *increase* the magnitude of the partial charges.

Potentials and forces; point charges, point dipoles, and multipoles. The force of interaction between two point charges, q_1 and q_2 at distance r is given by *Coulomb's law*, which states that two charges attract if they have opposite signs and repel if they have equal signs,[9]

$$\mathbf{F}_1 = \frac{-1}{4\pi\varepsilon_0}\frac{q_1 q_2}{r^2}\mathbf{e}_{12} = -\mathbf{F}_2. \tag{3.9}$$

The constant ε_0 is the permittivity of vacuum (see table of constants in Appendix I). By convention, the force $\mathbf{F}_1$ on the charge q_1 is along the unit vector $\mathbf{e}_{12}$ that points to the other charge, q_2. If the charges have the same sign, the force points away from the other charge, as in Fig. 3.6. The force $\mathbf{F}_2$ on the charge q_2 due to q_1 is always equal and opposite to $\mathbf{F}_1$. In general, the *potential energy of interaction* (a scalar) is the integral of the force of interaction over the distance needed to remove them to infinite separation; conversely, the force is the gradient of the potential energy (here written in three different ways, the last in terms of x, y, and z components).

$$V(\mathbf{r}) = -\int \mathbf{F}(\mathbf{r})d\mathbf{r}$$

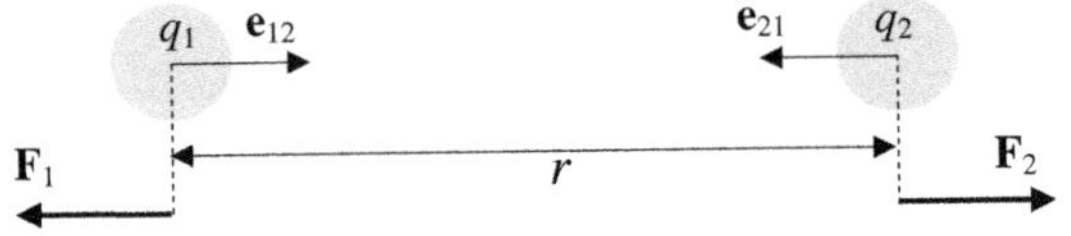

FIGURE 3.6 Illustration of the conventional statement of Coulomb's Law.

$$\mathbf{F}(\mathbf{r}) = -\nabla V = -\frac{\partial V}{\partial \mathbf{r}} = -\left(\frac{\partial V}{\partial x}, \frac{\partial V}{\partial y}, \frac{\partial V}{\partial z}\right). \tag{3.10}$$

The Coulomb energy is taken as zero at $r = \infty$

$$V(r) = -\int_{\infty}^{r} \mathbf{F} d\mathbf{r} = \frac{1}{4\pi\varepsilon_0}\frac{q_1 q_2}{r} = q_1\phi_2(r). \tag{3.11}$$

When charge is expressed in atomic units (au, having the same magnitude as the electronic charge), distance in Å and energy in kcal/mol, one obtains the value of V by multiplying the value of $q_1 q_2/r$ by a proportionality constant of 332 kcal mol^{-1} Å au^{-2}.

The *electrostatic potential*, $\phi_2(r)$ in Eq. 3.11 is the potential energy required to bring a *unit charge* (1) from infinity to some point at a distance r from another charge (2). The *electrostatic potential* at some point, **P**, due to a distribution of charges is a sum of terms, one for each charge (Fig. 3.5):

$$\phi = \frac{1}{4\pi\varepsilon_0}\sum_i \frac{q_i}{r_i}. \tag{3.12}$$

The sum over all charges is a complex function of the coordinates of all the point charges used to describe the molecule under consideration. When the distance of the point to the center of the charge distribution, R, is large relative to the dimensions of the charge distribution, it is convenient to expand this expression as a Taylor series in $1/R$, where R is the distance to the center. This gives the *multipole expansion*:

$$\phi = \frac{1}{4\pi\varepsilon_0}\sum_i\left(\frac{q_i}{R} + \frac{q_i\mathbf{d}_i \bullet \mathbf{R}}{R^3} + \cdots\right) = \frac{1}{4\pi\varepsilon_0}\left(\frac{Q}{R} + \frac{\boldsymbol{\mu} \bullet \mathbf{R}}{R^3} + \cdots\right), \tag{3.13}$$

where we have introduced the net molecular charge, Q, and the molecular *dipole moment*, μ, defined by

$$\boldsymbol{\mu} = \sum_i q_i\mathbf{d}_i; \tag{3.14}$$

d_i represents the vector from the center of charge to the charge q_i (Fig. 3.5). The dipole moment is a vector quantity, and $\mu \cdot \mathbf{R}/\mathrm{R}$ represents the projection of the dipole moment on a vector, **R** *from* the center of the charge distribution.

The higher order terms in Eq. 3.13 (whose coefficients are called *quadrupole*, *octupole*, etc.) become important at smaller distances or when both the net molecular charge, Q and the dipole moment are zero or very small, that is, for apolar and/or symmetrical molecules.

A *point dipole* is an abstraction used to describe the field due to a real charge distribution only at a large distance from the charge distribution. It has no net charge and all its higher multipoles are equal to zero. Thus, the field due to a pair of charges $(+q, -q)$ separated by a distance d can be described at some distance $R \gg d$ by a point dipole moment directed from the negative toward the positive charge and of magnitude $q^2 d$. For $R \sim d$, the point dipole description breaks down. Point quadrupoles, octupoles, etc., are also abstractions and are described by matrices of increasingly higher order (first order for dipole, second order for quadrupole, third order for octupole, etc.).

The potential energy for a point charge interacting with a (permanent) point dipole is

$$V_{q,\boldsymbol{\mu}} = q\phi_{\boldsymbol{\mu}} = \left(\frac{1}{4\pi\varepsilon_0}\right)\frac{q\boldsymbol{\mu}\cdot\mathbf{R}}{R^3} = \frac{1}{4\pi\varepsilon_0}\frac{q\mu\cos\alpha}{R^2}, \tag{3.15}$$

where the scalar product has been replaced with $\mu R\cos\alpha$, α being the angle between the μ and $\mathbf{R}$ vectors. Remember that for any angle ϑ, $-1 \le \cos\vartheta \le +1$. The potential energy for two interacting point dipoles is given by

$$V_{\boldsymbol{\mu},\boldsymbol{\mu}} = \left(\frac{1}{4\pi\varepsilon_0}\right)\frac{(\boldsymbol{\mu}_1\cdot\mathbf{R}_{12})(\boldsymbol{\mu}_2\cdot\mathbf{R}_{21})}{R^5}, \tag{3.16}$$

where $\mathbf{R}_{12}$ is the vector from dipole 1 to 2. One sees that this energy depends on the distance between the dipoles and their relative orientations. For any given arrangement the energy can be positive or negative; if one of the dipoles is flipped, the energy changes sign, and thus the energy averaged over all possible arrangements is zero. Given a preference for configurations with low energy, as in a system at thermal equilibrium, one can obtain the average, temperature-dependent dipole–dipole potential:

$$\langle V_{ab}\rangle_T = -\frac{2}{3}\left(\frac{1}{(4\pi\varepsilon_0)^2}\frac{\mu_a^2\mu_b^2}{k_\mathrm{B}T}\right)\frac{1}{r_{ab}^6}. \tag{3.17}$$

The assumptions of large distances and free rotation make this approximation useful mainly for molecules interacting in the gas phase, which is not usually the case in biology. Nonetheless, dipoles play an important role in physics, and one may come across the concept of a point dipole and the $1/r^6$ functionality in a number of instances (e.g., the London dispersion force, polarization and light scattering, fluorescence resonance energy transfer). In addition, dipoles are often discussed in biology. For example, it is recognized that the α-helix with its large dipole moment plays a role in enzyme catalysis, by locally affecting the electrostatic potential at the active site. This effect, however, is best described quantitatively in terms of the point charge approximation.

Quantum mechanical description. In the point charge approximation, polar interactions are treated in terms of collections of *fixed* point charges. While we might reasonably consider the nuclei to be fixed point charges (the Born approximation), the electrons are properly described by a probability distribution function. The dipole moment (or, for that matter, any one of the terms of the multipole expansion) is a property of this electronic distribution function and must, therefore, be calculated from the molecular wave functions. Quantum mechanically, the dipole moment (μ) is given as the expectation value of the dipole moment operator:

$$\mu = \int dr \Psi \left(\sum_i e_i r_i \right) \Psi^*, \tag{3.18}$$

where Ψ is the wave function for electrons *and* nuclei, e_i is the charge on each nucleus and electron and r_i is the distance of each charge from the center of charge of the molecule. Anything that alters the quantum state and electron distribution of the molecule also *alters* the molecular dipole moment (or *induces* a dipole where there was none before). There are no reliable classical approaches to describing electron redistribution in one molecule in response to another that is close by. QM is essential in this case.

Induced dipoles. Even if a molecule does not have a net charge or a dipole moment (i.e., a nonpolar molecule), it can interact with other charged, polar, or nonpolar molecules through *polarization*, the simplest manifestation of which involves *induced dipoles*. The origin of polarizability lies in electron redistribution in response to an external field. A positive charge q approaching a molecule exerts an attractive force on the electrons and a repulsive force on the nuclei. If the force is small, one may treat the electrons as bound elastically to the nucleus via harmonic springs, and then the resulting separation of center of electron charge and nucleus depends linearly on the force. The induced dipole, μ is the product of a polarizability tensor, α and the field, $\boldsymbol{E}$ produced by the charge distribution,

$$\mu = \alpha \boldsymbol{E}. \tag{3.19}$$

If the polarizability is isotropic, and the field is produced by a single charge of magnitude q at distance R, the magnitude of the dipole is given by

$$\mu = \alpha q / R^2; \tag{3.20}$$

if the charge is positive, the induced dipole points from the charge toward the molecule and if the charge is negative, the dipole points from the molecule toward the charge. The electrostatic potential due to the induced dipole is given by

$$\phi = \frac{-1}{4\pi\varepsilon_0} \frac{\mu}{R^2} = \frac{-1}{4\pi\varepsilon_0} \frac{\alpha q}{R^4}, \tag{3.21}$$

and the change in potential energy when the charge is changed by a small amount, dq is

$$dV = \phi dq = \frac{-1}{4\pi\varepsilon_0} \frac{\alpha q}{R^4} dq. \tag{3.22}$$

Thus, the potential energy for introducing the polarizing ion by growing its charge from 0 to q and inducing the dipole is found by integration of dV,

$$V(R) = \int dV = \int_0^q dq \frac{-1}{4\pi\varepsilon_0} \frac{\alpha q}{R^4} = \frac{-1}{4\pi\varepsilon_0} \frac{1}{2} \frac{\alpha q^2}{R^4} = \frac{-q\mu}{8\pi\varepsilon_0 R^2}. \quad (3.23)$$

This is just *half* the energy of interaction of a charge, q with a *permanent* dipole, μ (Eq. 3.15 with $\cos\alpha = -1$, as the induced dipole points always from the positive charge toward the molecule). This is because it takes energy to deform the molecular charge distribution to produce the induced dipole.

3.4.2 Hydrogen Bonds

A special type of interaction occurs between two electronegative atoms (such as O, N, F, S) when one (the donor, D) has a hydrogen atom bonded to it and this is situated between the donor and the other atom (the acceptor, A) to give an arrangement symbolized with A$\cdots$H-D. This is termed the *hydrogen bond.* When a hydrogen bond is present, the A$\cdots$H distance is significantly less than the minimum energy distance between nonbonded atoms of type A and nonpolar hydrogen, that is:

$$d_{\mathrm{A}\cdots\mathrm{H}} < R_{\mathrm{A}} + R_{\mathrm{H}} - 0.2\ \text{Å},$$

and similarly:

$$d_{\mathrm{A}\cdots\mathrm{D}} < R_{\mathrm{A}} + R_{\mathrm{D}} - 0.2\ \text{Å},$$

where R_{A}, R_{H} and R_{B} are van der Waals radii. In some cases of exceptionally strong hydrogen bonds, the equilibrium position of the hydrogen atom may be halfway between A and D, but usually one atom is clearly the donor with which the hydrogen atom shares a chemical bond. Ice is a special case in which the hydrogen atoms belong equally to two oxygen atoms but oscillate between two equilibrium positions, one closer to one oxygen atom and the other closer to the other oxygen atom.

The energy of a typical hydrogen bond, such as that between two water molecules, in the gas phase (~ -5 kcal/mol) is intermediate between those of covalent (of the order of -200 to -100 kcal/mol) or ionic (~ -50 kcal/mol) interactions and that of dispersion interactions (from -0.1 to -0.5 kcal/mol). However, when considering the contribution of any of these interactions to the folding stability of a protein, they are effectively much weaker, because formation of one intramolecular hydrogen bond occurs at the expense of disrupting interactions between protein and water. On the other hand, it is energetically extremely unfavorable to disrupt the interactions of a polar group, such as backbone C=O, with solvent and then *not* use this group in an intramolecular hydrogen bond in the folded protein! For this reason, the need to form as many hydrogen bonds as possible is an important determinant of the architecture of a folded protein.

The nature of the hydrogen bond has been widely argued. For some very short and thus very energetically favorable H-bonds, there is evidence for electron sharing between electronegative atoms *via* the shared H. For most H-bonds this is not likely, but there is surely electron polarization. Thus, the total potential energy of a cluster of H-bonded water molecules is not the sum of the energies of individual H-bonds between the water molecules, a form of "cooperativity." However, the simple description used in the MM force field, in terms of interaction of (effective) excess positive and negative *partial charges*, works surprisingly well for hydrogen bonds in proteins, in nucleic acids and even, to a first approximation, in water.

3.4.3 Nonpolar Forces: Electron Correlation, Dispersion and Repulsion

Dispersion forces. Even electrically neutral atoms interact via dipoles. The electrons are on average spherically distributed, but instantaneously this is not so; an atom has a fluctuating dipole moment (even if its average value is zero), and hence two atoms interact favorably even in the absence of the formation of a chemical bond as a result of a correlation of their instantaneous dipole moments. As two electron distributions approach each other, they perturb each other's wave functions, causing a redistribution of electrons in each molecule that lowers the energy. For a pair of identical neutral and otherwise noninteracting atoms (say, two He atoms), the average dipole moment of each remains zero, but the electron distributions are correlated so that the instantaneous dipole moments have, on average, a parallel alignment.

The resulting *correlated dipole interaction* can be described only with a quantum mechanical model. It varies as $1/r^6$, where r is the distance between atoms. The magnitude of the potential of interaction (called the *London dispersion potential*) can be estimated experimentally from second virial coefficients of dilute gases, and theoretically by its relation to the atomic polarizability.

Atoms and molecules also experience strong repulsive interactions as they approach to the point that their electronic wave functions begin to intermingle. When this occurs, MOs can form. If unoccupied valence orbitals are available, the occupancy of bonding MOs will exceed that of antibonding MOs and a covalent bond will form. If not (e.g., two He atoms), the two AOs will overlap. The *Pauli exclusion principle* states that two electrons in the same orbital cannot have the same spin quantum number. The result is a very sharply rising repulsive potential as nonbonded atoms approach (Fig. 3.7). This is customarily approximated in terms either of e^{-ar} or $1/r^n$ functionalities. With $n = 12$, this potential combined with the London dispersion interaction results in the familiar *Lennard-Jones 6–12 potential* (Fig. 3.7; Table 3.4),

$$V(r_{ij}) = 4\varepsilon_{ij}\left[\left(\frac{\sigma_{ij}}{r_{ij}}\right)^{12} - \left(\frac{\sigma_{ij}}{r_{ij}}\right)^{6}\right], \tag{3.24}$$

where σ is the value of the distance r_{ij} for which $V = 0$, often called the van der Waals distance, and ε is the depth of the potential at its minimum.

The combined molecular repulsions and dispersive attractions are often referred to as *van der Waals interactions*. Each atom pair contributes only a small amount of dispersion energy, but because there are so many nonbonded pairs, the aggregate

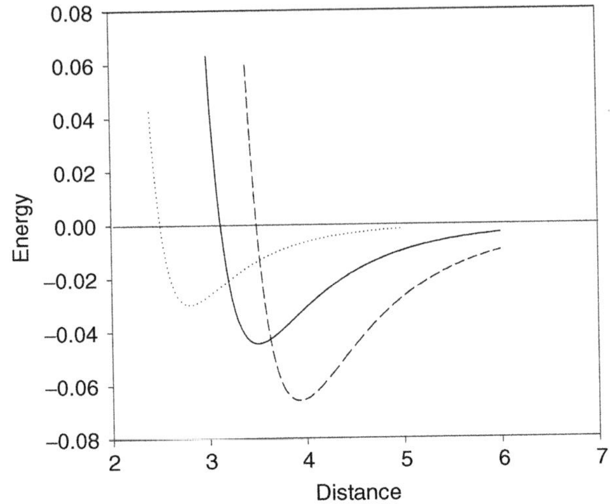

FIGURE 3.7 The Lennard-Jones energy function (left to right) of two H atoms, of a H and a C atom and of two C atoms (opls force field; atom types H_a and C_a of Table 3.4. Energy in kcal/mol, distance in Å.)

TABLE 3.4 Lennard-Jones parameters for the atoms in an alanine residue in a peptide (opls/amber force field; units of ε are kcal/mol)

Atom Type	ε	σ (Å)
H_N	0.000	
H_a	0.030	2.50
C	0.105	3.75
C_a	0.066	3.50
N	0.170	3.25
O	0.210	3.15

For mixed atom pairs, one uses the arithmetic mean of individual σ's and the geometric mean, $\sqrt{(\varepsilon_i \varepsilon_j)}$, for ε. The type C_a is used for C^α and C^β, and the type H_a is used for the hydrogens attached to these two carbon atoms.

dispersion energy is a significant contribution. Because the terms are very sensitive to interatomic distances, fine details of packing of a protein interior depend on these interactions, and play an important role in determining if a protein fold is globally stable with a given set of amino acid residues. Estimates of coefficients ε_{ij} and σ_{ij} can be obtained by systematic analysis of the properties of liquids and packing of crystals of small nonpolar molecules.

3.5 MOLECULAR MECHANICS

3.5.1 A "Complete" Empirical Force Field

We have now discussed all energy components of currently used force fields, including the terms for deformation of bonded geometry as well as those for nonbonded interactions. A complete expression has the following form:

$$
\begin{aligned}
V_{\text{MM}}(\mathbf{X}) &= \sum_{\text{bonds}} \frac{K_{l,i}}{2}(l_i - l_{i,0})^2 + \sum_{\text{angles}} \frac{K_{\theta,i}}{2}(\theta_i - \theta_{i,0})^2 \\
&+ \sum_{\text{torsions}} \frac{K_{\rho,i}}{2}\{1 - \cos[n_i(\rho_i - \rho_{i,0})]\} \\
&+ \sum_i \sum_{j>i} \left\{ \frac{1}{4\pi\varepsilon_0} \frac{q_i q_j}{r_{ij}} + 4\varepsilon_{ij} \left[\left(\frac{\sigma_{ij}}{r_{ij}}\right)^{12} - \left(\frac{\sigma_{ij}}{r_{ij}}\right)^{6} \right] \right\}. \qquad (3.25)
\end{aligned}
$$

The double sum represents the nonbonded interactions over all atom pairs (although atom pairs included in bond-length stretching and bond-angle bending terms are usually omitted). The potential energy depends implicitly on the atomic coordinates, **X**, because bond lengths, bond angles, dihedral angles and interatomic distances themselves can all be expressed in terms of atomic coordinates. **X** is a $3N$-dimensional vector, with elements the x, y, and z coordinates of each atom in the molecule. Some of the terms in this force field resemble expressions for fundamental forces (Coulomb) or forces derived from basic principles (dispersion and exclusion forces). However, this should not mislead us into thinking that these terms represent fundamental forces. They are totally empirical.

This expression can be improved with additional terms. The terms for bonded interactions (bond lengths and bond angles) are power series in the deviations from minimum energy structures and these series can be expanded with higher powers of deviations in bond length and angle and cross terms. Similarly, the cosine term for the torsion potential can be expanded with additional terms of a Fourier series. While representing atoms as partial point charges is certainly more reasonable than a multipole representation of nonbonded interactions, it is also a significant approximation. Clearly, atoms will behave differently depending on their chemical and physical environments, so that it is not possible to assign the same partial charge to the same atom in all situations, even in the same type of molecule. This may require that we have different force fields for different types of molecules, for example, for proteins, nucleic acids, or lipids, and at different temperatures. Thus, no matter how well we describe the "bonded" terms, these force fields remain *empirical*, that is, designed for particular situations. The "design process" involves assigning parameters to the various terms in the force field. A complete force field for use with proteins and nucleic acids in aqueous solution contains a not insignificant number of parameters, each of which must be determined by comparison with accurate quantum mechanical models or with experimental data (or both). The approach would not be as successful as it has proved to be, if one could not assume that chemically similar groups in different molecules share the same parameters.

Development of accurate parameters continues to be a very important undertaking. Probably the most important effect that is missed in the empirical approximation is the effect of polarization of electron distributions as ions approach atoms or highly polar atoms approach each other. Efforts are under way in several laboratories to incorporate appropriate polarization of atoms and molecular dipoles in MM force fields. Another defect is the use of the simple point charge model, which assigns a net charge to each atom, placed at the nucleus. This is inappropriate for highly nonspherical distributions of the outer shell electrons, as in the sp^2 hybridized oxygen atom of a C=O group, and this can in all likelihood be improved by placing some of the atomic charge off-center. The choice of model and parameters requires extensive calibration against results of more accurate quantum mechanical calculations.

As noted, the MM approximation is appropriate if the deviations from equilibrium are small, and can therefore be used to represent systems such as liquids and macromolecules in solution in thermal equilibrium and to represent conformation change by internal rotation about single bonds (where deviations of the energy of each degree of freedom are limited to a few times $k_B T$). As the potential function, Eq. 3.25, is an analytical function of the atomic coordinates, atomic forces can be expressed analytically (at small additional computational cost). Given energy and forces, it is a straightforward computational problem to minimize the energy (to a local minimum of the energy). At the minimum, first derivatives of the energy with respect to any coordinate are of course zero; the second-derivative matrix can be evaluated and, with a knowledge of the atomic masses, interpreted in terms of a set of so-called normal modes, which are harmonic vibrations, typically of small amplitude, about a conformation of minimum energy. Given a system of energy, forces and masses, the dynamic behavior of the system can be calculated by applying Newton's equations of motion. Molecular dynamics trajectories more fully explore conformation space in the neighborhood of a single energy minimum and, given enough time, also in the neighborhood of other local energy minima.

3.5.2 Molecular Structure and Motion: Normal Modes of a Rigid-Body Molecule

Normal mode analysis is concerned with molecular motion in the vicinity of an energy minimum. The molecule is considered as a collection of masses (the nuclei) that are connected into an elastic body by a variety of springs representing deformations of bonds, bond angles, torsions, and nonbonded contacts. When the internal displacements of the nuclei are small, these can be described in terms of three simple types of motion. Two of these are collective motions of all atoms, namely, translation of the molecular center of mass and rotation about the three principal molecular axes of inertia. For these motions, one uses as QM models the particle in a box and the rigid rotator. The third type of motion consists of internal vibrations, and these vibrations are modeled with the quantum mechanical harmonic oscillator model. For a molecule to behave as a harmonic oscillator, the energy would have to change quadratically with the extent of deformation away from the minimum energy structure. This is the case for real molecules only in the limit of small deformations (which means in practice at low temperatures). Under these conditions, the nuclei of a molecule may be considered as masses connected by *harmonic* springs.

For such a molecule, a classical analysis of the coupled motions of nuclei shows that there are $3N$ independent collective motions (see Appendix F). Six have roughly zero frequency and involve motions that leave the internal geometry of the molecule fixed. Three of these involve simultaneous motion of all the nuclei along three orthogonal directions in space and correspond to translations of the center of mass (*rigid-body translations*) along Cartesian coordinate axes. The other three zero-frequency motions involve coordinated movement of all nuclei around the three orthogonal principal rotation axes, p_1, p_2, and p_3 of the molecule that pass through the center of mass (*rigid-body rotations*). These three axes intersect at the center of mass. In Fig. 3.8, the Cartesian coordinate axes have been chosen to coincide with the principal axes.

The remaining $3N - 6$ independent collective motions are all internal and are termed *normal modes*. Normal modes are motions in *independent normal coordinates*. The equation of motion of each normal mode is the equation of motion of a simple harmonic oscillator with natural angular velocity ω_i. The Cartesian coordinates of each atom that contributes to a normal mode, change as functions of time, each with its own amplitude; in vector form this is expressed as

$$(\delta \mathbf{X})_i = a_i \boldsymbol{\xi}_i \cos \omega_i t, \tag{3.26}$$

where ξ_i is the normal coordinate vector describing the normal mode motion in terms of Cartesian coordinates and ω_i the angular frequency of the i-th normal mode. Normal modes are *independent* because motion along one normal coordinate does not influence nor is influenced by motion along other normal coordinates. The

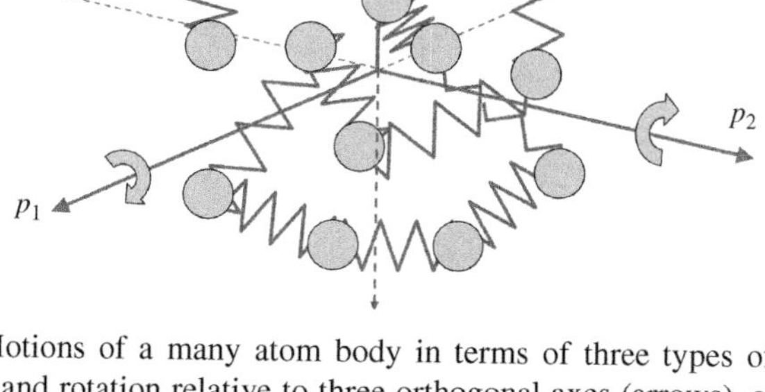

FIGURE 3.8 Motions of a many atom body in terms of three types of nuclear motion: global translation and rotation relative to three orthogonal axes (arrows), and internal vibrations. The axes intersect at the center of mass.

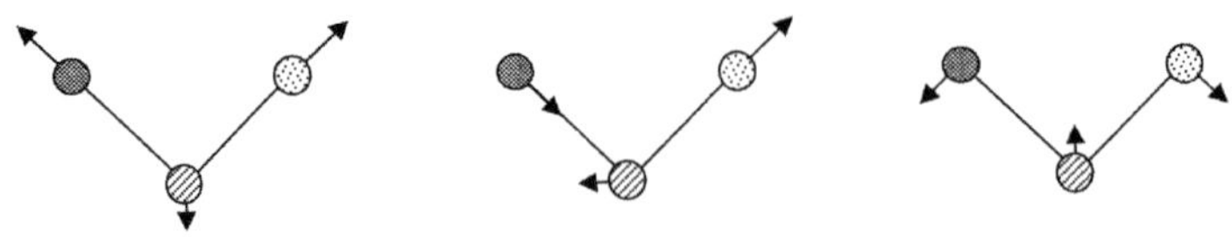

FIGURE 3.9 The three normal modes of a triatomic molecule; left: the two bonds stretch and shrink at the same time; middle: one bond stretches while the other bond shrinks and vice versa; right: the bond angle increases and decreases. Note that all three normal modes keep the center of mass and the three principal rotational axes fixed in space.

normal mode vectors are "orthogonal" (different normal mode vectors ξ have scalar products equal to zero), and they are (usually) normalized, that is,

$$\boldsymbol{\xi}_i \boldsymbol{\xi}_j = \delta(i - j), \tag{3.27}$$

where δ is the Kronecker delta function.

At any time, the motion is a superposition of the motions attributed to all normal modes (and to translation and rotation of the whole molecule), each with its own amplitude, a_i and phase. The normal modes of a triatomic molecule are illustrated in Fig. 3.9. (See discussion of normal modes in Appendix F.)

Normal mode vibrations vary widely in the amount of energy that corresponds to a deformation along normal coordinates. Each normal mode has a distinct effective force constant and an effective mass that is, for each normal mode, a different function of all atomic masses. Thus, each normal mode has a distinct frequency. Some normal modes involving coordinated motions of large numbers of atoms have very low force constants (and thus frequency) and are described as molecular "breathing" modes. Translational and rotational motions are limited by collisions with other molecules, for which the force constants are relatively small, and the frequencies of the corresponding motions are low. Low frequency motions are quite well treated by classical mechanics. However, local internal vibrations involving only two nuclei (bond stretches) or three nuclei (bond angle bends or torsions about double bonds) are much more energetic and require a quantum mechanical formulation.

By treating all normal motions (three translations, three rotations, and $3N - 6$ collective vibrations) with the corresponding simple QM model, one achieves a fairly accurate description of molecular motion and energy, if the deformations are not large (i.e., if the approximation that the energy is quadratic in the deformations remains valid).

The normal modes are a property of structure and interatomic forces in the neighborhood of a minimum of the potential energy. In the case of multiple energy minima, each has its own set of normal modes. This is a significant limitation when one seeks to apply this model to a polyatomic molecule (macromolecule), where many different conformations, each corresponding to a different minimum of the potential energy, can be achieved by rotations about chemical bonds.

3.5.3 Molecular Structure and Motion: Molecular Dynamics

An alternative, and widely used, application of the mechanical model of molecular structure is to simulate atomic motion in a molecular system as a function of time according to classical mechanics. This is called *molecular dynamics simulation.* In contrast to what is the case for a normal modes model, the explored region is not restricted to the immediate neighborhood of an energy minimum; in fact, given enough time, a molecular dynamics simulation explores an increasing segment of conformation space including additional structures of locally minimum energy. Molecular dynamics simulations explore the "energy landscape" and, if that landscape is explored sufficiently completely, this can lead to an equilibrium description in terms of the Boltzmann distribution. This is an enormous advantage, without which it would, for example, be impossible to represent the behavior of a liquid.

We here describe some key issues of molecular dynamics simulation; we have devoted an entire chapter to a more detailed description of this technique (Chapter 7). Newton's equations of motion are integrated numerically, and energy, forces, new velocities, and finally new coordinates are computed, essentially indefinitely, at small time intervals. The time step is set by the period of the fastest vibration, typically a bond vibration, with a period of around 10^{-14} s, so that an appropriate time step would be 10^{-15} s or 1 fs. The earliest application to biological macromolecules was a 20-ps simulation of a small protein (bovine pancreatic trypsin inhibitor with 58 residues) in vacuo.[10] Current computer technology and fast processing algorithms allow classical mechanics simulations of a small protein molecule in solution to reach into the tens of μs range, and their range can be expected to increase further. Nevertheless, the gap between the time scales attainable in simulations and those of key events that ultimately control biological function remains large, and will remain so for the foreseeable future.

NOTES

1. Chapter 11 of Eisenberg and Crothers (see Suggested Reading at end of chapter) discusses in detail the rules for forming valence bond of such molecular orbitals (MOs) and gives examples in terms of diatomic molecules, which can form bonding or antibonding MOs. This chapter might be useful reading for the student whose knowledge of quantum chemistry is sketchy.
2. Yang, W., Lee, T.-S. A density-matrix divide-and-conquer approach for electronic structure calculations of large molecules. *J. Chem. Phys.* **103**: 5674–5678 (1995).
3. Molecular mechanics methods are discussed in Section 3.5 and in Chapter 7. Calculation of the QM energy for a semiempirical QM method typically takes on the order of a factor of 1000 more computer time than calculation of the MM energy. This ratio increases to 10^6 and up for a high level QM method modified for linear scaling.
4. For the HBr bond stretch the force constant (for the quadratic potential) $K_l = 415$ J/m^2, the frequency $\nu = 8.0 \times 10^{13}$/s = 2650/cm (experimental value from the infrared spectrum), the wavelength $\lambda = 3.8$ μm, the energy level spacing $h\nu = 7.5$ kcal/mol.

5. Ponder, J.W., Case, D.A. Force fields for protein simulations. *Adv. Prot. Chem.* **66**: 27–85 (2003), Cheatham, III, T.E., Young., M.A. Molecular dynamics simulation of nucleic acids: Successes, limitations and promise. *Biopolymers* **56**: 232–256 (2001).
6. MacKerell, A. D., Bashford, D., Bellott, M., Dunbrack, R. L. Jr., Evanseck, J. D., Field, M. J., Fischer, S., Gao, J., Guo, H., Ha, S., Joseph-McCarthy, D., Kuchnir, L., Kuczera, K., Lau, F. T. K., Mattos, C., Michnick, S., Ngo, T., Nguyen, D. T., Prodhom, B., Reiher, W. E., Roux, B. III, Schlenkrich, M., Smith, J. C., Stote, R., Straub, J., Watanabe, M., Wiórkiewicz-Kuczera, J., Yin, D., Karplus, M. All-atom empirical potential for molecular modeling and dynamics studies of proteins. *J. Phys. Chem.* **B 102**: 3586–3616 (1998).
7. van Gunsteren, W.F., Billeter, S.R., Eising, A.A., Hünenberger, P.H., Krüger, P., Mark, A.E., Scott, W.R.P., Tironi, I.G. *Biomolecular Simulation: the GROMOS96 Manual and User Guide*. Zürich: Vdf Hochschulverlag AG an der ETH Zürich; 1996.
8. Jorgensen, W. L., Maxwell, D. S., Tirado-Rives, J. Development and testing of the OPLS all-atom force field on conformational energetics and properties of organic liquids. *J. Am. Chem. Soc.* **118**: 11225–11236 (1996).
9. Older texts may use a different definition of the unit of charge (as well as cgs units), in which case the factor $4\pi\varepsilon_0$ does not appear in Coulomb's equation and a factor 4π instead of $1/\varepsilon_0$ appears in Poisson's equation (Appendix G).
10. McCammon, J. A., Gelin, B. R., Karplus, M. Dynamics of folded proteins. *Nature (London)* **267**: 585–590 (1977).

SUGGESTED READING

Eisenberg, D.S., Crothers, D. *Physical Chemistry: With Applications to the Life Sciences*. Benjamin/Cummings: Menlo Park, CA, 1979.

Atkins, P., de Paula, J. *Physical Chemistry*. 7th ed., New York: Freeman, 2001.

Bohm, D. *Quantum Theory*, Courier Dover Publications, 1989

Karplus, M., Porter, R.N. *Atoms and Molecules*. Menlo Park, CA: Benjamin, 1970.

4

Water and the Hydrophobic Effect

4.1 INTRODUCTION

One cannot complete a description of forces within and between macromolecules without discussing the "hydrophobic force." Knowledge of properties of water and theories of water structure is a requirement for an understanding of this indirect force that has been used to refer to the apparent attraction between hydrocarbon molecules and between hydrocarbon side chains in aqueous solution, which is considerably larger than the direct attraction between hydrocarbon solutes or hydrocarbon side chains. In this chapter, we first review properties, in particular, the thermodynamic properties, of water. Next, we show how simulations with simple mechanical models can reproduce many physical properties of water and give insight into water structure. We then review the high energetic cost of forming interfaces between water and vacuum or water and hydrocarbon solutes, and the consequent "hydrophobic effect" that is responsible, among others for the poor solubility of hydrocarbons in water. The results of simulations are described that reproduce the association of small hydrophobic molecules in water and a more highly organized water structure in the interface between water and hydrocarbon. Thus, the term *force* is somewhat misleading, in that this strong interaction is indirect and results from interactions between water molecules. The hydrophobic force is not the negative gradient of a potential energy; although the attraction is real, the hydrophobic force is an average of a fluctuating force, thermodynamic in nature, the average force being the negative gradient of a *free* energy.

Equilibria and Kinetics of Biological Macromolecules, First Edition. Jan Hermans and Barry Lentz.

4.2 STRUCTURE OF LIQUID WATER

On the one hand, the high degree of randomness of gases and, on the other hand, the high degree of organization of crystalline solids are distinct advantages for theoretical and experimental structural studies of these states of matter. Liquids display a balance between disorder and order, and are more difficult to study than either one. And, among liquids, liquid water has distinct properties that place it in a class of its own, as summarized, for example, in review articles by Stillinger and Franks.[1]

4.2.1 Energy, Free Energy, and Entropy

The excess enthalpy and free energy (defined relative to those of a noninteracting set of the same number of molecules at the same V and T) are large and negative, at 300 K.

$$\Delta H_{\mathrm{x}} = -10 \text{ kcal/mol},$$

$$\Delta G_{\mathrm{x}} = -6.3 \text{ kcal/mol},$$

$$T\Delta S_{\mathrm{x}} = -3.7 \text{ kcal/mol}.$$

The comparison here is with small molecules of similar molecular weight, such as methane, for which ΔH_{x} is an order of magnitude smaller (and for which the boiling point at atmospheric pressure is 112 K). Quantum mechanical calculations all establish a high polarity (high dipole moment, large partial charges in the derived simple point charge description) for the water molecule. By contrast, interactions between methane molecules in the liquid phase are due to van der Waals forces, which are an order of magnitude smaller than the largely Coulombic interactions between water molecules.

Indeed at close separation, strong hydrogen bonds form between water molecules, as is clearly seen in the structure of ice (Fig. 4.1), in which each water molecule is surrounded by four nearest neighbors, to which it is connected via hydrogen bonds. Every hydrogen atom in ice lies between two oxygen atoms and, on an average, is closer to one oxygen atom than the other, but is shared by both. The water molecules form puckered six-membered rings ("chairs" and "boats"), and the overall structure is hexagonal and quite open, that is, a good deal of space remains within each "cage" of water molecules.

The excess energy (= observed energy − "thermal" energy; see Chapters 5 and 6) of ice is −12 kcal/mol. In a first approximation this establishes the energy of formation of each hydrogen bond as on the order of −6 kcal/mol, *many times* $k_B T$. The excess energy of the liquid is less in magnitude than that of ice but still large, leading to the inescapable conclusion that hydrogen bonds dominate the structure of liquid water and that the properties of liquid water must be intimately bound up with the hydrogen-bonded structure. While much progress has been made toward understanding this structure by a combination of experimental and theoretical approaches, some important aspects are still debated.

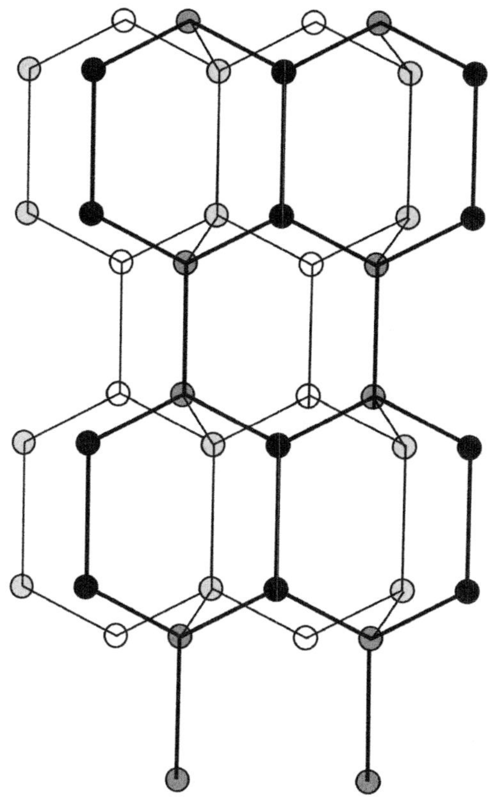

FIGURE 4.1 Schematic structure of ice. Darker circles represent oxygen atoms of layers of water molecules successively closer to the viewer. The four nearest neighbors of each molecule are arranged at the corners of a tetrahedron; one hydrogen atom is located between each pair of close neighbors.

4.2.2 Other Properties of Water

While the excess thermodynamic functions of water stand out simply by their magnitude, a number of properties of liquid water are quite extraordinary (always compared with those of other liquids). Most remarkable are

Density of the liquid is greater than that of the solid (at the melting point).

Density has a maximum at 4 °C.

Temperatures of melting, boiling, and critical point are anomalously high.

Mobility of H^+ and OH^- is high in both the solid and fluid states.

Heat capacity is anomalously large.

The heat capacity of liquid water (18 cal/K) is roughly twice than expected due to the increase with temperature of thermal motions (9 cal/K, see Chapter 5). This additional heat capacity is termed *configurational* or *structural* heat capacity. It is that part of the heat capacity that arises from a change in structure of a material, rather than from excitation of mechanical degrees of freedom (translation, rotation, vibration). Unlike the heat capacity of normal liquids, the heat capacity of water decreases with temperature due to a decrease in this configurational term.

4.2.3 Theories of Water Structure

Many theories have been developed in order to explain these unusual properties. The more successful theories applied to many elements, allowing all these theories to be condensed into two alternative views. Both views considered the structure of ice (Fig. 4.1) as a reference, and assumed that liquid water could be treated as a distorted or relaxed form of its hydrogen-bonded lattice.

One view (A) held that water structure is much like that of ice, but that the lattice geometry is deformed and contains distorted H-bonds. This is termed a *continuum view*. The other view (B) held that water molecules can be divided into classes according to the number of hydrogen bonds each molecule forms. These models were termed *discrete models*. Both these models explain the large heat capacity of water in terms of the H-bonded structure of water, but in fundamentally different ways: (A) as the temperature of liquid water rises, the higher energy of *bent and distorted hydrogen bonds* accounts for the configurational heat capacity; (B) alternatively, a steady *decrease in the number of hydrogen bonds* with increasing temperature causes the energy to rise. To put it in a simple sense, when ice melts, do hydrogen bonds break or do they deform? The debate over these alternatives raged for years.

In the early seventies, Rahman and Stillinger[2] produced a breakthrough by simulating liquid water with a molecular mechanics model (see also Chapter 7) with few adjustable parameters, with which they were able to make a large number of quantitative predictions, and show remarkable agreement with observed properties of liquid water. Their work shifted the emphasis of analysis of models of water structure from painstaking calculations of the properties of dimers, trimers, and larger clusters that were assumed to constitute the liquid phase, to consideration of an ensemble of many molecules, in many conformations together sampling a Boltzmann distribution, an approach that allows the calculation of physical properties as ensemble averages.

By then analyzing the behavior of individual water molecules in such a simulated sample, Rahman and Stillinger could follow hydrogen-bonding patterns as their simulation evolved. Figure 4.2 illustrates such a pattern. Water molecules form small polyhedra, with the distribution of polyhedron size dependent on the cut-off interwater potential energy, V^* used to define a hydrogen bond. For reasonable V^*, there are a few nonhydrogen-bonded waters, but most have one to four and sometimes up to six bonds. The bonds, however, can be distorted from the tetrahedral geometry observed in ice. Thus, Stillinger's water fits neither the discrete nor the continuum models.

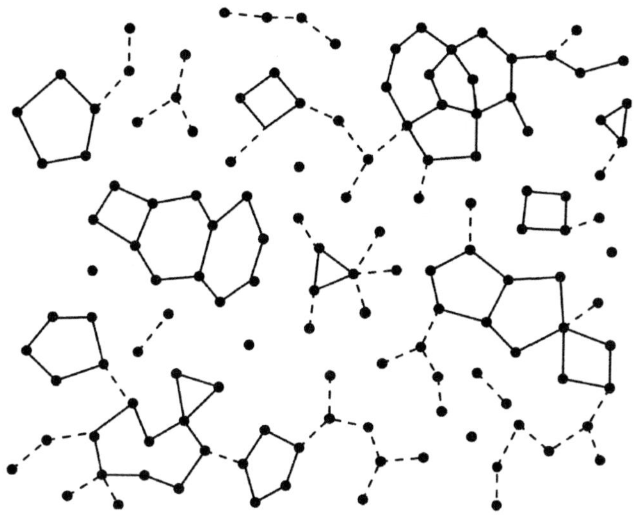

FIGURE 4.2 Two-dimensional representation of hydrogen-bond network in water simulation. Hydrogen-bonded polyhedra are emphasized by solid lines. From Rahman, A., Stillinger, *J. Am. Chem. Soc.* **95**: 7943–7948 (1973).

In molecular dynamics, the empirical potential assumed is a key to the success of the method. Rahman and Stillinger used a potential termed ST2 (Fig. 4.3). This potential reasonably captured the tetrahedral geometry of water in ice, but was computationally unwieldy because of the five centers of force. This potential has been superseded by two almost identical models having only three point charges: SPC,[3] and TIP3P.[4] Jorgensen and coworkers developed the TIP4P potential with four, and most recently the even more complex TIP5P model with five force centers.[5] All these models use a standard molecular mechanics energy function, Eq. 3.25 with intermolecular Lennard–Jones and electrostatic energy terms and forces, except that the geometry of the water molecule is held fixed. The distributions of point charges in these models are illustrated in Fig. 4.3 and their parameters are summarized in Table 4.1, where $A = 4\varepsilon_{\mathrm{OO}}\sigma_{\mathrm{OO}}^{6}$ and $B = 4\varepsilon_{\mathrm{OO}}\sigma_{\mathrm{OO}}^{12}$ are the coefficients of the $1/r^6$ and $1/r^{12}$ terms of the Lennard–Jones potential between oxygens.

The partial charges of H are unknowns for which a "best value" can be chosen. The long-range attraction of the water molecules is known from experimental studies and determines the value of the attractive parameter A for O ... O interactions. A second adjustable parameter, B determines the short-range repulsion. The best values of these parameters have generally been found by running a series of molecular dynamics simulations, and asking which simulation gave the best agreement with experiment for the internal energy and the density. These potentials have been compared in molecular dynamics simulations as summarized in Table 4.2. Not surprisingly, all do a reasonably good job of predicting the density, d and the energy of vaporization ΔH_{vap} against which they were parameterized. Surprisingly,

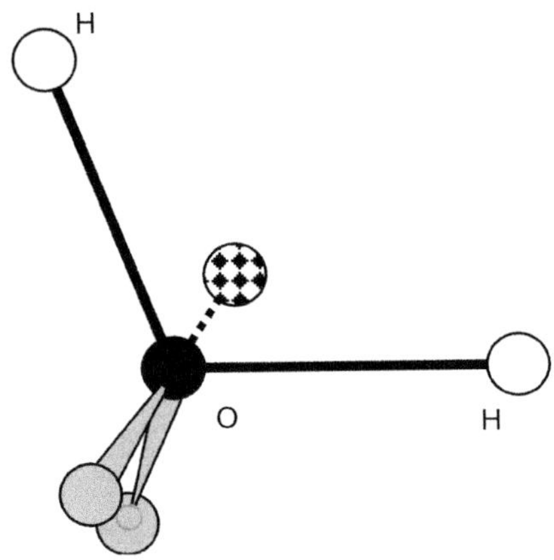

FIGURE 4.3 Diagram of different empirical water potentials. SPC and TIP3P have charges on the atom centers, H and O. ST2 and TIP5P have charges on the hydrogens and at the pseudo-atoms colored gray; TIP4P has charges on the hydrogens and on the "soccer ball." All models include a Lennard–Jones potential with a van der Waals radius somewhat greater than half the O – O distance in ice.

TABLE 4.1 Parameters for various water potentials[a]

	SPC	TIP3P	TIP4P	TIP5P
r_{OH}	1.0	0.9572	0.9572	0.9572
θ_{HOH}	109.47	104.52	104.52	104.52
$B \times 10^{-3}$	629	582	600	545
A	625	595	610	590
q_O	−0.82	−0.834	0	0
q_H	0.41	0.417	0.52	0.241
q_M	0	0	−1.04	−0.241
r_{OM}	0	0	0.15[b]	0.70
θ_{MOM}	0	0	0	109.47

Distances in Å, angles in degrees, energies in kcal/mol, charges in atomic units
[a]See review by Jorgensen, W.L., Tirado-Rives, J. *Proc. Natl. Acad. Sci. USA* **102**: 6665–6670 (2005).
[b]For TIP4P the point M lies on the bisectrix *toward* the hydrogens.

TABLE 4.2 Comparison of the performance of water potentials[a]

	SPC	TIP3P	TIP4P	ST2	TIP5P	Expt
d (g cm^{-3})	0.985	1.002	1.001	0.925	0.999	0.997
ΔH_{vap} (kcal/mol)	10.74	10.41	10.65	10.96	10.46	10.51
C_p (cal/mol/K)	20	20	20	20	29	18.0
$10^6\ \kappa$ (atm^{-1})	60	64	60	63	41	45.8
$10^5\ \alpha$ (K^{-1})	106	92	44	69	63	25.7
ε	60	88	60		82	78.3
D(M^{2}S^{-1}) $\times 10^9$	3.9	5.1	3.3		2.6	2.3

[a]See review by Jorgensen, W.L., Tirado-Rives, J. *Proc. Natl. Acad. Sci. USA* **102**: 6665–6670 (2005).

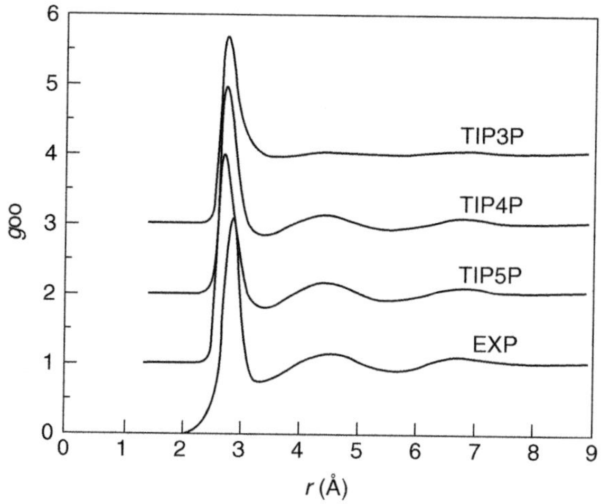

FIGURE 4.4 Radial distribution functions predicted by different water potentials. The SPC model performs quite well, also. From Jorgensen, W.L., Tirado-Rives, *J. Proc. Natl. Acad. Sci. USA* **102**: 6665–6670 (2005).

the simple, three-charge SPC model does an excellent job with the O – O radial distribution function (Fig. 4.4). None do well with the isothermal compressibility $\kappa = (\partial V/\partial P)_T$ or the expansivity $\alpha = (\partial V/\partial T)_P$ although these are usually within a factor of three of the experimental. The TIP5P model gives good values of the diffusion coefficient D and the dielectric constant ε (see Table 4.2). This model also excels by predicting within a few degrees the freezing behavior and the temperature of maximum density, at which the other models do poorly.

The O–O radial distribution function of liquid water (dashed curves in Fig. 4.4) has a very sharp maximum for near neighbors at $r = 2.8 - 2.9$ Å. The number of near neighbors does not quite have the value of 4 it has in ice. Contrast this with the number of near neighbors in a sample of closely packed spheres, which is 12. This is highly significant, and means that water is highly structured and that there is a great deal of free space in liquid water. This is reasonably well predicted by all the potential functions. An additional, less pronounced maximum indicates some long-range order, which is reasonably well predicted by the SPC, TIP4P, and ST2 potentials but not so well by TIP3P. It is surprising that the simple, three-charge SPC model does such a good job with the radial distribution function.

The two simplest models, SPC and TIP3P do a very reasonable job overall, and are widely used as "explicit water" in simulations of biological macromolecules with force-fields for proteins and nucleic acids that integrate one of these models of the water molecule. Experience shows that highest accuracy is obtained by simulating proteins and nucleic acids with explicit representation of all atoms and with a number of explicit water molecules sufficient for several layers adjacent to the macromolecule.

4.3 THE HYDROPHOBIC EFFECT

4.3.1 Role in Stabilizing Biological Structures

Walter Kauzmann proposed the hydrophobic effect as a key contributor to the stability of globular proteins. His ideas were based on a knowledge of, on the one hand, the presence of a significant proportion of hydrocarbon side chains in proteins and, on the other hand, the poor miscibility and unfavorable (positive) free energy of mixing of hydrocarbon molecules and water. He reasoned that this would produce an unfavorable free energy contribution to the process of unfolding a protein, as this would then result in the forced exposure to water of hydrocarbon side chains. The correctness of his ideas was amply demonstrated by the subsequent observation that, indeed, hydrocarbon side chains are predominantly buried in the protein interior, to an extent that correlates with the free energy of transfer of the side chain from a less polar solvent to water (an important series of measurements by Tanford and Nozaki; see Section 17.8).

Subsequently, it was realized that the hydrophobic effect is also responsible for the stability of lipid bilayers, which are formed by phospholipid molecules having a small polar headgroup and two long hydrocarbon tails. Depending on the exact structure, these molecules pack in double layers, with polar groups on each surface and hydrocarbon side chains in the center of the double layer to form a lamellar phase that contributes the basic structure of cell membranes. Similar amphiphilic detergent molecules (often with a single hydrocarbon tail) form spherical or cylindrical assemblies, called *micelles*, again with the polar head groups on the surface exposed to water and the hydrocarbon tails packed away in the center. (Fig. 4.5)

It turns out that the name hydrophobic effect is apt, as no other solvent approaches the peculiar properties of water, which is so apparently hostile to nonpolar solutes and so apparently welcoming to other, polar solutes. It should then not be surprising that in order to gain an understanding of what produces the hydrophobic effect one must refer to the structure of water, which in turn means considering how the hydrogen-bonded structure changes in the presence of a hydrocarbon solute.

4.3.2 Thermodynamic Description for Small Hydrocarbons

The hydrophobic effect is often called a *hydrophobic force* or a *force of mean field* that drives hydrophobic groups together or out of aqueous solution. However, it is not a force that the groups exert on each other, but it is the resultant of attractive forces between water molecules that tend to decrease the area of the hydrocarbon–water interface; the "force" is just (minus) the gradient of the resultant free energy change. The free energy, and therefore also the force, derive from the structural properties of liquid water and, at least for the present the hydrophobic effect finds its description best in thermodynamics, although molecular mechanics simulations can yield useful insights. We begin with the thermodynamic observations that do the most to define the phenomenon.

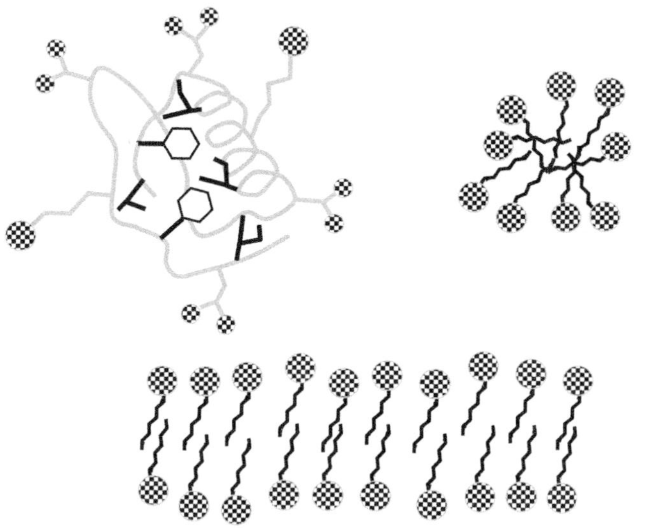

FIGURE 4.5 Protein with a polar side chains (black) in the hydrophobic core, polar side chains (checkerboard) on the surface. Micelle and lipid bilayer, each with hydrocarbon chains (black) in the hydrophobic core and polar head groups (checkerboard) on the surface.

Dissolution of simple aliphatic nonpolar molecules in water is of course unfavorable (positive free energy change). However, and this comes as a considerable surprise, the source of this positive free energy change is in the entropy, not the enthalpy (Table 4.3). Thus, the removal of small hydrocarbons from water to a nonpolar solvent is driven by a large increase in entropy, and accompanied by a nonnegligible increase of the enthalpy. In addition, the partial molar volumes of nonpolar molecules are often much lower in water than in nonpolar solvents (Table 4.4).

The molecular interpretation of these thermodynamic observations is that nonpolar solutes somehow fit within the open structure of water. This idea derives credence from crystalline hydrates of small organic compounds crystallized from water, which contain cages of water molecules surrounding the nonpolar molecules, called *clathrate* structures (see Fig. 4.6). On the basis of these observations, Frank

TABLE 4.3 Thermodynamic parameters for transfer of normal aliphatic hydrocarbons from the liquid (or nonpolar) solution to aqueous solution at 25 °C[a]

	ΔG° (kcal mol^{-1})	ΔH° (kcal mol^{-1})	ΔS° (cal mol^{-1} K^{-1})	$-T\Delta S^\circ$ (kcal mol^{-1})
Methane	2.5 to 3.1	−2.9 to −2.2	−18 to −17	5.0 to 5.5
Ethane	3.3 to 3.9	−2.4 to −1.3	−19 to −17	5.0 to 5.8
Propane	4.9	−2.1 to −1.4	−23 to −21	6.3 to 7.0
Butane	5.8 to 6.0	−1.0 to −0.7	−23 to −22	6.5 to 6.8

[a] Némethy, G. and Scheraga H.A. *J. Chem. Phys.* **36**: 3382–400 (1962) and **36**: 3401–3416 (1962).

TABLE 4.4 Partial molal volumes (cm³/mol) of gases in water or organic solvents[a]

	Ar	N_2	CH_4	C_2H_6
H_2O	32	33	37	51
C_6H_6	43	58	52	73
CCl_4	44	52	52	
CS_2	45		56	

[a] Némethy, G. and Scheraga H.A. *J. Chem. Phys.* **36**: 3382–400 (1962) and **36**: 3401–3416 (1962).

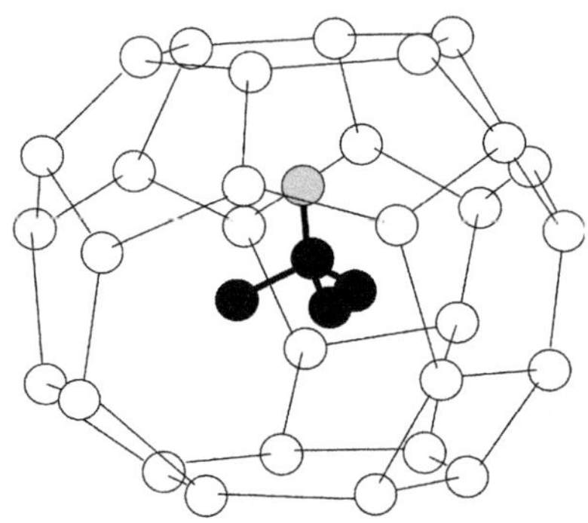

FIGURE 4.6 A water-hydrocarbon clathrate structure (crystal structure of *t*-butyl amine clathrate). Open circles: water oxygen atoms; thin lines: hydrogen bonds; filled circles: CH_3 groups; gray circle: NH_2 group; heavy lines: C—C and C—N chemical bonds.

and Wen[6] proposed a cluster model of water structure, and later Némethy and Scheraga developed a thermodynamic treatment based on this model.[7] In this model nonpolar solutes interact favorably (i.e., favorable ΔE) with the open structure of liquid water because they are able to fit into the large free volumes in water. According to this model, the water molecules surrounding the cavity assume a *more* ice-like arrangement, which indeed should result in the large negative entropy change associated with hydration of nonpolar solutes. This molecular interpretation is useful as a thinking tool, but this is still an interpretation. In the following section, we examine the hydrophobic effect from the perspective of molecular dynamics to see if the simple picture of Frank and Wen holds up.

A more empirical, and straight-forward way to view the hydrophobic effect is in terms of the free energy of transfer of a single CH_2-group from water to a nonpolar solvent (or alternatively, from water to vacuum), as estimated from the free energies of transfer of aliphatic hydrocarbons of increasing chain length, according to

$$\Delta G^{\mathrm{o}}_{\mathrm{tr}}(\text{—CH}_2\text{—}) = \Delta G^{\mathrm{o}}_{\mathrm{tr}}[\mathrm{H}(\text{—CH}_2\text{—})_{n+1}\mathrm{H}] - \Delta G^{\mathrm{o}}_{\mathrm{tr}}[\mathrm{H}(\text{—CH}_2\text{—})_n\mathrm{H}].$$

The transfer of a single CH_2-group from water to nonpolar solvent is favorable ($\Delta G^{\mathrm{o}} \approx -1.4\,\mathrm{kcal\,mol^{-1}}$ for solvent hexadecane[8]), with a large positive ΔS^{o} and a small positive ΔH^{o} and thus the process is said to be *entropy-driven*.

4.3.3 Thermodynamic Description for Hydrophobic Surfaces

At times, there is merit in considering the transfer of a solute into water as a two-step process (see also discussion in Chapter 13). In the first step, a cavity is created of the right size to accept the solute, and in the second step the solute is introduced and solute–solvent interactions are established. The free energy in the first step is invariably unfavorable, and can be roughly estimated as the product of the inner surface of the cavity and the surface tension of water, which is on the order of 79 cal/Å^2/mol (and depends on the somewhat arbitrary choice of locating the interface between solute and solvent). The free energy of the second step can vary from relatively weak Lennard–Jones interactions for a nonpolar solute to very significant polarization free energy for a polar molecule or ion.

This is referred to as the *surface tension model*, and has merit as the size of the cavity that accepts a solute grows to macroscopic proportions, the interface becomes essentially flat and thus the specific free energy of cavity formation should approach the surface free energy or surface tension of water. However, the enthalpy of formation of the flat, macroscopic interface is equal to +175 cal/mol/Å^2, and thus in going from the microscopic to the macroscopic situation, the nature of the driving force for hydrophobicity changes from entropic to enthalpic.

The small-hydrocarbon model is taken as the more appropriate model for describing hydration of nonpolar amino acid side chains.

4.3.4 Molecular Dynamics Description: Water Structure, Free Energy, and the "Hydrophobic Force"

There have been several molecular dynamics treatments of the interaction between two small hydrophobic molecules in water using the potentials outlined in Section 4.2.3. We will consider here the first such attempt, one that modeled the interaction between neon atoms.[9]

In this simulation, water molecules in the first shell next to a neon atom reorient so that their hydrogen-bond directions all point toward other water molecules and never toward the neon (Fig. 4.7). Roughly eight nearest neighbor and six next-nearest neighbor water molecules surround the neon atom. The calculated reorientational correlation times of molecules in this 14-member first shell increase by 1.2–1.7-fold over bulk water molecules. The observed distribution corresponds roughly to a distorted clathrate-like structure similar to those seen in crystalline hydrates of organic molecules.

When two neon atoms are initially placed in contact, the neon–neon distance oscillates around about 3 Å for roughly 2.5 ps, then switches to about 6 Å. Analysis of hydrogen-bonding patterns shows that this corresponds to a shift from two neon atoms encased in a single cage to each neon having its own cage (i.e., a single layer of water being shared by the two neon atoms—see Fig. 4.8).

According to the hydrophobic effect, the free energy of a pair of neon atoms or of hydrocarbon molecules in aqueous solution is lowest when the two molecules are in close proximity, essentially van der Waals contact, with a minimum of contacts with water molecules. With the advent of new simulation techniques and more

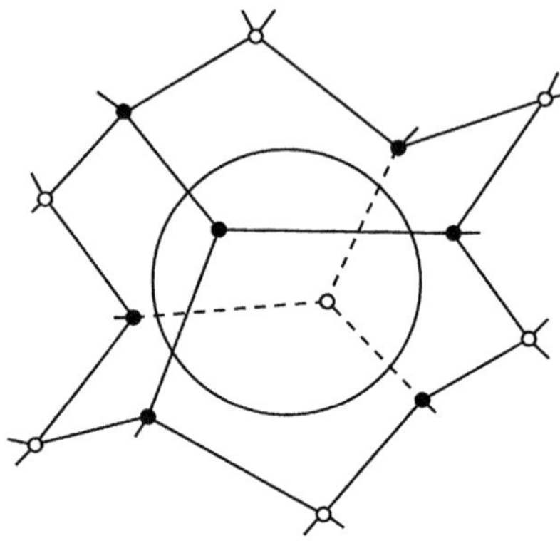

FIGURE 4.7 Typical cage of ST2 water around a neon atom. From Geiger, A., Rahman, A., Stillinger, F.H. *J. Chem. Phys.* **70**: 263–276 (1979).

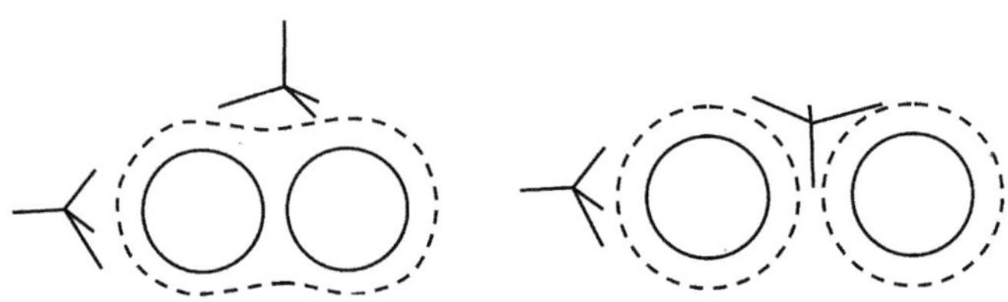

FIGURE 4.8 Two configurations of a pair of neon atoms and two nearby water molecules in ST2 water. From Geiger, A., Rahman, A., Stillinger, F.H. *J. Chem. Phys.* **70**: 263–276 (1979).

powerful computers, it has become possible to calculate such free energy differences by molecular dynamics simulation. This requires a thorough sampling of the Boltzmann distribution, that is, producing several hundred thousand configurations of the system. The free energy change along a reaction path ("reaction coordinate") can be estimated by numerical integration of the mean force between the two solute molecules as these are brought together in the solution, and is termed a *potential of mean force*. This is illustrated by the results of a simulation of two methane molecules in water represented with the TIP3P potential. This study[10] is discussed in detail in Section 7.8.2. (Simulation techniques are also discussed in detail in that chapter.) The computed free energy for the pair of methane molecules has a minimum when the two molecules are separated by about 3 Å; these short distances are favored over the fully separated state by a free energy difference of nearly −0.9 kcal/mol (Fig. 7.5); the direct interaction between the model methane molecules (a Lennard–Jones potential) has a minimum of only −0.29 kcal/mol.

Thus, molecular dynamics simulations support the general view of the hydrophobic interaction first suggested by Frank and Wen, namely, that nonpolar solutes appear to enhance the structure of already structured liquid water, with a concomitant increase of the free energy.

NOTES

1. (a) Stillinger, F. H. Water revisited. *Science*: 451–457 (1980). (b) Franks, F., *Water, A Matrix of Life*, 2nd Ed., The Royal Society of Chemistry, 2000. For a table comparing properties of four liquids, see http://hyperphysics.phy-astr.gsu.edu/hbase/tables/liqprop.html#c1.
2. Rahman, A., Stillinger, F.H. Hydrogen-bond patterns in liquid water *J. Am. Chem. Soc.* **95**: 7943–7948 (1973).
3. Berendsen, H.J.C., Postma, J., van Gunsteren, W.F., Hermans, J. In *Intermolecular Forces*. (Pullman, B., Ed.) pp 331–342, Dordrecht, Holland: Reidel, 1981.
4. Jorgensen, W.L., Chandrasekhar, J., Madura, J.D., Impey, R.W., Klein, M.L. Comparison of simple potential functions for simulating liquid water. *J. Chem. Phys.* **79**: 926–935 (1983).
5. See review by Jorgensen, W.L., Tirado-Rives, J. Potential functions for atomic-level simulations of water and organic and biomolecular systems. *Proc. Natl. Acad. Sci. USA* **102**: 6665–6670 (2005).
6. Frank, H.S., and Wen, W.-Y. Ion-solvent interaction. Structural aspects of ion-solvent interaction in aqueous solutions: a suggested picture of water structure. *Discuss. Faraday Soc.* **24**: 133–140 (1957).
7. Némethy, G. and Scheraga H.A. Structure of water and hydrophobic bonding in proteins. (a) A model for the thermodynamic properties of liquid water. *J. Chem. Phys.* **36**: 3382–400 (1962). (b) Model for the thermodynamic properties of aqueous solutions of hydrocarbons. *J. Chem. Phys.* **36**: 3401–3416 (1962).
8. Kyte, J. The basis of the hydrophobic effect. *Biophys. Chem.* **100**: 193–203 (2003).
9. Geiger, A., Rahman, A., Stillinger, F.H. Molecular dynamics study of the hydration of Lennard-Jones solutes. *J. Chem. Phys.* **70**: 263–276 (1979).
10. Czaplewski, C., Rodziewicz-Motowidlo, S., Liwo, A., Ripoll, D.R., Wawak, R.J., Scheraga, H.A. Molecular simulation study of cooperativity in hydrophobic association. *Protein Sci.* **9**: 1235–1245 (2000).

SUGGESTED READING

1. Tanford, C. *The Hydrophobic Effect: Formation of Micelles and Biological Membranes*. New York: John Wiley & Sons, Inc., 1980

Part 2

Statistical Mechanics: The Molecular Basis of Thermodynamics

In the first part of this book we have reviewed two pillars of physical chemistry, thermodynamics, and quantum mechanics. Thermodynamics relates such concepts as energy, work, heat, entropy, and chemical energy of macroscopic systems on a strictly empirical basis. On the other hand, quantum mechanics provides a knowledge of molecular energetics on the basis of first principles, but does not provide useful descriptions of macroscopic systems. Quantum mechanics teaches that, rather than having a continuum of energy, a system can occur in any one of infinitely many states of different energy and gives, at least in principle, a complete description of atomic, molecular, and intermolecular energy states.

The link between molecular and thermodynamic energetics is established by statistical mechanics (more precisely statistical thermodynamics), which are the subject of this second part. Statistical mechanics is concerned with how energy is distributed on a macroscopic scale over quantum states of many molecules (or many systems), or in a single system over a fairly long periods of time in order to determine its average properties. A key postulate is that the quantum state of a system randomly evolves with time so that over a long time all quantum states will appear with the same probability as observable in a large ensemble of systems at a given time (the so-called ergodic hypothesis).

The second key postulate is that average system properties reflect the most probable distribution over quantum states. We will discover the intuitively appealing result that the most probable distribution of systems or molecules over quantum states is consistent with the Second Law of Thermodynamics. Once we know the probability distribution over quantum states, each with its associated energy, we can

Equilibria and Kinetics of Biological Macromolecules, First Edition. Jan Hermans and Barry Lentz.

(of course) calculate average system properties, including thermodynamic functions strictly on the basis of the energy levels predicted by quantum theory (see Figure on this page).

This probability distribution (actually, its normalizing factor) is called *the partition function*, and is determined by how we constrain the system by our choice of independent variables. We begin by dealing with an ensemble at constant energy and composition (an isolated system; Chapter 5); in particular, we treat a special case of this ensemble, a system containing independent molecules (which represents the ideal gas state), completely in terms of quantum mechanical energy levels. We shall call this the molecular ensemble.

Two additional ensembles are treated in Chapter 6. We show first how an ensemble kept at constant temperature and composition, a so-called canonical ensemble, can be used to good effect with classical energy functions, and we apply this ensemble to the energetics of molecular interactions.

This chapter finishes with a grand ensemble; this is an ensemble of open systems that can exchange energy and contents, and are (thereby) kept at constant temperature and chemical potentials. This ensemble provides a theoretical description of concentration fluctuations of solutions, and we apply this to light scattering by (macromolecular) solutions.

In Chapter 7, we describe how molecular simulation methods using classical, empirical force fields can be used to explore the Boltzmann distribution of systems of hundreds or thousands of atoms and can be used to compute system averages and thermodynamic functions by application of statistical mechanics.

The canonical and grand canonical partition functions and simulation studies will, in subsequent chapters, be key to describing a variety of macromolecular equilibria that underlie ligand binding, conformation change, allosteric effects, the helix-coil transition of polypeptides, and protein unfolding. It is fair to say that these properties cannot be fully understood without knowledge of statistical thermodynamics.

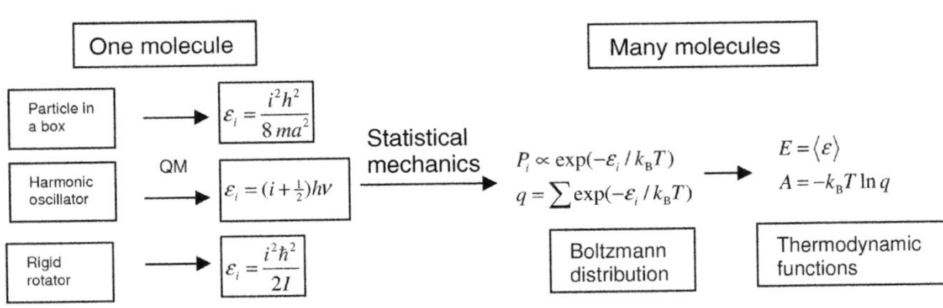

From quantum mechanics to thermodynamics.

5

The Molecular Partition Function

5.1 INTRODUCTION

Statistical mechanics deals with how the internal energy (a thermodynamic quantity) partitions between the energy states available to a system comprised of any number of molecules. The quantity that describes this is called a *partition function*. Not only do statistical approaches require a large number of observational elements but thermodynamics also involves large or macroscopic systems. We therefore consider first a collection of a very large number of identical and *independent* (i.e., noninteracting) molecules, call this the "molecular ensemble" and derive its "molecular partition function" by seeking a description of the most likely or average molecule or set of molecules. The key to describing the average properties of a single molecule is our assumption that the molecules are independent and identical.

5.2 THE MAXWELL–BOLTZMANN DISTRIBUTION

5.2.1 Statistical Distribution of Energy: an Example

In order to simplify the discussion, we consider first an isolated system of defined energy, E containing a fixed number, N of *independent and distinguishable* molecules[1] confined to a volume V, each of which is characterized by a *molecular quantum state* ψ_i *of energy* ε_i. Unless the total energy, E is small, there are usually many ways in which these molecules can be distributed over the available

Equilibria and Kinetics of Biological Macromolecules, First Edition. Jan Hermans and Barry Lentz.

molecular quantum states (n_1, in a state ψ_1 with ε_1, n_2 in ψ_2 with ε_2, ... n_i in ψ_i, etc.), consistent with fixed total energy E.

This is illustrated in Table 5.1 for the simple case of four molecules distributed between five possible molecular quantum states (ψ_i with $\varepsilon_i = [i-1]e$), such that the total system energy is $E = 4e$. Each distribution is described by a set of numbers, $\{n_i\}$. The possible distributions for the example given in Table 5.1 are the sets of numbers: {2,0,2,0,0}, {1,2,1,0,0}, and so on. The notation {2,0,2,0,0} means 2 molecules in quantum state 0, 2 in quantum state 2, 0 molecules in quantum states 1, 3, and 4. Each distribution (denoted by index j) has the same number of molecules (N), energy ($E = 4e$, because the system is isolated) and volume (V) but defines a distinct system quantum state, $\Psi(\{n_i\})$.

Distinguishable molecules can be assigned in many ways to the set of molecular quantum states established by a given distribution $\{n_i\}$. Thus, there can be a large number, $\Omega(\{n_i\})$ of distinct configurations or arrangements of *particular* molecules in molecular quantum states consistent with each system quantum state $\Psi(\{n_i\})$. The system quantum state is said to have a *configurational or statistical degeneracy* of $\Omega(\{n_i\})$.[2] In our simple example, Ω is 6, 12, 1, 12, or 4 (Table 5.1) for the five possible distributions of molecules over molecular quantum states, each distribution corresponding to a different system quantum state. Table 5.2 illustrates how we obtained a value of 12 for Ω ({1,2,1,0,0}).

For this example, we can obtain the total number of configurations consistent with $E = 4e$, which is equal to 35, by summing over the degeneracy of all possible quantum states. In general, any ensemble can be configured in

$$\sum_j \Omega(\{n_i\}_j) = \mathcal{W} \tag{5.1}$$

TABLE 5.1 Five physically distinct ways to distribute four particles ($N = 4$) over five quantum states with energy levels 0, e, $2e$, $3e$ and $4e$ to give a total energy of $4e$

# In Level 0	# In Level 1	# In Level 2	# In Level 3	# In Level 4	$\Omega(\{n_i\})$	$\{n_i\}$
2	0	2	0	0	6	{2,0,2,0,0}
1	2	1	0	0	12	{1,2,1,0,0}
0	4	0	0	0	1	{0,4,0,0,0}
2	1	0	1	0	12	{2,1,0,1,0}
3	0	0	0	1	4	{3,0,0,0,1}

See text for definition of Ω and $\{n_i\}$.

TABLE 5.2 Twelve distinct ways to distribute four particles (numbered 1 through 4) in four equally spaced energy stated (marked with different symbols) to give the same configuration, $\{n_i\} = \{1, 2, 1, 0\}$, with $E = 4e$

❶ ②③ **4**	❶ ②④ **3**	❶ ③④ **2**	❷ ③④ **1**
❷ ①③ **4**	❷ ①④ **3**	❸ ①④ **2**	❸ ②④ **1**
❸ ①② **4**	❹ ①② **3**	❹ ①③ **2**	❹ ②③ **1**

For example: in quantum state 3 1 4 2, in row 2 of column 3, particle 3 is in the ground state, particles 1 and 4 are in the state with energy level e, and particle 2 is in the state with energy level $2e$.

ways, where the sum is over all possible quantum states $\Psi(\{n_i\}_j)$ of the system. Different molecular energy levels occur with frequencies proportional to $\sum_j \Omega(\{n_i\}_j) n_{i,j}$, where $n_{i,j}$ is the number of molecules in molecular quantum state i in the system quantum state defined by the distribution $\{n_i\}_j$. This gives a ratio 60 : 40 : 18 : 12 : 4 for the molecular energy levels $i = 0$ through 4 in the system of Table. 5.2. Even this tiny system shows a decrease in the likelihood of observing quantum sates with higher energies, which, as we shall see next, becomes the exponential Boltzmann distribution for very large systems.

5.2.2 An Ensemble of Systems

We shall now extend this simple example to the general case where the set of molecular energy levels is much more complex, as it is for real molecules, and the number of molecules is very large, as in a macroscopic system.

Given a system of many particles, we can construct system quantum states in a very large number of ways, $\mathcal{W}$, but each consistent with the constraints that it has energy E, volume V, and N particles. Note that $\mathcal{W}$ gets extremely large as N and E get large. In fact, $\mathcal{W}$ becomes so large that in any practical time period not all possible configurations can be observed. Just as it is not possible to observe a system for an "infinite" time (much more than the lifetime of the observer), it is not possible to sample all possible configurations of the system. In statistical mechanics, we assume that an average of a *statistical sample* (a very large random sampling) of all possible quantum states describes the macroscopic physical state of a system. Such a collection of $\mathcal{M}$ systems, all having the same number of molecules, volume, and energy is denoted as a *microcanonical ensemble* (Table 5.3). We continue to discuss a special case of this ensemble in which the molecules are independent and distinguishable. This ensemble will allow us to derive the average properties of any molecule as it evolves over time.

TABLE 5.3 A microcanonical ensemble of systems of independent molecules. All systems have the same composition, volume, and energy, while the distribution of molecules over molecular quantum states varies

N, V, E $\{n_i\}_1$	N, V, E $\{n_i\}_2$	N, V, E $\{n_i\}_3$	...
N, V, E $\{n_i\}_j$	...		
N, V, E $\{n_i\}_m$	...		
N, V, E $\{n_i\}_t$	...		N, V, E $\{n_i\}_{\mathcal{M}}$

We can obtain the average value of some mechanical system variable, X, in the system of interest if we make two fundamental postulates of statistical thermodynamics[3]:

- All allowed distinct quantum states of a system are *intrinsically* equally probable; however, as we shall see, the *actual* probability of each quantum state is limited by the available energy.
- Any given system followed over time visits all quantum states with the same probability with which they occur in the ensemble.[4]

On the basis of these postulates, we obtain the average of any mechanical system variable, X, by finding the average of X over the values X_j that it has in all configurations of the ensemble,[5]

$$\langle X \rangle = \frac{\sum_{j=1}^{\mathcal{M}} \Omega(\{n_i\}_j) X_j}{\sum_{j=1}^{\mathcal{M}} \Omega(\{n_i\}_j)}, \tag{5.2}$$

where X_j is the value of the mechanical variable in system quantum state, j. For independent molecules, X_j will reflect an average over this mechanical variable for all the independent molecules in the system:

$$\langle X \rangle_j = \frac{\sum_i n_{i,j} X_i}{\sum_i n_{i,j}} = \frac{\sum_i n_{i,j} X_i}{N}.$$

The summation in the expression for $\langle X \rangle$ is over all system quantum states $\{n_i\}_j$, and the summation in the expression for $\langle X \rangle_j$ is over molecular quantum states in the jth system quantum state.

The reader will recognize from the example in Table 5.2 that $\Omega(\{n_i\})$ is the number of ways of putting N different (i.e., distinguishable) balls (molecules) into differently numbered boxes (molecular quantum states), such that n_1 are in box 1, n_2 in box 2, etc., which is given by the following combinatorial expression,

$$\Omega(\{n_i\}) = \frac{N!}{\prod_i n_i!}. \tag{5.3}$$

Of course, the allowed choices of $\{n_i\}$ are limited by the constraints (rules or canons) imposed by the properties of the microcanonical ensemble,

$$E = \sum_i n_i \varepsilon_i; \; N = \sum_i n_i. \tag{5.4}$$

5.2.3 Maximizing the Likelihood

A property of the combinatorial expression (Eq. 5.3) is that, as N increases, fewer terms dominate the sum in Eq. 5.2, that is, some configurations have many more distinct quantum states in the ensemble than the others. When N becomes very large, one configuration becomes enormously more probable than *all* other configurations, such that these other possible configurations of the system can be ignored, and the sum in Eqs. 5.1 and 5.2 is approximated by its maximum term. When that is the case, the problem of finding the average value of X becomes the problem of identifying the set $\{n_i\}$ that maximizes $\Omega(\{n_i\})$ subject to the constraints shown in Eq. 5.4, and calculating the appropriate value of $\langle X \rangle$ for this configuration, as shown in the second sum of Eq. 5.2.

For a finite system, the probability that the distribution of molecules in the system over the energy levels is slightly different from the most probable distribution can be substantial; in a real system, the distribution will fluctuate over time, and two similar systems will rarely have an identical distribution. However, if the number of molecules in the system is very large, these deviations will be very small and will not have a measurable effect.

Mathematically, it is easier to seek the maximum of $\ln \Omega$, which of course occurs for the same set $\{n_i\}$ as the maximum of Ω. The function $\ln \Omega$ depends on the entire set of molecular quantum state occupancies, $\{n_i\}$. In order to find the maximum value of $\ln \Omega$ that satisfies the constraints, Ludwig Boltzmann (1844–1906) used a procedure ascribed to Joseph Lagrange (1736–1813), in which the constrained functions, multiplied by constants, are added to the quantity to be maximized, and the maximum of this sum is obtained by setting its derivatives with respect to all variables equal to zero.

$$\begin{aligned} B &= \ln \Omega(\{n_i\}) + \alpha \sum_i n_i - \beta \sum_i n_i \varepsilon_i \\ \frac{\partial B}{\partial n_i} &= 0, i = 1, 2, \ldots . \end{aligned} \tag{5.5}$$

The identification and evaluation of the so-called Lagrange multipliers α and $-\beta$ is part of the solution (we use a negative multiplier, $-\beta$, so that the solution for n_i given subsequently will converge). The solution is simple when each of the factorials occurring in the expression for Ω is replaced with $n^n e^{-n}$, the leading terms of Stirling's formula for $n!$,[6]

$$\begin{aligned} B &= N \ln N - N - \sum_i n_i \ln n_i + \sum_i n_i + \alpha \sum_i n_i - \beta \sum_i n_i \varepsilon_i \\ \frac{\partial B}{\partial n_i} &= -\ln n_i + \alpha - \beta \varepsilon_i = 0, \quad i = 1, 2, \ldots \\ n_i &= \exp \alpha \exp(-\beta \varepsilon_i). \end{aligned} \tag{5.6}$$

Setting the sum of all the n_i equal to N assigns a value to α, leading to

$$\begin{aligned} n_i &= \frac{N \exp(-\beta \varepsilon_i)}{\sum_i \exp(-\beta \varepsilon_i)} \\ P_i &= \frac{n_i}{N} = \frac{\exp(-\beta \varepsilon_i)}{\sum_i \exp(-\beta \varepsilon_i)}, \end{aligned} \tag{5.7}$$

where P_i is the probability of molecules in the system being in the ith quantum state with energy ε_i. This probability function is called the *Boltzmann distribution function*.

The Boltzmann distribution defines the most probable distribution of independent molecules over molecular quantum states, that is, the one that maximizes the number of quantum states associated with the distribution consistent with the imposed constraints. We will call this distribution $\{n_i\}_{\text{B}}$. Thus, the Boltzmann probability function defines a distribution that *optimizes the randomness of the energy distribution* of an isolated system of independent molecules. Recall that, at equilibrium, the Second Law of Thermodynamics requires that the entropy of an isolated system is maximized. Thus, if we identify the *configurational or statistical degeneracy* of the system as related to the entropy, the Boltzmann distribution satisfies the Second Law. Statistical mechanics *replaces* the Second Law of Thermodynamics with the statement that the number of ways, Ω, of putting a large number of different molecules into different molecular quantum states (of different energy) is maximized in an isolated system. Conservation of energy, the First Law, is of course maintained in statistical mechanics, which determines only the most likely way in which this energy is distributed over quantum states.

Note that by ignoring all distributions $\{n_i\}_j$ except that for which $\Omega(\{n_i\}_j)$ is maximal, we imply that, while no two systems in the ensemble need be identical at the molecular level, if we randomly pick any system, it will with high probability (approaching 1 as N approaches infinity) have the distribution defined by the Boltzmann probability function, and the distributions of individual systems in the ensemble will increasingly become indistinguishable from one another as N increases.

To further establish the connection to thermodynamics, we must correctly choose the Lagrange multiplier, β. To gain insight into the meaning of β, we first note from Eq. 5.7 that P_i decreases exponentially with increasing ε_i. In Fig. 5.1, we plot P_i as a function of ε_i for a model system with $\varepsilon_i = i \cdot e$ where e is a constant energy level spacing. Our physical intuition suggests that fewer high-energy states would be occupied if the thermal energy (temperature) *decreases*. This is predicted by the Boltzmann distribution as β *increases*, as is evident from Fig. 5.1. Hence, we might expect β to be related to the reciprocal of the temperature. This says that to obtain a reasonable probability that the system exists in other than the lowest energy state, β must be much larger than the energy level spacing. For $\beta \gg e$, high energy states are highly unlikely but not forbidden, and the probability that a molecule exists in a high energy state increases with increasing temperature (decreasing β).

Boltzmann chose β to be inversely proportional to T, with the proportionality constant $1/k_{\text{B}}$ ($k_{\text{B}} = R/N_{\text{A}}$, Boltzmann's constant, equals Boyle's gas constant divided by Avogadro's number). As we shall see, this makes a connection between statistical mechanics and thermodynamics and fixes the temperature scale as the Kelvin scale inherent to the experimental value of the gas constant, and sets the

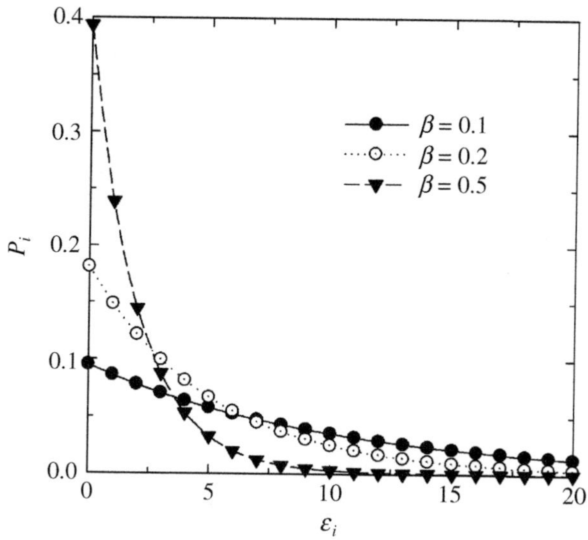

FIGURE 5.1 The Boltzmann probability, P_i as a function of energy, ε_i for the case of equally spaced energy levels ($\varepsilon_i = 0, 1, 2, \ldots$) is plotted at three values of β, that is, at three temperatures. P_i decreases with energy in all cases, but at high temperature (low β; solid line), the high-energy states are relatively more populated. The state with lowest energy is always the most populated.

thermal energy to 0 at $T = 0\,\mathrm{K}$,[7]

$$\beta = \frac{1}{k_\mathrm{B}T}. \tag{5.8}$$

We see later that this choice leads to a statistical definition of free energy and entropy.

5.3 THE MOLECULAR PARTITION FUNCTION AND THERMODYNAMIC FUNCTIONS

The denominator in Eq. 5.7 is called the *molecular partition function*,

$$q = \sum_i \exp[-\varepsilon_i/(k_\mathrm{B}T)] = \sum_n \omega_n \exp[-\varepsilon_n/(k_\mathrm{B}T)], \tag{5.9}$$

where the sum over i is over *quantum states*. As noted, each molecular quantum state may be degenerate such that ω-fold degenerate states occur ω times in the sum, so that we can perform the sum over *energy levels*, n, if we take into account this quantum degeneracy.[2]

It should be evident that a single molecule does not sample all possible quantum states according to a Boltzmann distribution on any time scale. However, if we could observe a single molecule for a long enough time (many times the characteristic lifetimes of the slowest motions experienced by nuclei and electrons in the molecule—i.e., μs), it would pass through most of its available quantum states. If over this long time, we insist that the molecule has an average energy, $\langle E \rangle$, then the probability that it exists in any one molecular quantum state during this time interval is defined by the Boltzmann distribution. Instead of doing this experiment, we have established the average behavior of a molecule on this very long time scale by considering an ensemble of a very large number of independent molecules. This is the meaning of the molecular partition function.

The configurational energy and entropy of a system of Boltzmann molecules can be defined in terms of q. We will initially make the simplifying assumption that a thermodynamic property of N molecules is the sum of that property over all N molecules, and just N times the average of that property for a single molecule.[8] Thus, the thermodynamic or average energy, U, is equated with the sum of the molecular energies for all molecules in the system, or simply N times the average energy for a single molecule. It is easy to derive from Eq. 5.9 that the total *configurational energy* (due to distribution over available states) is related to the temperature derivative of the molecular partition function,

$$U = E = N\sum_i P_i\varepsilon_i = N\langle\varepsilon\rangle = -N\frac{\partial \ln q}{\partial\beta}. \tag{5.10}$$

In order to find an expression for the free energy, we use an expression (Eqs. 5.1–5.30) from the chapter on thermodynamics

$$U = \left(\frac{\partial (A/T)}{\partial (1/T)}\right)_{V,N}.$$

Making use of Boltzmann's assumption that $\beta = 1/k_{\mathrm{B}}T$, Eq. 5.10 identifies A as

$$A = -Nk_{\mathrm{B}}T \ln q. \tag{5.11}$$

This is the key expression that links Boltzmann's statistical description of an isolated system of N independent molecules with the thermodynamics of such a system. We rewrite the above-mentioned definition of ε_i in terms of P_i and q as follows:

$$\varepsilon_i = -\frac{1}{\beta}\ln(P_i q). \tag{5.12}$$

and substitute this into the expression for the internal energy, U (Eq. 5.10). With Eq. 5.11 for the free energy, A, this gives

$$U = E = -\frac{N}{\beta}\sum_i P_i \ln(P_i q) = -\frac{N}{\beta}\sum_i P_i \ln P_i - \frac{N}{\beta}\ln q = -\frac{N}{\beta}\sum_i P_i \ln P_i + A. \tag{5.13}$$

With the entropy, S, defined by $A = U - TS$, this gives

$$S = \frac{E}{T} - \frac{A}{T} = -Nk_B \sum_i P_i \ln P_i. \tag{5.14}$$

The expression $N\sum P_i \ln P_i$ can be equated with $\ln \Omega$ via Eqs. 5.3 and 5.7[9] to give the following famous expression[10]

$$S = k_B \ln \Omega. \tag{5.15}$$

Thus, Boltzmann's identification of β as $1/(k_B T)$ identifies the *configurational entropy* of an isolated system of independent, distinguishable molecules as $k_B \ln \Omega$. The justification of this fundamental identification is given in terms of the consistency between statistical and thermodynamic expressions for energy and entropy. Because each individual $P_i \leq 1$, the entropy is positive (as it should be). As $\{n_i\}$ defined by the Boltzmann distribution always maximizes $\Omega(\{n_i\})$, it follows that $\ln \Omega$ must increase if the system undergoes a spontaneous change to a different thermodynamic state characterized by $\{n_i\}$, meaning that $dS \geq 0$, which agrees with the Second Law for a system having constant energy.

A system can have entropy that is not configurational. When the absolute temperature is zero, and only the ground state is occupied, the entropy is $-k_B\, \omega_0 \ln \omega_0$, where ω_0 is the degeneracy of the ground state. This is very small, but not exactly zero, as had been assumed in the Third Law. In these expressions, the reference state for the internal energy, $U(0)$ is taken as the energy at 0 K (the sum of electronic ground state energy and zero point vibrational energy). The reference state for the entropy is $S(0)$, the entropy at 0 K, which, except for degeneracy, is 0. Thus, at 0 K, $A(0)$ is almost entirely in the form of internal energy, $U(0)$.

5.4 APPLICATION TO MACROMOLECULES

5.4.1 Independent Motions, Independent Quantum States

Exact evaluation of the energies of the quantum states of a macromolecule is not currently possible. The problem simplifies greatly if the quantum states associated with different modes of motion of the molecule do not depend on the quantum states of other modes of motion, that is, different modes of motion are *independent*. This is actually *a reasonable approximation*, as the motions of electrons are largely independent of the motions of much heavier nuclei (Born–Oppenheimer approximation). Recall from the discussion in Section 2.5 that the motion of the N covalently linked atoms making up a molecule can be approximated in terms of the motions of an (elastic) "rigid body" that corresponds to the minimum-energy structure of the molecule.

1. The translational motion of the center of mass of the molecule in three *orthogonal* directions

2. Rotational motion about three *orthogonal* rotational axes and
3. $3N$–6 internal and *independent* normal mode vibrations involving coordinated motions of the atoms (*normal coordinates*).

In this way, the total energy of a macromolecule can be described by independent electronic, translational, rotational, and vibrational energy levels

$$\varepsilon_{i,j,k,l} = \varepsilon_{\text{elec},i} + \varepsilon_{\text{trans},j} + \varepsilon_{\text{vib},k} + \varepsilon_{\text{rot},l}, \tag{5.16}$$

with each set of energy levels being independent of the energy levels for the other motions. In writing the partition function of a molecule, we sum the Boltzmann probability (an exponential of the total energy) over all possible quantum states. Because of the properties of the exponential function and the fact that the indices of the sum over quantum states are independent, we can replace the sum of products of exponentials by the product of the sums of exponentials,[11]

$$\sum_{i,j,k,l,m} \exp\left(-\frac{\varepsilon_{i,j,k,l}}{k_B T}\right) = \left[\sum_i \exp\left(-\frac{\varepsilon_{\text{elec},i}}{k_B T}\right)\right]\left[\sum_j \exp\left(-\frac{\varepsilon_{\text{trans},j}}{k_B T}\right)\right]$$
$$\times \left[\sum_k \exp\left(-\frac{\varepsilon_{\text{vib},k}}{k_B T}\right)\right]\left[\sum_l \exp\left(-\frac{\varepsilon_{\text{rot},l}}{k_B T}\right)\right] \tag{5.17}$$

Thus, we can write a partition function for each independent motion and the molecular partition function becomes the product of these

$$q = q_{\text{trans}}\, q_{\text{elec}}\, q_{\text{vib}}\, q_{\text{rot}} = q_{\text{trans}}\, q_{\text{int}}, \tag{5.18}$$

where q_{int} represents the partition function for the electronic, vibrational, and rotational energy levels. The partition functions for each of the nuclear motions can be obtained from the quantum mechanical description of the particle in a box, the harmonic oscillator, and the rigid rotator. But first, we address the motions of electrons.

5.4.2 Electronic Partition Function

At normal temperatures, only the ground state electronic level is occupied, so there is only one state that is occupied, and

$$q_{\text{elec}} = \omega_0 \exp\left[-\varepsilon_0/(k_B T)\right], \tag{5.19}$$

where ω_0 is the degeneracy and ε_0 is the energy of the electronic ground state. As there is only one state occupied, this partition function contributes only minimally to the entropy of a system ($k_B \ln \Omega$), but contributes mightily to the internal energy because of the magnitude of ε_0. Chemical reactions, ligand binding, conformation changes, formation of substrate-enzyme transition state, and formation of macromolecular complexes all alter ε_0.

5.4.3 Translational Partition Function

We have seen that the energy levels for the translational motion of a molecule are

$$E_n = \frac{n^2 h^2}{8ma^2}, \qquad n = 1, 2, 3, \ldots \tag{5.20}$$

$$q_{\text{trans},x} = \sum_n \exp\left(-\frac{1}{k_B T}\frac{n^2 h^2}{8ma^2}\right). \tag{5.21}$$

Because $E_n - E_{n-1} \ll k_B T$, even for quite large n, very many states contribute to the sum, and we can replace this sum by an integral (this is termed the *classical limit*). There are essentially no situations that will concern us as biophysicists where the classical limit for the translational partition function would not hold.

$$q_{\text{trans},x} \approx \int_0^\infty dn \exp\left(-\frac{n^2 h^2}{8ma^2 k_B T}\right) = \frac{1}{2}\left(\frac{8ma^2 k_B T\pi}{h^2}\right)^{\frac{1}{2}} = \frac{(2\pi m k_B T)^{\frac{1}{2}}}{h} a, \tag{5.22}$$

and combining the partition functions for the three perpendicular directions gives

$$q_{\text{trans}} = q_{\text{trans},x}\, q_{\text{trans},y}\, q_{\text{trans},z} = \frac{abc}{\left(\dfrac{h^2}{2\pi m k_B T}\right)^{\frac{3}{2}}} = \frac{V}{\Lambda^3}; \quad \Lambda = \left(\frac{h^2}{2\pi m k_B T}\right)^{\frac{1}{2}}. \tag{5.23}$$

Λ is termed the *de Broglie wavelength* of a particle and represents a classical (temperature-dependent) interpretation of its wave dimension. Thus, we see that the translational partition function is a measure of how many positions a particle can occupy in the box of interest.

Application of Eq. 5.11 gives the free energy, A, and this gives an expression for the pressure P

$$A_{\text{trans}} = -k_B T \ln q_{\text{trans}} = -k_B T \ln V + 3k_B T \ln \Lambda$$

$$P = -\left(\frac{\partial A}{\partial V}\right)_{N,T} = \frac{k_B T}{V}. \tag{5.24}$$

If we are describing one mole of particles, this is exactly Boyle's experimental gas law, $PV = RT$, where R is the gas constant, $R = k_B \times$ Avogadro's number. We derived this expression from "first principles," that is, on the basis of quantum mechanics and statistical mechanics. This agreement with experiment provided Boltzmann's justification for equating β with $1/k_B T$ in Eq. 5.8 and sets the Kelvin temperature scale (inherent in Boyle's law) as the scale used in statistical thermodynamics.

5.4.4 Vibrational Partition Function

The $3N$–6 internal degrees of freedom of a macromolecule can be described in terms of independent normal modes, each of which behaves as a harmonic oscillator.[12] Each normal mode vibrational partition function has the energy level distribution of a harmonic oscillator,[13] leading to an expression for the partition function of a single normal mode,

$$q_{\text{vib,j}} = \sum_{n=0} \exp\left[-\frac{\left(n+\frac{1}{2}\right)h\nu_j}{k_{\text{B}}T}\right] = \frac{\exp[-h\nu_j/(2k_{\text{B}}T)]}{1-\exp[-h\nu_j/(2k_{\text{B}}T)]}. \tag{5.25}$$

When $k_{\text{B}}T \gg h\nu_j$ (in the so-called *classical limit*), a simpler expression for the vibrational partition function can be used,

$$q_{\text{vib,j}} \approx \int_0^\infty dn \exp\left[-\frac{nh\nu_j}{k_{\text{B}}T}\right] = \frac{k_{\text{B}}T}{h\nu_j}. \tag{5.26}$$

However, the frequencies of many vibrations are too high to fall in this limit (generally, oscillators with $k_{\text{B}}T \gg 10h\nu_j$ are reasonably well described by the classical partition function), and for these the more complex expression must be used. The frequency spacings of many vibrational modes can be determined by accurate spectroscopy. It is customary in spectroscopy to expresses radiation frequencies in units of cm^{-1};[14] one then estimates vibrational energy level spacings by multiplication with $h = 2.86\ldots$ cal mol^{-1} cm^{-1} and may then compare this to $k_{\text{B}}T$. The error in using the classical form of q_{vib} is close to 1% for $h\nu/(k_{\text{B}}T) = 10^{-2}$. At $T = 300$ K, this corresponds to a frequency of $\nu = 2.1\ \text{cm}^{-1} = 63\ \text{ns}^{-1}$, or a period of oscillation of circa 16 ps.

The energy of a vibrational degree of freedom is easily shown to be $k_{\text{B}}T$ in the classical limit but when $k_{\text{B}}T \sim h\nu_j$, a more complex expression is needed:

$$E_{\text{vib},j} = \frac{h\nu_j}{2}\left(\frac{1+\exp\left(-\beta h\nu_j\right)}{1-\exp(-\beta h\nu_j)}\right) = \coth(\beta h\nu_j), \tag{5.27}$$

where β replaces $1/(k_{\text{B}}T)$. (The reader may wish to derive this result and compare the classical and exact quantum mechanical energy associated with a single vibrational motion with $h\nu = 0.01k_{\text{B}}T, 0.1k_{\text{B}}T$, and $0.5\ k_{\text{B}}T$.)

For a macromolecule with $3N$–6 normal modes, we can write the vibrational partition function as a nested sum over all possible quantum numbers of all $3N$–6 modes

$$q_{\text{vib}} = \sum_{n_1, n_2, \ldots, n_{3N-6}} \cdots \sum \prod \exp\left[-\frac{\left(n_1+\frac{1}{2}\right)h\nu_1}{k_{\text{B}}T}\right]\exp\left[-\frac{\left(n_2+\frac{1}{2}\right)h\nu_2}{k_{\text{B}}T}\right]\cdots.$$

Because the quantum number n_i varies independently of any other quantum number n_j, this nested sum can again be written as a product of sums,

$$q_{vib} = \left(\sum_n \exp\left[-\frac{(n+\frac{1}{2})\,h\nu_1}{k_B T}\right]\right)\cdots\left(\sum_n \exp\left[-\frac{(n+\frac{1}{2})\,h\nu_{3N-6}}{k_B T}\right]\right).$$

Thus, the vibrational partition function for a macromolecule is given by the product over the vibrational partition functions of all $3N-6$ normal modes of the molecule,

$$q_{\text{vib}} = \prod_{i=1}^{3N-6} q_{\text{vib},i}. \qquad (5.28)$$

5.4.5 Rotational Partition Function

Global rotation: In Chapter 2, we showed that the energy levels for a body rotating about one axis (the one-dimensional rigid rotator) are

$$E_m = m^2 k_B \Theta_{\text{rot}};\quad m = 0, \pm 1, \pm 2, \ldots,$$

where $\Theta_{\text{rot}} = \hbar^2/(2Ik_B)$, k_B is the Boltzmann constant, and I is the moment of inertia of the rotator. Θ_{rot} is called the *rotational temperature*. Because the energy level spacing for most rotational motions is always much less than $k_B T$, the rotational partition function is always treated in the classical limit, which leads to the classical partition function for such a simple one-dimensional rotator,

$$q^1_{\text{rot}} = \pi^{1/2}\frac{T^{1/2}}{(\Theta_{\text{rot}})^{1/2}}.$$

The rotational energy levels for a three-dimensional rigid rotator are more complex. This is due to the fact that two successive rotations about any two of the three independent axes of a macromolecule can produce the effect of a single rotation about the third, making the three motions not truly independent. If we take the product of the q_{rot}s for rotation about the three principal axes, this leads to overcounting of configurations by a factor of π. If we correct for this overcounting by dividing by π, the rotational partition function for the three-dimensional rigid rotator becomes

$$q_{\text{rot}} = \frac{\sqrt{\pi}}{\sigma}\left(\frac{T^3}{\Theta_1\Theta_2\Theta_3}\right)^{\frac{1}{2}};\quad \Theta_i = \frac{h^2}{8\pi^2 I_i k_B}, \qquad (5.29)$$

Here, σ is the symmetry number ($\sigma = 1$ for asymmetric molecules).

Internal rotation: Many molecules have internal freedom that corresponds to rotation of one part of the molecule with respect to the remainder, with a chemical bond as axis. This is common in proteins, where such motions about backbone peptide bonds[15] lead to different global rigid-body structures or *conformations*. The folded conformation of a biological macromolecule is generally viewed in first approximation as a semirigid configuration of atoms.[16] As such, the definition of a rigid body (one without major internal conformational freedom) applies roughly to each conformation. Our treatment thus far, which correctly accounts only for the internal motions of a rigid body, does not apply to motions that change the conformation. We deal with this problem in two ways. When the rotation is very hindered (the barrier being many times greater than k_BT, as for a double bond), the macromolecule is not free to reach other conformations through rotation about this bond. In this case, rotations about this bond can be treated as small-amplitude oscillations using a vibrational model and partition function. Large chemical groups attached to the atoms connected by a bond often create substantial obstacles to free rotation and large energy barriers. When the barrier to rotation about a chemical bond is small (on the order of k_BT), a more complex partition function must be used. When such rotations lead to different structures with different basic mechanical properties (center of mass, rotational axes, and normal modes), this leads to multiple conformational states, each one of which can be described in terms of the molecular partition function of a rigid body; the molecular partition function is then the sum of such partial contributions for all states.

5.4.6 Use of the Molecular Partition Function

The model of independent molecular motions that we have developed in this section can be combined with the methods of computing the molecular electronic energy E_{elec} as a function of the atomic coordinates according to the quantum mechanics described in Section 3.2. Taking as an example a two-atom molecule, one proceeds by minimization of E_{elec} with respect to the interatomic distance r_{12}, calculating E_{elec} using any of the various methods, such as Hartree–Fock and density functional theory. This gives the electronic energy ε_0 and interatomic distance $r_{12,0}$ of the ground state. If, as is typically the case, the minimization employs the Newton method,[17] a value of the second derivative at the minimum $(d^2E_{\text{elec}}/dr^2)_0$ is available, and can be equated with the force constant for stretching the bond. As the atomic masses are known, a knowledge of $r_{12,0}$ gives us the principal moment of inertia needed to evaluate the rotational partition function, while the masses and the force constant for stretching the bond together define the vibrational frequency needed to evaluate the vibrational partition function. Evaluation of the translational partition function requires only the atomic masses.

This is easily generalized to molecules containing more than two atoms; in that case, the internal vibrations must be evaluated in terms of normal modes calculated using the full (Hessian) matrix of partial second derivatives of the electronic energy with respect to the atomic coordinates. Thermodynamic functions of (simple) molecules in the gas state calculated by this approach agree well with experiment, provided the quantum mechanical calculations are done at a high level of accuracy.

Larger molecules, and *a fortiori*, macromolecules, have access to multiple distinct conformations of minimum energy, and for each the ground state energy, principal moments of inertia, and spectrum of normal mode vibrations can be evaluated separately, as we have seen for a general rigid body. If the minima are separated by energy barriers many times $k_B T$,[18] the molecular partition function can be closely approximated by writing it as a sum over minimum-energy states k,

$$q = \sum_k q_k = \sum_k q_{\text{trans},k}\, q_{int,k}$$

$$q = q_{\text{trans}} \sum_k q_{int,k} = q_{\text{trans}}\, q_{\text{int}}. \tag{5.30}$$

The second equation recognizes that the factor q_{trans} depends only on the mass, temperature, and volume of the system and is identical for all states; q_{int} is a sum over all rigid body conformations.

5.4.7 Equipartition of Energy in the Classical Limit

With the exception of motions under the influence of a strong potential (e.g., stiff internal vibrations), as described, contributions from translation, vibration, and rotation to the molecular partition function are all simple expressions in the classical limit, and it is thus easy to derive the contributions of each molecular degree of motional freedom to the system thermodynamics. In particular, the mean internal molecular energy (which is also the thermodynamic energy, U) is found with Eqs. (1.30) and (5.11)

$$A = -k_B T \ln q$$

$$\langle E \rangle = U = \partial(A/T)/\partial(1/T), \tag{5.31}$$

and application to the high-temperature limiting expressions for the partition functions for translation, vibration, and rotation gives

$$\langle E_{\text{trans}} \rangle = \frac{3}{2} k_B T$$

$$\langle E_{\text{vib}} \rangle \approx k_B T$$

$$\langle E_{\text{rot}} \rangle \approx \frac{3}{2} k_B T. \tag{5.32}$$

(The reader is invited to derive these expressions from the corresponding terms in the molecular partition function.)

One sees that each of the three translational degrees of freedom contributes $\frac{1}{2} k_B T$ to the molecule's internal energy, that each vibrational degree of freedom contributes $k_B T$, and that each rotational degree of freedom contributes $\frac{1}{2} k_B T$ to this energy. This is the very useful *principle of equipartition of thermal energy.*

The difference between the contributions of rotational/translational motions and that of vibrational motions is due to the fact that the former motions contribute only kinetic energy, while the latter contribute both kinetic and potential energy. In general, any motion (even a hindered rotation or translation) that experiences some potential, even a weak one, must be treated with a vibrational partition function and contributes more to the total internal energy of a macromolecule than does a potential-free motion. Note that q_{rot} and q_{trans} are each written for 3 degrees of freedom, while q_{vib} is written for $3N-6$ independent vibrational degrees of freedom for a molecule with N atoms. If a single atom moves in a potential, its motions contribute $k_{\mathrm{B}}T$ per degree of freedom in the classical limit; this amounts to roughly 1.8 kcal/mol, a significant contribution to the internal energy.

5.4.8 Limitations of the Molecular Partition Function

Constancy of system energy. The molecular partition function was developed for a system containing N independent molecules, with a constant total energy. This limits the microcanonical ensemble as a model of thermodynamic systems, for which, in many experimental situations, the temperature, rather than the energy is given or controlled.

Harmonic approximation. In the model presented here, internal motions are treated as harmonic oscillations, and, in fact, *all* molecular deformations have been treated as producing energy increases that are proportional to the square of the deformations (internal motions in a potential). For molecular systems, this is only an approximation. For example, the energy for changing the length of a chemical bond follows an anharmonic potential (approximated by a Morse potential; see Section 3.3.1). The treatment of molecular structure as a collection of harmonic oscillators is in general a good approximation only if the deformations from the minimum-energy conformation are small, that is, the approximation will work at low temperature, but will progressively fail as the temperature is raised.

Complex molecular systems. Complex molecular systems cannot be described with energy levels given by quantum mechanics, not because the quantum approach is incorrect, but because the Schrödinger equation cannot be solved without making the simplifying assumption of independent energy levels. Even a system seemingly as simple as liquid water is vastly too complex.[19]

In order to address these problems, we turn to Chapter 6, in which we introduce the canonical ensemble. In this ensemble, the temperature is the controlled variable; also, Section 6.3 contains a formulation of the partition function in terms of classical mechanics.

NOTES

1. In Section 6.2.5, we shall discuss how to modify the theory for a system of particles that are not distinguishable.

2. Our simple example assumes molecular quantum states with no degeneracy. Real molecular quantum states can have their own *quantum degeneracies*, ω_i, which can increase the degeneracy of system quantum states. Quantum degeneracy is ignored at this stage of our treatment of statistical mechanics.
3. A postulate is a key assumption that cannot be proven upon whose validity the whole field rests. Many fields of science have such postulates; postulates can be proved wrong *only* by experimental evidence. (See also our introduction to Chapter 1.)
4. As a consequence, long-time-averaged properties equal ensemble-averaged properties as t and $N \to \infty$. This is termed the *ergodic hypothesis.*
5. In statistical thermodynamics, it is essential to define and keep track of indices over which one does sums. We use here j for system quantum states, i for molecular quantum states, n_i for the number of molecules in molecular quantum state i.
6. Stirling's formula, $N! \approx N^N e^{-N}\sqrt{2\pi N}$ is an excellent approximation for large N. See Section AI.1.
7. The symbol β is frequently used (also in many places in this book) as an abbreviation for $1/(k_B T)$.
8. We will see later in this chapter that this requires that the molecules are not only independent but also distinguishable.
9. From Eq. 5.3 and Sterling's approximation, we get

$$\Omega = N! \Big/ \prod_i n_i! \approx N^N \Big/ \prod_i n_i^{\,n_i} = \prod_i (n_i/N)^{-n_i}$$

$$\ln \Omega = -N \sum_i (n_i/N) \ln (n_i/N) = -N \sum_i (P_i \ln P_i).$$

10. This is the fundamental postulate that links Boltzmann's statistical description of molecules to thermodynamics, although the more familiar form of this linkage is given by Eqs. 5.5–5.11 for the free energy. Boltzmann clearly saw this as his essential discovery; his tomb in Vienna bears the inscription "$S = k \ln W$" (with W for our Ω or $\mathcal{W}$ of Eq. 5.1, cf. "Wahrscheinlichkeit" = probability).
11. This is easily demonstrated by gathering all the terms with $i = 0$ and factoring out $\exp(-\beta\varepsilon_{\text{elec},0})$. Because of the properties of the exponential, the term $\exp(-\beta\varepsilon_{\text{elec},0})$ now multiplies a sum over all other indices. This process can be repeated until one has a sum over i of terms, each of which has $\exp(-\beta\varepsilon_{\text{elec},i})$ multiplying a sum over all other indices. By this process, one ends up with a sum over i of $\exp(-\beta\varepsilon_{\text{elec},i})$ (this sum equals q_{elec}) times a sum over the other three indices. This can be continued for the remaining indices until the appropriate product of four sums is the result.
12. Note that a similar, but somewhat more complex functional form, exists for an anharmonic oscillator (e.g., the Morse potential, Eq. 5.3–5.6)—a case that we will not treat so as to avoid complication.
13. See Section 3.5.2 and Appendix F for additional discussion of normal mode vibrations.
14. With frequency ν in cm^{-1}, wavelength $\lambda = 1/\nu$ cm. To obtain frequency in s^{-1}, multiply frequency in cm^{-1} by the velocity of light *in vacuo*, $c = 3.00 \ldots \times 10^{10}$ cm/s.
15. Rotations about covalent bonds within side chains of a folded protein molecule can also lead to different conformations; many side chains packed in the molecule's interior

remain in a single, well-defined conformation, while surface side chains have a choice of conformations (the unique conformer set of each type of side chain).

16. Inherent to this definition is the understanding that a conformation exists at an energy minimum and that an energy barrier at least of the order of $k_B T$ exists between it and other conformation that are reached by concerted backbone bond rotations.
17. Newton's method uses a function's first and second derivatives and is especially powerful in the neighborhood of a minimum. For a function $f(x)$ of a single variable, the minimum is approached by changing the estimate x to a new estimate by adding $\Delta x = -f'/f''$. Newton minimization of a function of many variables employs the partial first derivatives $\partial f/\partial x_i$ and the Hessian matrix $\partial^2 f/(\partial x_i \partial x_j)$.
18. *Exactly* how many times $k_B T$ is a matter of judgment. If only 1 or 2 $k_B T$, the higher energy conformation is easily accessible from the lowest energy one. We must then consider that the molecular partition function is of an average conformation that is defined not just by energy but also by entropy. We shall see that the molecular partition function can be written as the exponential of a free energy multiplied by β. As such, the same formal expression will hold. Obviously, situations such as this can become quite complex and will involve additional assumptions and approximations. These will not be dealt with in this book.
19. How to represent the partition function of liquid water and, hence, of the solvent surrounding a biological macromolecule, remains a problem of enormous magnitude.

SUGGESTED READING

Davidson, N., *Statistical Mechanics.* New York: McGraw-Hill. 1962, (*republished by Courier Dover Publications*; *2003*).

Hill, T.L., *An Introduction to Statistical Mechanics*. Reading, MA: Addison-Wesley, 1960.

Hill, T.L., *Statistical Mechanics. Principles and Selected Applications,* New York: McGraw-Hill, 1964.

McQuarrie, D.A., *Statistical Mechanics*. Sausalito, CA: University Science Books, 2000.

Widom, B., *Statistical Mechanics. A Concise Introduction for Chemists.* Cambridge: Cambridge University Press, 2002.

6

System Ensembles and Partition Functions

Mathematics is *a language.*
—Reportedly said by Gibbs at a Yale faculty meeting

6.1 INTRODUCTION

In the preceding chapter, the molecular partition function has been defined for a system containing a large number of *noninteracting* copies of the same molecule. Therefore, the only macroscopic system that can be represented by this ensemble is the ideal gas. Another shortcoming of describing a molecular system in terms of a microcanonical ensemble is that the temperature T, which along with the volume V and the number of particles N is a commonly controlled experimental variable, was introduced only by way of the Maxwell–Boltzmann derivation of the most probable configuration $\{n_i\}_\mathrm{B}$, not as an independent variable. It is much more common and convenient to work experimentally with systems with given V (or P), T, and N and to relate thermodynamic functions, such as E, to these independent system variables. In addition, binding problems, which are of considerable biological importance, require that the number of molecules in a system be free to vary so that a balance is reached between the favorable entropy associated with ligands free in solution and the favorable internal energy associated with ligands bound to a macromolecule. In such cases, N of at least one species varies, and the system is open.

Equilibria and Kinetics of Biological Macromolecules, First Edition. Jan Hermans and Barry Lentz.

In this chapter we describe two new ensembles. The *canonical ensemble* contains closed systems able to exchange energy, with *average* system energy defined by temperature. The *grand canonical ensemble* contains open systems able to exchange energy and particles, with average system contents defined by chemical potentials as independent variables. In addition, we shall (independently) remove the restriction that systems contain noninteracting molecules.

6.2 CLOSED SYSTEMS: THE CANONICAL ENSEMBLE

6.2.1 An Ensemble of Systems at Thermal Equilibrium

The canonical ensemble by definition is used to describe a closed system with V, T, and N as independent system variables. N is fixed but not necessarily large; molecules may or may not interact; and molecules need not be all the same species. The systems are at thermal equilibrium (T is the same for all systems) but need not be described by the same quantum state or have the same energy. We now develop an *ensemble* of these *systems* that we will use to describe the average or thermodynamic system behavior. This ensemble of systems is called the *canonical ensemble.*[1] Unlike in Chapter 5, we allow the molecules in one system to interact physically, although the individual systems, or elements in a canonical ensemble, are independent except that they exchange energy (i.e., are at thermal equilibrium).

Just as in Chapter 5, an ensemble of systems is again used to describe the thermodynamic or time average properties (e.g., P or E) of a single real system (ergodic hypothesis). However, in the microcanonical ensemble developed in Chapter 5, the Boltzmann description was approached only as the number of molecules in the system, N, became very large. Because we limited ourselves to independent and distinguishable molecules in developing the microcanonical ensemble and the molecular partition function, this condition ensured that the total statistical degeneracy of the system quantum states $\mathcal{W}$ was very large. However, the molecules in the system we now treat are not assumed to be independent and we do not insist that N is very large. Thus, we must now insist that the size of the canonical ensemble $\mathcal{R}$ (Table 6.1) be large enough to provide an adequate random sampling of all possible system quantum states.

Another significant difference between the microcanonical ensemble and the ensemble developed here is that we now construct an ensemble of many systems having the same composition and volume but *varying* energy. By doing that, we replace the difficult problem of developing a time average for a dynamic system exchanging energy with its constant-temperature environment (i.e., the system quantum state Ψ evolves with time through all the possible system quantum states Ψ_j,[2] consistent with the system having a *fixed thermal* energy) with the much simpler problem of describing a large number of identical systems in thermal equilibrium with each other, but otherwise independent (i.e., having different quantum states Ψ_j and energies E_j).

TABLE 6.1 **The canonical ensemble on the left is here subdivided into a superensemble of $\mathcal{U}$ canonical ensembles on the right. Each sub-ensemble in the superensemble contains m_j systems in the jth system quantum state, such that $\sum_j \{m_j\}_k = \mathcal{M} = \mathcal{R}/\mathcal{U}$, and has the same total energy $\mathcal{M}\overline{E}$, where $\overline{E}$ Is the average or thermodynamic internal energy of the system**

N, V, T E_1, Ψ_1	N, V, T E_2, Ψ_2	N, V, T E_3, Ψ_3	...		N, V, T $\mathcal{M}\overline{E}$ $\{m_j\}_1$	N, V, T $\mathcal{M}\overline{E}$ $\{m_j\}_2$	...	...
...				$\Rightarrow$	...			
N, V, T E_r, Ψ_r	...				N, V, T $\mathcal{M}\overline{E}$ $\{m_j\}_k$	...		
...	...		N, V, T $E_{\mathcal{R}}, \Psi_{\mathcal{R}}$		...	...		N, V, T $\mathcal{M}\overline{E}$ $\{m_j\}_{\mathcal{U}}$

Each system in the canonical ensemble has its own unique system quantum state Ψ_j and energy E_j that, in general, are functions of the positions and momenta of all the particles in the ensemble element. While the molecules within each element are not assumed to be independent, each ensemble element is independent of all other elements. The total number of elements in the ensemble, the total energy of the ensemble, and the average energy in the ensemble are given by

$$\mathcal{R} = \sum_j m_j;\ E_{\text{tot}} = \sum_j E_j m_j;\ \langle E \rangle = \frac{\sum_j E_j m_j}{\mathcal{R}}.$$

Here, m_j is the number of times that a quantum state with energy E_j appears in the ensemble.

We begin by noting the analogy between the system of independent and distinguishable molecules discussed in Chapter 5 and the independent elements (systems) of the current, canonical ensemble. In the microcanonical ensemble, independence of the molecules allowed us to determine by statistics the most likely distribution of molecules in the ensemble over molecular quantum states, which was then, for a system containing a large number of molecules, found to be the Boltzmann distribution $\{n_i\}_{\text{B}}$. In the current canonical ensemble, the elements are also independent and distinguishable. At this point, many textbooks simply state that because $\mathcal{R}$ is indefinitely large, one distribution over system quantum states will dominate the ensemble, and simply state that this distribution is the same as that found in a system of a large number of independent molecules, that is, the Boltzmann distribution.

In order to clarify and mathematically justify this argument, we divide the $\mathcal{R}$ systems in the canonical ensemble into $\mathcal{U}$ subensembles, each containing $\mathcal{M} = \mathcal{R}/\mathcal{U}$ systems. The right side of Table 6.1 illustrates this new superensemble. We can do this because $\mathcal{M}$, $\mathcal{R}$, and $\mathcal{U}$ are all indefinitely large. Each subensemble (index k) contains $\mathcal{M}$ independent systems distributed over their possible quantum states, such that m_1 are in state 1, m_2 in state 2, ..., m_j in state j, etc. Each of these elements is itself an indefinitely large canonical ensemble defined by T, V, N and a set $\{m_j\}_k$ that defines the number of times that a particular system quantum state with energy E_j appears. Because the subensembles are in thermal equilibrium (same T), each has the same average or thermodynamic internal energy of a system and satisfies the condition that each subensemble has energy $\mathcal{M}\overline{E}$, where $\overline{E}$ is the average or thermodynamic internal system energy related to T. It is now easy to see that this is the equivalent of the situation we treated in deriving the Boltzmann distribution over molecular quantum states in Chapter 5 (indefinitely large number of independent molecules N, fixed energy E, and fixed V for each element in the microcanonical ensemble), and we can use exactly the same mathematical method to determine the most likely set $\{m_j\}_{\text{B}}$ leading to the thermodynamically defined energy, E, as the one having the greatest statistical degeneracy $\Omega(\{m_j\}_{\text{B}})$. Consequently,

$$m_j = P_j \cdot \mathcal{M};\ P_j = \frac{\exp\left[-E_j/\left(k_B T\right)\right]}{Q};\quad \text{where } Q = \sum_j \exp\left[-E_j/\left(k_B T\right)\right], \tag{6.1}$$

Q is called the *canonical partition function*. Here, E_j is the energy associated with a *system quantum state* Ψ_j that reflects all N molecules in the system as well as interactions between them. If we wish to determine the probability of finding a system with a particular energy, the sum, of course, becomes over *system* energy levels, j, but system quantum degeneracies must multiply each term of the sum.

6.2.2 Relation of the Partition Function to Thermodynamic State Functions

The canonical partition function describes a *system* of N molecules in volume V and at temperature T, that is, N, V, and T are the independent system variables. The microcanonical ensemble has independent variables N, V, and E. However, E is linked to T by the choice of k_B and by the choice of the Boltzmann probability distribution function (Eq. 5.7). Hence, both ensembles have the same set of independent variables N, V, and T. The thermodynamic state functions defined by these independent variables are A, E, and S, and the two ensembles bear similar relations between these state functions and their respective partition functions,

$$\begin{aligned}
A - A(0) &= -k_B T \ln Q \\
U - U(0) &= \langle E \rangle = \frac{\partial(A/T)}{\partial(1/T)} = -\frac{\partial \ln Q}{\partial\left(1/k_B T\right)} \\
S - S(0) &= -\frac{\partial A}{\partial T} = -k_B \ln \Omega.
\end{aligned} \tag{6.2}$$

In these expressions, the reference state for the internal energy $U(0)$ is taken as the internal energy at $T = 0\,\text{K}$ (the sum of electronic energy and zero-point vibrational energy). The temperature dependence of the partition function provides the internal energy due to thermal distribution over available energy levels, called the *configurational energy* $\langle E \rangle$. The reference state for the entropy is $S(0)$, the entropy at $T = 0\,\text{K}$, which, except for degeneracy, is 0. Thus, at $T = 0\,\text{K}$, $A(0)$ is totally in the form of internal energy $U(0)$.

Using the expression for Q in Eq. 6.1 to obtain U according to this last equation, we obtain

$$-k_B T \ln Q = \langle E \rangle - k_B T \ln \Omega_B = A, \tag{6.3}$$

which shows that $A(T)$ is the total configurational energy reduced by the energy devoted to increasing the configurational entropy of the system, as this energy is unavailable to do work. This identifies the entropy with $k_B T \ln \Omega_B$. Recall that a spontaneous change must have $\Delta S \geq 0$. Thus, for a change in a system to occur spontaneously, it must result in an increase in the system partition function, that is, in the average number of states that contribute to the average system energy. Again, as the Boltzmann distribution maximizes the entropy of the system, it satisfies the second law of thermodynamics.

One important reason for introducing the canonical ensemble and its partition function is that the mean energy of a thermodynamic system is not a convenient independent variable for defining the thermodynamic state of a system. As the temperature, T, and the thermal energy are related, the canonical partition function simply switches these conjugate variables, making T the independent variable and U or $\langle E \rangle$ the dependent one. Thus, the temperature is set, but the energy fluctuates about a mean value equal to that of the dominant system in the ensemble. If this partition function can be defined on the basis of the molecular quantum states of molecules in the system, then the above-mentioned expressions show that all the thermodynamics of the system can be obtained from knowledge of the molecular properties of the system. This is possible only in the case of independent molecules.

6.2.3 Multiplication of Partition Functions of Independent Systems

As we know, the terms in partition functions are exponentials in the system energy. The partition function of a system obtained by the combination of two *independent* systems is then a double sum of the products of two Boltzmann factors, one from each system. Independence means that the state and energy of either system is not affected by the state of the other system; one sees that the double sum equals the partition function of the combined system,

$$\begin{aligned} Q_{ij} &= \sum_{i,j} e^{-\beta E_{ij}} = \sum_i \sum_j e^{-\beta E_{ij}} = \sum_i \sum_j e^{-\beta(E_i + E_j)} = \sum_i \sum_j e^{-\beta E_i} e^{-\beta E j} \\ &= \sum_i e^{-\beta E_i} \sum_j e^{-\beta E_j} = Q_i Q_j. \end{aligned} \tag{6.4}$$

Therefore, the partition function of the combined noninteracting systems equals the *product of the partition functions of the individual systems.*

6.2.4 Application: Free Energy is an "Extensive" Function

In order to obtain an expression for the partition function of solvent (or, for that matter, *any* homogeneous macroscopic system), we apply Eq. 6.4 to the case of such a liquid (or solid or gas), by dividing a sample into two equal parts at identical P and T, to give two systems that are energetically independent. Because these systems are macroscopic, contributions to the energy from the surface and interface can be ignored. The partition function of the combined system then equals the product of the partition functions of the parts, and the free energy of the whole is twice the free energy of each part:

$$\begin{aligned} Q_{\text{whole}} &= Q_{\text{half}}^2 \\ A_{\text{whole}} &= -k_B T \ln Q_{\text{whole}} = -2k_B T \ln Q_{\text{half}} = 2A_{\text{half}}. \end{aligned} \tag{6.5}$$

This shows, from principles of statistical mechanics, that the free energy is an extensive thermodynamic function (proportional to system size). It follows that related functions such as entropy and specific heat share this property. This derives an important statement of thermodynamics from principles of statistical mechanics.

6.2.5 Distinguishable and Indistinguishable Molecules

Derivation of the canonical partition function does not require the assumption that molecules in each element of the ensemble are independent. However, we shall see that treatment of interacting molecules can become quite complex, and independence is often assumed for convenience or when it might be a reasonable approximation (e.g., in dilute solution). According to Eq. 6.4, the canonical partition function of N noninteracting molecules at T and V can be expressed as a product over molecular partition functions for all distinct molecules in the system.[3] Thus, the canonical partition function for a system of *independent and distinguishable molecules*, N_1 of type 1, N_2 of type 2, etc., is

$$Q = \prod_i q_i^{N_i} \left[\text{distinguishable and independent}\right]. \tag{6.6}$$

The average energy of the system is simply N times the average energy for each molecule $\langle \varepsilon \rangle$, which is related to T as defined by the Boltzmann distribution of molecules over molecular quantum states (i.e., n_1, in a state ψ_1 with ε_1, n_2 in ψ_2 with ε_2, ... and n_i in ψ_i with ε_i; see Eq. 5.10.). This partition function might describe an ideal crystal for which the unique lattice position of each particle makes it distinguishable from all other particles. This physical model is the basis for the Einstein crystal, in which each molecule in the lattice is a three-dimensional quantum harmonic oscillator attached to *its own* preferred position in space by a spring, and for the treatment of adsorption to an ideal crystal, originally developed by Irving Langmuir.

However, *identical molecules in an ideal gas are not distinguishable*, in that the system is physically unchanged if we exchange two identical molecules. In this case, we would count too many configurations by using the above-mentioned expression, which treats each molecule as unique. Thus, in order to arrive at the partition function for an ideal gas containing a mixture of *independent, indistinguishable molecules* (N_i of each kind), the preceding result must be divided by a combinatorial term reflecting the number of ways in which N identical molecules can be exchanged,

$$Q = \prod_i \frac{q_i^{N_i}}{N_i!} \left[\text{Indistinguishable and Independent}\right]. \tag{6.7}$$

This equation, including the division by $N_i!$, applies formally to ideal mixtures or to dilute gases (as close to an ideal mixture as we come in reality) but also to *independent solute molecules in dilute solution*, although in that case the function q_i has a different significance than in an ideal mixture, as it takes account also of the interactions of the solute with solvent molecules and the perturbation of solvent–solvent interactions due to the presence of a solute molecule. (This is discussed further in Chapter 10.)

6.2.6 Application: Chemical Potential of the Ideal Gas

A common example of the use of the canonical partition function is to obtain the chemical potential for an ideal gas mixture. Eq. 6.7 is readily applied to this system, as particles in an ideal gas do not interact. [Alternatively, an equivalent result can be obtained using the phase-space integral, Eq. 6.16 with potential energy $E_{\text{pot}}(\mathbf{X}) = 0$.]

We consider N_0 molecules of type 0 and N_1 molecules of type 1, contained in volume V at temperature T, and write the molecular partition function of each component as a product of electronic through translational partition functions (Eq. 5.18),

$$q_k = q_{k,\text{int}} q_{k,\text{tran}} = q_{k,\text{int}} \frac{V}{\Lambda_k^3} \text{ with } q_{k,\text{int}} = q_{k,\text{elec}} q_{k,\text{rot}} q_{k,\text{vib}}; \quad \text{for } k = 0, 1.$$

(Recall from Eq. 5.23 that $\Lambda = h/(2\pi m k_B T)^{1/2}$ is the de Broglie wavelength.) The partition function for the entire system of noninteracting molecules is then

$$Q = \frac{1}{N_0! N_1!} q_0^{N_0} q_1^{N_1}. \tag{6.8}$$

Using Stirling's approximation for $N!$ $(N^N e^{-N})$, and substituting for q_k, we obtain

$$Q = \left(\frac{q_{0,\text{int}}}{\Lambda_0^3} \frac{eV}{N_0} \right)^{N_0} \left(\frac{q_{1,\text{int}}}{\Lambda_1^3} \frac{eV}{N_1} \right)^{N_1}, \tag{6.9}$$

and the chemical potential of either component with

$$A = -k_B T \ln Q$$

$$\mu_k = \left(\frac{\partial A}{\partial N_k} \right)_{T,V,N_{i \neq k}} = -k_B T \ln \left(\frac{q_{k,\text{int}}}{\Lambda_k^3} \right) + k_B T \ln \frac{N_k}{V}. \tag{6.10}$$

We can compare this with an expression derived on the basis of thermodynamics (Eq. 1.39)

$$\mu = \mu^\bullet + k_B T \ln \frac{N}{V},$$

where $\mu^\bullet$ is the standard chemical potential. Combining the two gives for each component

$$\mu_k^\bullet = -k_B T \ln \left(\frac{q_{k,\text{int}}}{\Lambda_k^3} \right). \tag{6.11}$$

We should remember that Eq. 1.39 was derived on an empirical basis, the ideal gas law; in contrast, the derivation given here is based on first principles. (You can easily derive Boyle's ideal gas law from the expression for A, by equating P with $-(\partial A/\partial V)_{T,N_k}$, as was done in Chapter 5 for a one-component system; see Eq. 5.24.)

6.2.7 Interacting Molecules

Unlike for the molecular ensemble, for which each subsystem was expressible in terms of independent molecular quantum states, each subsystem of the canonical ensemble *can* contain *interacting* molecules, which makes it usually impossible to express system quantum states in terms of molecular quantum states. However, in order to determine an expression for the average of some mechanical variables in this system, we do not need to know exactly how intermolecular interactions alter the quantum states of the system. We need only know that the quantum states exist and can be chosen in such a way that they have defined energies and maximal degeneracies, even though identification of the wave functions and thus energies and degeneracies of these quantum states is at this point largely impossible.

Molecules that interact will have energies related not just to their quantum states but also to the quantum states of neighboring molecules. We can estimate how an interaction might alter the energy of a pair of molecules: $E_{\mathrm{AB}} \approx E_{\mathrm{A}} + E_{\mathrm{B}} + \Delta E_{\mathrm{AB}}$, where ΔE_{AB} can be estimated experimentally, quantum mechanically, or using the types of approximate intermolecular potentials discussed in Chapter 3. However, interactions will also influence quantum state degeneracies. In simple terms, favorable interactions are constraints that limit spatial redistributions of molecules. Even unfavorable interactions (e.g., van der Waals contacts) limit molecular redistributions. This means that the number of configurations of the most likely system quantum state will be less than Ω_{B} for independent molecules. Intuitively, the more the interactions lower quantum state energies, the more they will limit configurational freedom and reduce Ω. Thus, we expect a complex interrelationship between the energies and degeneracies of quantum states in interacting systems. However, the distribution over system quantum states that maximize $\Omega(\{E_i, \omega_i\})$ will still satisfy the Maxwell–Boltzmann statement of the Second Law, and will thus define the system's thermodynamics, but we have no way of predicting the interrelationship between ΔE_{AB} and Ω. Approximate approaches to this problem are at the crux of modern theories of liquids and solutions.

In the following section, we present a formulation of the canonical partition function that applies to systems that are described with classical mechanics and a classical energy function, that is, an energy function that is continuous in atomic positions and velocities.

6.3 THE CANONICAL PARTITION FUNCTION OF SYSTEMS WITH CONTINUOUS ENERGY DISTRIBUTIONS: THE PHASE-SPACE INTEGRAL

As we have seen, the description of molecular energies by quantum mechanics, together with the theory of statistical mechanics, leads to a complete description of the equilibrium properties of molecular *ensembles* based on first principles. We have also indicated the difficulties inherent in the quantum theory that make application to complex systems well-nigh impossible. In many cases, the only available description of the molecular energetics is an approximation to the underlying quantum mechanical energies. Frequently, the approximate energy function is continuous, rather than quantized; this is the case, for example, for the empirical molecular mechanics energy function introduced in Chapter 3.

When one uses a continuous energy function, the summation of the Boltzmann factor over quantum states required to evaluate the partition function will have to be replaced by an integral. Just as an integral may be approximated by a sum, a sum can, in many cases, be approximated by an integral. In the previous chapter (Section 5.7), we have already seen that, in the high temperature limit, integration can replace summation of a partition function expressed in terms of quantum states.

Classical mechanics decomposes the energy of a molecular system into kinetic and potential energy, the former a function of the atomic velocities, and the latter a function of the atomic coordinates, both functions continuous, without reference to quantum states. The correspondence between the quantum and classical representations is not immediately obvious.

It turns out (see subsequent text for a justification) that summation of the Boltzmann factor over enumerable quantum states can be approximated by integration of the Boltzmann factor over atomic momenta ($\mathbf{p} = m\mathbf{v}$) and coordinates. The first step is to approximate the summation of the Boltzmann factor over quantum states (having energies $E_{q,i}$) with a summation of the Boltzmann factor for a continuous energy E_c which is a function of momenta p and coordinates x, over discrete values of momenta and coordinates. The second step is to approximate the latter sum as an integral. For a single degree of freedom (one momentum and one coordinate) this gives

$$\begin{aligned} q = \sum_i \exp\left(-\frac{E_{q,i}}{k_B T}\right) &\approx \sum_i \sum_j \exp\left(-\frac{E_c\,(i\,\Delta p, j\,\Delta x)}{k_B T}\right) \\ &= \frac{1}{\Delta p\,\Delta x} \sum_i \sum_j \Delta p\,\Delta x \exp\left(-\frac{E_c\,(i\,\Delta p, j\,\Delta x)}{k_B T}\right) \\ &\approx \frac{1}{\Delta p\,\Delta x} \iint dpdx \exp\left(-\frac{E_c\,(p,x)}{k_B T}\right). \end{aligned} \tag{6.12}$$

The magnitude of the element $\Delta p \Delta x$ must be set to ensure that the number of elements in the two summations is equal, and this is achieved by invoking the Heisenberg uncertainty principle, $\Delta p \Delta x = h$, Eq. 2.3, which suggests that it does not make sense to consider phase–space volumes smaller than this, so that we count just one state for each element of size $\Delta p \Delta x = h$. This gives the classical partition function q_c, as

$$q_c = \frac{1}{h} \iint dpdx \exp\left(-\frac{E_c\,(p,x)}{k_B T}\right). \tag{6.13}$$

In classical (Newtonian) mechanics, kinetic energy E_{kin} and potential energy E_{pot} are independent, the kinetic energy is a function only of momentum and mass and the potential energy is a function only of position, and there is a simple relation between kinetic energy and momentum p and mass m:

$$E_c(p, x) = E_{kin}(p) + E_{pot}(x),$$
$$E_{kin}(p) = \frac{p^2}{2m}. \quad (6.14)$$

This gives for the partition function

$$\begin{aligned} q_c &= \frac{1}{h}\int dp \exp\left(-\frac{E_{kin}(p)}{k_B T}\right)\int dx \exp\left(-\frac{E_{pot}(x)}{k_B T}\right) \\ &= \frac{1}{h}\int dp \exp\left(-\frac{p^2}{2mk_B T}\right)\int dx \exp\left(-\frac{E_{pot}(x)}{k_B T}\right) \\ &= \frac{1}{h}(2\pi m k_B T)^{1/2}\int dx \exp\left(-\frac{E_{pot}(x)}{k_B T}\right). \end{aligned} \quad (6.15)$$

For a multidimensional system containing N particles, this so-called *phase–space integral*[4] is generalized to

$$\begin{aligned} Q_c &= \frac{1}{h^{3N}}(2\pi m k_B T)^{3N/2}\int\ldots\int d\mathbf{X} \exp\left(-\frac{E_{pot}(\mathbf{X})}{k_B T}\right) \\ &= \frac{1}{\Lambda^{3N}}\int\ldots\int d\mathbf{X} \exp\left(-\frac{E_{pot}(\mathbf{X})}{k_B T}\right) \end{aligned} \quad (6.16)$$

where $\Lambda = h/(2\pi mkT)^{1/2}$ is the de Broglie wavelength, which earlier appeared in the partition function for a particle in a box (Section 5.4). The first two factors are constant for all problems in which the composition, that is, the numbers of different kinds of atoms, does not change. Thus, a chemical equilibrium or a conformational equilibrium of a macromolecule is determined by the potential energy term in the phase–space integral, which is termed the *configuration(–space) integral*:

$$z = \int\ldots\int d\mathbf{X} \exp\left(-\frac{E_{pot}(\mathbf{X})}{k_B T}\right) = \int d\mathbf{x}_1\int d\mathbf{x}_2\ldots\int d\mathbf{x}_N \exp\left[-\frac{E_{pot}(\mathbf{X})}{k_B T}\right]$$
$$A_z = -k_B T \ln z, \quad (6.17)$$

where $\mathbf{x}_i$ represents the coordinates of atom i, and $\mathbf{X}$ the entire coordinate set of all N atoms. This configuration integral also behaves as a partition function and has a free energy A_z associated with it. However, this free energy does not include the kinetic energy (i.e., the symbol z indicates a partition function in which the kinetic component of the energy has been omitted from the Boltzmann factor.)

The configuration integral does not contain Planck's constant h and also does not depend on the atomic and molecular masses. This results from the fact that momenta and coordinates are not linked in classical mechanics, while the Heisenberg Uncertainty Principle links them in quantum mechanics. Because changes in the state of a system do not involve changes in mass when systems are treated with classical mechanics, changes in their equilibrium thermodynamic properties do not depend on the atomic and molecular masses (at constant T.) However, the masses do appear in the thermodynamic functions of quantum mechanical systems, meaning that state changes in such systems are dependent on mass. When quantum mechanical systems are treated in the high temperature or classical limit, terms containing the mass and Planck's constant still appear but cancel whenever *differences* in values of thermodynamic functions for two states are evaluated.

Many useful treatments are based on the classical formalism. It should, however, be pointed out that the classical treatment is increasingly inadequate at low temperatures. More precisely stated, the classical treatment breaks down when the spacing between successive energy levels is not small compared with $k_{\mathrm{B}}T$. (See also Section 5.4.) At ambient temperatures, this is, for example, the case for the energy levels for stretching chemical bonds, and for many other relatively local modes of molecular deformation, such as bending of bond angles and out-of-plane twisting of aromatic rings. (See Eq. 5.26 and accompanying text.) This is not surprising, as we clearly need quantum mechanics to describe any change involving significant rearrangements of electrons. As an example, it would be a very bad approximation to describe the equilibrium for a reaction in which one or more chemical bonds is made or broken with a fully classical partition function.

Simple examples. Our description of the phase–space and configuration integrals did not constitute a rigorous derivation. One can easily satisfy oneself that the classical formalism agrees with the high temperature limits of the particle in a box, harmonic oscillator, and rigid rotator that we have introduced in the previous chapter. (This will establish the origin of the term *classical limit* as used in this book.)

For example, for the *particle in a box*, E_{pot} is zero inside the box (volume V) and infinite outside it, and Eq. 6.16 gives

$$q_c = \frac{1}{h^3}\left(2\pi m k_{\mathrm{B}} T\right)^{3/2} V, \tag{6.18}$$

which one can easily verify is identical to the expression given in Eq. 5.22.

For a *harmonic oscillator* with force constant K, the potential energy is a quadratic function of the coordinate

$$E_x = \frac{1}{2} K x^2. \tag{6.19}$$

Substitution of this energy in Eq. 6.16 and integration gives a simple expression for the classical partition function, which is identical to that given in Eq. 5.26 for the condition that $k_{\mathrm{B}}T \gg h\nu$. The reader may wish to verify this identity.

For a one-dimensional *rigid rotator* consisting of two equal masses held at a fixed distance d, the atomic position is restricted to a circle with radius d, and the value of the configuration integral is just $2\pi d$, which gives the partition function as states in Eq. 5.29 for $T \gg \Theta_{rot}$, as can be verified by the reader using the definition of Θ_{rot}.

6.4 APPLICATION: RELATION BETWEEN BINDING AND MOLECULAR INTERACTION ENERGY

As an example of a serious application of the configuration integral, Eq. 6.17, we use that equation to describe the thermodynamics of a simple system of interacting molecules in terms of the interaction energy function.

The easiest system to treat contains one molecule of type A and N_B molecules of type B. The molecules of type B do not interact with one another, that is, the energy of a B–B pair is equal to zero for any intermolecular distance. In practice, this corresponds to the limit of high dilution of B. Furthermore, at high dilution of B, encounters of A with several B at a time are negligible. Thus, except for the interaction between A and B, the system is an ideal gas.

The energy of an A–B pair is described by a function $E_{AB}(r_{AB})$ of the intermolecular distance r_{AB}. Molecular interactions are short range, and thus $E_{AB}(r_{AB})$ is equal to zero at long distance. The configurational partition function is then

$$z = \frac{1}{N_B!}\int_V d\mathbf{x}_A \int_V d\mathbf{x}_{B,1} \ldots \int_V d\mathbf{x}_{B,N} \exp\left[\sum_{i=1}^{N} -\beta E_{AB}\left(r_{AB,i}\right)\right], \tag{6.20}$$

where the factor $N_B!$ is needed because the N_B molecules of B are indistinguishable; the A–B interaction energy is summed over all molecules of B and the integrals are all over the volume, V. Because there are no cross terms, the integrals can be separated

$$z = \frac{1}{N_B!}\int_V d\mathbf{x}_A \int_V d\mathbf{x}_{B,1} \exp\left[-\beta E_{AB}\left(r_{AB,1}\right)\right] \ldots \int_V d\mathbf{x}_{B,N_B} \exp\left[-\beta E_{AB}\left(r_{AB,N_B}\right)\right], \tag{6.21}$$

Integration over $\mathbf{x}_A$ gives a factor of V, and the N_B integrals over $\mathbf{x}_B$ are all the same, so that

$$z = \frac{1}{N_B!} V \left\{\int_V 4\pi r_{AB}^2 dr_{AB} \exp\left[-\beta E_{AB}\left(r_{AB}\right)\right]\right\}^{N_B}, \tag{6.22}$$

where each integral over the coordinates of a B molecule has been replaced by an integral over a spherical shell centered on the one A molecule.

What interests us is how the partition function has changed relative to that of a system without interaction (i.e., zero interaction energy) between A and B molecules z_0

$$z_0 = \frac{1}{N_B!} V^{N_B+1}$$

$$z/z_0 = \left\{ \frac{1}{V} \int_V 4\pi r_{AB}^2 dr_{AB} \exp\left[-\beta E_{AB}\left(r_{AB}\right)\right] \right\}^{N_B}. \tag{6.23}$$

The change of the free energy of this system due to the interaction of the A and B molecules is then

$$\Delta A = -\frac{1}{\beta} \ln\left(z/z_0\right) = -\frac{N_B}{\beta} \ln \frac{\int_V 4\pi r_{AB}^2 dr_{AB} \exp\left[-\beta E_{AB}\left(r_{AB}\right)\right]}{V}. \tag{6.24}$$

The integral, $\Upsilon = \int_V 4\pi r_{AB}^2 dr_{AB} \exp\left[-\beta E_{AB}\left(r_{AB}\right)\right]$ is nearly equal to V; hence, we can write the ratio Υ/V as $1+\varepsilon$, where $\varepsilon = \left(\Upsilon - V\right)/V$ is arbitrarily small (as there is no obvious upper limit to our choice of V), and then equate $\ln\left(1+\varepsilon\right)$ with ε itself

$$\Delta A = -\frac{N_B}{\beta} \frac{\int_V 4\pi r_{AB}^2 dr_{AB} \exp\left[-\beta E_{AB}\left(r_{AB}\right)\right] - V}{V}$$

$$\Delta\mu_A^o = -\frac{N_B}{\beta V} \int_V 4\pi r_{AB}^2 dr_{AB} \left\{\exp\left[-\beta E_{AB}\left(r_{AB}\right)\right] - 1\right\}. \tag{6.25}$$

This change of the free energy results from the addition of B to a system per molecule of A, which is therefore the change in standard chemical potential of A, $\Delta\mu_A^o$. If the interaction of A and B molecules is a strong preferential attraction then this represents the standard free energy of binding.

Note that the exponential gives the probability of observing B at some distance r_{AB} from A relative to the probability of observing B at the same distance from B; subtraction of 1 from the exponential produces the *excess* relative concentration (excess if $E_{AB} < 0$); multiplication by the concentration of B, N_B/V produces the excess concentration of B near A; finally, integration changes this to the excess *number* of molecules of B near A, that is, if the association is strong and specific, the number of AB dimers (for a single molecule of A), N_{AB} that is,

$$\frac{N_B}{V} \int_V 4\pi r_{AB}^2 dr_{AB} \left\{\exp\left[-\beta E_{AB}\left(r_{AB}\right)\right] - 1\right\} = N_{AB} = K_b \frac{N_B}{V}, \tag{6.26}$$

where the right-most expression is obtained if we define the equilibrium constant for the process of bringing excess B into contact with A as $K_b = N_{AB} \times V/N_B$.

In more complicated situations (e.g., the interaction energy depends on relative orientation of the interacting molecules or the interaction is modulated by the presence of solvent) the energy in Eq. 6.26 may be replaced by a free energy. Thus, we have here derived a relation allowing *the estimation of a binding constant given an energetic model* of the interaction of macromolecule and ligand. In Section 7.7, this model has been applied to the self-association of methane molecules in water modeled with molecular dynamics simulation.

6.5 APPLICATION: BINDING OF LIGAND TO A MACROMOLECULE

The result derived in the preceding section, while it covers all cases of weak and strong favorable interactions, and even unfavorable interactions (so-called *exclusion*; see Section 9.5), was developed for a spherically symmetrical model of the interactions, and does not *per se* cover the case of tight binding of a ligand to a preorganized binding site in which the bound ligand is a well-defined part of the structure, and which is ubiquitous in biology. For such a case, the energy for transferring the ligand molecule from the solution into the binding site ΔE^* is negative for only a small range of the $3N$ degrees of freedom (positions, orientations, internal rotations) of the ligand molecule relative to the position and orientation of the macromolecule. As an approximation, we represent the range of energy and positions in the bound state by a square-well potential: the energy difference ΔE^* is constant ($\Delta E^* < 0$) in a volume v_b (Fig. 6.1).

One can consider two closed systems, one with a ligand bound to a site on a macromolecule (M·L), the other with the macromolecule having an empty site (M) and the ligand in solution (L). The energy of the two systems differ by a binding energy ΔE^*, and the entropy of the two systems differ because of the fact that one ligand molecule moves freely over the solution volume V in one case, and is confined to a much smaller volume v in the other case. Accordingly, the ratio of the partition functions of the two systems, one containing a single macromolecule with bound ligand and $N_L - 1$ free ligand molecules, and the other containing the free macromolecule and N_L free ligand molecules, is equal to the ratio of the two volumes multiplied by the Boltzmann factor for the energy difference

$$\frac{Q_{\mathrm{ML},N_L-1}}{Q_{\mathrm{M},N_L}} = \frac{v_b}{v} \exp\left(-\beta \Delta E^*\right).$$

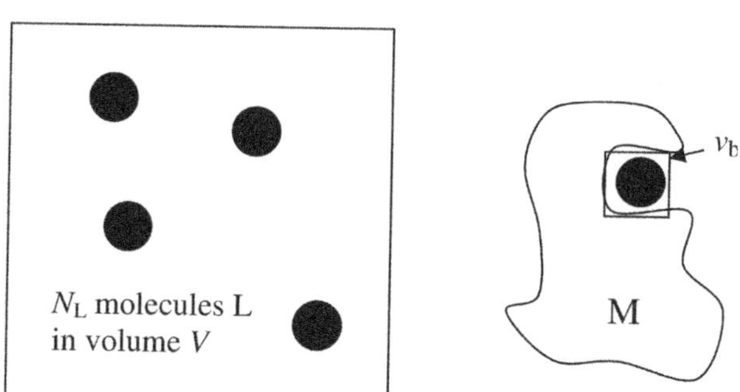

FIGURE 6.1 Entropic component of ligand binding: one ligand molecule, L moves from a large volume to a small volume in the binding site of macromolecule M.

The free energy change for transferring one ligand molecule from dilute solution to macromolecule M is then

$$\Delta A_{\rm b} = \Delta E^* + k_{\rm B}T \ln \left(v/v_{\rm b}\right). \tag{6.27}$$

The first term on the right, ΔE^* is independent of concentration, while the second term depends on concentration and is large and positive as $v \gg v_b$; the binding process consists of moving a ligand from a large effective volume v in solution to a much smaller volume, $v_{\rm b}$ of a binding site on the macromolecule. For example, if the concentration of B in solution is 10^{-6} M, $v \approx 4.5 \times 10^9$ Å^3, while 10 Å^3 is a reasonable estimate of $v_{\rm b}$. This gives about 13 kcal/mol for the second term; accordingly, the binding (free) energy ΔE^* must be less than -13 kcal/mol for $\Delta A_{\rm b}$ to be negative (and, thus, for binding to occur). This illustrates the point that ligand binding is a playoff between an unfavorable concentration-dependent, entropic term (the cratic free energy; Chapter 1) and a favorable, concentration-independent term (the unitary or standard free energy term; Chapter 1.) As the concentration of the ligand increases, the cratic term becomes less positive and is overcome by the negative unitary terms and the binding site becomes saturated with ligand.

An interesting observation has been made by Jencks. He considered a pair of binding sites to which two small ligands, B and C, could bind. He then asked what would be the effect on the affinity of linking B and C into a single molecule, allowing $\Delta E^*_{\rm BC}$ to be the sum of $\Delta E^*_{\rm B}$ and $\Delta E^*_{\rm C}$, and $v_{\rm b,B} = v_{\rm b,C} = v_{\rm b,BC}$. The answer is that $\Delta A_{\rm b,BC}$ is lowered relative to $\Delta A_{\rm b,B}$ and $\Delta A_{\rm b,C}$ by removal of one unfavorable entropic term $k_{\rm B}T \ln \left(v/v_{\rm b}\right)$. This results in a considerable affinity for BC, even if the individual ligands B and C bind weakly. A consequence of this is that an interaction that is too weak to be expected to be of biological significance can be important (e.g., in a regulatory system) if accompanied by a strong interaction of part of a ligand molecule with a neighboring site. The strong interaction provides most of the free energy to overcome the unfavorable cratic contribution to $\Delta A_{\rm b}$, allowing the weaker interaction to occur.

Using the concepts developed in Chapter 3, one can think of the energy $E_{\rm AB}$ as actually a free energy resulting from the sum of many contributions from interactions between macromolecule and ligand, minus similarly constructed sums for interactions between ligand and macromolecule and solvent, dominated by hydrogen bonds and other charge–charge interactions and hydrophobic interactions. Setting this energy constant inside a small volume $v_{\rm b}$ is a gross simplification of the actual, complex free energy profile centered on the conformation of minimum free energy, and involving many degrees of freedom, not only translational and rotational motions of the ligand but also internal degrees of freedom of the ligand and the macromolecule. In addition, all terms must be considered for the free macromolecule and free ligand in solution. In principle, these terms can be evaluated given a realistic (and computationally tractable) model of the energy function, but, in practice, this is so difficult that no reliable method exists for relating affinities of biological macromolecules to first principles. Despite the magnitude of the approximations made in the simple treatment presented here, it still captures the essence of the binding process: a balance between a favorable standard or unitary free energy change and the unfavorable cratic contribution to the overall free energy of the process.

6.6 OPEN SYSTEMS: THE GRAND CANONICAL ENSEMBLE OR GRAND ENSEMBLE

The canonical ensemble applies to closed systems in which N is fixed, that is, the composition of all species in each member system is fixed. The single most important molecular event in biology is binding of a ligand to a recognition site, where the concentrations (or, more generally, activities) of bound and free ligands are not fixed. For binding experiments, the ligand concentration is varied by the experimentalist. Cellular events modulate concentrations of ligands *in vivo*, and these regulate binding of ligand to macromolecules.

A system that can exchange material with a reservoir is an *open system*. Such a system can have a large number (M) of components (labeled Λ), where N_Λ is variable for each of the M components. Instead, the chemical potential, μ_Λ is fixed for each species, Λ. The thermodynamic condition of compositional equilibrium establishes equilibrium of all components between systems.[5]

In order to study open systems in a statistical thermodynamics framework, we must consider another ensemble: the *grand canonical ensemble* or *grand ensemble*, which is illustrated in the left panel of Table 6.2. This ensemble consists of a very large number of systems at *thermal and compositional equilibrium* such that each system has a defined V, T, and a set of chemical potentials $\{\mu_\Lambda\}$, but *both the energy, E and composition $\{N_\Lambda\}$ vary between systems* within the ensemble. In Table 6.2, the set $\{N_\Lambda\}_k$ represents a particular composition, with that composition being one determinant of the quantum state (Ψ_r) of the system. As there are other determinants of system quantum state (e.g., distributions of molecules over molecular quantum states and interactions between molecules), there can be many systems that have the same composition but still have different quantum states

On the right side of Table 6.2, we have *rearranged* the ensemble to include in the kth row all systems with the same composition $\{N_\Lambda\}_k$. Each row represents a canonical ensemble of systems that all have the same composition, temperature and volume, but of which the quantum state, and thus the energy, varies. As the quantum state depends on the set $\{N_\Lambda\}_k$, we can use the methods outlined in Section 6.2 and determine the most likely distribution $\{m_j\}_k$ that defines the average properties and thus average energy for the kth row $\langle E\rangle_k$ in terms of the Boltzmann distribution. Of course, $\{N_\Lambda\}_k$ need not be the most likely or equilibrium composition of the system. To obtain this, we need to find the distribution $\{m_j\}_k$ that maximizes the statistical degeneracy $\Omega(\{m_{j,k}\})$, of the *entire* ensemble, as was done for the canonical ensemble in Section 6.2. To accomplish this, we use the same method of Lagrange multipliers that we used to optimize the degeneracy and obtain the Boltzmann distribution for the canonical ensemble, except that the addition of composition as a variable makes the problem a bit more mathematically complex.

TABLE 6.2 The grand canonical ensemble of systems

Grand ensemble

$\{\mu_\Lambda\}, V, T$ E_1, Ψ_1 $\{N_\Lambda\}_1$	$\{\mu_\Lambda\}, V, T$ E_2, Ψ_2 $\{N_\Lambda\}_2$	$\{\mu_\Lambda\}, V, T$ E_3, Ψ_3 $\{N_\Lambda\}_3$	…
…	…		
$\{\mu_\Lambda\}, V, T$ E_r, Ψ_r $\{N_\Lambda\}_r$			
…	…		$\{\mu_\Lambda\}, V, T$ $E_{\mathcal{R}}, \Psi_{\mathcal{R}}$ $\{N_\Lambda\}_{\mathcal{R}}$

$\Rightarrow$

Rearranged grand ensemble

$\{\mu_\Lambda\}, V, T$ $\Psi_{1,1}$ $\{N_\Lambda\}_1$	$\{\mu_\Lambda\}, V, T$ $\Psi_{1,2}$ $\{N_\Lambda\}_1$	…	$\{\mu_\Lambda\}, V, T$ $\Psi_{1,l}$ $\{N_\Lambda\}_1$	…
…	…			
$\{\mu_\Lambda\}, V, T$ $\Psi_{k,l}$ $\{N_\Lambda\}_k$	…		$\{\mu_\Lambda\}, V, T$ $\Psi_{k,l}$ $\{N_\Lambda\}_k$	
…	…			$\{\mu_\Lambda\}, V, T$ $\Psi_{\mathcal{L},\mathcal{M}}$ $\{N_\Lambda\}_{\mathcal{L}}$

Once again, we restrict the choice of $m_{j,k}$ to those that satisfy the constraints that they yield the thermodynamic average energy and composition, E and $\{N_\Lambda\}$. Thus, we have $2+\mathrm{M}$ constraints for the maximization of $\Omega(\{m_j\}_k)$. As in Section 6.1, the first of these is an artificial constraint, as the value of $\mathcal{R}$ is not a defined quantity but is needed to define the total energy in terms of $\langle E\rangle$ and overall composition in terms of $\langle\{N_\Lambda\}\rangle$.

$$\sum_{j,k} m_{j,k} = \mathcal{R},$$

$$\sum_{j,k} m_{j,k} E_{j,k}\left(V, \{N_\Lambda\}_k\right) = \mathcal{R}\langle E\rangle,$$

$$\sum_{j,k} m_{j,k} N_{\Lambda,k} = \mathcal{R}\langle N_\Lambda\rangle; \text{ for } \Lambda = 1, 2 \ldots \mathcal{M}.$$

In addition to $\beta = 1/k_\mathrm{B}T$ and the normalizing multiplier that derives from the arbitrary $\mathcal{R}$ constraint, this introduces a Lagrange multiplier for each component, which is then identified with the chemical potential μ_Λ of that component. We give here only the result of this optimization process, but present it in Appendix C for the case of a single component, with a variable content of N molecules.

The probability that a system quantum state $\Psi_{k,j}$ with a given energy $E_{k,j}$ and composition $\{N_\Lambda\}_k$ contributes to the system's thermodynamics is given by

$$P\left[E_{k,j}, \{N_\Lambda\}_k\right] = \frac{\exp\left(-E_{k,j}/k_\mathrm{B}T\right)\prod_\Lambda \exp\left(N_{\Lambda,k}\mu_\Lambda/k_\mathrm{B}T\right)}{\Xi}, \tag{6.28}$$

where $N_{\Lambda,k}$ is the number of molecules of type Λ in this system quantum state, and Ξ is the *grand canonical or grand partition function*,

$$\Xi = \sum_k \sum_j P\left[E_{k,j}, \{N_\Lambda\}_k\right]. \tag{6.29}$$

All terms for a given composition share a common factor $\prod_\Lambda \exp(N_{\Lambda,k}\mu_\Lambda/k_\mathrm{B}T)$, and thus we first carry out the summation over quantum states j for each particular composition k, that is, we sum over all elements in a given row of the ensemble with a given fixed $\{N_\Lambda\}_k$. This yields a convenient and common form of the grand canonical partition function shown on the right side of the following equation[3]

$$\begin{aligned}\Xi &= \sum_{\{N_\Lambda\}_k} \sum_j \exp\left[\left(-\frac{E_j}{k_\mathrm{B}T}\right)\prod_\Lambda e^{N_{\Lambda,k}\mu_\Lambda/k_\mathrm{B}T}\right]\\ &= \sum_{\{N_\Lambda\}_k}\left[Q\left(\{N_\Lambda\}_k, V, T\right)\prod_\Lambda e^{N_\Lambda\mu_\Lambda/k_\mathrm{B}T}\right].\end{aligned} \tag{6.30}$$

The remaining sum is over all possible compositions, and each term in the sum is the canonical partition function for a given composition $\{N_\Lambda\}_k$ times a weighting factor that is the Boltzmann factor of the Gibbs free energy of the system in equilibrium with its surroundings $G = \sum_\Lambda (N_\Lambda \mu_\Lambda)$. As the canonical partition function is the negative exponential of the Helmholtz free energy of a species, the weighting factor for each term of the sum is an exponential of a difference between the Helmholtz and Gibbs free energies of the system, which is PV for a given composition $\{N_\Lambda\}$.

The utility of the grand ensemble is that it allows us to define the probability of observing any composition $\{N_\Lambda\}$ in the open system,

$$P\left[\{N_\Lambda\}\right] = \frac{Q_{\{N_\Lambda\}} \prod_\Lambda \exp\left(N_\Lambda \mu_\Lambda / k_\mathrm{B} T\right)}{\Xi}$$

$$= \frac{\exp\left[-A\left(\{N_\Lambda\}, T, V\right)/k_\mathrm{B} T\right] \prod_\Lambda \exp\left(N_\Lambda \mu_\Lambda / k_B T\right)}{\Xi}. \qquad (6.31)$$

The probability of an open system having composition $\{N_\Lambda\}$ can be thought of as the Boltzmann factor of the work required to move components into a system kept at constant volume from an open reservoir with chemical potentials $\{\mu_\Lambda\}$ in order to achieve this composition. Alternatively, the probability $P[\{N_\Lambda\}]$ can be thought of as the product of the probability of the system having Helmholtz free energy A with $\{N_\Lambda\}$ fixed,[6] and the probability of "importing" each of the components Λ from a reservoir with chemical potential μ_Λ. The latter probability is just the Boltzmann factor of the Gibbs free energy of the components in the reservoir $G = \sum_\Lambda N_\Lambda \mu_\Lambda$.

For a given $\{N_\Lambda\}$, the difference $G(\{N_\Lambda\}) - A\{N_\Lambda\}$ is just $PV(\{N_\Lambda\})$, and each term in the sum is just $\exp[\beta PV(\{N_\Lambda\})]$. Because A and G are both state function, PV is also a state function with T, V, and $\{\mu_\Lambda\}$ as independent variables. This is the state function defined by the logarithm of the grand partition function,

$$k_\mathrm{B} T \ln \Xi = -A\left(T, V, \{\langle N_\Lambda \rangle\}\right) + \sum_\Lambda \langle N_\Lambda \rangle \mu_\Lambda = -A + G = PV, \qquad (6.32)$$

where the average composition $\{\langle N_\Lambda \rangle\}$ corresponds to the thermodynamic distribution of components between systems. As we have seen for the microcanonical and canonical partition functions, this distribution satisfies the condition that the entropy of the system be maximized. The distribution $\mathrm{P}\{N_\Lambda\} = \exp(\beta PV\{N_\Lambda\})/\Xi$ also satisfies this condition. The composition that "wins out," $\langle\{N_\Lambda\}\rangle$, is the composition that maximizes the playoff between a favorable broad compositional distribution (maximal entropy) and one that produces the greatest decrease in the Helmholtz free energy (normally the lowest internal energy).

Of great importance is the following relation between the grand partition function, composition, and chemical potential (which is easily derived from Eqs. 6.30 and 6.31,)

$$\langle N_\Lambda \rangle = k_B T \left(\frac{\partial \ln \Xi}{\partial \mu_\Lambda} \right)_{V,T,\mu_{j \neq \Lambda}} = \frac{\partial \ln \Xi}{\partial \ln a_\Lambda} \approx \frac{\partial \ln \Xi}{\partial \ln [L_\Lambda]}, \quad (6.33)$$

where the third equality may be used in the case of an ideal solution.

Fluctuations in composition and phase equilibria are two physical phenomena most appropriately analyzed in terms of the grand partition function. Other ensembles can be thought of that may be more appropriate for other problems. In general, the choice of a partition function, and corresponding state function, depends on the choice of independent thermodynamic variables to describe a system, and this choice is one of convenience, that is, according to the variables we have used experimentally to define the system. Hill's Statistical Mechanics provides an excellent treatise on this issue for the student seeking a more in-depth treatment.

6.7 FLUCTUATIONS

One of the major insights from the statistical treatment of thermodynamics is that no state (allowed by quantum mechanics) is impossible, but that each state is assigned a probability. Many possible states have a low probability, and occur rarely. Consequently, system state functions that have not been fixed in order to determine the thermodynamic ensemble fluctuate about mean values.

Fluctuation of the energy. For example, in a system with N, V, and T fixed (i.e., described by the canonical partition function), the energy has a *most probable* value, which is also its average value $\langle E \rangle$, but the system exchanges energy with a heat bath, and as a result its instantaneous energy fluctuates about the mean. Thus, there is a distribution of energies, and of other such system properties (Fig. 6.2). Such distributions are, in a first approximation, characterized by their width, which

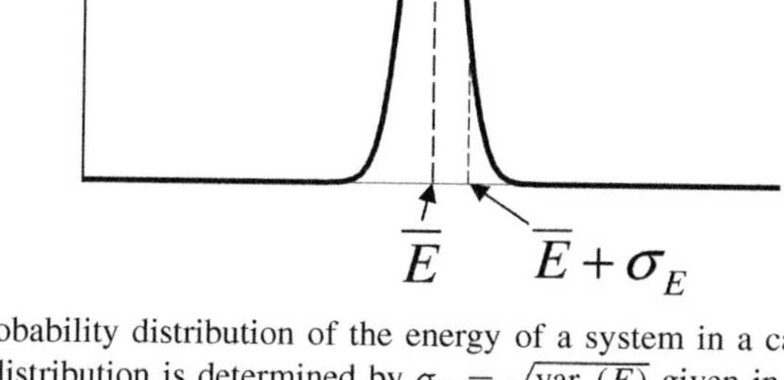

FIGURE 6.2 Probability distribution of the energy of a system in a canonical ensemble. The width of the distribution is determined by $\sigma_E = \sqrt{\mathrm{var}\ (E)}$ given in Eq. 6.35.

is related to the variance (or root-mean-square deviation from the mean),

$$\mathrm{var}\,(x) = \sigma_x{}^2 = \left\langle (x - \langle x \rangle)^2 \right\rangle = \left\langle x^2 \right\rangle - \langle x \rangle^2 . \tag{6.34}$$

It turns out that the variance of the energy is proportional to the heat capacity (related to the second temperature derivative of the partition function) and increases with the square of the temperature. We can show this using the canonical partition function,[7] which gives the probability of states of different energy, and use this to evaluate the mean square and mean energies,

$$\mathrm{var}\,(E) = \left\langle E^2 \right\rangle - \langle E \rangle^2 = \frac{\sum_i E_i{}^2 \exp\left(-\beta E_i\right)}{Q} - \left(\frac{\sum_i E_i \exp\left(-\beta E_i\right)}{Q}\right)^2 . \tag{6.35}$$

As we have seen before, the partial derivative of the partition function with respect to one of the factors in the exponent, here β, gives the ensemble average of the other factor of the exponent, here E,

$$\frac{\partial \ln Q}{\partial \beta} = \frac{1}{Q}\frac{\partial Q}{\partial \beta} = \frac{-\sum_i E_i \exp\left(-\beta E_i\right)}{Q} = -\langle E \rangle , \tag{6.36}$$

and, if we repeat the differentiation,

$$\begin{aligned}
\frac{\partial^2 \ln Q}{\partial \beta^2} &= \frac{\partial\left(\frac{1}{Q}\frac{\partial Q}{\partial \beta}\right)}{\partial \beta} = \frac{1}{Q}\frac{\partial^2 Q}{\partial \beta^2} + \frac{\partial Q}{\partial \beta}\frac{\partial\left(\frac{1}{Q}\right)}{\partial \beta} = \frac{1}{Q}\frac{\partial^2 Q}{\partial \beta^2} - \frac{1}{Q^2}\left(\frac{\partial Q}{\partial \beta}\right)^2 \\
&= \frac{\sum_i E_i{}^2 \exp\left(-\beta E_i\right)}{Q} - \left(\frac{1}{Q}\frac{\partial Q}{\partial \beta}\right)^2 \qquad (6.37) \\
\frac{\partial^2 \ln Q}{\partial \beta^2} &= \left\langle E^2 \right\rangle - \langle E \rangle^2 = \mathrm{var}\,(E) .
\end{aligned}$$

Thus, the magnitude of the fluctuations in E is given by the second derivative of the partition function with respect to β. This can be related to derivatives of the free energy with respect to the temperature,

$$\begin{aligned}
&\frac{\partial(A/T)}{\partial(1/T)} = \langle E \rangle = -\frac{\partial \ln Q}{\partial \beta} \\
&\frac{\partial}{\partial T}\frac{\partial(A/T)}{\partial(1/T)} = \frac{\partial\langle E \rangle}{\partial T} = C_V = -\frac{\partial}{\partial T}\frac{\partial \ln Q}{\partial \beta} = \frac{1}{k_B T^2}\frac{\partial^2 \ln Q}{\partial \beta^2} \qquad (6.38)
\end{aligned}$$

and hence

$$\mathrm{var}\,(E) = k_B T^2 C_V \tag{6.39}$$

that is, the specific heat C_V is a measure of the energy fluctuation. For a macroscopic system, the root-mean-square fluctuation of E is an extremely small fraction of the system's thermal energy, and cannot be detected.

Fluctuation of concentration. Similarly, for an *open system* described by fixed T, V, and chemical potential, the number of particles, and hence the concentration fluctuates. Such a system can be realized in practice by studying a volume V that is part of a much larger system that can absorb the fluctuations in the contents of V without measurable changes in its own content. Small volumes within much larger solution volumes can be studied readily by a variety of spectroscopic methods that rely on narrow light beams, as in the light scattering experiments described in the following section.

Following a similar derivation as used for fluctuations of the energy, double differentiation of the grand partition function with respect to the chemical potential μ_Λ gives the extent of *fluctuation of the number of particles* N_Λ as

$$\text{var}\left(N_\Lambda\right) = \left\langle N_\Lambda^2\right\rangle - \left\langle N_\Lambda\right\rangle^2 = \left(k_B T\right)^2 \frac{\partial^2 \ln \Xi}{\partial \mu_\Lambda^2}$$

$$\text{var}\left(N_\Lambda\right) = k_B T \frac{\partial \left\langle N_\Lambda\right\rangle}{\partial \mu_\Lambda}. \tag{6.40}$$

In the special case of an ideal gas or dilute solution,

$$\mu_\Lambda = \mu_\Lambda^{\text{o}} + k_B T \ln\left(\left\langle N_\Lambda\right\rangle / V\right)$$

$$\text{var}\left(N_\Lambda\right) = \frac{k_B T}{\partial \mu_\Lambda / \partial \left\langle N_\Lambda\right\rangle} = \left\langle N_\Lambda\right\rangle$$

$$\sqrt{\text{var}\left(N_\Lambda\right)} = \sqrt{\left\langle N_\Lambda\right\rangle}. \tag{6.41}$$

Thus, the root-mean-square deviation is equal to the square root of the (mean) number of particles in the volume. This shows that, while the size of fluctuations increases with system size, the *relative* fluctuation becomes smaller when the system size is increased.

For any system, one can show using the grand partition function that

$$\text{var}\left(N\right) = \frac{N k_B T \kappa}{V}; \quad \frac{\text{var}\left(N\right)}{\left\langle N_\Lambda\right\rangle^2} = \frac{k_B T \kappa}{\left\langle N_\Lambda\right\rangle V} \tag{6.42}$$

where κ is the compressibility

$$\kappa = -\frac{1}{V}\left(\frac{\partial V}{\partial P}\right)_{N,T}. \tag{6.43}$$

6.8 APPLICATION: LIGHT SCATTERING AS A MEASURE OF FLUCTUATIONS OF CONCENTRATION

The intensity of light scattered by a liquid or gas depends on the magnitude of the (local) fluctuations in refractive index. Light scattering intensities are measured by passing a narrow incident beam (typically from a red or blue laser) through a solution and observing the light scattered by a small volume V illuminated by the incident beam, as shown in Fig. 6.3.

The refractive index of a dilute solution, n differs from that of the pure solvent, n_0 by a small amount that is proportional to the concentration of solute, c_1 (expressed in units of mass/volume) and a proportionality constant that depends on the nature of both solute and solvent,

$$n - n_0 = c_1 \frac{dn}{dc_1}. \tag{6.44}$$

Thus, the fluctuation of the refractive index of a dilute solution containing a single solute, expressed as its variance, is given by

$$\begin{aligned} \operatorname{var}(n) &= \left(\frac{dn}{dc_1}\right)^2 \operatorname{var}(c_1) = \left(\frac{dn}{dc_1}\right)^2 \left(\frac{dc_1}{dN_1}\right)^2 \\ \operatorname{var}(N_1) &= \left(\frac{dn}{dc_1}\right)^2 \left(\frac{dc_1}{dN_1}\right)^2 \frac{k_B T}{\partial \mu_1 / \partial N_1} \end{aligned} \tag{6.45}$$

and in an ideal solution (i.e., at low concentration, c_1) this becomes with Eq. 6.41

$$\operatorname{var}(n) = \left(\frac{dn}{dc_1}\right)^2 \left(\frac{dc_1}{dN_1}\right)^2 N_1. \tag{6.46}$$

Here, the ratio $dc_1/dN_1 = c_1/N_1$ is just the molecular mass of the solute molecules m_1

$$\operatorname{var}(n) = \left(\frac{dn}{dc_1}\right)^2 {m_1}^2 \frac{c_1}{m_1} = \left(\frac{dn}{dc_1}\right)^2 m_1 c_1. \tag{6.47}$$

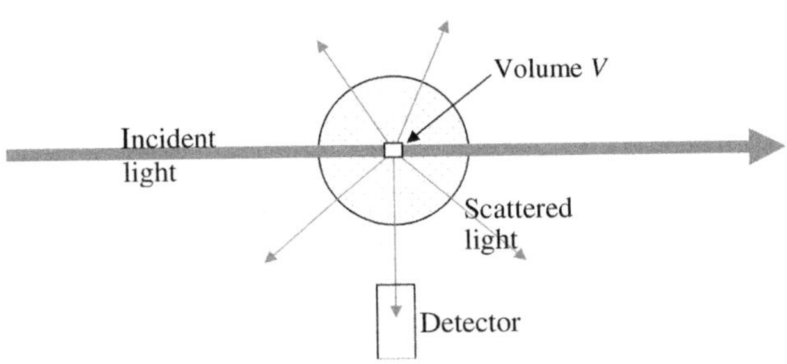

FIGURE 6.3 Schematic of instrumentation for measurement of light scattering in solution.

Thus, the fluctuations in refractive index depend on the molecular mass of the solute and light scattering intensity provides a quantitative measure of the molecular weight of the solute. The intensity scattered per unit of volume is given by

$$\frac{R}{I} = \frac{2\pi^2 n^2}{\lambda^4}\left(\frac{dn}{dc_1}\right)^2 m_1 c_1, \qquad (6.48)$$

where I is the intensity of the incident beam and λ the wavelength. (This is a simplified form of the scattering equation, as the scattered intensity contains a background of light scattered by the solvent and depends also on the direction of polarization of the incident light relative to the direction in which the scattering is observed.)

If the molecules have dimensions that are not negligible relative to the wavelength, the scattered intensity drops off as the angle between incident and scattered beam increases; this effect can be used to measure molecular dimensions, principally the radius of gyration.

If the solution is nonideal, it is incorrect to substitute Eq. 6.41 in Eq. 6.45. Very concentrated protein solutions are very nonideal, and in these solutions $\partial\mu_1/\partial N_1$ is very large. (This effect will be discussed in some detail in Section 9.5.2.) Accordingly, the fraction of light scattered, and hence not transmitted, by very concentrated protein solutions (such as the solutions that transmit and focus light in the eye) is very low.

Measurements of light scattering intensity remain useful in studies of formation and size of supermolecular complexes in solution, but have otherwise largely been superseded by more convenient measures of molecular mass. So-called dynamic light scattering experiments measure the shift in the frequency of the scattered light due to molecular motion, that is, a molecular Doppler effect, and can be interpreted in terms of diffusive motion.

NOTES

1. Canonical ≈ "following the law or canon" (related to "cane," a stick used as a ruler). Also known as a Gibbsian ensemble, after Josiah Willard Gibbs, 1839–1903, professor at Yale, a great contemporary of Boltzmann who is considered the father of chemical thermodynamics.
2. While it would surely be a currently intractable challenge to define the quantum states and their energies for a system of interacting molecules, it is in principle possible, and we will not concern ourselves with the difficulty of this task at this point.
3. This is because the sum of a product of exponentials over several independent indices is equal to the product of sums over each index.
4. The $6N$-dimensional space of $3N$ p-coordinates and $3N$ x-coordinates is called the phase space.
5. Note that one of the most annoying things about the grand canonical partition function is keeping track of indices of the many quantities that change. The subscript Λ is used here to label the M different chemical species present in the system. We use j to index

system quantum states, and l to index the different possible "compositional distributions" in the ensemble.

6. This term defines the thermodynamic properties of the subensemble that contains a fixed number of each type of molecule (N_Λ). It does not depend on the thermodynamically defined concentration of each species at equilibrium.
7. We note that this treatment of fluctuations has been successfully given in terms of statistical, that is, molecular, thermodynamics, but can not be formulated in terms of standard macroscopic thermodynamics.

SUGGESTED READING

Davidson, N., *Statistical Mechanics*. New York: McGraw-Hill, 1962, (republished by Courier Dover Publications; 2003).

Hill, T.L., *An Introduction to Statistical Mechanics*. Reading, MA: Addison-Wesley, 1960.

Hill, T.L., *Statistical Mechanics. Principles and Selected Applications*. New York: McGraw-Hill, 1964.

McQuarrie, D.A., *Statistical Mechanics*. University Science Books: Sausalito, CA, 2000.

Widom, B., *Statistical Mechanics. A Concise Introduction for Chemists*. Cambridge: Cambridge University Press, 2002.

7

Sampling Molecular Systems with Simulations

Today's powerful tool is tomorrow's scrap.

7.1 INTRODUCTION

Simulations of molecular structures and dynamics have become widely used as ways to investigate the behavior of interacting molecules, including biological macromolecules in solution, and can be expected to remain widely used as techniques improve and, especially, as computational facilities become ever more powerful. In an earlier chapter devoted to the theory of molecular energies and molecular forces (Chapter 3), we introduced the concept of the empirical molecular mechanics (MM) force field with which to rapidly evaluate the potential energy as a sum of a very large number of inter- and intramolecular energies. In two subsequent chapters (Chapters 5 and 6), we introduced the theory of statistical thermodynamics for the study of molecular ensembles. There we saw how correlation of positions and momenta of all molecules due to intermolecular potentials makes evaluating the partition function of a system impossibly complex. Here we describe how computers and the laws of mechanics can help deal with this complexity.

In this chapter, we discuss solutions to some of the technical issues of simulations of molecular systems; however, the focus of this chapter is the application of the statistical thermodynamics theory to the problem of estimating a variety of ensemble (and bulk) physical properties from simulations of molecular systems. The discussion will end with calculations of free energy differences in terms of potentials of mean force and "computer alchemy."

Equilibria and Kinetics of Biological Macromolecules, First Edition. Jan Hermans and Barry Lentz.

7.2 BACKGROUND

We have seen how knowledge of the potential energy levels of a molecular system obtained with quantum mechanics (QM) can be combined with the theory of statistical mechanics to give a complete statistical description that allows calculation of exact physical properties, including thermodynamic functions (Chapter 5, Section 4). However, in practice, the application is limited by the difficulty of evaluating the energy levels of any except very simple systems.

We have also described briefly how this method can be extended to continuous energy functions, with good accuracy in many cases of interest (Chapter 6, Section 3). Summation over the energy levels is now replaced by an integral over conformation space ("phase-space integral," Eq. 6.15), and this allows more complicated systems to be treated, but at the expense of having to assume an energy function that approximates the distribution of energy levels of QM. The next great difficulty that is encountered is the large number of independent variables (the atomic coordinates) and the complexity of the energy functions of systems of many atoms, which characteristically display very many local minima, any of which may make a significant contribution to the partition function. Consequently, integration over all phase space for systems with many particles, such as (samples of) liquids and macromolecules, is feasible with only the simplest potential functions.

Instead, equilibria and kinetics of many-atom systems are dealt with by simulation approaches that *sample configuration space according to a Boltzmann distribution*. The reason is simple: if the sample of conformations obtained by the simulation is sufficiently large and representative, it allows estimation of any physical property, g, that is expressible in terms of the atomic coordinates, as an ensemble average according to

$$\overline{g} = \frac{\int g(\mathbf{X}) e^{-E(\mathbf{X})/k_B T} d\mathbf{X}}{\int e^{-E(\mathbf{X})/k_B T} d\mathbf{X}} \approx \langle g \rangle_{\text{sample}} = \frac{1}{\mathcal{N}} \sum_{j=1}^{\mathcal{N}} g(\mathbf{X}_j), \tag{7.1}$$

where $\mathbf{X}_j$ represents the coordinate set of one of the $\mathcal{N}$ configurations[1] of the system in the sample. The convergence of $\langle g \rangle_{\text{sample}}$, or lack thereof, with increasing sample size is used to establish if the sample is adequate. We note that the integral in the denominator is the partition function (the configuration integral of Eq. 6.16), which cannot be evaluated by simulation methods discussed in this chapter.

Such samples can be obtained by two methods. The first, *molecular dynamics*, simulates the thermal motion of the system by integrating the equations of motion. The second, *Monte Carlo*, constructs the sample by stepping randomly through conformation space and using the Metropolis criterion to ensure that the sample, if it is large enough, converges to a Boltzmann distribution.

The most common type of energy function used in simulations is the classical *molecular mechanics* energy (Eq. 3.25). However, there are numerous, and widely used, implementations of QM, which, in effect, estimate continuous energy functions that can be used to produce energy and forces that can be used in dynamics simulations according to otherwise classical mechanics, that is, using Newton's equations of motion. As mentioned, calculation of the QM energy typically takes much more computer time than calculation of the MM energy, and the vast majority of simulations of biological macromolecules is carried out with MM energy and forces.

In this chapter, we describe some of the methods for doing molecular simulations and for analysis of results, with emphasis on applications to equilibria and kinetics.

7.3 MOLECULAR DYNAMICS

This method for sampling the Boltzmann distribution is based on the integration of Newton's equations of motion (i.e., according to classical mechanics). These relate potential energy, V, force, F, and mass, m, to acceleration, a, which itself is the first derivative of the associated velocity, v, and the second derivative of the associated coordinate, X, with respect to time[2]

$$F_i = -\frac{\partial V}{\partial X_i}$$

$$a_i = \ddot{X}_i = \dot{v}_i = \frac{F_i}{m_i}, \tag{7.2}$$

and these equations are solved by numerical integration. In a numerical integration, the velocities and positions are extrapolated in small time steps on the basis of forces calculated for each new set of coordinates. The key assumption is that the time step is small enough that the forces, and therefore also the accelerations, do not change significantly as a result of the small displacements that happen in that time interval. The velocity at the end of the next time step is then set equal to the velocity at the current time augmented by the product of the time step and the acceleration, and the position is updated with both a term linear in the time step and a quadratic term to account for the fact that the velocity is changing linearly during the time step, according to the following equations

$$F_i(t) = -\frac{\partial V}{\partial X_i(t)}$$

$$v_i(t + \delta t) = v_i(t) + a_i \delta t$$

$$X_i(t + \delta t) = X_i(t) + v_i \delta t + \frac{1}{2} a_i (\delta t)^2. \tag{7.3}$$

Repetition of this algorithm produces "trajectories" of arbitrary length (as permitted by available computer resources).

Figure 7.1 shows trajectories for the following force equation

$$a = \ddot{x} = -4\pi^2 \cos(2\pi t) = -4\pi^2 x \tag{7.4}$$

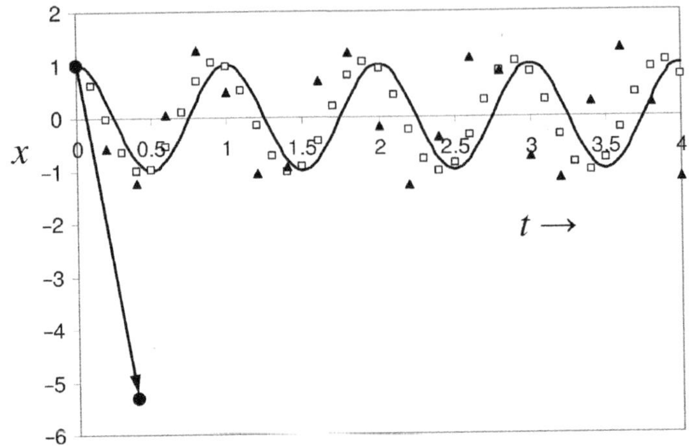

FIGURE 7.1 Numerical integration of x as a function of time according to $\ddot{x} = -4\pi^2 \cos(2\pi t) = -4\pi^2 x$, with starting conditions $x = 1$, $\dot{x} = 0$. Smooth line, no symbols: the analytically integrated function $x = \cos(2\pi t)$; squares: time step = 0.1; triangles: time step = 0.2; circles: time step = 0.4 (additional points are off scale).

with initial values of $x = 1$, $\dot{x} = 0$ at time $t = 0$. Analytical integration gives the figure's smooth curve according to

$$\dot{x}(t) = -2\pi \sin(2\pi t)$$
$$x(t) = \cos(2\pi t). \tag{7.5}$$

Trajectories computed by numerical integration according to Eq. 7.3 are shown as sets of points. One sees that if the time step is small, these trajectories follow the analytical result reasonably accurately; on the other hand, a long time step is desirable to save computer time and extend the length of the trajectory. With longer time steps, the period of the oscillation decreases, but the energy is conserved. Finally, when the time step is chosen too large, at 0.4 times the period or longer, the system is unstable. The value of the largest time step that can be used in a molecular simulation thus depends on the period of the most rapid oscillation, typically the period of bond stretch oscillation of a bond to a hydrogen atom. With Equation 2.11 and force constant from Table 3.1 this period is on the order of

$$\begin{aligned} \text{period} &= \frac{1}{\nu} = 2\pi\sqrt{\frac{m}{k}} \approx 2\pi\sqrt{\frac{10^{-3}}{4.182 \times 644 \times 10^3 \times 10^{20}}} \\ &= 12 \times 10^{-15}\text{s} = 12\,\text{fs}. \end{aligned} \tag{7.6}$$

Dynamics simulations of molecular structure are, therefore, carried out with time steps of around 1 fs. Force fields in which hydrogen atoms are not explicitly represented or, better, in which bond lengths do not vary, can be used with slightly longer time steps.

A rough scale of time constants for some dynamic events in biological macromolecules is as follows:

Vibration of bond length	10^{-14} s
Rotation of side chain at protein surface	10^{-11}–10^{-10} s
Libration of buried side chain	10^{-11}–10^{-9} s
Relative motion of domains	10^{-11}–10^{-7} s
Rotation of buried side chain	10^{-14}–10 s
Folding/unfolding	10^{-6}–10 s

Molecular dynamics simulations of macromolecules in solution are typically carried out on a relatively small set of molecules in a small, properly scaled volume; this volume is repeated in all directions in order to create an effectively infinite system having no boundary with vacuum (periodic boundary conditions).

There are two implicit *limitations* of this procedure that deserve mention. First, all molecules in the system are assumed to behave according to classical mechanics. This is reasonable for a description of thermal motion and conformation change but is unable to deal with processes such as electron transfer and making and breaking of chemical bonds. Approaches to such problems have been developed (the QM/MM method, see Section 3.2) in which only a small region requires a high level quantum treatment.

Nevertheless, as available computer resources grow, treatment of an entire many-atom system with forces computed according to QM becomes increasingly feasible. A simulation of an entire protein (crambin) has been done in which the protein's energy and forces were computed over a 300 ps time interval with a semiempirical QM method with the "divide and conquer" approach.[3] Also, more accurate density functional methods have been applied over much shorter simulation times.[4]

A second limitation is that, even if the use of classical mechanics (i.e., Newton's equations of motion) is a reasonable approximation for the molecules or atoms in the system, the potential energy function, $V(\mathbf{X})$, is empirical and very approximate, making this easily the most severe limitation of the method. Many choices have to be made in developing an MM force field. This accounts for the variety of available force fields and for the different results that may be obtained when treating the same system with different force fields.

It is reasonable to ask if an integration scheme as highly predictive as that followed in molecular dynamics simulations will properly (i.e., randomly and thoroughly) sample a Boltzmann ensemble. Another way to phrase this is to ask if the time average ensemble produced in the simulation approaches the canonical ensemble (the ergodic hypothesis). This fundamental question has not been completely answered, although both theory and experience point in that direction.

In the next section, we digress from our discussion of dynamics simulations by mentioning an alternative simulation method that is guaranteed to sample the Boltzmann distribution.

7.4 METROPOLIS MONTE CARLO

This alternative method for sampling the Boltzmann distribution generates a random sequence of structures and rejects or retains these on the basis of relative energies according to the Metropolis Monte Carlo scheme,[5] which acts in the following manner:

Step 1. Given a conformation, $\mathbf{X}_i$ calculate its energy, E_i.

Step 2. Generate a new conformation, $\mathbf{X}_i'$ by a random change in the structure, and calculate its energy, E_i'.

Step 3. The next conformation added to the growing sample is either $\mathbf{X}_i$ or $\mathbf{X}_i'$, depending on two successive comparisons: the first comparison checks if the energy of the new conformation is lower, and if that is the case, the new conformation is added to the sample and becomes the next starting point for another random change of the structure. The second comparison is required if the energy of the new conformation is higher. In that case, the Boltzmann factor for the (positive) energy difference (this Boltzmann factor has values in the range 0–1) is compared with a function R_U (0:1), which generates a random number obeying a uniform distribution over the interval 0–1.[6] If the random number is the lesser of the two values, then the new conformation is added to the sample; otherwise, the *old* conformation is added to the sample. Whichever is the case, this last conformation becomes the starting point for the next random change of the structure. This is summarized in the following code

$$
\begin{aligned}
&(E_i' \leq E_i)?\, \mathbf{X}_{i+1} = \mathbf{X}_i' \\
&\text{else} \\
&\qquad \{\exp[-(E_i' - E_i)/k_B T] > R_U(0:1)\}?\, \mathbf{X}_{i+1} = \mathbf{X}_i' \\
&\qquad \text{else}\ \mathbf{X}_{i+1} = \mathbf{X}_i\ \text{(i.e., replicate conformation } i\text{)}.
\end{aligned}
\tag{7.7}
$$

A final requirement is that a positive change of a coordinate in step 1 be as likely as a negative change.

The scheme can be repeated indefinitely (again, to the limit of available computer time).

The Metropolis Monte Carlo scheme generates a distribution in which each accessible state is represented in proportion to its Boltzmann factor. In principle, it is possible to widely sample conformation space in a shorter time than is possible with molecular dynamics simulation. In practice, *successful* large changes of conformation are extremely rare. The reason is that it is difficult to make a large change in coordinates of many atoms without deforming the geometry or causing intermolecular clashes. Most changes will then cause large increases of the energy (many times $k_B T$), and as a consequence the probability is very small that such a structure will be accepted in the Metropolis scheme.

7.5 SIMULATION OF A CONDENSED SYSTEM

It has long been recognized that an isolated set of a small number of molecules is a poor system with which to simulate a liquid, as many molecules in such a system are at the interface with vacuum and thus theirs is not an environment fully characteristic of the liquid. It turns out that the solution–vacuum boundary can be eliminated by the use of so-called *periodic boundary conditions*; in its simplest case, a rectangular simulation volume is surrounded in all directions by exact replicas. Such a system has no boundaries, at the cost of the introduction of an artificial periodicity (Fig. 7.2).

A major disadvantage of a system with periodic boundaries is that the number of atom pairs is infinite. In early simulations, the infinite sum that represents the

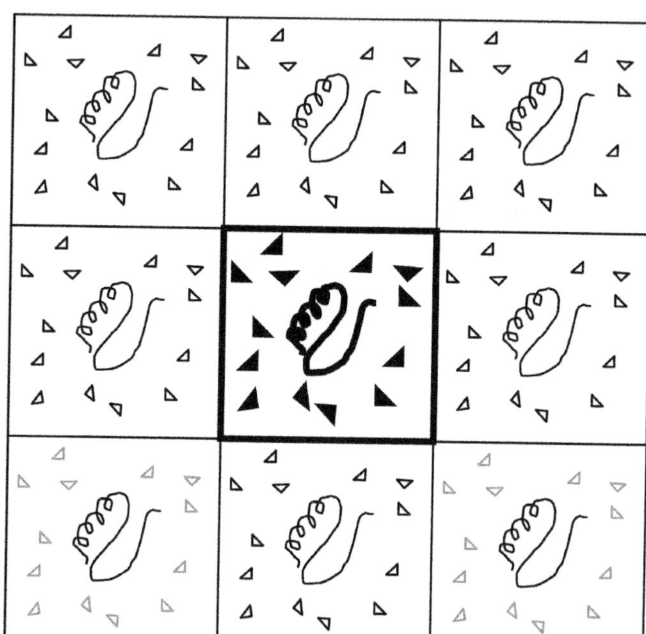

FIGURE 7.2 Simulation of a protein molecule in explicit solvent with periodic boundary conditions. In three dimensions, the central cell is surrounded by a first layer of 26 identically filled cells, and so on, indefinitely.

nonbonded energy in the MM energy function, Eq. 3.25),

$$V_{\text{nonbond}}(\mathbf{X}) = \sum_i \sum_{j>i} \left\{ \frac{1}{4\pi\varepsilon_0} \frac{q_i q_j}{r_{ij}} + 4\varepsilon_{ij} \left[\left(\frac{\sigma_{ij}}{r_{ij}} \right)^{12} - \left(\frac{\sigma_{ij}}{r_{ij}} \right)^{6} \right] \right\}, \tag{7.8}$$

was reduced to a finite sum by truncation, that is, by omitting terms for all pairs at separation greater than a preset cutoff distance. This is not such a bad approximation for terms in $(1/r_{ij})^{12}$ and $(1/r_{ij})^{6}$, but it is a very poor approximation for the term in $1/r_{ij}$, which represents the Coulomb energy, and it was found that this approach gives rise to serious artifacts in systems with polar molecules, that is, systems with large Coulomb energies. (The errors of using cutoffs on Coulomb interactions in simulation of a periodic system containing positive and negative ions are particularly severe.)

Many years ago, the calculation of Coulomb energy and forces for a periodic system was cast in terms of the sum of two rapidly converging functions by Ewald, and relatively fast-computing codes for such "*Ewald summation*" have since been developed. A fast Ewald scheme developed by Darden is used in many simulation programs.[7]

The combination of periodic boundary conditions with Ewald summation is an elegant and practical solution to the problem of representing condensed systems. Of course, the artificially imposed periodicity does lead to artifacts, but these pale in significance relative to the errors that are avoided by using these techniques.

In practice, Ewald summation requires more computation than does a direct summation with a typical cutoff. However, the extra cost of using Ewald summation can be greatly reduced by the use of a so-called *multiple time step scheme*: forces that change rapidly (such as bond stretch forces) are computed frequently, and forces that change slowly (such as forces between well-separated electrostatic charges) are computed only every few time steps. It turns out that this can be accomplished by a fairly simple adaptation of the standard integration scheme of Eq. 7.3.[8]

7.6 CONNECTING MICROSCOPIC AND MACROSCOPIC SYSTEM PROPERTIES

In the first part of this chapter, we have briefly outlined how molecular simulations are carried out. As discussed earlier (Eq. 7.1), a simulation will produce a sample of system configurations according to a Boltzmann distribution, and system properties can be averaged over this sample. We now consider the crucial step of connecting properties that can be calculated for the simulated molecular system, with measurable properties of the corresponding real molecular system, and it is here that results from statistical mechanics obtained in Chapters 5 and 6 prove invaluable.

As a first, rather trivial, example, one can estimate the system energy, E with

$$\langle E\rangle = \langle E_{\text{pot}}\rangle + \langle E_{\text{kin}}\rangle, \tag{7.9}$$

where $\langle E_{\text{kin}}\rangle$ and $\langle E_{\text{pot}}\rangle$ are potential and kinetic energy, respectively, averaged over the time course of the simulation. The former is calculated with the energy function used as the basis for the simulation and the latter with

$$E_{\text{kin},i} = \sum_{j=1}^{N} \frac{1}{2} m_j v_{j,i}^2, \tag{7.10}$$

with m_j the atomic mass of atom j and $v_{j,i}$ its velocity in structure i (N atoms).

A value of the specific heat, C_V, can be estimated from the energies obtained in a series of simulations at different temperatures. A more sophisticated approach uses the results of a single trajectory and estimates C_V from the fluctuation of E, measured as the mean square deviation (msd), with Eq. (6.34), that is, according to

$$C_V = \frac{\text{var}(E)}{k_B T^2} \approx \frac{\text{msd}(E)}{k_B T^2}. \tag{7.11}$$

One thermodynamic function that cannot be estimated from simulations is the free energy, A. The free energy is defined in terms of the integral of the Boltzmann factor over the entire phase or configuration space (the partition function) and not as an ensemble average, and, as mentioned, this integral cannot be computed in a simulation.[9] Consequently, the entropy also is not accessible by this approach.

Given a dynamics trajectory of a sample of N molecules of a liquid, a value of the *diffusion coefficient* can be estimated from a knowledge that diffusive motion results in molecular displacements, s, that increase with time according to

$$\langle s^2\rangle = 6Dt, \tag{7.12}$$

so that the diffusion coefficient, D, can be obtained from a plot of mean square displacement versus simulation time, t, which is found to be nonlinear at short times, but becomes linear at longer simulation times (Fig. 20.3).

Simulation temperature in a system containing N particles is determined by the kinetic energy. Each translational degree of freedom contributes $\frac{1}{2}k_B T$ to the kinetic energy, so that one has for a single particle (see Section 5.4.7)

$$\langle E_{\text{kin}}\rangle_1 = \frac{1}{2} m\langle v^2\rangle = \frac{3}{2} k_B T, \tag{7.13}$$

for motion in three dimensions, where $\langle v^2\rangle$ is the mean thermal velocity. For motion in a single dimension

$$\langle v_x^2\rangle = \frac{k_B T}{m}. \tag{7.14}$$

One also has for a system of many particles that

$$T = \frac{2}{3}\frac{\langle E_{\mathrm{kin}}\rangle_N}{Nk_{\mathrm{B}}}. \tag{7.15}$$

One may use the average of either the molecular or the atomic kinetic energy, with corresponding values of N for number of atoms or number of molecules. As the kinetic energy, E_{kin} is known for each configuration, one can use this equation to compute a value of the temperature for each configuration.

Equation 7.15 can be used also as the basis for a practical mechanism to control the temperature of a simulated system, by gradually increasing all velocities if the temperature is too low or decreasing them if it is too high. Often, E rises steadily as a result of imperfections in the calculation of the potential energy (such as use of a finite cutoff on nonbonded interactions; see preceding text). In that case, it is necessary to use a scheme that artificially maintains the kinetic energy at the correct value or at least approximately at the correct value. For example, the "Berendsen thermostat" at each time step multiplies each of the 3N velocities, v_{ij} by the same (small) factor determined by the ratio between the actual and the desired kinetic energy. The new value of each velocity component, $v_{ij}{}'$ is determined by the following equation

$$v_{ij}{}' = v_{ij}\left[1 + \frac{\delta t}{\tau}\left(\frac{\frac{3}{2}Nk_{\mathrm{B}}T}{E_{\mathrm{kin}}} - 1\right)\right]^{1/2}, \tag{7.16}$$

where τ is a "relaxation time" chosen by the user of the simulation program, that determines the magnitude of the fluctuations of the kinetic energy.

Another approach to maintaining a constant mean temperature (constant mean kinetic energy) is by the use of a combination of friction and random forces according to the Langevin equation, as discussed in Chapter 20 (Eq. 20.26).

7.7 AN EXAMPLE: DYNAMICS OF Ace-Ala-Nme IN SOLUTION

We illustrate simulation results by analyzing the time dependence of several internal coordinates in a simulation of a small peptide (Figs. 7.3 and 7.4). The simulated molecule (Fig. 7.3) is best known by its nonsystematic name, alanine dipeptide. It is properly called *N*-acetyl-alanyl-*N*-methyl amide, abbreviated as Ace-Ala-Nme and often is more loosely described as terminally blocked alanine.

The structure of the alanine dipeptide possesses flexibility as a result of relatively unhindered internal rotation about the bonds marked φ and ψ (symbols that we shall use for the torsion angles). The amide groups at each end have a planar equilibrium structure and are not very flexible. This molecule has been much studied in simulations as a model of the polypeptide backbone, ever since the pioneering work of Ramachandran in the 1960s.

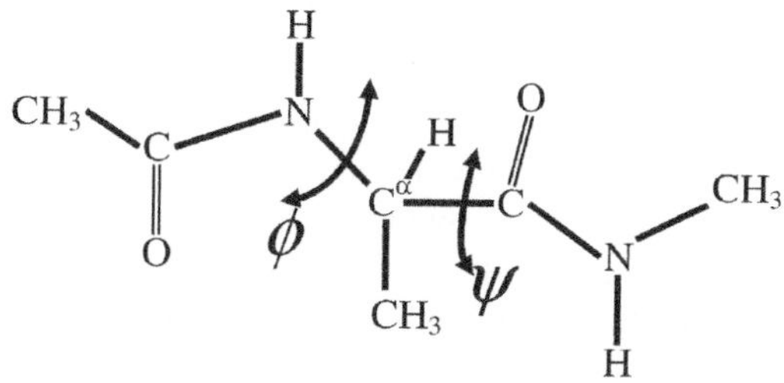

FIGURE 7.3 Structure of Ace-Ala-Nme ("alanine dipeptide") with principal torsion angles φ and ψ.

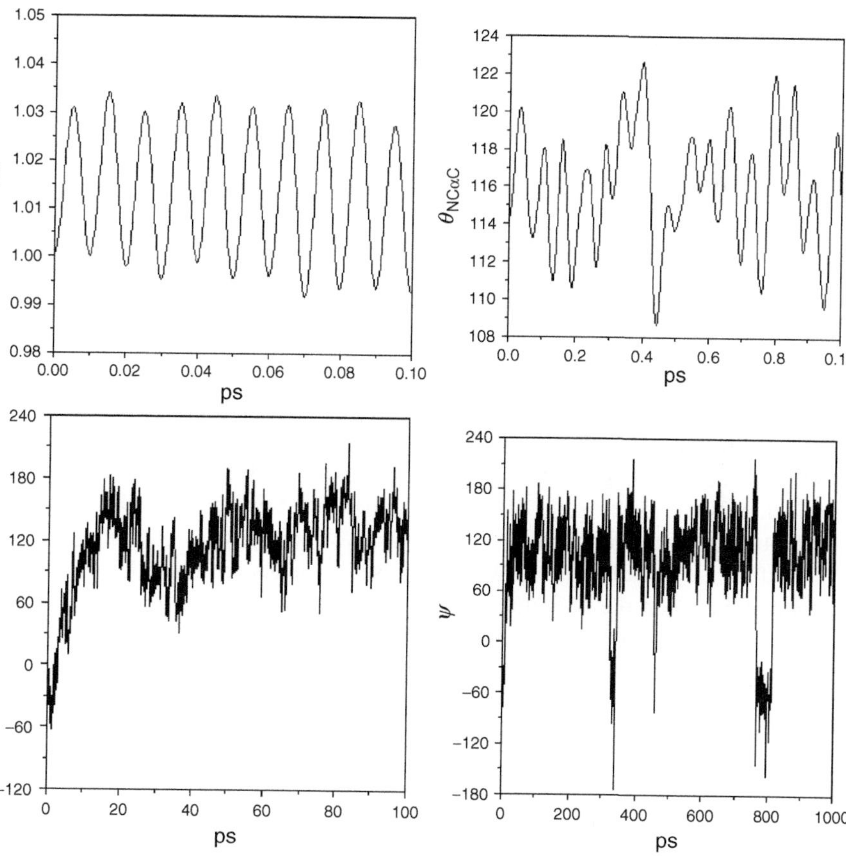

FIGURE 7.4 Variation with time of three internal coordinates of Ace-Ala-Nme (alanine dipeptide).

We discuss here some of the results of a 1 ns long simulation of the alanine dipeptide in a box of explicit water molecules (modeled as SPC; see Section 4.7) with periodic boundary conditions, using graphs of time series of three internal coordinates, one of the two N–H bond lengths, the N–C^{α}–C bond angle and the torsion angle ψ (Fig. 7.4). To begin with, the behavior of the N–H bond length is highly dominated by an oscillation according to a single frequency of 100 ps^{-1}, that is, with a period of 10 fs, although neither the amplitude of the motion nor the midpoint of the oscillations is precisely constant, presumably in response to other motions of the solute and solvent. The equilibrium value of the bond length is circa 1.015 Å, slightly more than the ideal value of this bond length of 1.00 Å, due to stretching as a result of forming a hydrogen bond to water. Next, the variation of the bond angle is dominated, although not nearly as strongly, by an oscillation with a frequency of approximately 16 ps^{-1}, but motions with other frequencies contribute significantly to the deviations from the mean value of circa 115°.

In the first approximation, amplitude and frequency of these dominant high frequency motions are related to amplitude and frequency of corresponding normal modes (see Appendix F); one sees, for example, that the frequency of N–H stretching is much larger than that of bending the bond angle, in agreement with predictions from the normal mode treatment, because of the difference in force constant (Tables 3.1 and 3.2) and in reduced mass (Eq. 2.11). However, normal mode vibrations are defined for systems in which the potential energy is a strictly quadratic function of the coordinates, which is not the case for the potential used in this and most other simulations.

At first glance, the variation of the torsion angle, ψ, appears to behave similarly, with lower frequency and larger amplitude. However, if the simulation is performed for a long time (1 ns), relatively sharp transitions appear between *two stable states* having distinctly different values of the coordinate, namely, around 120° and around −60°. We denote these states as β and α, respectively. The results of the 1 ns simulation (lower right panel of Fig. 7.4) allow us to estimate equilibrium and rate constants for the transition from α to β as approximately $K_{\alpha\beta} = 0.1$, $k_{\beta\rightarrow\alpha} = 4\times10^{9}$ s^{-1}, and $k_{\beta\rightarrow\alpha} = 4\times10^{10}$ s^{-1}. (More precise estimates require a larger number of transitions between the two states, that is, a longer simulation.)

Experimental studies have only recently been able to shed light on the conformation of the alanine residue in short peptides in solution.[10] There appears to be a consensus for a propensity for PPII-type structures [with dihedral angles (φ, ψ) near (−75°,145°)], for beta-type structures [(φ, ψ) near −125°,120°)], and α-helix-type conformations [(φ, ψ) near (−60°,−60°)], and that this third conformation accounts for 20% of the total population in Ala-Ala-Ala; thus, the observation in this simulation of the alanine dipeptide that 10% of the structures have $\psi < 0$, is not unreasonable. Corresponding experimental kinetic results appear unavailable.

This example shows that under the right circumstances, simulations can give thermodynamic (and kinetic) information on conformation changes, the conditions being that the equilibrium constant must not be too far removed from unity and the rate constant large enough to observe several transitions in the time allotted for the simulation. However, there are many interesting cases in which the transitions are much slower and not even a single transition is likely to occur in the time interval achievable in a simulation. Similarly, there are many situations in which the transition is unfavorable, something that will make the occurrence of a transition even less likely. In order to study these cases, more sophisticated simulation methods have been developed; in these methods the transitions are forced.

7.8 FORCED TRANSITIONS

7.8.1 Potential of Mean Force

Potential of mean force (PMF) is simply another term for a free energy profile along a reaction coordinate. We can explain the origin of the term PMF by referring to a basic principle of mechanics, according to which a potential, $V(\mathbf{X})$ (a scalar) is related to a force, $\mathbf{F}$ (a vector), the force being equal to the negative gradient (a vector) of the potential

$$\mathbf{F} = -\nabla V(\mathbf{X}). \tag{7.17}$$

For a one-dimensional problem

$$F_x = -\frac{\partial V}{\partial x}. \tag{7.18}$$

In a molecular system (real or simulated), atomic and molecular forces fluctuate as a result of thermal motion. In the absence of a potential gradient, the force will fluctuate about a mean value of zero. For a system experiencing a nonzero mean force the value of the potential can be recovered from measurements of the mean force at many different values of x if one numerically integrates the force,

$$V(x) \approx -\int \langle F(x)\rangle dx. \tag{7.19}$$

Hence, we call this a *potential of mean force*.

If done very slowly, such a process approximates a "reversible" process (or a "quasi-static" process), and, according to the basic principles of thermodynamics, the work in a reversible, *isothermal* process is equal to the free energy change

$$\Delta A \approx \int \langle F(\lambda)\rangle d\lambda, \tag{7.20}$$

where λ represents a generalized reaction coordinate and F the (external) force along it.

Four main techniques have been widely used for computing potentials of mean force in simulations, and these are discussed in Appendix D.

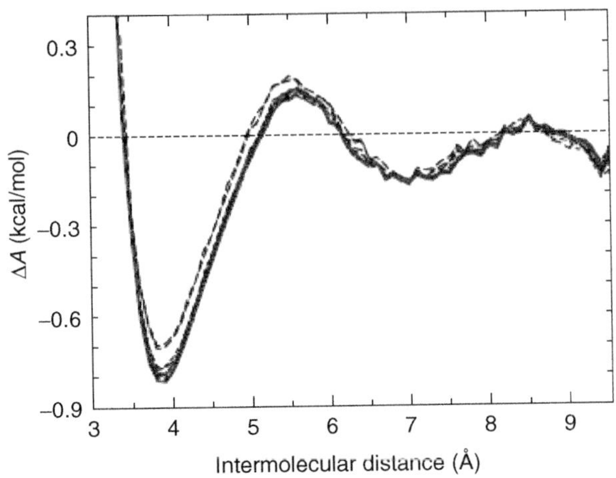

FIGURE 7.5 Potential of mean force of a pair of methane molecules in water, showing convergence with increasing simulation time. From Czaplewski, C., Rodziewicz-Motowidlo, S., Liwo, A., Ripoll, D.R., Wawak, R.J., Scheraga, *Protein Sci.* **9**: 1235–1245 (2000).

7.8.2 PMF for Molecular Association–Dissociation

As a first example, we take a system containing two interacting solute molecules, and seek to determine the free energy for bringing these molecules into close proximity. This free energy is, in effect, the PMF for a reaction coordinate along the intermolecular distance, r. The system chosen for this example consists of two methane molecules in solution in water, and has been simulated with a united-atom potential for methane,[11] explicit water molecules modeled with the TIP3P potential and periodic boundary conditions.[12] Results are shown in Fig. 7.5. One sees that the curve is not perfectly smooth: just as with physical experiments, simulations are subject to statistical error, which can be systematically reduced by using longer simulations, or by averaging over a large number of independent simulations.

A free energy minimum at a separation of 4 Å is evidence of a significant attraction between the two methane molecules. At distances between 4 and 8 Å, the two modeled methane molecules in water are effectively attracted in this simulation, while at close separation the molecules repel one another owing to the $1/r^{12}$ term in the intermolecular potential energy. The model produces a clear representation of the *hydrophobic effect*. (This example has also been discussed in Chapter 4.) The last section of this chapter shows how these model data can be used to estimate a predicted association equilibrium constant for the interaction of two methane molecules.

7.8.3 PMF for Conformation Change

In this example, we determine the free energy for changing the conformation of the alanine dipeptide between two stable conformation states, α and β. This free

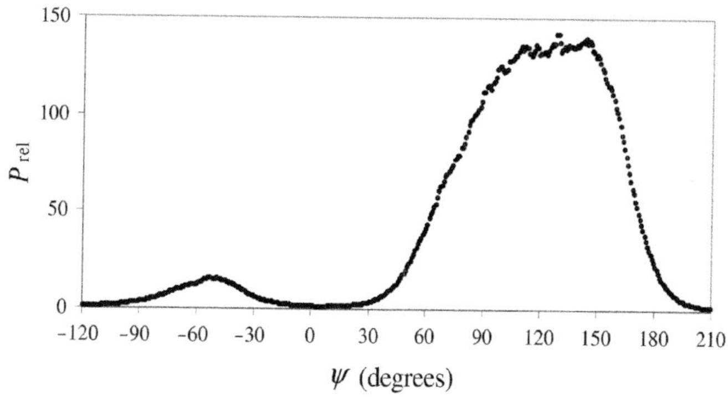

FIGURE 7.6 Conformational (relative) probability distribution based on the potential of mean force for torsion of the alanine dipeptide, Ace-Ala-Nme about the C^{α}—C bond of the alanine residue, from simulation. The small peak corresponds to the α-conformation and the large peak to the β-conformation.

energy is, in effect, the PMF for a reaction coordinate along the torsion angle ψ. The PMF has been computed for the same energy function that was used in the free simulations discussed in Section 7.7; the corresponding probability distribution is shown in Fig. 7.6. A minimum of the probability P (i.e., a transition state) occurs at 10° and another, at a much lower value of P, between -150 and $-120°$. Summing the probability over each of the two ranges separated by the minima gives the relative probabilities of the two conformations, and thus an estimate of the equilibrium constant $K_{\alpha\beta} = 20$. This is in good agreement with the value of 10 determined in a free simulation, which was based on barely adequate statistics (Section 7.7).

7.8.4 PMF for Formation of a Cavity in Water

An early application of the SPC water potential was a determination of the free energy for formation of a cavity as a PMF.[13] The cavity was formed by introducing, into a system of water molecules, a "cavity particle" X that interacted with the water molecules only via a repulsive force between it and the oxygen atoms, according to

$$V_{X,O} = B_{X,O}/r_{X,O}^{12}. \tag{7.21}$$

The value of $r_{X,O}$ at which $V_{X,O} = k_B T$ is taken as the thermal or effective radius, R_{eff} of the cavity. Simulations were done over a range of values of effective radius, by varying the value of the reaction coordinate, $B_{X,O}$. Results, recalculated with a larger system and longer simulations, are shown in Fig. 7.7. The free energy for growing the cavity is closely proportional to the square of the effective radius, that is, to the effective cavity surface; from the value of the slope follows an estimate of the surface tension of 63 cal/(mol Å^2), somewhat less than the experimental value of 79. In the solvent shell lining the cavity, the water molecules were found

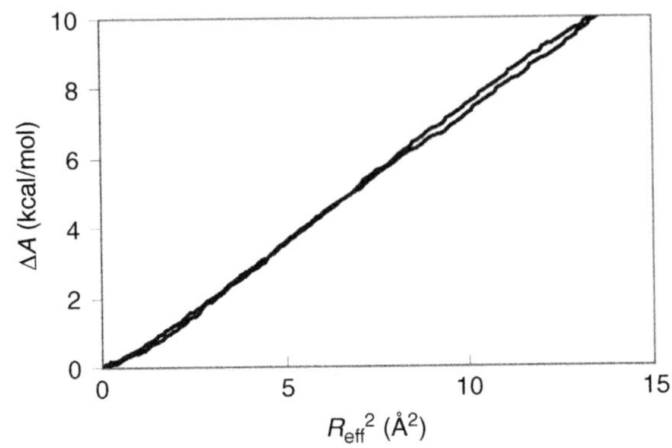

FIGURE 7.7 Potential of mean force for formation of a cavity in water, from molecular dynamics simulations. Duplicate traces represent results calculated with increasing and decreasing radius.[13]

to prefer orientations compatible with formation of hydrogen bonds with other water molecules, in that OH vectors tended to avoid directions pointing toward the center of the cavity, in agreement with what had been found in simulations of Ne (Figs. 4.7 and 4.8 and related discussion).

7.9 POTENTIAL OF MEAN FORCE FOR CHANGES OF CHEMISTRY: "COMPUTER ALCHEMY"

7.9.1 Free Energy of Transfer from Vacuum to Water

The preceding section gave examples of PMFs for displacement along geometric reaction coordinates: an intermolecular distance, an angle for torsion about a chemical bond and the radius of a cavity. Clearly, this approach can be extended to any reaction coordinate that can be defined in terms of molecular geometry. In practice, these conformation changes are accomplished by a *perturbation of the energy function* in order to force the conformation to adequately sample all states between the two endpoints. The perturbation can be a force that maintains the reaction coordinate at specified fixed values, or an umbrella potential that restrains the reaction coordinate to a particular interval, and is an addition to the potential energy (see Appendix D).

We can now introduce an important generalization of the concept of PMF to allow any arbitrary change of the potential function used in the simulation, *including* a change of the energy function that describes the molecular physics. As physics and chemistry of a simulated system are determined by the form and parameters of the energy function used in the simulation (and, of course, temperature and volume), a simulation in which one systematically changes the potential function can give an estimate of the PMF for a change of the system's physics or chemistry.

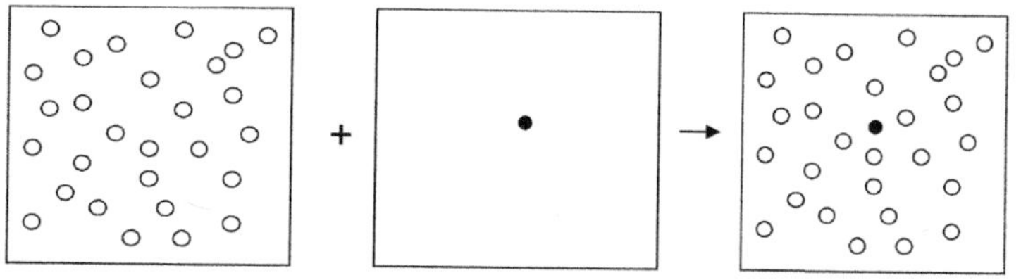

FIGURE 7.8 The process of transferring a solute molecule (•)from vacuum to solvent (◦).

As a first example, we describe how the *free energy for transfer of a solute from vacuum to solvent* (solvation free energy; Fig. 7.8) can be calculated with simulations of a system in which the interactions between solute and solvent are controlled, and allowed to vary between nonexistent and fully effective interactions.

It is simplest to describe how this calculation is done for a typical MM energy function, which is largely, although not necessarily entirely, a sum of many *local* terms. However, the approach can be applied with some modifications also when the energy is computed via QM. The system contains one solute molecule and an adequate number of solvent molecules. The potential energy function of this system contains a sum of terms for intramolecular solvent and intermolecular solvent–solvent interactions, V_{00}, another sum for intramolecular solute energy, V_{11} and, finally, a term for solute–solvent interactions, V_{01}. The total energy function is the sum of these three terms,

$$V = V_{00} + V_{01} + V_{11}. \tag{7.22}$$

If we now transfer the solute to vacuum, the energy *functions* for solvent and solute remain the same, while the energy function for the solute–solvent term, V_{01} is now equal to 0. Thus, by modifying the potential energy by multiplying the term for solute–solvent interaction with a coupling parameter, λ, one obtains *a new energy function that can describe either system,* (the solution when $\lambda = 1$, and the separated systems when $\lambda = 0$),

$$V = V_{00} + \lambda V_{01} + V_{11}. \tag{7.23}$$

Furthermore, we have written the energy as a continuous function of λ, so that we can obtain the change in free energy due to the change of λ. The transfer free energy of the solute from vacuum to solution is then given by

$$\Delta A_{\text{transfer}} = \int_0^1 \frac{\partial A}{\partial \lambda} d\lambda = \int_0^1 \langle V_{01} \rangle d\lambda. \tag{7.24}$$

This shows that the free energy is calculated as a PMF, where the force results from a perturbation of the energy function; the one difference is that the perturbation consists of a change of the potential energy function that would be normally used for the simulation, rather than the addition of a term to control the conformation.

As an example, we report duplicate *estimates* of free energy of *transfer of methane from vacuum to water* ($\Delta A_{\mathrm{v}\to\mathrm{w}} = 3.12$ and 3.20 kcal/mol) and from water to vacuum ($\Delta A_{\mathrm{w}\to\mathrm{v}} = -3.06$ and -3.18 kcal/mol) from simulations in which the transfer was accomplished in 200 ps with the slow-growth method described in Appendix D. The small difference between duplicate simulations indicates a rather good precision, as does the near-equality of the magnitude of $\Delta A_{\mathrm{v}\to\mathrm{w}}$ and $\Delta A_{\mathrm{w}\to\mathrm{v}}$.[14]

As a closely related example, we mention a Monte Carlo simulation of the transfer of a water molecule at 298 K from liquid water to vacuum, in which the free energy of transfer was calculated with the thermodynamic perturbation method described in Appendix D.[15] The calculated Gibbs free energy of transfer from water to vacuum was 6.1 kcal/mol, and compares well with the experimental value $\Delta G_{\mathrm{w}\to\mathrm{v}} = 6.3$ kcal/mol. (Similar good agreement between theory and experiment has been found for the SPC and TIP3P models.)

7.9.2 Molecular Transformation of Ligand Bound to a Macromolecule

As a second case, we consider the example of transforming a ligand molecule bound to a protein in solution (Fig. 7.9). Call the starting state ligand $\mathrm{L_A}$, and the final state ligand $\mathrm{L_B}$. In this case, one splits the energy function into six parts, for solvent V_{00}; for ligand V_{11}; for protein V_{22}; and cross terms V_{01}, V_{02}, and V_{12},

$$V = V_{00} + V_{11} + V_{22} + V_{01} + V_{02} + V_{12}. \tag{7.25}$$

The form of only three of these, V_{01}, V_{11}, and V_{12}, depends on the chemistry of the ligand molecule, and if one now writes these as follows

$$V_{01} + V_{11} + V_{12} = (1-\lambda)(V_{01} + V_{11} + V_{12})_{\mathrm{L_A}} + \lambda(V_{01} + V_{11} + V_{12})_{\mathrm{L_B}}, \tag{7.26}$$

one has a potential function that describes the complex with ligand $\mathrm{L_A}$ when $\lambda = 0$, and the complex with ligand $\mathrm{L_B}$ when $\lambda = 1$. The function is computable and continuous for intermediate values of λ, although the corresponding intermediate states are not *physically* realizable, that is, there exist no molecules in nature that are part ligand A and part ligand B.

The free energy of transforming the complex is then obtained with

$$\begin{aligned}\Delta A^{\mathrm{c}}_{\mathrm{L_A}\to\mathrm{L_B}} &= \int_0^1 \frac{\partial A}{\partial\lambda} d\lambda = \int_0^1 \left\langle \frac{\partial V}{\partial \lambda}\right\rangle d\lambda \\ &= \int_0^1 -(V_{01} + V_{11} + V_{12})_{\mathrm{L_A}} + (V_{01} + V_{11} + V_{12})_{\mathrm{L_B}} d\lambda. \end{aligned}\tag{7.27}$$

An analogous expression gives the free energy of transforming the ligand molecule free in solution by simulation of the same system but without a protein

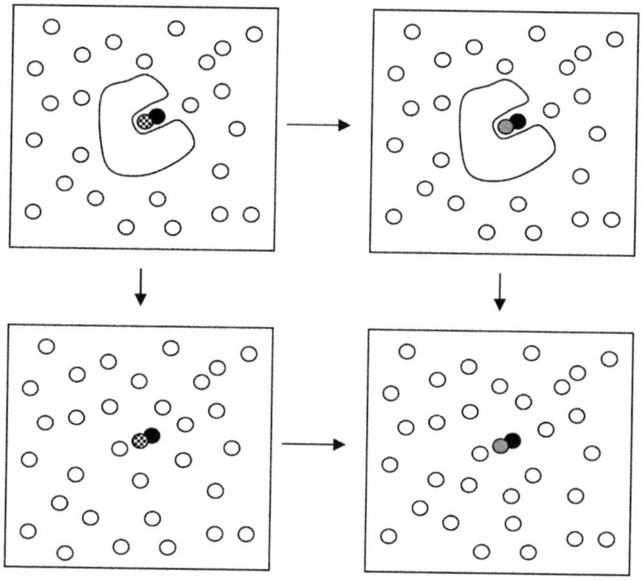

FIGURE 7.9 The difference in affinity of two different ligands is studied in terms of two processes (horizontal arrows), one for changing the chemistry of the ligand molecule with the ligand bound to the macromolecule (top) and the other for changing the ligand free in solution (bottom). The processes on the left and right (vertical arrows) correspond to the binding of the ligands to the macromolecule.

molecule,

$$\Delta A^{\mathrm{s}}_{\mathrm{L_A}\to\mathrm{L_B}} = \int_0^1 \langle -(V_{01}+V_{11})_{\mathrm{L_A}} + (V_{01}+V_{11})_{\mathrm{L_B}} d\lambda \rangle . \tag{7.28}$$

Figure 7.9 illustrates the two processes for which the free energy differences are estimated by simulations according to Eqs. 7.27 and 7.28.

These two free energy differences cannot be measured experimentally. Instead, measurements of the affinities of the two ligands for the protein can give the (standard) free energies for the binding of the two ligands to the protein, $\Delta A^{\circ}_{\mathrm{LA}}$ for the reaction

$$\mathrm{P} + \mathrm{L_A} \to \mathrm{PL_A},$$

and $\Delta A^{\circ}_{\mathrm{LB}}$ for the reaction

$$\mathrm{P} + \mathrm{L_B} \to \mathrm{PL_B}.$$

Each of these two processes can be represented in terms of the reaction illustrated in Fig. 7.10.

The relationships between the two free energies of binding and the two free energies obtained by the simulations are evident if one considers the following

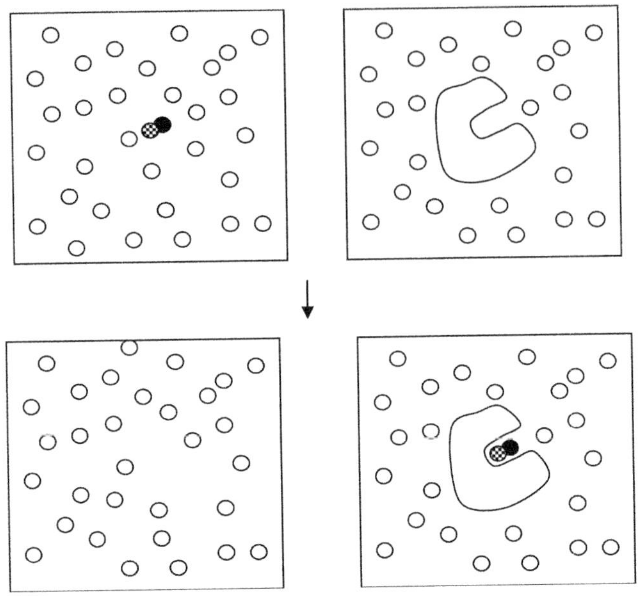

FIGURE 7.10 The complete binding process of one kind of ligand (vertical arrow) involves removing the ligand from water and inserting it in the binding pocket, and is illustrated here in terms of four systems: ligand in water, ligand-free protein in water, water without solute, and protein–ligand complex in water.

thermodynamic cycle

$$\begin{array}{ccc} & \Delta A^{\mathrm{s}}_{\mathrm{L_A}\to\mathrm{L_B}} & \\ \mathrm{P}+\mathrm{L_A} & \leftrightarrow & \mathrm{P}+\mathrm{L_B} \\ \Delta A^{\circ}_{\mathrm{LA}}\ \downarrow\uparrow & & \downarrow\uparrow\ \Delta A^{\circ}_{\mathrm{LB}} \\ \mathrm{PL_A} & \leftrightarrow & \mathrm{PL_B} \\ & \Delta A^{\mathrm{c}}_{\mathrm{PL_A}\to\mathrm{PL_B}} & \end{array} \tag{7.29}$$

for which the four free energy differences are related according to

$$\Delta A^{\mathrm{s}}_{\mathrm{L_A}\to\mathrm{L_B}} + \Delta A^{\circ}_{\mathrm{L_B}} - \Delta A^{\mathrm{c}}_{\mathrm{L_A}\to\mathrm{L_B}} - \Delta A^{\circ}_{\mathrm{L_A}} = 0. \tag{7.30}$$

This can be rearranged to equate the difference between the computable transformation free energies of the two types of ligand molecule (left of the equal sign) and the measurable difference in standard free energy of binding (right of the equal sign),

$$\Delta A^{\mathrm{s}}_{\mathrm{L_A}\to\mathrm{L_B}} - \Delta A^{\mathrm{c}}_{\mathrm{L_A}\to\mathrm{L_B}} = \Delta A^{\circ}_{\mathrm{L_A}} - \Delta A^{\circ}_{\mathrm{L_B}}. \tag{7.31}$$

Note that two of the four systems shown in Fig. 7.10 are not represented in the simulations (ligand-free protein in water and pure water). Hence, these simulations, which do not involve all four systems, cannot yield estimates of the free energy of binding. However, these two systems do not contain the ligand, and hence their contributions cancel when the results of two simulations are used to estimate the *difference* of the binding free energies for different ligands (see Eq. 7.32).

Excellent *examples* are given by the first two published articles to use this approach.[16] Both articles address the same problem, namely, which of two stereoisomers, S or R of an inhibitor of HIV protease binds more tightly. The reason for doing these simulations was created by a crystallographic structure determination of a complex of the protease formed from a racemic mixture of the inhibitor, in which the stereochemistry was not adequately resolved. The articles produce almost identical values of the affinity difference. The second cited simulation study gave 3.95 $\pm$ 0.3 kcal/mol for transformation in the complex and 1.05 $\pm$ 0.1 for transformation in the solution, which then gave with Eq. 7.32 a value of 2.9 kcal/mol for the difference in standard free energies of binding. Subsequently, solution experiments gave a value of 2.6 kcal/mol. The agreement with experiment is quite good, and it was at one time thought that application of this simulation method had promise for the design of enzyme inhibitors as drugs. However, the rather long simulation times (on the order of nanoseconds) needed to obtain converged estimates have proved an obstacle to widespread application.

In Chapter 18 we discuss how these computer alchemy calculations are applied to the problem of estimating rates of enzymatic reactions from first principles, with a high level QM treatment of reactive site and substrate, while the remainder of the enzyme and the solvent are treated with an MM potential.

7.10 THE POTENTIAL OF MEAN FORCE AND THE ASSOCIATION EQUILIBRIUM CONSTANT OF METHANE

We conclude this chapter with another example of how to connect microscopic and macroscopic molecular properties. We consider briefly how the equilibrium constant for dimerization of methane in water (as represented by the force field used in the molecular dynamics simulation) can be estimated from a PMF (as in Fig. 7.5). The principles have been laid out in the previous chapter (Section 6.4). Differences are, first, that in this case the interaction of two molecules is described in terms of a *free* energy (PMF, Fig. 7.5), and, second, that the interaction is between molecules of the same species.

We shall base the treatment on Eq. 6.25 which relates the number of AB dimers per A monomer for a given concentration of B. Here, the free energy simply replaces the energy, E_{AB}, the concentration of A replaces that of B, and, to make the equation give the number of AA dimers per monomer of A, division by a factor of 2 is needed to avoid double counting of dimers. This gives

$$N_{\mathrm{AA}}/N_{\mathrm{A}} = \frac{N_{\mathrm{A}}}{2V}\int_V 4\pi r_{\mathrm{AA}}^2 ds_{\mathrm{AA}}\{\exp[-\beta A_{\mathrm{AA}}(r_{\mathrm{AA}})] - 1\}$$
$$= K_b\frac{N_{\mathrm{A}}}{V}. \tag{7.32}$$

The derivation of Eq. 6.25 made use of the fact that molecular interaction energies are short-ranged, and A_{AA} is zero at large distances. Accordingly, we must shift the entire PMF so that its value is zero at large separation (as has already been done for the data in Fig. 7.5).

When this equation is applied to the data of Fig. 7.5, the resulting association equilibrium constant of methane in water (for the MM model) turns out to be circa 6 L/mol ($K_{\mathrm{d}} = 160$ mM; $\Delta A^{\circ} = -1$ kcal/mol relative to a standard state of 1 M), which is of course very weak by the standards of powerfully associated biological macromolecules, whose affinity is based on a great many relatively weak interactions acting in concert. We are unaware of any experimental measurement of this equilibrium constant.

We note here that when one says that two molecules bind to one another, this provides no insight into the mode of the interaction. All one can say is that the two molecules are likely to be near one another. From the PMF shown in Fig. 7.5, it is clear that a higher probability exists to find the second molecule in the range from, say 3 to 12 Å than at very long distances (where the free energy is slightly higher) or at very short distances (where the free energy is much higher). Observable binding is the result of a favorable balance of these two effects; by contrast, an unfavorable balance would result in observable exclusion. This statement anticipates the discussion of the equilibrium dialysis experiment in Chapter 8, in which evidence of binding of small molecule L to macromolecule M is the simple observation of excess L in the compartment containing M (Fig. 8.3), and where it is seen how the equilibrium constant is calculated from a series of quantitative observations of this excess.

NOTES

1. We use "configuration" to refer to the collective coordinates of all atoms in a *system* containing one or more molecules, and "conformation" for the structure of a single molecule.
2. Here we use a notation common in physics: $\dot{x} = dx/dt$.
3. Yang, W., Lee, T.-S. A density-matrix divide-and-conquer approach for electronic structure calculations of large molecules. *J. Chem. Phys.* **103**: 5674–5678 (1995).
4. Car, R., Parrinello, M. Unified approach for molecular dynamics and density-functional theory. *Phys. Rev. Lett.* **55**: 2471–2474 (1985).
5. Metropolis, N., Rosenbluth, A. W., Rosenbluth, N. M., Teller, A. H., Teller, E. Equation of state calculations by fast computing machines. *J. Chem. Phys.* **21**: 1087–1092 (1953). Edward Teller is the father of the hydrogen bomb.

6. See Appendix AI.8 for a discussion of random number generators.
7. Darden, T.A., York, D.M., Pedersen, L.G. Particle mesh Ewald: An *N*.log(*N*) method for Ewald sums in large systems. *J. Chem. Phys.* **98**: 10089–10092 (1993). For a description of the Ewald method, see, for example, the book by Leach cited at the end of this chapter.
8. Tuckerman, M.E., Berne, B.J., Martyna, G.J. Reversible multiple time scale molecular dynamics. *J. Chem. Phys.* **97**: 1990–2001 (1992)
9. Numerical integration of a multidimensional function, $\int \ldots \int \exp[-\beta E(\mathbf{X})] d\mathbf{X}$ is not feasible for technical reasons.
10. See, for example, Schweitzer-Stenner, R., Eker, F., Huang, Q., Griebenow, K. Dihedral angles of trialanine in D_2O determined by combining FTIR and polarized visible Raman spectroscopy. *J. Am. Chem. Soc.* **123**: 8628–9633 (2001).
11. The methane molecule is represented as a single Lennard-Jones particle.
12. Czaplewski, C., Rodziewicz-Motowidlo, S., Liwo, A., Ripoll, D.R., Wawak, R.J., Scheraga, H.A. Molecular simulation study of cooperativity in hydrophobic association. *Protein Sci.* **9**: 1235–1245 (2000).
13. Postma, J.P.M., Berendsen, H.J.C., Haak, J.R. Thermodynamics of cavity formation in water. *Faraday Symp. Chem. Soc.* **17**: 55–67 (1982).
14. Experiment gives the free energy of transfer of methane from vacuum to water as $\Delta G_{v \to w} = 1.94$ or $\Delta A_{v \to w} = 2.54$ kcal/mol to account for the change in PV. The parameters used in the simulation are from a "first generation" force field. Transfer energies and free energies should be used to improve the accuracy of force field parameters, as was done in the opls force field.
15. Jorgensen, W.L., Blake, J.F., Buckner, J.K. Free energy of TIP4P water and the free energies of hydration of CH_4 and Cl^- from statistical perturbation theory. *J. Chem. Phys.* **129**: 193–200 (1989).
16. (a) Ferguson, D.M., Radmer, R.J., and Kollman, P.A. Determination of the relative binding free energies of peptide inhibitors to the HIV-1 protease. *J. Med. Chem.* **34**, 2654–2659, 1991. (b) Tropsha, A., and Hermans, J. Application of free-energy calculations to HIV protease/inhibitor binding. *Protein Eng.* **5**: 29–33 (1992).

SUGGESTED READING

Allen, P.M., Tildesley, D.J., *Computer Simulation of Liquids*. Oxford: Clarendon Press, 1987.

Leach, A.R., *Molecular Modeling: Principles and Applications*. Essex, England: Addison Wesley, Longman, (1996)

Brooks, C.L., Karplus, M., Pettitt, B.M., *Proteins: A Theoretical Perspective of Dynamics, Structure, and Thermodynamics*. New York: Wiley-Interscience, 1988.

van Gunsteren W.F., Weiner P.K., eds., *Computer Simulation of Biomolecular Systems*. Leiden: Escom, 1989.

Schlick, T., *Molecular Modeling and Simulation*. New York: Springer, 2002.

Part 3

Binding to Macromolecules

Proteins are molecular machines that transduce the free energy of intermolecular interactions (binding) into a variety of other events (such as catalysis, formation or change of structure, force generation, active transport). It is not possible to study the molecular aspects of biology without being well versed in the phenomenon of ligand binding.

With some exceptions, proteins (and probably also other biological macromolecules) do not have unique stable conformations but exist in a dynamic equilibrium between a number of conformations that depends on conditions including the presence of bound ligand molecules. Again with important exceptions, biological macromolecules interact with one another and with small molecules (ligands) to form noncovalent complexes, that can be thought of as stabilized by *numerous* individually relatively weak interactions. This in turn allows for a high degree of *specificity* of the interactions, as the interacting groups on the macromolecule have a specific configuration that is complementary to a particular ligand, but frequently not to a chemically very similar one. Furthermore, the affinity for each ligand species depends on the macromolecule's conformation, and, as we shall see, binding of ligand can be accompanied by a switch of the conformation, in which case the binding and the conformation change are said to be coupled. Finally, often the macromolecule (frequently a complex of like subunit macromolecules) has multiple binding sites and, as we shall see, this can be a cause of unexpected behavior.

In this part of the book, we review first the equilibrium treatment of binding in terms of simple equilibria (Chapter 8). In the following chapter, we discuss binding in terms of thermodynamics; this includes a thermodynamic treatment of

Equilibria and Kinetics of Biological Macromolecules, First Edition. Jan Hermans and Barry Lentz.

molecular interactions that is capable of dealing simultaneously with "binding" and "exclusion" or "crowding". We develop the concept of "free energy of binding" and its relation to the binding isotherm and discuss models and experiments of crowding in macromolecule solutions (Chapter 9). In Chapter 10 we develop the statistical mechanics of macromolecules in ideal solution in terms of a molecular partition function, and extend this to ligand binding in Chapter 11. In Chapter 12 these ideas are applied to the case of interdependent binding and conformational equilibrium and ligand binding and interdependent self-association (the so-called coupled equilibria). This is followed by Chapter 13 on allosteric function, with emphasis on the cooperative binding of oxygen to hemoglobin, on which the theory of coupled equilibria had first been tested (and on which research has continued over the years). Part 3 is completed by Chapter 14 on charge-dependent interactions: binding of ions, interactions between charged side chains, and interactions of charged molecules with solvent, which are all complicated by the relatively long range of these interactions.

8

Binding Equilibria

8.1 INTRODUCTION

This chapter begins with a treatment of one-on-one binding, that is, to a sole site for a particular ligand and reviews approaches to measurement of binding. Following that, we discuss how to approach binding to multiple sites on the same macromolecule for the same ligand species; this section makes the important distinction between analysis in terms of stoichiometric binding (mole of ligand per mole of macromolecule) and analysis in terms of site binding (saturation of individual binding sites). The chapter ends with some recommendations for use of experimental data in terms of specific models.

8.2 SINGLE-SITE MODEL

8.2.1 Formalism

One may write the binding of ligand to a single site on a macromolecule (protein or other) as a bimolecular reaction

$$\mathrm{M} + \mathrm{L} \leftrightarrow \mathrm{ML},$$

and the rules of mass action then imply the existence of an equilibrium constant, the binding or association constant K_b, that is related by the following equation

Equilibria and Kinetics of Biological Macromolecules, First Edition. Jan Hermans and Barry Lentz.

to the activities of the three species (where for the protein species M and ML we have replaced activities with concentrations)

$$K_b = \frac{[\mathrm{ML}]}{[\mathrm{M}]a_\mathrm{L}} = \frac{1}{K_d}. \tag{8.1}$$

In this simple case, K_b is both the stoichiometric binding constant and the site binding constant. (See subsequent text for definitions of these terms.) The dissociation constant (K_d) is for the opposite reaction and is the reciprocal of the association constant. The average fraction of sites on the macromolecule occupied by ligand, or saturation f_b is given by

$$f_b = \frac{[\mathrm{ML}]}{[\mathrm{M}] + [\mathrm{ML}]} = \frac{K_b a_\mathrm{L}}{1 + K_b a_\mathrm{L}}. \tag{8.2}$$

It is important to note that this *formulation* is not changed when a conformation change of the macromolecule accompanies binding of the (single) ligand (although the value of the binding constant will be affected by this).

Many macromolecules (including complexes of identical subunits) have multiple independent binding sites for the same ligand. In that case, the average number of ligand molecules bound per macromolecule ν_L is found by summation of the saturation of all n sites,

$$\nu_\mathrm{L} = \sum_{i=1}^{n} f_{b,i} = n f_b, \tag{8.3}$$

where the second equality is true if the n sites are equivalent.

8.2.2 Plots

Several options are available for plotting binding isotherms. Linear and semilog plots are recommended. The once popular Scatchard plot was a convenient method to estimate binding constants, but has become obsolete now that computer fits are available to everyone. A fourth possibility, the Hill plot, is discussed in the chapter on oxygen binding to hemoglobin (Chapter 13, Section 12).

1. *Linear plot*: This is simply a plot of saturation f_b as a function of the ligand activity, a_L (Fig. 8.1). It is rational to design a series of binding experiments with many measurements at low ligand activity, where the saturation changes rapidly, and fewer at activities at which the binding is almost saturated. In that case, data obtained at low values of a_L tend to be bunched together in a linear plot, and will not equally contribute to the visual fit.
2. *Semilog plot*: In a sensibly designed series of binding experiments, the values of ligand activity might be distributed evenly on a logarithmic rather than a linear scale. It is then preferable to plot saturation or ν_L as a function of the logarithm of ligand activity, as in Fig. 8.2, because the points will be evenly spaced along the abscissa.

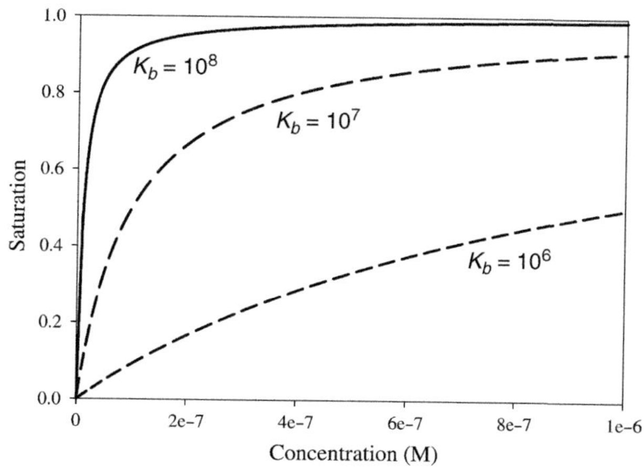

FIGURE 8.1 Saturation of binding to a site (or n equivalent sites) as a function of *free* ligand concentration, [L], for (left to right) $K_b = 10^8$, $K_b = 10^7$, $K_b = 10^6$.

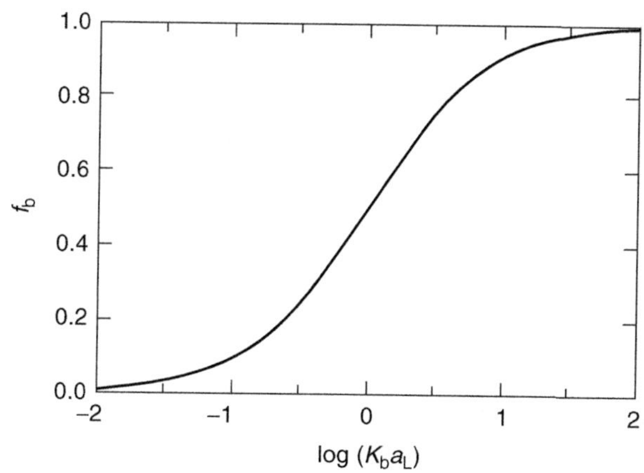

FIGURE 8.2 Saturation of binding to a site as a function of *free* ligand activity, a_L on a semilogarithmic scale.

The semilogarithmic binding curve for binding to a single site has a slope

$$\frac{df_b}{d\ln a_L} = \frac{K_b a_L}{\left(1 + K_b a_L\right)^2}, \tag{8.4}$$

which has a maximum value of 1/4 for $K_b a_L = 1$. (On a $\log_{10}$ scale, as in Fig. 8.2, the slope is slightly less than 1/2.)

3. *Scatchard plot*: Given a series of measurements of f_b as a function of a_L, one can use linearized forms of the single-site equation Eq. 8.1. For example,

the Scatchard form plots f_b/a_L as a function of f_b, and the double reciprocal form plots $1/f_b$ as a function of $1/a_L$. A problem with this approach is that these linearized forms overweight data at the beginning of the titration, which are the most difficult to obtain accurately.

8.3 MEASURING LIGAND ACTIVITY AND SATURATION

When using these (and other equations) for estimating binding constants, one has the quite general practical problem of how to measure the *free* ligand concentration and the fraction of empty (and full) sites. We describe briefly some of the methods that one may use to measure *saturation of binding sites* as a function of *concentration of free ligand.*

8.3.1 Equilibrium Dialysis

Equilibrium dialysis is a venerable method for measuring binding of small molecules to macromolecules. The solution is divided into two compartments separated by a semipermeable membrane, with pores large enough to freely pass ligand and solvent, but not the macromolecule. The concentration of free ligand, c_{free} is (assumed to be) equal in both compartments; this is also the total concentration of ligand in the solvent compartment, $c_{\text{L},2}$. The other compartment contains an additional ligand bound to the macromolecule (Fig. 8.3), so the concentration of bound ligand can be found as the difference of the concentrations in the two compartments.

Disadvantages of such classical equilibrium dialysis experiments are the long equilibration times and rather large amounts of material that they require. The equilibration time can be enormously reduced by using the inside of small beads of

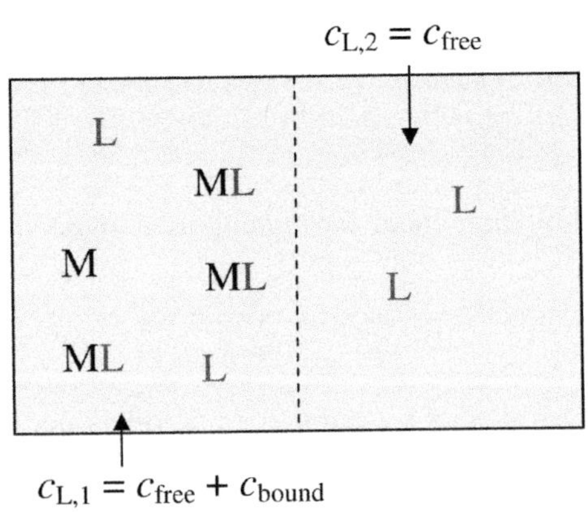

FIGURE 8.3 Measurement of binding with a fixed macromolecule concentration while varying activity of ligand, by equilibrium dialysis.

highly cross-linked hydrophilic polymer (such as Sepharose) as the compartment into which the macromolecule cannot penetrate ("solvent" compartment); more common use of these gel beads is in columns on which to separate macromolecule and bound ligand. Unfortunately, equilibrium dialysis measurements, even with modern microchambers, consume much time and material.

A "perfect" equilibrium dialysis system will be used to advantage in subsequent chapters in which interactions between different kinds of molecules are analyzed with thermodynamic and statistical mechanical theory.

8.3.2 Potentiometric Titration

An alternative setup for accurate binding studies is possible if the concentration of free ligand can be measured in terms of an electric potential. This is, for example, the case for hydrogen ion (and is widely used in measuring pH; Fig. 8.4). Even better, the potentiometric measurements directly report the ligand's activity, rather than its concentration. (While the macromolecule is typically dilute and can be assumed to behave as in an ideal solution, this is not always the case for the ligand. For example, the activity of H^+ ion in 0.1 M HCl is only 0.079 M.)

In the case of potentiometric titration, the amount of bound ligand is found by subtracting the amount of ligand needed for titration of solvent, from that needed for titration of an identical volume of the macromolecule solution, using, in both cases, a relatively concentrated solution of the ligand as titrant.

8.3.3 Indirect Observation

The methods described in the preceding two sections provide the most detailed information about ligand binding to macromolecules because exact values of "ligand bound" and "free ligand" are directly measured (equilibrium dialysis) or are

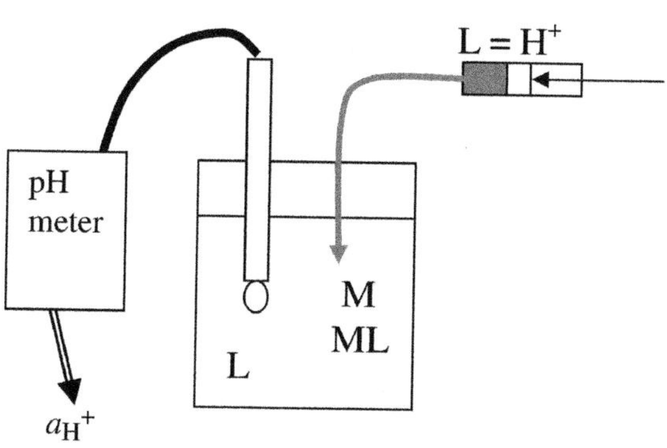

FIGURE 8.4 Measurement of binding with a fixed amount of macromolecule while varying activity of ligand, by potentiometric titration.

available using a reasonable assumption and subtraction. The average number of ligands bound, ν_L, is then found by dividing by the concentration or amount of macromolecule in the sample. The number of independent equivalent sites, binding sites, n is obtained unambiguously at saturation ($f_b = 1 = \nu_L/n$). When saturation cannot be easily reached experimentally, n is still quite well estimated by performance of experiments over a range of macromolecule concentrations and analyzing these with K_b and n as adjustable parameters. The saturation f_b then follows by division by the number of binding sites, n, which will have to be determined separately.

An alternative and very practical way to measure the *saturation of site(s) on the macromolecule* is to use an observable "signal" X that changes in a monotonic and saturable manner with the extent of ligand binding. For example, X can be absorbance or fluorescence intensity of the solution containing a given concentration of protein at a particular wavelength, or the heat evolved or taken up by the sample on mixing solutions of the components.

The only requirements are that the value of X is different for the ligand-free (X_0) and the ligand-bound forms of the protein (X_{sat}) and that it changes linearly with the saturation. With this assumption, the relation between the saturation f_b and value of the observed quantity X becomes

$$f_b = \frac{\nu_L}{n} = \frac{X - X_0}{X_{sat} - X_0} = \frac{K_b a_L}{1 + K_b a_L}$$

$$X = \left(X_{sat} - X_0\right) \frac{K_b a_L}{1 + K_b a_L} + X_0. \tag{8.5}$$

Note that this method of determining the saturation can be combined with measurement of free ligand concentration or activity as in Sections 8.3.1 and 8.3.2. Alternatively, the concentration of free ligand $c_{L,free}$ can be found by subtracting the concentration of bound ligand $c_{L,bound}$ from the total ligand concentration c_L according to

$$c_{L,free} = c_L - c_{L,bound} = c_L - c_M f_b, \tag{8.6}$$

where c_M is the concentration of the macromolecule. This leads to a quadratic expression for $X - X_0$ as a function of c_L that can be easily solved for K_b. This can give very good results, unless one is in a regime where the difference between total and bound concentrations is very small, that is, at low ligand/macromolecule ratio. It is worth noting that, if the concentration of L is large relative to M, then $c_{L,free} \approx c_{L,tot}$ and the analysis becomes much easier.

This is illustrated in Fig. 8.5, where saturation computed according to Eq. 8.2 is plotted as a function of total ligand concentration. If the binding is weak (lower curve), then the data will not be adequate to determine the binding constant K_b with good accuracy, nor the saturating value of X, X_{sat} or the stoichiometry n.

If the affinity is not very low (middle curve), and the concentration of bound ligand cannot be neglected, one can determine the concentration of free ligand, and hence a_L, by relating it to c_M and c_L using the conservation of mass of protein and ligand, resulting in a quadratic form of the binding equation. (This

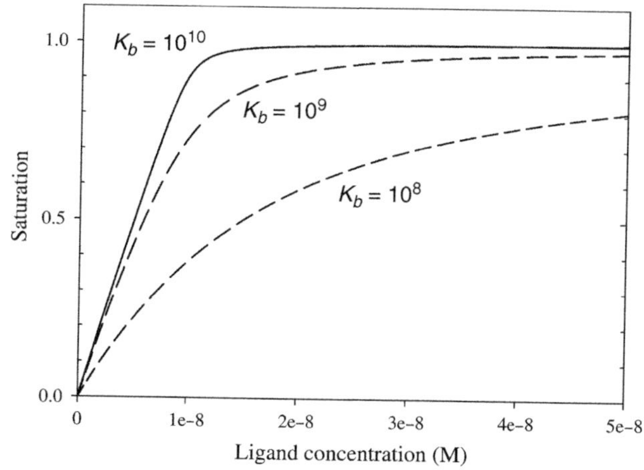

FIGURE 8.5 Saturation of binding to a site ($c_M = 10^{-8}$) as a function of *total* ligand concentration, c_L, for (left to right) $K_b = 10^{10}$, $K_b = 10^9$, $K_b = 10^8$.

is discussed further in Section 8.3.5.) In this instance, nc_M is roughly equal to the dissociation constant $K_d\ (1/K_b)$, an ideal condition for performing a binding experiment, as all three parameters are accessible from a multiple (at least two) experiments performed at somewhat different macromolecule concentrations.

Finally, the top curve in Fig. 8.5 shows what happens if the binding is extremely tight (which, in practice, means that the concentration of macromolecule, c_M is small relative to the dissociation constant K_d). In that case, it may not be possible to determine the concentration of free ligand, as essentially all added ligand is bound to the protein (top curve). In this instance, K_b cannot be determined with accuracy, but n and X_{sat} can be.

We mention that, if multiple sites are present, this method of analysis can yield estimates of all three parameters (K_b, X_{sat}, n) only if the sites are equivalent and change the value of X equally as they are occupied. Otherwise, even for the simple case of multiple independent sites binding the same ligand, the fit of the experimental data requires, besides the protein molarity c_M, the determination of two unknown parameters per site, namely, a signal increment $\delta X_{sat,i}$ and a *site* binding constant K_i for each site. This will be addressed in more detail in Sections 8.4 and 8.5.

8.3.4 Measuring Interactions Between Macromolecules

Neither equilibrium dialysis nor potentiometric titration is a useful technique for studying interactions of two macromolecular species. However, changes of spectra or heat exchange on mixing can be applied quite well to study interacting macromolecules. Interaction of macromolecules can also be studied uniquely by equilibrium ultracentrifugation and (less commonly) by measuring the intensity of scattered light (for which see Section 6.12).

The use of sedimentation measurements in biophysical studies of macromolecules is extensively discussed in earlier textbooks. Briefly, in *sedimentation equilibrium*, the concentration, c of a macromolecule species, i assumes a Boltzmann distribution in the centrifugal field (angular velocity ω), and consequently varies with distance from the axis of centrifugation, R according to

$$c_i = a_i \exp\left[\frac{\frac{1}{2}m_i\left(1-\overline{v}_i\rho\right)\omega^2 R^2}{\left(k_B T\right)}\right]$$

$$N_i = \int_V c_i dV, \tag{8.7}$$

where m is the solute's molecular mass, $\overline{v}$ its partial specific volume, and ρ solvent density. The integral of c over the solution volume determines the constant a in terms of the total number of molecules N by conservation of mass. For two associating macromolecular species 1 and 2, such an expression applies for each species and for the 1–2 complex. In addition, the concentrations are related by the association equilibrium constant K_b and conservation of mass determines relations between N_1, N_2, and N_{1-2} and the total concentration of species 1 and 2 as sums of bound and free,

$$K_b = \frac{c_{1-2}}{c_1 c_2}$$

$$N_{1-2} + N_i = N_{i,0} \quad \text{for } i = 1, 2. \tag{8.8}$$

A best value of K_b can be determined by fitting these equations to measurements of macromolecule concentration as a function of R in solutions containing both macromolecules.

The dependence of protein concentration with R in the analytical ultracentrifuge is determined by measuring UV absorption or refractive index; values of the so-called buoyant mass, $m\,(1-\overline{v}\rho)$ of species 1 and 2 can be determined in separate experiments with a single solute. Sedimentation equilibrium experiments are run at relatively modest rotor speed; their main disadvantage is the long times needed to reach equilibration.

A relatively new technique for measuring binding of macromolecules to other macromolecules has an advantage of speed. In this setup, species 1 is attached (preferably tethered, so as to reduce interactions with the surface) to the metal-covered face of the window (glass or quartz) of a flow cell (Fig. 8.6). The detection principle relies on *surface plasmon resonance* (SPR), an electron charge density wave phenomenon that arises at the surface of a metallic (typically gold) film when light is reflected at the film under specific conditions. The resonance is a result of energy and momentum being transferred from incident photons into surface plasmons, and is sensitive to the refractive index of the medium on the opposite side of the film from the reflected light; thus, the presence of a macromolecular layer changes the intensity of the reflected signal. If this seems obscure it is useful to note that this obscurity does not matter in the context of determining binding affinities.

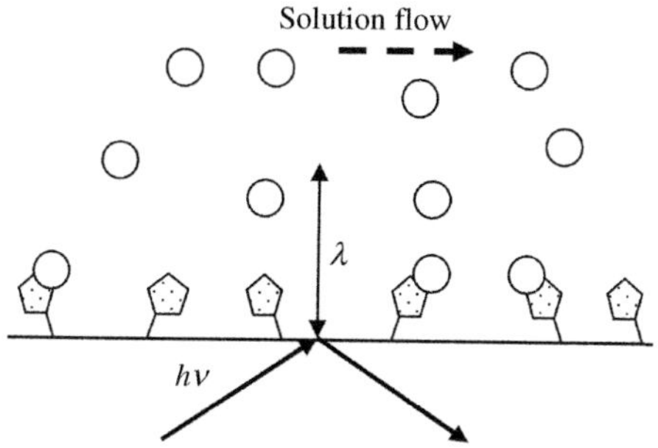

FIGURE 8.6 Schematic of a surface plasmon resonance experiment. The horizontal line represents a metal (gold) coating of the cell window.

If one now replaces the original solvent with a solution of species 2, which binds to species 1, the refractive index of the surface layer again changes with a concomitant change in the signal. In fact, the signal changes linearly with the saturation, and is thus a perfect instance of a signal, X that can be used to determine, with Eq. 8.5 saturation, f_b as a function of concentration of free ligand (simply the concentration of the solution of species 2, which is flowed past the bound species 1 until the signal stabilizes, indicating that equilibrium has been established). Once properly attached to the metal surface, the same sample of protein (species 1) can be used to measure interaction with a series of solutions of ligand (species 2) to determine the entire binding isotherm, and experiments with a series of ligand species can often be performed before the sample deteriorates. SPR is also suitable for measuring the on- and off rates of macromolecular interactions if these do not exceed 1 s^{-1}.

Commercial SPR instruments advertized as "biosensors" may not all be equally suitable for accurate determination of binding constants.[1] In any case, the fact that one species is (indirectly) attached to a metal surface requires parallel solution studies with an ultracentrifuge or calorimeter to establish that this attachment does not affect the affinity.[2]

8.3.5 Example: Isothermal Calorimetric Titration

As an example, we discuss results of isothermal calorimetric titration of a protein with an octapeptide ligand (Fig. 8.7).[3]

The experiment is done as follows: At equal time intervals, small aliquots of peptide solution are successively added to the protein solution, and the heat produced is recorded as a function of time. (For instrumentation, see Section 1.9.) The heat evolution is reported in the top panel. Each square in the bottom panel represents the incremental integrated heat change per mole of added octapeptide for one aliquot, $\delta q/\delta N_2$, as a function of the amount of added peptide ligand, expressed as molar ratio of ligand (component 2) to protein (component 1) $r_2 = N_2/N_1$.

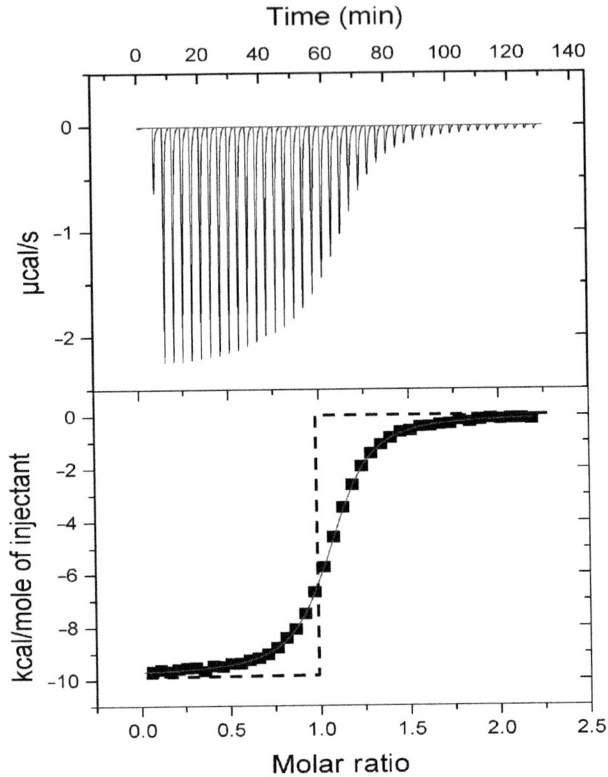

FIGURE 8.7 Isothermal titration data for an exothermic reaction, binding of an octapeptide to a PDZ3 domain (concentration = 0.1 mM). (Top) Raw data: the heat produced is recorded as a function of time. (Bottom) Incremental integrated heat change per mole of added octapeptide for each aliquot. The dashed curve describes the sharp titration that would be observed for extremely tight binding; the solid curve is the result of a computer fit for a simple binding isotherm.[3]

As mentioned in Section 8.3.3, analysis in terms of a binding model introduces a number of unknowns. For binding to a single site, these are the binding constant K_b, the observable "signal" at saturation X_{sat}, and the molarity of protein c_M. A curve fitting procedure was used to determine precise values of X_{sat}, which here represents the standard enthalpy change ($\Delta H^\circ = -9.8$ kcal/mol), and of the binding constant ($K_b = 9.0 \times 10^5\ M^{-1}$) for a simple binding isotherm according to Eqs. 8.1 and 8.5. In this case, the affinity is not very low, and the concentration of bound ligand cannot be neglected, so that the concentration of free ligand, and hence a_L, must be determined by conservation of mass of protein and ligand, according to Eq. 8.6. Substitution into the expression for the equilibrium constant gives a quadratic equation, one of whose roots gives $X(= q)$.[4]

The resulting fit of computed values to the experimental data is shown as a solid curve of δq as a function of molar ratio ($r_2 = c_L/c_M$) in the bottom panel

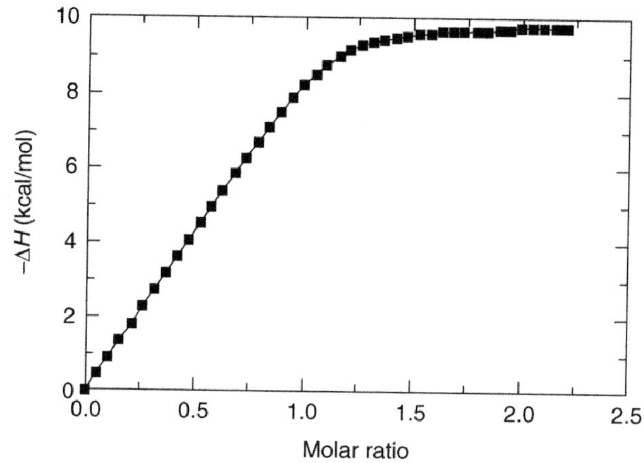

FIGURE 8.8 Data in bottom panel of Fig. 8.7 plotted as total, rather than incremental, heat change.

of Fig. 8.7. For comparison, the dashed curve indicates the results expected in the limit of extremely tight binding, where every added ligand molecule binds to a protein molecule, but only until equal concentrations of ligand and protein are present (molar ratio $r_2 = 1$).

The fit of model to data is excellent, which suggests that a simple one-site binding model is appropriate. This is confirmed by a plot of the summed incremental change per mole of macromolecule, $\Delta H = q/N_{\mathrm{M}}$ as a function of r_2 (Fig. 8.8). Comparison with curves of saturation as a function of total ligand concentration of Fig. 8.5 further indicates that this is a system with relatively tight binding of ligand to a single binding site.

8.4 MULTIPLE SITES FOR A SINGLE LIGAND

In many cases, several sites on a macromolecule are able to bind the same ligand. The situation is complicated by a number of factors. Such sites can be *equivalent* (indistinguishable) or all *different* (distinguishable); the site binding constants need not be identical (as for binding of hydrogen ion to proteins; cf. Chapter 14). Also, although multiple binding sites for the same ligand can be *independent*, this is often not the case.

Independent sites are invariant during the binding process, that is, binding at one site is independent of whether other sites are occupied. Dependent or *coupled sites* have affinities that change with the extent of occupancy of other sites by ligand. Coupling can result from direct interactions between ligands or because the binding of the ligands induces conformation changes in the macromolecule that alter the nature of other sites. The functional properties of many if not most proteins depend on coupled binding equilibria. The simple formalism described thus far is thus insufficient to describe most of the functional consequences of binding in biology, and needs to be extended.

8.4.1 Stoichiometric Binding Constants

If we consider a macromolecule with n sites, there is only one species with no bound ligand, and only one species with ligand bound to all n sites. However, there are *multiple* species with intermediate saturation. For example, in molecule ML, the ligand can be in any one of n available sites, and ML collectively represents n distinct *molecular* entities. Thus, [ML] is actually the sum of the concentrations of n possible molecular species, each having exactly one bound ligand molecule

$$[\mathrm{ML}] = \left[\mathrm{ML}^1\right] + \left[\mathrm{ML}^2\right] + \cdots + \left[\mathrm{ML}^k\right] + \cdots + \left[\mathrm{ML}^n\right] = \sum_k \left[\mathrm{ML}^k\right], \tag{8.9}$$

where the superscript k indicates to which of n sites the ligand is bound.

For addition of the second ligand to form ML_2, the situation is more complex. We can add the second ligand to any of $n-1$ sites, but half of the species created in this way are identical because it does not matter which of the two occupied sites is filled first. Thus, there are a possible $n\,(n-1)/2$ species of ML_2. Following this line of reasoning, we see that the *stoichiometric* species ML_i corresponds to $C_i^n = n!/\left[(n-i)!i!\right]$ molecular species. This number is equal to the number of ways of distributing i equivalent balls over n equivalent boxes, which is a binomial coefficient. Summation of this binomial coefficient over *all* values of i gives a total of $\sum_i C_i^n = 2^n$ molecular species with different occupancy of the binding sites; as each site can have two states, free and bound, making 2^n different possible species, when there are n sites. We will use the index j to label these species.

For each species, one can write a *stoichiometric equilibrium binding constant* as for each step in which the number of bound ligands, i increases by one, so that

$$\mathcal{K}_i = \frac{\left[\mathrm{ML}_i\right]}{\left[\mathrm{ML}_{i-1}\right] a_\mathrm{L}} = \frac{\sum_j \mathrm{ML}_i^j}{\sum_j \left[\mathrm{ML}_{i-1}^j\right] a_\mathrm{L}}. \tag{8.10}$$

In this last equation, the summation over j refers to all different molecular species in which exactly i binding sites are occupied. These *incremental stoichiometric binding constants*, $\mathcal{K}$ relate to equilibria in which the number of bound ligand molecules increases by one, without reference to particular sites at which the binding occurs.

In an analogous manner, one can define a *full stoichiometric equilibrium constant*, κ_i for binding exactly i ligand molecules to the macromolecule, regardless of which sites are occupied,

$$\kappa_i = \frac{\left[\mathrm{ML}_i\right]}{[\mathrm{M}]\, a_\mathrm{L}^i} = \frac{\sum_j \left[\mathrm{ML}_i^j\right]}{[\mathrm{M}]\, a_\mathrm{L}^i}. \tag{8.11}$$

For $i = 1$ and for $i = 2$,

$$\kappa_1 = \frac{[\mathrm{ML}_1]}{[\mathrm{M}]\, a_\mathrm{L}} = \frac{\sum_j \left[\mathrm{ML}_1^j\right]}{[\mathrm{M}]\, a_\mathrm{L}} = \mathcal{K}_1$$

$$\kappa_2 = \frac{[\mathrm{ML}_2]}{[\mathrm{M}]\, a_\mathrm{L}^2} = \frac{\sum_j \left[\mathrm{ML}_2^j\right]}{[\mathrm{M}]\, a_\mathrm{L}^2},$$

where we recognize that a species ML_k consists of all possible species with k bound ligands, that is, $\sum_j [\mathrm{ML}_k^j]$. Elimination of [M] from the expressions for κ_1 and κ_2 in this last equation gives

$$\kappa_2 = \frac{\sum_j \left[\mathrm{ML}_2^j\right]}{\frac{1}{\mathcal{K}_1}\sum_j \left[\mathrm{ML}_1^j\right] a_\mathrm{L}} = \mathcal{K}_1 \frac{\sum_j \left[\mathrm{ML}_2^j\right]}{\sum_j \left[\mathrm{ML}_1^j\right] a_\mathrm{L}} = \mathcal{K}_1 \mathcal{K}_2, \tag{8.12}$$

and, in general,

$$\mathcal{K}_i = \frac{\sum_j \left[\mathrm{ML}_i^j\right]}{\sum_j \left[\mathrm{ML}_{i-1}^j\right] a_L}$$

$$\kappa_i = \frac{[\mathrm{ML}_i]}{[\mathrm{M}]\, a_L^i} = \frac{\sum [\mathrm{ML}_i]_j}{[\mathrm{M}]\, a_L^i} = \mathcal{K}_1 \mathcal{K}_2 \ldots \mathcal{K}_i$$

$$\mathcal{K}_i = \frac{\kappa_i}{\kappa_{i-1}}. \tag{8.13}$$

An important general conclusion is that the binding of ligands to n sites is governed by n equations with (at most) n independent (sequential or full stoichiometric) binding constants.

Stoichiometric binding constants are suitable for a phenomenological description of binding data, but cannot reveal microscopic details about the arrangement of L on M. Even so, if n is greater than 3 or 4, even very accurate measurements of the binding isotherm do not suffice to determine all n full stoichiometric constants with some accuracy (although these full constants are most easily tied to experiment), unless the affinities of individual sites are dramatically different. In general, such complex binding equilibria can be unraveled only by a series of different experiments and some simplifying assumptions. We leave discussion of such problems to Chapter 11, wherein we treat the statistical approach to binding equilibria.

8.4.2 Site Binding Constants

Mechanistic descriptions of binding are typically done in terms of so-called site binding constants. A single-site binding constant refers to the binding of one molecule of ligand to a particular site, while all other sites that can bind ligand are kept as they are (i.e., occupied by specific ligands or empty). A site binding constant relates the concentration of one species of macromolecule having a unique choice of bound and free sites to the concentration of the macromolecule with the one additional ligand bound in a specific site. Binding affiinity at one site will often depend on the state of occupancy of other sites. Sites can interact for a number of reasons. The site binding constant of a given site may depend on the conformational state of the macromolecule and this again may depend on the state of occupancy of all other sites (see also Chapters 11–13). Binding sites may be in close proximity and bound ligands may interact directly, favorably, or unfavorably. Ionic surface-bound ligands interact significantly over distances on the order of the size of protein molecules, and, if the ligands have the same charge, this will destabilize particular arrangements of bound ligand more than others (see also Chapter 14).

The site binding constant for binding to a particular site m, is defined according to

$$\left[\mathrm{ML}_i^{m,\{i\}_j}\right] = K_m^{m,\{i\}_j}\left[\mathrm{ML}_{i-1}^{\{i-1\}_j}\right] a_{\mathrm{L}}. \tag{8.14}$$

The subscript j refers to a particular set of occupied sites consistent with the stoichiometric species ML_i, that is, to a particular distribution $\{\mathrm{ML}_i\}_j$ of ligands over distinct sites or in other words, to different molecular species. For the special case of a single ligand, as discussed here, we will often use $\{i, j\}$ to represent $\{\mathrm{ML}_i\}_j$. The index m refers to a particular site among all those that can be occupied in $\mathrm{ML}_{i,j}$, and the equilibrium described is one in which this particular site moves from an unoccupied to an occupied state.

We note that, in general, K_m depends on the particular distribution in which site m exists as the occupancy of one site may depend on the state of occupancy of other sites ("coupling or linkage", Chapters 12 and 13). As mentioned, the number of ways in which it is possible to configure the species ML_m is the binomial coefficient C_m^n, that is, there are C_m^n unique species with m sites occupied by a single type of ligand, and $n - m$ sites vacant. Thus, there are $n - m$ distinct paths between any one of the C_m^n species with occupancy m and species with occupancy $m + 1$, for a total of $C_m^n\,(n - m)$ for all species with m sites occupied; summation over all values of m gives the total number of distinct binding steps as $\sum C_m^n(n - m) = n \cdot 2^{n-1}$.

For $n = 2$, there are four possible site binding constants, but we will see in Section 8.4.5 that only three of these are independent. For $n = 4$ (as for hemoglobin), this increases dramatically to 32, but again not all of these can be independent. This dependence arises from the existence of multiple paths for filling binding sites between, say, species M and species $\mathrm{ML}_{m_j}^{(j)}$ (if $m_j > 1$). (Actually, the twofold symmetry of hemoglobin reduces the number of nonequivalent oxygenation states to 10.)

Because site binding constants reflect individual binding reactions, they offer the greatest insight into molecular events associated with binding and ideally are the objective of careful biophysical studies of association processes. It is easy to see from the discussion here how quickly an analysis in terms of site binding constants

TABLE 8.1 Definitions of three types of binding constants

Symbol	Type	Formula
$\mathcal{K}_i$	Incremental stoichiometric binding constant: stepwise addition of the m-th ligand molecule $L + ML_{i-1} \rightleftarrows ML_i$	$\mathcal{K}_i = \dfrac{\sum_j \left[ML_i^j\right]}{\sum_j \left[ML_{i-1}^j\right]_j a_L} = \dfrac{\kappa_i}{\kappa_{i-1}}$
κ_i	Full stoichiometric binding constant: simultaneously add the first m ligand molecule $M + iL \rightleftarrows ML_i$	$\kappa_i = \dfrac{\sum_j [ML_i]_j}{[M]\, a_L^i} = \prod_{i=1}^{m} \mathcal{K}_i$
$K_i^{(j)}$	Single-site binding constant: bind one ligand molecule to site number m, while the other ligands in set $\{i\}_j$ remain in place.	$\left[ML_i^{m,\{i\}_j}\right] = K_m^{m,\{i\}_j} \left[ML_{i-1}^{\{i-1\}_j}\right] a_L$

becomes very complex, and nearly intractable. All the sites constants for a complex system generally cannot be determined independently from binding data alone. One may address this problem by obtaining additional experimental data; for example, if the protein changes conformation as a result of binding, the conformation change may be characterized by spectroscopy (such as circular dichroism, fluorescence, or NMR). One can take advantage of the fact that different methods may reflect events at different sites. Alternatively, or in addition, one may seek to make simplifying assumptions about the binding process (e.g., exclusion of some molecular species, equivalence of some sites, etc.). For as few as two or three sites, this can become a challenging problem even to set up, let alone solve. This problem is taken up again in Chapter 11.

The definitions of the three types of binding constants are summarized in Table 8.1.[5] The relations between the stoichiometric binding constants and the site binding constants are complex, as will become clear in what follows.

8.4.3 Independent, Nonequivalent Sites

The simplest case of binding to multiple sites is that in which the sites are independent (i.e., each site binding constant is independent of occupancy of other sites).

If the sites are independent, the total saturation, ν_L will simply be the sum of the saturations, f_b for the n sites, each calculated with the single-site model of Eq. 8.2,

$$\nu_L = \sum_{i=1}^{n} i \cdot f_i = \sum_{i=1}^{n} \frac{K_i a_L}{1 + K_i a_L}; \quad \text{independent nonequivalent sites,} \qquad (8.15)$$

with K_i the site binding constant for site i. When multiple sites are present, a semilog plot of saturation will clearly show distinct "steps" at different $\log a_L$,

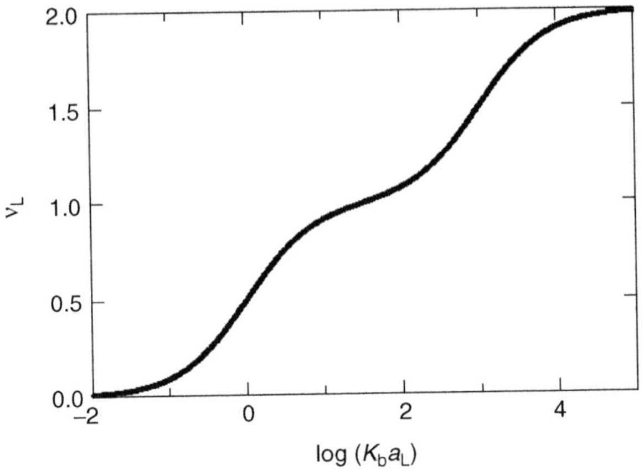

FIGURE 8.9 Saturation of binding to two sites, the second site with 1000-fold lower affinity, as a function of ligand activity (i.e., $K_{b,2} = 10^{-3} \cdot K_{b,1}$).

as in Fig. 8.9, but only if the affinities are sufficiently different. Depending on precision and number of experimental data, K_i values differing by an order of magnitude would be needed.

This simple description in terms of site binding constants seemingly belies the underlying complexity due to the presence of 2^n species with different numbers of bound ligands or different distributions of ligands over the sites; that is, one should consider the presence of multiple species having the same number of bound ligands (except for the species M and ML_n),

$$[\mathrm{ML}] = \sum_i K_i\,[\mathrm{M}]\,a_\mathrm{L}$$

$$[\mathrm{ML}_2] = \sum_i K_i \sum_{j>i} K_j\,[\mathrm{M}]\,a_\mathrm{L}^2 \tag{8.16}$$

etc.

Using this approach, the expression for the *sum of concentrations of all species* can be greatly simplified by the following factorization

$$\sum_i [\mathrm{ML}_i] = [\mathrm{M}] \prod_{i=1}^{n} (1 + K_i a_\mathrm{L}); \quad \text{independent nonequivalent sites.} \tag{8.17}$$

8.4.4 Independent, Equivalent Sites

If the sites are equivalent (multimeric protein; synthetic polymer; binding to backbone of polynucleotide), the analysis given in the preceding subsection remains applicable, except that all K_b now have the same value, that is,

$$\nu_L = \sum_{i=1}^{n} i \cdot f_i = \frac{nK_b a_L}{1 + K_b a_L} = \frac{na_L}{K_d + a_L}; \quad \text{independent equivalent sites.} \quad (8.18)$$

The expression on the far right shows the form when K_d is used instead of K_b. Also, the sum of concentrations of all species becomes

$$\sum_i [ML_i] = [M]\left(1 + K_b a_L\right)^n; \quad \text{independent equivalent sites.} \quad (8.19)$$

In this case, there is *another, more roundabout, way to formulate the problem*, by taking into account the number of ways it is possible to configure the species ML_i, which we have seen is a binomial coefficient

$$P_{i,n} = \frac{n!}{i!\,(n-i)!} = C_i^n.$$

As each of the species in ML_i is obtained in i successive binding steps each with the same K, the concentration of ML_i is given by

$$\left[ML_i\right] = P_{i,n}\,[M]\,K_b{}^i a_L{}^i$$

$$\frac{\left[ML_i\right]}{[M]} = C_i^n \left(K_b a_L\right)^i. \quad (8.20)$$

We can then derive the expression for the average number of ligands per macromolecule to be

$$\nu_L = n f_b = \frac{\sum_{i=1}^{n} i C_i^n \left(K_b a_L\right)^i}{\sum_{i=0}^{n} C_i^n \left(K_b a_L\right)^i} = \frac{nK_b a_L}{1 + K_b a_L}, \quad (8.21)$$

where we have replaced the sums by the finite expressions given in any mathematics handbook. This more circuitous route arrives at the same expression by using a simple statistical analysis.

When there are n equivalent and independent sites, there are n ways of forming a species ML, but there is only one species M. Therefore, the stoichiometric binding constants for the first step ($\mathcal{K}_1 = \kappa_1$) are n times that for the individual step, $\mathcal{K}_1 = nK_b$, the values of $\mathcal{K}_i$ for different values of m are not the same and generally $\mathcal{K}_i \neq K_b$. From Eq. 8.20 and the definitions of stoichiometric binding constants in terms of concentrations of macromolecular species it follows that for this simple model

$$\kappa_i = \frac{\left[ML_i\right]}{[M]\,a_L^i} = C_i^n \left(K_b\right)^i$$

$$\mathcal{K}_i = \frac{\left[ML_i\right]}{\left[ML_{i-1}\right] a_L} = \frac{n-i+1}{i} K_b, \quad (8.22)$$

which is an unnecessarily complex description of a very simple problem, but is nonetheless instructive to consider.

8.4.5 Coupled Sites

The problem is much more difficult when the sites are not independent, but coupled or linked. We illustrate this for the case of coupling between two sites (Fig. 8.10; see also the cited articles by Klotz and Hunston), the affinity of each site depends on whether its companion site is occupied, that is, $K_{b,12} \neq K_{b,2}$, even though both refer to binding of ligand to site 2. From the description of stoichiometric binding constants, we can write the concentration of any species as $[\mathrm{ML}_m] = \kappa_m a_\mathrm{L}^m$, meaning that we can write the average number of ligands bound as

$$\nu_\mathrm{L} = \frac{\sum_m m \cdot \kappa_m a_\mathrm{L}^m}{1 + \sum_m \kappa_m a_\mathrm{L}^m}.$$

For this simple system, m can be only 1 or 2, that is, there are only *two* full stoichiometric constants (κ_1 and κ_2), related to the two incremental stoichiometric binding constants ($\kappa_1 = \mathcal{K}_1$ and $\kappa_2 = \mathcal{K}_1\mathcal{K}_2$). If we can measure ν_L or a signal proportional to it, we can obtain estimates of these two stoichiometric binding constants, but only if they are sufficiently different in magnitude. However, our explanation of site binding constants suggested that there were four site constants, although we noted that these would not all be independent. The reason is that the condition of equilibrium for the *thermodynamic cycle* in Fig. 8.10 requires that $K_{b,1}K_{b,12} = K_{b,2}K_{b,21}$, resulting in only *three* independent site constants, still making it impossible for site binding constants to be obtained from measurements of ν_L without additional information.

On the other hand, if we know the independent site constants, we can easily obtain the stoichiometric constants. Thus, for our example:

$$\mathcal{K}_1 = K_{b,1} + K_{b,2}$$
$$\mathcal{K}_2 = \frac{K_{b,1}K_{b,12}}{K_{b,1} + K_{b,2}}. \tag{8.23}$$

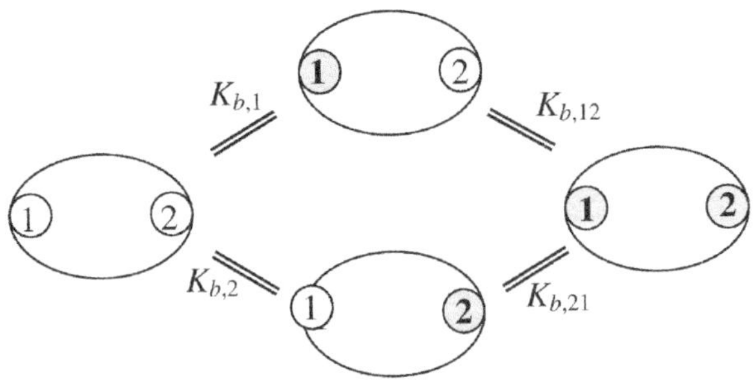

FIGURE 8.10 Two-site binding model; filled-in sites indicate sites with bound ligand.

Note that we do not include in the expression for κ_2 (or $\kappa_2 = \mathcal{K}_1\mathcal{K}_2$) a term in $K_{b,2}K_{b,21}$, as this product of site binding constants defines the same distinct molecular species as does the term $K_{b,1}K_{b,12}$. Thus, if one has a way of independently determining any one of the three independent site binding constants, one can use this and the measurement of ν_L to obtain a complete microscopic description of binding for this simple case. We have seen earlier that one can write the average number of bound ligands per macromolecule in the following manner

$$\nu_{\mathrm{L}} = \frac{\sum_i i\,[\mathrm{ML}_i]}{\sum_i [\mathrm{ML}_i]} = \frac{\kappa_1 a_{\mathrm{L}} + 2\kappa_2 a_{\mathrm{L}}^2 + \cdots}{1 + \kappa_1 a_{\mathrm{L}} + \kappa_2 a_{\mathrm{L}}^2 + \cdots}, \tag{8.24}$$

where all possible molecular species with exactly i bound L are included in $[\mathrm{ML}_i]$. One can see from this discussion that the problem of expressing all stoichiometric binding constants in terms of the site binding constants for a particular model becomes very tedious even for $n = 3$.

An example: in the classic case of oxygen binding to hemoglobin ($n = 4$) mentioned earlier, there are 10 different molecular species, related by nine independent binding constants. However, we can use full stoichiometric binding expressions to simplify the problem if we recognize that the ith stoichiometric species contributes i bound ligands, leading to the seemingly simple equation

$$\nu_{\mathrm{L}} = \sum_{i=1}^{4} i \cdot f_i = \sum_{i=1}^{4} \frac{i \cdot \kappa_i a_{\mathrm{L}}^i}{1 + \kappa_i a_{\mathrm{L}}^i}. \tag{8.25}$$

With minor modification, this becomes the Adair model for oxygen binding to hemoglobin (see Chapter 13, Eq. 13.9). While this is a marvelously simple expression, it provides us with no insight into the binding mechanism. We will see in Chapter 13 that obtaining our current best guess about the mechanism of hemoglobin action has required a combination of many ingenious thermodynamic and kinetic experiments.

Coupling is the rule, not the exception in cases of great physiological importance and interest. This section has given a formal description of coupled binding, independent of the mechanism by which coupling occurs (direct ligand interaction or binding site interaction). In Chapters 12 and 13. we discuss mechanisms by which coupling may occur *via* indirect interactions between binding sites, *via* a binding-dependent conformation change.

8.4.6 Partitioning of Macromolecule: Relation to Statistical Thermodynamics

Many of the equations derived in this section contain sums of concentrations of different species of macromolecule distinguished by how many ligand molecules are bound and to which specific binding sites. The partitioning over these species is, of course, obtained by dividing each term in the sum by the total concentration of macromolecule.

We shall see in Chapters 10 and 11 that the terms in any such sum of concentrations of macromolecular species $\sum_m[ML_m]$ correspond also to the terms for different molecular species, M, ML, etc. in a statistical mechanical partition function for macromolecules in solution containing dilute ligand L. Consequently, by application of the relation between partition functions and thermodynamics, the sum can give information about thermodynamic functions, in the first instance the free energy. The important correspondence between thermodynamics and statistical mechanics of ligand binding to macromoleules will be developed and explored in detail in Chapters 10 and 11. These chapters use the methods of statistical thermodynamics to develop and apply a "molecular grand partition function," the use of which helps considerably in setting up analyses for complex binding problems so that appropriate approaches can be taken to solve this challenging problem.

8.5 A FEW PRACTICAL RECOMMENDATIONS

We have here described several formulations of multiple binding equilibria, any of which may be encountered in the literature. Multiple formulations can be valid; the choice of which to use in new work will depend on circumstances. Currently, because it is so easy to fit even complicated theoretical expressions containing adjustable parameters to experimental data by least squares criteria, we believe that the following guidelines should be adopted in the first instance.

1. Carefully choose measurement conditions. For example, when dealing with experimental data, one may seek to fit observed values of "signal" used to detect saturation X versus ligand concentration c_L according to the single-site model. This fit involves adjusting three parameters, stoichiometry of binding n, signal at saturation X_{sat}, and binding constant K_b. The molarity of protein, c_M may be precisely known, but more often cannot be determined nearly as closely as the concentration of ligand. The experiment must be carefully planned if it is to give accurate values of all three parameters. It is necessary for the site concentration to be comparable to the dissociation constant ([site] $\approx K_d = 1/K_b$) and for the ligand concentration to be sufficiently high in order to define X_{sat}. Normally, [site] $= c_M$ if only one site is present, but can be $n \cdot c_M$ when n independent and equivalent sites are present. If this is not the case, then, if $K_d \ll$ [site], nearly all sites will have ligand bound, giving a good value of X_{sat} (and n if this is not 1), but not a useful estimate of K_b, and, if $K_d \gg$ [site], the measurements will not extend insufficiently close to saturation, and give a poor estimate of X_{sat}. If one does not have a good estimate of K_b (thus K_d), it is wise to perform experiments using at least two values of c_M to ensure reasonable estimates of binding parameters. If more than a single site is present, it is best to perform equilibrium dialysis experiments, but, if another technique is used, multiple experiments over a wide range of c_M can give a rough estimate of n if the response (X_i) is the same for all sites.

2. Fit the parameters in the equation to the data, and *do not* transform the data to fit the equation (as is done to make a Scatchard plot). For example, given a set of measurements of an observable X that reports occupancy of a binding site, as a function of free ligand concentration, it is asked if these data support the single-site binding model, according to the expression given in Eq. 8.5. This equation has three independent variables, X_0, X_{sat}, and K_b. A least-squares fit is then done to minimize $\Sigma(X_{model} - X_{obs})^2$ summed over all observations. This is preferable because, if all observations of X are equally precise, they should all contribute equally, which is not the case for a Scatchard-type analysis.

   As X_0 is nearly always well defined, and a hyperbolic function generally defines two parameters, it is clear that X_{sat} must be well established experimentally to obtain a decent estimate of K_b. The computer fit of model to data must include a statistical analysis that defines the standard error of parameters. If the data do not adequately limit the model, one or both parameters may be ill defined. If one or more of the parameter uncertainties is very large, the model is underdetermined by the data. In such instances, one must use additional experimental data or independent information to place constraints on parameter values.

3. Formulate the model in terms of site binding equilibrium constants. Site binding constants of equivalent independent sites are all the same, and by using these, one avoids (i) the factorials that enter into the formulation with stoichiometric binding constants K_1, K_2, etc., (ii) the order-of-magnitude differences between stoichiometric binding constants κ_1, κ_2, etc. and (c) a third advantage is that site binding constants can be related to the binding constants of model systems with only a single site, and differences can often be interpreted mechanistically.

4. Compare plots of experimental data and observables predicted by the model. Systematic deviations of the calculated from the observed values will indicate that attempts should be made to improve the model; randomly scattered deviations indicate statistical error. The most straightforward way to accomplish this is with a plot of the relative deviation ($[X_{model} - X_{obs}]/X_{obs}$) *versus* the independent variable. Such plots should accompany any complex fit of experimental data.

NOTES

1. Navratilova, I., et al. Thermodynamic benchmark study using Biacore technology. *Anal. Biochem.* **364**: 67–77 (2007).

2. Day, Y.S.N., Baird, C.L., Rich, R.L., Myszka, D.G. Direct comparison of binding equilibrium, thermodynamic, and rate constants determined by surface- and solution-based biophysical methods. *Protein Sci.* **11**: 1017–1025 (2002).

3. The heat evolved as a result of the addition of the i-th aliquot of volume δV_i containing $\delta n_{L,i}$ moles of ligand, is $\Delta H_i = \Delta H^\circ(n_{ML,i} - n_{ML,i-1})$, where $n_{ML,i}$ is the number

of moles of protein–ligand complex, $n_{ML} = Vc_{ML}$. Total volume is $V = V_0 + \sum \delta V_i$. The concentrations of ML, M, and L are related by $c_{ML} = K_b c_M c_L$. Total amount of protein $n_M + n_{ML} = Vc_{M,0}$ is constant. Total amount of ligand is $V(c_{ML} + c_L) = i\delta n_L$. This defines a set of simultaneous equations with unknowns K_b and ΔH°, for each experimental point. The values of the unknowns are found by a least-squares fit to the experimental values of ΔH_i.

4. For single-site binding, the solution is $X = X_0 + (X_{sat} - X_0)\{y - (y^2 - 4c_L c_M)^{1/2}\}/(2c_M)$, where $y = 1/K_b + c_L + c_M$, X_0 being the signal in the absence of ligand.
5. The symbol k, when used in connection with reactions, is reserved for rate constants throughout this book. Accordingly, the meaning of the symbols given in this table is not the same as in the papers by Klotz and Hunston cited at the end of this chapter.

SUGGESTED READING

Klotz, I.M., Hunston, D.L. Protein interactions with small molecules, *J. Biol. Chem.* **200**: 3001–3009 (1975).

Klotz, I.M., Hunston, D.L. Protein affinities for small molecules: Conceptions and misconceptions, *Arch. Biochem. Biophys.* **193**: 314–328 (1979).

Van Holde, K.E., Johnson, W.C., Ho, P.S., *Principles of Physical Biochemistry*. Upper Saddle River, NJ, 07458: Prentice Hall, 1998.

Wyman, J., Gill, S. J., *Binding and Linkage*, Mill Valley, CA 94941: University Science Books, 1990.

9

Thermodynamics of Molecular Interactions

9.1 INTRODUCTION

In the preceding chapter, we have treated the interaction of macromolecule and ligand using chemical (or stoichiometric) equilibria as a model. While this is appropriate for strongly and specifically interacting species, important situations exist in which this model should not be applied. The stoichiometric equilibrium model presumes a strong favorable, and in biological systems, specific interaction, and is, as a result, unsuited for treating repulsive molecular interactions. Interactions with solvent are typically not specific and relatively weak, and each macromolecular solute molecule interacts with many solvent molecules. Accordingly, the chemical equilibrium model is inappropriate also in mixed solvents, when different solvent components interact differentially, but weakly, with the macromolecule.

The typically high solubility of biological macromolecules indicates that for these, water is an excellent solvent and that, effectively, the macromolecular solutes repel one another, and as a consequence can coexist in a crowded cellular environment, and can closely approximate ideal behavior in dilute solution. In fact, it seems reasonable to suppose that this mutual repulsion is no less a selective advantage than the specific favorable interactions that exist between macromolecules that are required for biological function.

In this chapter, we show how to treat molecular interactions, including weak, nonspecific interactions, in terms of thermodynamics. Interaction of two solutes implies a change of the free energy of the solution, and can be described by the dependence of the chemical potential of either component on the concentration

Equilibria and Kinetics of Biological Macromolecules, First Edition. Jan Hermans and Barry Lentz.

of the other component. This principle can be used as the basis for deriving some completely general results; general in the sense that these are independent of the nature of the interactions (as should be the case for any treatment that is strictly based on thermodynamics). This results in a formulation that with equal success describes *favorable and unfavorable interactions* and a formulation of the change in the chemical potential of the macromolecule due to the interaction with another molecular species, in effect a formulation of the *free energy of binding*. This is applied in some detail to excluded volume interactions between macromolecules, and to interaction of macromolecules with solvent components in mixed solvents.

9.2 RELATION BETWEEN BINDING AND CHEMICAL POTENTIAL: UNIFIED FORMULATION OF BINDING AND "EXCLUSION"

Consider again the model equilibrium dialysis experiment illustrated in Fig. 8.3. The macromolecule, M is present only in the compartment on the left, while the small molecule L can pass freely from one compartment to the other. At equilibrium, the chemical potential of L, μ_L is the same in both compartments (and is determined by the concentration of L in the right-hand compartment). Let us now add a small number of molecules of M, dN_M to the left compartment, while keeping μ_L constant. If the macromolecule interacts with the small molecule, then N_L in the left compartment changes as a result of that addition, that is, if

$$\left(\frac{\partial N_L}{\partial N_M}\right)_{\mu_L} \neq 0, \tag{9.1}$$

then the two species interact. We speak of *binding* when

$$\left(\frac{\partial N_L}{\partial N_M}\right)_{\mu_L} > 0, \tag{9.2}$$

and this is, certainly in biology, the most common and most relevant case. In that case, we can equate this quotient with the average number of molecules of L that bind to one molecule of M,

$$\left(\frac{\partial N_L}{\partial N_M}\right)_{\mu_L} = \nu_L (\text{in the case of binding}). \tag{9.3}$$

However, it is also possible that

$$\left(\frac{\partial N_L}{\partial N_M}\right)_{\mu_L} < 0, \tag{9.4}$$

in which case we speak of *exclusion* of species L by species M. (These effects are mutual; i.e., if M attracts or excludes L, then L also attracts or excludes M.)

9.3 FREE ENERGY OF BINDING

We next relate the dependence of μ_M on μ_L to the saturation of binding (or, alternatively, the exclusion), that is, to $\partial N_L/\partial N_M$. One reason for doing this is to be able to relate shifts in conformational equilibria (as expressed in changes in chemical potentials of different species) to differences in ligand binding, and thereby obtain a general formalism for describing coupled binding and conformation change.

We begin by remembering that the Gibbs free energy, G is related to the composition and the chemical potentials of the components according to Eq. 1.32), $G = \sum N_i \mu_i$. Let us for the moment consider two solutes, M and L. Differentiation with respect to the chemical potential of either component gives

$$\langle N_M \rangle = \left(\frac{\partial \ln G}{\partial \mu_M}\right)_{V,T,\mu_L}$$
$$\langle N_L \rangle = \left(\frac{\partial \ln G}{\partial \mu_L}\right)_{V,T,\mu_M}. \qquad (9.5)$$

If we differentiate a second time, but with respect to the chemical potential of the other component, and omit the < > symbols denoting the thermodynamic average, we obtain

$$\left(\frac{\partial N_M}{\partial \mu_L}\right)_{\mu_M} = \frac{\partial^2 \ln G}{\partial \mu_L \partial \mu_M} = \left(\frac{\partial N_L}{\partial \mu_M}\right)_{\mu_L}. \qquad (9.6)$$

We can now make use of a general equality, called the *Euler chain rule* (Eq. AI.14),

$$\frac{(\partial N_M/\partial \mu_L)_{\mu_M}}{(\partial N_M/\partial \mu_M)_{\mu_L}} = -\left(\frac{\partial \mu_M}{\partial \mu_L}\right)_{N_M}, \qquad (9.7)$$

and substituting Eq. 9.6, we obtain

$$\frac{(\partial N_L/\partial \mu_M)_{\mu_L}}{(\partial N_M/\partial \mu_M)_{\mu_L}} = \left(\frac{\partial N_L}{\partial N_M}\right)_{\mu_L} = \nu_L = -\left(\frac{\partial \mu_M}{\partial \mu_L}\right)_{N_M}, \qquad (9.8)$$

and therefore

$$\left(\frac{\partial \mu_M}{\partial \mu_L}\right)_{N_M} = -\nu_L. \qquad (9.9)$$

Writing both μ_L and μ_M in the usual way, we obtain

$$\frac{1}{k_B T}\frac{\partial \mu_M^o}{\partial \ln a_L} = -\nu_L. \qquad (9.10)$$

(The amount of ligand bound or excluded per macromolecule ν_L is independent of the concentration of the macromolecules.) Integration of this expression gives

$\mu_{\mathrm{M}}^{\mathrm{o}}$ as a function of ligand activity, given the binding curve, that is, given ν_{L} as a function of $\ln a_{\mathrm{L}}$,

$$\mu_{\mathrm{M}}^{\mathrm{o}} = -k_{\mathrm{B}}T \int \nu_{\mathrm{L}} d \ln a_{\mathrm{L}}$$

$$\mu_{\mathrm{M},a_{\mathrm{L}}}^{\mathrm{o}} - \mu_{\mathrm{M},a_{\mathrm{L}}=0}^{\mathrm{o}} = -k_{\mathrm{B}}T \int_{a_{\mathrm{L}}=0}^{a_{\mathrm{L}}} \nu_{\mathrm{L}} d \ln a_{\mathrm{L}}. \tag{9.11}$$

The second form gives the change in standard chemical potential of M (due to the interaction of M with L) that results from a change of the ligand's activity from 0 to a value of a_{L}. The difference between standard chemical potentials of a macromolecule with ligands bound and in the absence of interacting ligand, according to Eq. 9, can be called the *standard free energy of interaction*. We note that this becomes the unitary binding free energy ($\mu_{\mathrm{ML}}^{\mathrm{o}} - \mu_{\mathrm{M}}^{\mathrm{o}}$) described in Chapter 10, if a well-defined stoichiometric complex ML is formed in the limit of high a_{L}. Interaction curves are accessible to direct experimental determination and thus this equation gives us a method to determine the dependence of the chemical potential of the macromolecule on the ligand activity directly from experiment, without assumption of a model of the binding (or, alternatively, the exclusion). According to this general result, binding of ligands ($\nu_{\mathrm{L}} > 0$) *lowers* the free energy of the macromolecule while exclusion ($\nu_{\mathrm{L}} < 0$) raises it. Thus, the conformation of a macromolecule to which ligand binds most favorably will be stabilized relative to other conformations. This point is again taken up in the following chapters.

9.4 MUTUAL RESPONSE

Consider now a system with at least three interacting solutes; we are especially interested in cases where two ligand species, 1 and 2 interact with a macromolecular species M, but not directly with one another. We can again apply Eq. 9.6 and this gives

$$\left(\frac{\partial N_{L_1}}{\partial \mu_{L_2}}\right)_{\mu_{L_1}} = \left(\frac{\partial N_{L_2}}{\partial \mu_{L_1}}\right)_{\mu_{L_2}}. \tag{9.12}$$

This states that if binding of one species to M is affected by a change in the activity (i.e., the concentration) of the other species, the reverse is also true, that is, the response is mutual. This effect is completely general and one cannot arrive at conclusions as to a specific mechanism on the basis of observing this equality.

A well-known example is the binding of oxygen to hemoglobin, which is accompanied by release of hydrogen ions, that is, $\partial N_{\mathrm{H+}}/\partial \mu_{\mathrm{O2}} < 0$. (Bohr effect, see Chapter 13). From structural studies, we understand that the H^+ ions are released by certain titratable groups of the protein as a result of the conformation change that accompanies oxygen binding. The resulting acidification is beneficial because it promotes the release of $\mathrm{CO_2}$ as oxygen is absorbed in the lungs. According to

Eq. 9.12, also $\partial N_{O2}/\partial \mu_{H^+} < 0$, and the metabolic production of H^+ ion promotes the release of oxygen in the body.

9.5 VOLUME EXCLUSION

9.5.1 Overview

So-called volume exclusion or crowding is a common form of exclusion; this is the effect, quite simply, of molecules of L not being able to occupy that part of the volume that is occupied by molecules of M, and vice versa. This was first analyzed by Ogston, who reasoned that two spherical particles would mutually exclude one another (more precisely, exclude each other's centers) from a volume equal to that of a sphere with radius equal to the *sum* of the radii of the two particles (Fig. 9.1a),

$$V_{s,s} = \frac{4\pi}{3}(R_1 + R_2)^3. \tag{9.13}$$

It is easy to also estimate the excluded volumes of spheres interacting with rod-like particles. The excluded volume of a sphere of radius R and a very thin rod of length L equals the volume of a cylinder of length L and radius R plus that of two half spheres of radius R

$$V_{s,r} = \pi R^2 L + \frac{4\pi}{3} R^3. \tag{9.14}$$

Again, the effect is mutual and the particles exclude equal volumes to one another.

The excluded volume of a pair of rods depends on the mutual orientation; it is smallest when the rods are parallel and largest when they are perpendicular, and this can lead to a spontaneous alignment in concentrated solutions of approximately cylindrical molecules such as particles of tobacco mosaic virus, polymers of sickle cell hemoglobin, or of liquids containing asymmetric molecules (liquid crystals).

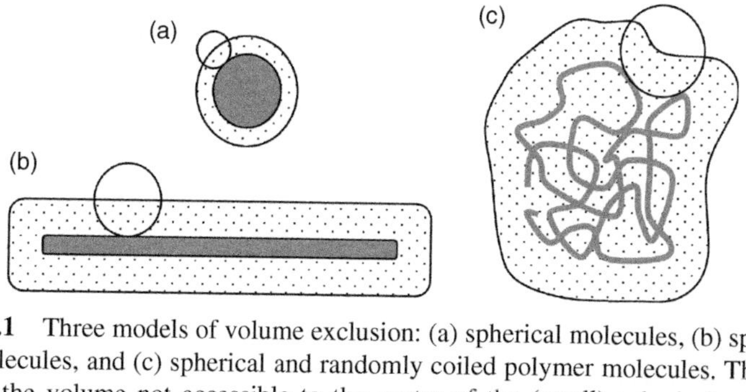

FIGURE 9.1 Three models of volume exclusion: (a) spherical molecules, (b) spherical and rod-like molecules, and (c) spherical and randomly coiled polymer molecules. The excluded volume [= the volume not accessible to the center of the (small) spherical solute] is the sum of stippled and solid parts.

Irregularly coiled individual (water-soluble) polymer molecules (such as the polyether polyethylene glycol, PEG) occupy large volumes, mostly filled with water. (For a discussion of polymer solutions see Chapter 15.) Coiled molecules can interpenetrate and do not exclude each other very effectively, but solid spherical particles of sufficiently large radius are excluded from the inside of coiled molecules (Fig. 9.1c). This leads to the well-known exclusion of PEG from membrane surfaces and to aggregation of membrane vesicles with increasing PEG concentration.

9.5.2 Mutual Exclusion of Solid Particles

As a model of mutual exclusion of solid particles, one may consider a (concentrated) solution of hard spheres. This model is presumably relevant to the interior of a cell, which is typically quite rich in compact polymers (proteins, nucleic acids, etc.); the magnitude of the excluded volume in the living cell can be considerable. Macromolecular structures of various shapes and sizes use up from 30% to 40% of the available volume, approximately one half of the limiting volume fraction, $\varphi_{\max}$ occupied by "closely packed" spheres ($\varphi_{\max}$ equals $0.740\cdots$). Figure 9.2 shows the (mean) free volume fraction, φ_{free} in a solution of spheres, as a function of the volume fraction, φ occupied by the spheres. Free volume is defined as the volume of the solution not excluded for insertion of additional spheres. The free volume depends on the size of the "solute" spheres relative to that of the "solvent" spheres, obviously being much smaller for larger inserts than for smaller ones. Smaller spheres can be accommodated much more readily than larger ones in the spaces between the largest spheres.

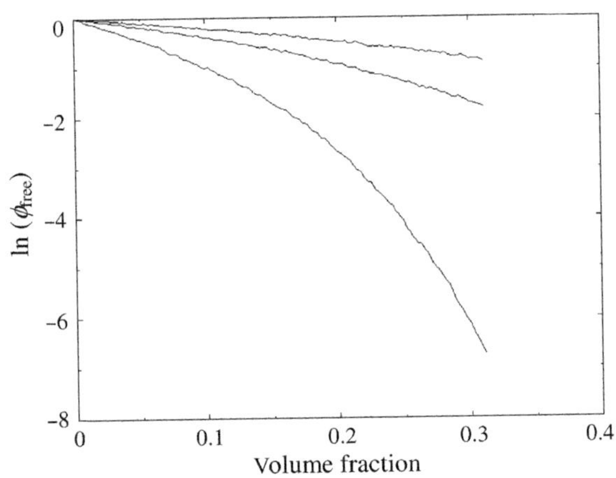

FIGURE 9.2 Free volume, that is, the fraction of total volume accessible for inserting an additional molecule in a solution of spherical molecules. The radius of the inserted molecule is (from top to bottom) 1/4 of, 1/2 of, and the same as, the radius of the molecules in the solution. (From a numerical simulation of a hard-sphere model.) Volume fraction cannot exceed that of closely packed spheres ($=0.740\cdots$); at that particle density, the free volume is zero.

One sees from the data reported in Fig. 9.2 that $\ln \varphi_{\text{free}}$ and $\partial \ln \varphi_{\text{free}}/\partial \varphi$ become very large and negative as the volume fraction, φ increases. The chemical potential of the solute, μ_1 is changed (increased) from that in an unencumbered volume according to[1]

$$\mu_1 = \mu_1^{\text{o}} - k_B T \ln \varphi_{\text{free}} + k_B T \ln(N_1/V). \qquad (9.15)$$

Hence, μ_1 (and also $\partial \mu_1/\partial N_1$) become extremely large as φ increases, approaching infinity as φ approaches the volume fraction of closely packed spheres. We see that when the spheres occupy 30% of the solution volume (the concentration of hemoglobin in human erythrocytes exceeds 300 g/l), free volume is equal to 1/400 and the chemical potential is raised by $k_B T \ln 6 = 3.6\,\text{kcal/mol}$. Nevertheless, hemoglobin molecules do not aggregate under these conditions. The results of such simulation experiments are reasonably well reproduced by statistical mechanical approaches (based on the Percus–Yevick equation as briefly discussed in Section 10.4, or based on scaled particle theory[2]).

The chemical potential of a macromolecular solute, μ_1 can be measured as a function of concentration by a variety of techniques. In *sedimentation equilibrium* the solute assumes a distribution according to a potential which is the sum of the solute's chemical potential, μ_1 and a potential due to the centrifugal field that increases as the square of the distance from the axis of rotation of the centrifuge. We have seen that the intensity of excess *light scattering* due to fluctuations of the solute concentration is proportional to $\partial N_1/\partial \mu_1$ (see Section 6.8). Measurements of the *osmotic pressure* report the chemical potential of solvent, μ_0 which is related to that of solute, μ_1 via the Gibbs–Duhem equation (see Section 1.5). Such measurements of the chemical potential of concentrated solutions of highly soluble globular proteins, such as hemoglobin[3] and serum albumin, show good agreement with theoretical predictions for a model of solute particles interacting via a hard-sphere potential with equivalent spheres of volumes close to molecular volumes estimated on the basis of the proteins' X-ray structures.

An example of experimental results is given in Fig. 9.3.[4] The smallest excluded volume effect is seen at pH 4.7, at which the average net charge of the protein molecules is zero (isoelectric point). Apparently, differences in molecular interactions (weak repulsion between negatively charged molecules at the higher pH values and possibly some attraction at the isoelectric point) can be adequately modeled simply with the use of slightly larger and smaller effective spheres.

Volume exclusion favors compact structures. As the unfolded form of many proteins is from experiment known to have a larger radius of gyration than the folded form, some cosolutes stabilize the folded structure as a result of volume exclusion. It has indeed been found that in concentrated solutions of globular proteins, protein folding equilibria are shifted in favor of the more compact folded state; also, equilibria of protein self-association are shifted in favor of the oligomeric or polymeric states, which typically have lower exposed surface per unit than the free monomers.[5] In the case of a dimerization equilibrium of a globular monomeric protein, an excluded volume effect raises the chemical potential of both monomer and dimer, with the net favoring the dimer. Estimates based on extensions of the

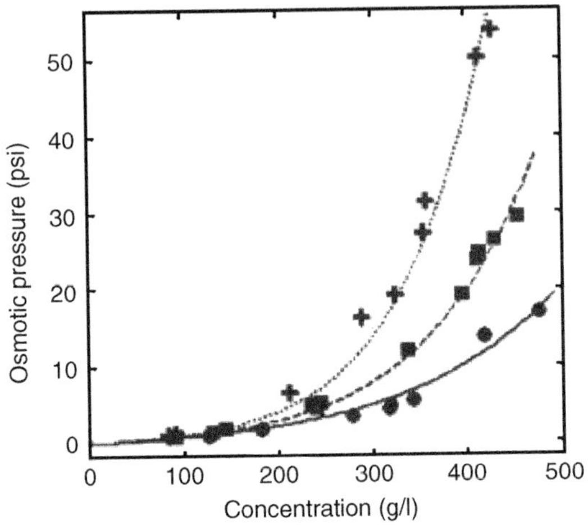

FIGURE 9.3 Osmotic pressure versus concentration of solutions of bovine serum albumin in 0.15 M NaCl at (Left to right) pH 7.4, 5.4, and 4.7. Points represent experimental data; curves are drawn according to a hard-sphere model with use of a slightly different sphere size at each pH. From Minton, A.P. *J. Pharm. Sci.* **96**: 3466–3469 (2007).

hard-sphere model to include, for example, a lozenge-shaped dimer predict an increase of the association equilibrium constant by a factor of around 10 when the volume fraction occupied by spheres of monomer size is 0.3.[5,6]

9.5.3 Mutual Exclusion of Protein Molecules and Coiled Macromolecules

Exclusion of protein molecules and coiled macromolecules can be an effective way to crowd protein molecules together. The addition of polymer such as PEG to a protein solution thereby raises the protein's chemical potential (see below) and thus favors phase separation (precipitation or crystallization; Fig. 9.4). The same effect excludes PEG from surfaces and accounts for the fact that PEG produces an "exclusion force" that is capable of driving membranes into near molecular contact.

When favorable interactions are absent between the polymer and a protein or a membrane surface, the effectiveness of PEG as a crowding agent (measured on a *mass*/volume basis) increases with the polymer's molecular weight. This is as expected from the basic properties of polymer molecules in solution (see Chapter 15). On an average, a randomly coiled polymer molecule has a radius that is roughly proportional to the square root of its molecular mass, M and hence occupies an effective volume, V_{eff} (much of which is solvent) proportional to $M^{1.5}$, and consequently the ratio of V_{eff}/M scales as $M^{0.5}$. The excluded volume depends of course on V_{eff}, but also on the size of the excluded molecule, which has to be large enough not to penetrate the coiled polymer molecules.[7]

As mentioned, volume exclusion can favor the native, folded state of a globular protein. Volume exclusion *may* play a role in the case of glycerol and various sugars

FIGURE 9.4 Mutual volume exclusion of coiled polymer molecules and globular protein molecules can lead to separation into polymer-rich and protein-rich phases.

(several mono- through tetra-saccharides) and other so-called osmolytes that have been found to convey greater stability on folded protein conformations. Among water-soluble polymers, PEG (a polyether) has been found to slightly destabilize native protein structures, while polyvinyl pyrrolidone (PVP) instead provides stabilization. These examples illustrate the fact that cosolute effects are often more complex than allowed for with simple excluded volume arguments.

9.6 ACCOUNTING FOR INTERACTIONS OF MACROMOLECULE AND SOLVENT COMPONENTS

Dissolved macromolecules and ligands typically interact strongly with water, and, presumably, the binding of ligand molecules replaces some water molecules, which become part of the bulk solvent. In order to formally recognize the participation of the solvent, the reaction of macromolecule and ligand to form a complex should be written as

$$M(H_2O)_{\nu_{0,M}} + L(H_2O)_{\nu_{0,L}} + (\nu_{0,ML} - \nu_{0,M} - \nu_{0,L})H_2O \rightleftarrows ML(H_2O)_{\nu_{0,ML}}, \tag{9.16}$$

to indicate the participation of $\Delta\nu_0 = \nu_{0,ML} - \nu_{0,M} - \nu_{0,L}$ water molecules. The energies for adding or removing these water molecules are far from negligible; in spite of this, the formulation of macromolecule–solute interactions so far has progressed without explicit consideration of water. This has been possible because the energy contributions for exchanging the $\Delta\nu_0$ interacting water molecules with bulk solvent and their standard chemical potential are both included in the standard free energy for the reaction. No explicit reference to the released water molecules is needed, as long as the chemical potential of water in the solvent, μ_0, does not change, that is, if this is negligibly affected by the presence of the ligand, a condition that is readily met in the typically dilute solution of a ligand species for which the macromolecule has a high affinity. (See Eq. 1.54.)

The condition of a constant μ_0 is frequently *not* met in mixed solvents, such as aqueous solutions containing high concentrations of cosolvents/cosolutes such as urea, guanidine, ethanol, glycerol, sugars, and salts, any of which change the stability of native proteins, as will be discussed in Chapter 17. When the concentration of the cosolute (component 2) changes, not only its own chemical potential, but also that of the solvent (component 0) changes, while the chemical potential of the macromolecule (component 1) depends on both.

Unfortunately, one cannot disentangle the effects of solvent and cosolute by a strictly thermodynamic approach, as indicated by the following simple analysis. When a solute molecule is introduced into a (mixed) solvent at constant volume V, some small volume v_d of the mixed solvent is trivially displaced from the volume V to the bulk. Specific interactions between the macromolecule and the solvent components result in additional adjustments of solvent content in the volume V in order to maintain constant chemical potentials of both solvent components (as in an equilibrium dialysis experiment in which the macromolecule is confined to one compartment, but solvent can equilibrate between both compartments; see Section 8.3.1). However, as long as the macromolecule interacts equally well or poorly with both solvent components, the displaced volume will still have composition equal to that of bulk solvent. Thus, an effect of interactions between macromolecule and solvent components is detectable by this measurement only if the interaction is unequal, as in that case the composition of the liquid in the displaced volume v will differ from that of the bulk solvent mixture.

For a more precise analysis, we make use of the Gibbs–Duhem relation between the chemical potentials of solvent and cosolute. This equation (Eq. 1.35) gives us $N_0 d\mu_0 + N_2 d\mu_2 = 0$ for the mixed solvent, and we then find that the chemical potentials of macromolecule and cosolvent are related by

$$\frac{d\mu_1}{d\mu_2} = \frac{\partial\mu_1}{\partial\mu_2} + \frac{\partial\mu_1}{\partial\mu_0}\frac{\partial\mu_0}{\partial\mu_2} = \frac{\partial\mu_1}{\partial\mu_2} - \frac{\partial\mu_1}{\partial\mu_0}\frac{N_2}{N_0}, \tag{9.17}$$

which with Eq. 9.9 becomes

$$\frac{d\mu_1}{d\mu_2} = -\nu_2 + \frac{N_2}{N_0}\nu_0, \tag{9.18}$$

with ν_2 cosolute molecules and ν_0 water molecules added to the volume V due to specific interactions. Clearly, the chemical potential of component 1 is affected by a change in chemical potential (i.e., in concentration) of component 2 only when the ratio ν_2/ν_0 differs from N_2/N_0, the ratio of cosolvent and solvent in the bulk solvent.

Some remarks are in order. First, if the cosolute binds preferentially, that is, $\nu_2 > 0$ then it is likely that the water is preferentially excluded, so that $\nu_0 < 0$ and participation of water enhances the effect on μ_1. Second, even in a relatively concentrated solution of cosolute, the ratio N_2/N_0 is quite small; because water is such a small molecule, the number of moles in a liter of water is over 55. Taking as an example 8 M urea, one finds that this contains over 60% water, so that N_2/N_0 is on the order of 0.25. Unless the number of water molecules released per "bound" urea molecule is very large, the term in ν_2 dominates. Third, to form a contact between macromolecule and cosolute *requires* removal of solvent from that location, and this makes it qualitatively understandable why a thermodynamic measurement of binding such as equilibrium dialysis cannot provide both ν_2 and ν_0.

As this is the case, it turns out to be equivalent, but more convenient, to express results in terms of constant number of molecules of solvent component 0, rather than constant volume, as by doing this one avoids having to include in the analysis the contents of the solvent volume v_d displaced by the introduction of the macromolecule. The interaction is then quantitated by the partial differential quotient $(\partial N_2/\partial N_1)_{N_0,\mu_2}$.

Accurate measurement of small amounts of cosolute bound or excluded in an equilibrium dialysis type experiment is quite difficult against a high background concentration of cosolute. Timasheff and Lee performed extensive measurements of preferential interactions of proteins with cosolutes, including chloroethanol, guanidine hydrochloride, and urea.[8] In these necessarily very accurate measurements, the preferential association of cosolute with protein is found by comparing solution densities at constant chemical potential and constant molality of cosolute.[9]

Other types of measurement are needed to distinguish even between mechanistically different scenarios. Consider, for example, the following two mechanistically quite different special cases. In the first case, cosolute molecules are excluded from the neighborhood of the macromolecule via volume exclusion, so that $\nu_2 < 0$, Eq. 9.18 gives $d\mu_1/d\mu_2 < 0$, when analyzed in terms of constant water content, $(\partial N_2/\partial N_1)_{N_0,\mu_2} < 0$. This result was to be expected, as in this case cosolvent molecules are mechanistically excluded.

In the second case, water molecules bind tightly to the macromolecule, and $\nu_0 > 0$. But here also, Eq. 9.18 gives $d\mu_1/d\mu_2 < 0$, and, when analyzed in terms of constant water content, $(\partial N_2/\partial N_1)_{N_0,\mu_2} < 0$. In this second case, some water molecules bind preferentially to the macromolecule, but if the total number of water molecules is kept constant, then cosolvent molecules must be removed in order to maintain a constant chemical potential of cosolute, that is, to maintain a constant solvent composition. Given the underlying mechanisms, one could say that in the first case the exclusion of cosolute is direct, and that in the second case it is indirect.

The effect of cosolutes on the stability of proteins is discussed in more detail in Chapter 17.

NOTES

1. Call the partition function [configuration integral, Eq. (6.17)] for N particles z_N. The partition function for $N+1$ particles is then $z_{N+1} = z_N \int_V \exp(-\beta E_{N+1}) dV$ where $E_{\mathrm{N+1}} = 0$ when the $N+1^{\mathrm{st}}$ particle is placed in the free volume, and otherwise infinite. Hence, $z_{N+1}/z_N = \varphi_{\mathrm{free}} V$, which one can then easily show to correspond to a change in μ_1, relative to a system with no volume exclusion, of $-k_{\mathrm{B}}T \ln \varphi_{\mathrm{free}}$.

2. Reiss, H., Frisch, H.L., Lebowitz, J.L. Statistical mechanics of rigid spheres. *J. Chem. Phys.* **31**: 369–380 (1959).
3. Ross, P.D., Minton, A.P. Analysis of non-ideal behavior in concentrated hemoglobin solutions. *J. Mol. Biol.* **112**: 437–452 (1977)
4. Minton, A.P. The effective hard particle model provides a simple, robust, and broadly applicable description of nonideal behavior in concentrated solutions of bovine serum albumin and other nonassociating proteins. *J. Pharm. Sci.* **96**: 3466–3469 (2007).
5. Minton, A.P. Molecular crowding: Analysis of effects of high concentrations of inert cosolutes on biochemical equilibria and rates in terms of volume exclusion. *Methods in Enzymology* **295**: 127–149 (1998).
6. Zhou, H.X., Rivas, G., Minton, A.P. Macromolecular crowding and confinement: biochemical, biophysical, and potential physiological consequences. *Annu. Rev. Biochem.* **37**: 375–397 (2008).
7. (a) Hermans, J. Excluded-volume theory of polymer-protein interactions based on polymer chain statistics. *J. Chem. Phys.* **77**: 2193–2203 (1982). (b) Knoll, D.A., Hermans, J. Polymer-protein interactions: comparison of experiment and excluded volume theory. *J. Biol. Chem.* **258**: 5710–5715 (1983).
8. (a) Timasheff, S.N. Protein-solvent interactions and protein conformation. *Acc. Chem. Res.* **3**: 62–68 (1970). (b) Lee, J.C., Timasheff, S.N. Partial specific volumes and interactions with solvent components of proteins in guanidine hydrochloride. *Biochemistry* **13**: 26–49 (1979).

9. Densities were measured with a precision density meter (Anton Paar, Gratz), based on the variation of the natural frequency of a hollow oscillator when filled with liquids of different density and, thus, mass. Molality is defined as concentration per kilogram of water (component 0).

SUGGESTED READING

Minton, A.P. Molecular crowding: analysis of effects of high concentrations of inert cosolutes on biochemical equilibria and rates in terms of volume exclusion. *Methods in Enzymology* **295**: 127–149 (1998).

Davis-Searles, P.R., Saunders, A.J., Erie, D.A., Winzor, D.J., Pielak, G.J. Interpreting the effects of small uncharged solutes on protein folding equilibria. *Ann. Rev. Biophys. Biomol. Struct.* **30**: 271–306 (2001).

10

Elements of Statistical Mechanics of Liquids and Solutions

10.1 INTRODUCTION

In contrast to the situation in an ideal gas, the molecules in a liquid interact favorably as the result of a variety of molecular attractive forces. Formation of a stable liquid state from a vapor is *favored* by the net negative energy of attractive intermolecular forces, but *opposed* by the loss of entropy because of the confinement of the molecules from the larger volume of the vapor into the smaller volume of the liquid plus the crowding (or volume exclusion) in the liquid resulting from intermolecular repulsion at short separation.

The statistical mechanical theory of liquids and, by extension, solutions has been studied extensively, but many questions remain unanswered. As we shall discuss in more detail in section 10.6, very large difficulties are encountered when one tries to describe the properties of model liquids with even very simple intramolecular forces, using a statistical mechanics *analysis*. These difficulties are then further magnified in the case of liquid water, with its not very simple, and very strong, intermolecular forces; rather, important insights into the equilibrium properties of liquid water have been obtained via the *numerical* approaches of molecular dynamics and Monte Carlo simulations. (See discussion in Chapter 4.)

As water is the solvent used in *all* experimental studies of biological macromolecules, much of the material discussed elsewhere in this book requires only an empirical knowledge of solvent properties such as, free energy, enthalpy and specific heat, dielectric coefficient, or viscosity. Most experimental studies in molecular biophysics are carried out in dilute solutions, and statistical mechanics does have

Equilibria and Kinetics of Biological Macromolecules, First Edition. Jan Hermans and Barry Lentz.

important things to say about the thermodynamics of dilute (ideal) solutions. The results of this theory are not very complicated, and are easily rationalized in terms of thermodynamics; however, statistical mechanics provides additional important insights into the nature of a dilute solution as well as a link to the behavior of nonideal liquids and solutions.

Accordingly, we have organized this chapter as follows. We begin by giving a simple derivation of an expression for the partition function of the ideal solution from well-known results of thermodynamics and, in Section 10.3, present a more complex derivation of this same expression on the basis of principles of statistical mechanics.

We similarly address the problem of writing a partition function for site-specific binding of ligand to macromolecule in two sections. In the first of these (Section 10.4) we follow Wyman's development based on stoichiometric binding theory, that is, on thermodynamics; in the second (Section 10.5) we give a more complex treatment based on statistical mechanics. The usage of this partition function is further elaborated in Chapter 11.

The chapter finishes with a short Section 10.6 on statistical mechanical theory of liquids.

10.2 PARTITION FUNCTION OF IDEAL SOLUTION FROM THERMODYNAMICS

On general principles, the canonical partition function of a (macroscopic) sample containing N_0 molecules of pure liquid, Q_0 can be written as

$$Q_0 = (q_0)^{N_0}. \tag{10.1}$$

Here we may call q_0 a unitary partition function for solvent; the *value* of q_0 reflects energetics both of intramolecular forces and of interactions between molecules and, as stated in the introduction to this chapter, is not readily accessible from first principles. One sees that this expression *must* have this form, if it is to agree with both thermodynamics and statistical mechanics (cf. Section 6.2.4): as the free energy, which is given by $A_0 = -k_B T \ln Q_0 = -N_0 k_B T \ln q_0$, is an "extensive" function, that is, it is proportional to the size of the system, in this case, proportional to N_0.

If we now add a single solute molecule to this solvent system, the partition function is modified, and we can write the new canonical partition function as

$$Q_{\text{soln}} = q_1 (q_0)^{N_0} = V q_1^{\text{o}} (q_0)^{N_0}, \tag{10.2}$$

where q_1 and q_1^{o} depend on a combination of intra- and intermolecular energetics and V is the volume of the solution. (Introduction of a factor V into this equation may seem arbitrary, but it is necessary in order to obtain the correct form of the chemical potential in Eq. 10.5. In Section 10.3.2 we shall base this and the following equation firmly on statistical mechanics.)

If another molecule is added to the solution, but this molecule interacts only with solvent, and not with other solute molecules (the definition of an ideal solution), then this will introduce another factor $q_1^\circ V$, and so on. However, we must also take into account that the solute molecules are indistinguishable, and for a solution containing N_1 independent solute molecules this gives

$$Q_{\text{soln}} = \frac{1}{N_1!}\left(q_1\right)^{N_1}\left(q_0\right)^{N_0} = \frac{1}{N_1!}\left(Vq_1^\circ\right)^{N_1}\left(q_0\right)^{N_0}. \tag{10.3}$$

By comparing this with the equation for the canonical partition function of an ideal gas mixture (Eq. 6.9), replacing $\left(eq_{0,\text{int}}\right)/\left(\Lambda_0^3 N_0\right)$ with q_0°, and ditto for component 1,

$$Q_{\text{gas mix}} = \frac{1}{N_1!}\left(Vq_1^\circ\right)^{N_1}\frac{1}{N_0!}\left(Vq_0^\circ\right)^{N_0}, \tag{10.4}$$

we see that solute (component 1) in the solution is represented similarly as component 1 in the ideal gas mixture, but that the term for solvent (component 0) is written differently from component 0 in the gas mixture. This difference in form between the expressions for Q_{soln} and $Q_{\text{gas mix}}$ agrees with what we know from thermodynamics, namely, the expressions for the chemical potential of solute in an ideal solution and in an ideal gas have the same form, while the expressions for chemical potential of solvent in solution and vapor are qualitatively different.

All effects of the solvent on solute and solute on solvent are included as part of the intrinsic properties of a solute molecule surrounded by solvent, but solvent is treated as pure solvent.[1]

An expression for the chemical potential of the solute is obtained with

$$A = -k_B T \ln Q$$

$$\mu_1 = \left(\frac{\partial A}{\partial N_1}\right)_{T,V,N_0} = -k_B T \ln q_1^\circ + k_B T \ln \frac{N_1}{V}. \tag{10.5}$$

This expression has the same form as the one derived on the basis of thermodynamics (Eq. 1.54, which we reproduce here as)

$$\mu_1 = \mu_1^\circ + k_B T \ln c_1,$$

which then gives

$$\mu_1^\circ = -k_B T \ln q_1^\circ. \tag{10.6}$$

In the following section, we examine how the same can be derived on the basis of statistical mechanics, alone, that is, avoiding the empirical aspects introduced by the use of thermodynamic principles.

10.3 STATISTICAL MECHANICS OF THE IDEAL SOLUTION

10.3.1 Based on the Ideal Gas Model

An ideal gas mixture is of course a special case (*all* interactions, not only those between solute molecules are zero) of an ideal solution. We can seek to apply a canonical partition function of the form that this has for a dilute ideal gas mixture (Eqs. 6.8 and 6.9) also to the ideal solution model. We rewrite the equation for this partition function as follows:

$$Q = \frac{q_1^{N_1}\, q_0^{N_0}}{N_1!\, N_0!} = \frac{\left(q_1^{\mathrm{o}} V\right)^{N_1}}{N_1!} \frac{\left(q_0^{\mathrm{o}} V\right)^{N_0}}{N_0!}$$

$$Q \approx \frac{\left(q_1^{\mathrm{o}} V\right)^{N_1}}{N_1!} \left(\frac{q_0^{\mathrm{o}} V}{N_0}\right)^{N_0} \approx \frac{\left(q_1^{\mathrm{o}} V\right)^{N_1}}{N_1!} (q_0)^{N_0}, \quad \text{for } N_0 \gg N_1. \tag{10.7}$$

In arriving at the last approximation, we have replaced $N_0!$ with N^N, the first factor in Stirling's formula, and recognized that the ratio of volume to number of molecules of a liquid, V/N_0 is in practice a constant. For the ideal gas the factors q_0 and q_1 represent partition functions which equal the Boltzmann sums over all quantum mechanical *molecular* energy levels. However, no analytic form exists with which to express the wave functions and energy levels of systems containing many interacting molecules.

We retain the symbols q_0 and q_1 for the ideal solution to represent effective contributions to the partition function. We shall no longer use the quantities $q_{1,int}$ and $q_{0,int}$ of Eq. 6.9. The functions q_0 and q_1 and the new functions q_0^{o} and q_1^{o} now take into account also the interaction of solute and solvent with surrounding solvent.[2] In an ideal solution all solute molecules experience interactions only with solvent, while interactions between solute molecules can be ignored, and solute molecules rearrange independently of one another. Thus, Eq. 10.7 expresses the partition function of a dilute solution as a product of identical terms for all molecules of the same species, thereby reflecting the fact that each molecule experiences the same large average interactions with its surrounding solvent molecules. In order to represent a solution, the factors q_0, q_1, q_0^{o}, and q_1^{o} must account not only for internal molecular contributions to the energy and entropy, but also for the changes in energy and entropy resulting from the interactions with solvent molecules. We shall use the term *molecular partition function* for both q and q^{o} also when these apply to the ideal solution.

10.3.2 Based on the Configuration Integral

As the partition function of the system we may instead use the configuration integral z of Eq. 6.17, which is a multiple integral of the Boltzmann factor of (only) the potential energy over the positions of all $N_0 + N_1$ atoms in the system, with the usual prefactors because the molecules of any one type are indistinguishable

$$z\left(N_0, N_1\right) = \frac{1}{N_0!}\frac{1}{N_1!}\int \cdots \int d\mathbf{X} \exp\left[-\beta E_{\text{pot}}\left(\mathbf{X}, N_0, N_1\right)\right]. \tag{10.8}$$

This partition function does not include a kinetic energy contribution, and neither do the expressions for energy, free energy and chemical potential that are based on it. However, all these contributions depend only on temperature and atomic mass, and are independent of the system's configuration. For the moment we shall ignore these contributions, and distinguish the expressions for partition functions lacking them with the subscript z.

We first express the effect of introducing a single solute molecule in a system of pure solvent by considering the partition function

$$z\left(N_0, 1\right) = \frac{1}{N_0!}\int \cdots \int d\mathbf{X} \exp\left[-\beta E_{\text{pot}}\left(\mathbf{X}, N_0, 1\right)\right]. \tag{10.9}$$

We can write the potential energy of this system as the sum of the potential energy for the solvent system and the difference due to the addition of the solute molecule

$$\Delta E_{\text{pot}} = E_{\text{pot}}\left(\mathbf{X}, N_0, 1\right) - E_{\text{pot}}\left(\mathbf{X}_0, N_0, 0\right), \tag{10.10}$$

and we can then rewrite the expression for $z\left(N_0, 1\right)$ as

$$z\left(N_0, 1\right) = \frac{1}{N_0!}\int \cdots \int d\mathbf{X}_0 \left\{\int \cdots \int d\mathbf{X}_1 \exp\left[-\beta \Delta E_{\text{pot}}\right]\right\} \exp\left[-\beta E_{\text{pot}}\left(\mathbf{X}_0, N_0, 0\right)\right].$$

Dividing both sides by $z\left(N_0, 0\right)$ we get

$$\frac{z\left(N_0, 1\right)}{z\left(N_0, 0\right)} = \frac{\int \cdots \int d\mathbf{X}_0 \left\{\int \cdots \int d\mathbf{X}_1 \exp\left[-\beta \Delta E_{\text{pot}}\right]\right\} \exp\left[-\beta E_{\text{pot}}\left(\mathbf{X}_0, N_0, 0\right)\right]}{\int \cdots \int d\mathbf{X}_0 \exp\left[-\beta E_{\text{pot}}\left(\mathbf{X}_0, N_0, 0\right)\right]}, \tag{10.11}$$

where on the right hand side we have written out the integral for $z\left(N_0, 0\right)$. One sees that this ratio is an ensemble average: the ensemble over which is averaged is that of the solvent system, and the function that is averaged is the integral of the Boltzmann factor for the incremental energy for *inserting a particle* (a solute molecule) *in any conformation, position, and orientation* into the system. (This is frequently referred to as *Widom particle insertion.*[3])

We can then rewrite this last equation as

$$z\left(N_0, 1\right) = \left\langle \int \cdots \int d\mathbf{X}_1 \exp\left[-\beta \Delta E_{\text{pot}}\right]\right\rangle z\left(N_0, 0\right). \tag{10.12}$$

Given a potential energy function, the ensemble average $\langle \int \ldots \int d\mathbf{X}_1 \exp[-\beta \Delta E_{\text{pot}}]\rangle$ can be evaluated numerically using simulation methods described in Chapter 7, in principle for any system, and, in practice, for simple systems. As the solute molecule can be placed anywhere in the volume V with equal probability, the expression is clearly proportional to the volume $V = N_0 v_0$, and thus we can, in analogy with Eq. 10, write it as $q^{o}_{1,z} V$, which gives

$$z\left(N_0, 1\right) = q^{o}_{1,z} V \cdot z\left(N_0, 0\right). \tag{10.13}$$

We can now in the same way insert a second solute molecule. The system volume over which is to be averaged is slightly larger, while the value of the ensemble average $\langle \exp[-\beta \Delta E_{\text{pot}}] \rangle$ differs slightly because of contributions from placement in the neighborhood of the first solute molecule. The resulting deviations are on the order of N_1/N_0; for the ideal solution, these deviations can in first approximation be ignored, and, if that is done, the partition function z contains a constant factor $q^{o}_{1,z} V$ for each solute molecule, so that

$$z\left(N_0, N_1\right) = \frac{1}{N_1!}\left(q^{o}_{1,z} V\right)^{N_1} \cdot z\left(N_0, 0\right). \tag{10.14}$$

(The change in content due to addition of N_1 solute molecules does need to be considered if one is to recover the slight dependence of solvent chemical potential on N_1, Eq. 1.54). If we then include the terms for kinetic energy, we may omit the subscript z. Furthermore, we have already shown that the partition function for the pure liquid can be written as $\left(q_0\right)^{N_0}$ according to Eq. 10.1, and this gives

$$Q\left(N_0, N_1\right) = \frac{1}{N_1!}\left(q^{o}_{1} V\right)^{N_1}\left(q_0\right)^{N_0},$$

which is identical with Eq. 10.3.

One sees here that the contribution to a solution's partition function made by independent solutes is equal to the product of the contribution that each solute makes to the system. We have encountered this obviously very *general principle of multiplication of partition functions of independent components* before, and applied it in such varied situations as the parts of a homogeneous macroscopic system and the independent molecules in an ideal gas (Sections 6.2.3 and 6.2.4).

In summary, we find that Eq. 10.7 correctly represents the partition function of an ideal solution, with factors q_0 and q^{o}_{1} that are independent of composition as long as interactions between solute molecules can be ignored. The mathematical definition of q^{o}_{1} in terms of the Widom integral is a welcome addition to the physical meaning we assigned to it in Eq. 10.7.

10.4 FORMULATION OF MOLECULAR BINDING INTERACTIONS IN TERMS OF A PARTITION FUNCTION: EMPIRICAL APPROACH BASED ON THERMODYNAMICS

We next extend these concepts to develop molecular partition functions for molecules that form complexes in solution with other solutes and for molecules that self-associate to form oligomers and polymers in solution. By definition, these systems are not ideal solutions. However, in most cases of interest, the interactions ("bonds") responsible for formation of the complexes are tight and specific, and each complex has a well-defined structure. Apart from forming these specific bonds, the interactions between complexes are very weak. Thus, when such solutions are sufficiently dilute, we may still apply the ideal solution model, if we explicitly account for the strong and specific interactions by considering the equilibrium distribution of macromolecules over distinct species, both ligand-free macromolecule and macromolecule with various combinations of bound ligand or monomer and oligomers of different sizes.

As discussed in great detail in Chapter 8, stoichiometric binding theory gives a familiar description of the partitioning of macromolecule over species with different number of bound ligand molecules. We follow here an approach pioneered by Wyman (see references to Wyman and Wyman and Gill at end of chapter), and capture this behavior by describing the partitioning of macromolecule in terms of the sum of concentrations over all stoichiometric species c_{ML_x},

$$c_{\mathrm{ML}_x} = c_{\mathrm{M}} + c_{\mathrm{ML}_2} + c_{\mathrm{ML}_2} + \cdots c_{\mathrm{ML}_n}. \tag{10.15}$$

Applying the usual equations for stoichiometric equilibrium (which have their theoretical foundation in thermodynamics, see Section 1.7), this can be written in terms of the activity of ligand a_{L} and stoichiometric binding constants κ_1, κ_2, and so on, introduced in Chapter 8, so that

$$c_{\mathrm{ML}_i} = c_{\mathrm{M}}\kappa_i a_{\mathrm{L}}^i, \tag{10.16}$$

and

$$c_{\mathrm{ML}_x} = c_{\mathrm{M}}\left(1 + \kappa_1 a_{\mathrm{L}} + \kappa_2 a_{\mathrm{L}}^2 + \kappa_3 a_{\mathrm{L}}^3 + \cdots + \kappa_n a_{\mathrm{L}}^n\right). \tag{10.17}$$

The terms in these equations describe the partitioning of macromolecule (expressed in terms of a probability distribution P) with *any* number of bound ligands ML_x over different free and bound species M, ML, ML_2, and so on in terms of a set of binding constants and the chemical potential of ligand

$$\frac{P_i}{P_0} = \frac{c_i}{c_0} = \kappa_i a_{\mathrm{L}}^i. \tag{10.18}$$

The expression in Eq. 10.17 does describe the partitioning of macromolecule over different species, but it does *not* represent a partition function as defined in statistical mechanics, which equals the (exponential) Boltzmann factor of a free energy. While it turns out that this is unimportant for many of the applications of Eq. 10.17 described by Wyman and Wyman and Gill in the cited references and in Chapter 11, it is actually quite simple to reconfigure this equation and thereby obtain an expression that is a partition function as defined in statistical mechanics.

This is accomplished by inserting in Eq. 10.17 in place of the concentration of macromolecule c_M, the corresponding molecular partition function q_M (as defined in Sections 10.2 and 10.3), which gives a new function

$$q_{ML_x} = q_M \left(1 + \kappa_1 a_L + \kappa_2 a_L^2 + \kappa_3 a_L^3 + \cdots + \kappa_n a_L^n\right). \qquad (10.19)$$

It is easy to see that the function q_{ML_x} represents the partition function of the macromolecule in the presence of bound ligand: The terms on the right are proportional to the probabilities P_i, and, in the absence of L, Eq. 10.19 gives $q_{ML_x}\left(a_L = 0\right) = q_M$, which is the correct partition function when there is no ligand in the solution.

In the following section we present a derivation of this expression strictly in terms of statistical mechanics. Applications will be presented in subsequent chapters.

10.5 A PURELY STATISTICAL MECHANICAL FORMULATION OF MOLECULAR BINDING INTERACTIONS

10.5.1 Introduction

In Section 10.3 of this chapter, we have presented a simple description of solvent and solutes in ideal solution based on statistical mechanics. In what follows we extend this description to include interactions between solute molecules, with main emphasis on site-specific binding, in which discrete complexes of a macromolecule with other molecules are formed at equilibrium.

Chapter 8 has treated site-specific binding equilibria in terms of chemical (stoichiometric) equilibria. A treatment based on statistical mechanics has the distinct advantage of being based on a *molecular* or microscopic view of events rather than on a thermodynamic view in which molecular details are ignored. It thus provides the possibility of deriving molecular insights from appropriate sets of observations. In addition, this treatment provides a direct connection to thermodynamic functions, in this case the chemical potential of the macromolecule. In its most general formulation, this treatment can be extended to such phenomena as exclusion, weak association, and solvent depletion as described in Chapter 9. However, these cases will not be included in the discussion that follows here. Of necessity, the formal statistical mechanical treatment developed here includes the thermodynamic description of binding in terms of stoichiometric (chemical) equilibria developed in Chapter 8.

In a typical experimental study on the binding of small molecules to a macromolecule, the number of macromolecules and volume (not precisely but approximately) in the system are fixed, while the chemical potential of the ligands is changed by titration, as discussed in Chapter 8. Binding experiments ideally involve determining the concentration (or activity) of bound and free macromolecule or bound and free ligand through such methods as potentiometric measurement (e.g., pH meter for H^+ ion) or equilibrium dialysis experiments (Figs. 8.3 and 8.4). Practically, however, one often must be satisfied with some spectroscopic or other means of detecting bound macromolecule (and consequently bound ligand, which gives bound when subtracted from total).

10.5.2 Model System

As we have seen, the first step in developing a partition function is to define the system, which in turn defines the ensemble, which defines the partition function.

First consider the simplest case of equilibrium dialysis experiment (Fig. 8.3). In this, a fixed number of macromolecules N_M is contained in a volume V of solution while small molecules L, that can interact with (bind to) the macromolecules, can freely pass through a semipermeable membrane separating this compartment from a very large reservoir of solution free of macromolecules. The compartment containing the solution of macromolecules is open to ligand L but closed to macromolecule. At equilibrium, the chemical potential μ_L is constant throughout the system and its value is known, as this is defined by the concentration of L in the reservoir.

In the absence of the reservoir, the experimental system has a volume V that contains a fixed number of macromolecules N_M and a fixed number N_L of molecules L that can interact with M. Given the system's composition, μ_L being a state function, is *defined*, although its *value* must be determined experimentally.

We divide the system conceptually into two compartments; one compartment is associated with the macromolecules, and consists of the collection of small volumes v_0 surrounding each macromolecule. Each volume v_0 is sufficiently large that it contains the macromolecule plus any affected ligand molecules and a shell of n_S solvent molecules, including all those affected by M. While macromolecules are known to define such small associated volumes, we are concerned here not with a real v_0 of defined size, but rather with a conceptual construct that reflects the reality that other solute molecules can partition between bulk solution and the distinct regions associated with M. For this reason, the actual value of v_0 is immaterial for our purposes. In the absence of any interaction between L and M, the concentrations of L in the two compartments will be the same; it is the interaction that alters this equality. Molecules of L in the bulk solvent compartment are referred to as free. The compartment(s) containing macromolecule have both free and interacting L molecules.

The solution can thus be conceptually, if not physically, separated into two volumes, one containing macromolecules and solvent, the other only solvent. Ligand L is at equilibrium between the two volumes, and the chemical potential of L, μ_L has the same value in both, although in this case, the concentration of free L and hence the chemical potential μ_L must be determined experimentally. Seen in this way, the solution containing M and L is an internal equilibrium dialysis experiment, and the same type of ensemble and partition function are applicable in both cases.

10.5.3 Ensemble

If we choose a canonical ensemble in which N_L and N_M are all fixed, this is inadequate to define the distribution of L between the compartments. If we choose a grand canonical ensemble in which each element is open with respect to *all* components, this fails to properly apply the constraint that the number of macromolecules M is fixed in each compartment (none in the bulk; N_M in solution). Thus, we use instead an ensemble with properties of both ensembles, the *semigrand ensemble*. Hill's *Statistical Mechanics*[4] discusses this ensemble as part of a family of related ensembles.[5] Just as the canonical ensemble is a subensemble of a grand ensemble, containing only those terms that pertain to one particular choice of content of *all* components, the semigrand ensemble is a subset of the grand ensemble limited to those terms that pertain to a fixed content of a *single* component, here the macromolecule M. (This is discussed in more detail in Appendix C).

10.5.4 Partition Function

The *grand partition function* Ξ for a system containing macromolecule M and a single type of ligand L is given by (see Chapter 6, Eq. 6.30)

$$\Xi = \sum_{N_M} \sum_{N_L} \left[Q\left(N_M, N_L, V, T\right) e^{\beta N_M \mu_M} e^{\beta N_L \mu_L} \right] e^{\beta N_S \mu_s},$$

where Q is the canonical partition function of a system containing N_M macromolecules and N_L ligand molecules, the sum being over all possible compositions. In addition to these, the system contains N_S solvent molecules that introduce a factor $e^{\beta N_S \mu_S}$ in each term.

We restrict what follows to ideal solutions, in which solute molecules do not interact, except for discrete, site-specific binding between macromolecules and ligand molecules; and, for the sake of simplicity, we restrict the discussion to a single type of ligand. In that case, a number $N_{L,bound}$ of the N_L ligand molecules is bound to macromolecule, while the remaining $N_{L,free}$ are free in solution. The number of bound ligand molecules varies with the number of macromolecules and their affinity for ligand; however, the concentration of free ligand is fixed (at the concentration of the reservoir, i.e., by the value of the chemical potential μ_L), so that the number of free ligand molecules is in a fixed proportion to the number of solvent molecules N_S.

In order to describe a system containing a *fixed* number of macromolecules, we select from Ξ only terms for one particular value of N_M, and, when we also separate the ligand molecules into bound and free, this gives

$$\Gamma\left(N_M\right) = \sum_{N_{L,bound}} \left[Q\left(N_M, N_{L,bound}, V, T\right) e^{\beta N_{L,bound} \mu_L} \right] e^{\beta N_{L,free} \mu_L} e^{\beta N_S \mu_s}. \quad (10.20)$$

This is then the semigrand partition function Γ for this system. (The common constant factor $\exp\left(\beta N_M \mu_M\right)$ has been dropped.)

We apply this equation to specifically describe the simple case of a solution containing a *single* macromolecule with n binding sites for ligand L. This macromolecule can then occur in any one of $n + 1$ states, with $N_{L,bound}$ ranging from

$i = 0$ to n bound ligand molecules, M, ML, ML_2 … ML_n. The semigrand partition function for this case is a sum of $n + 1$ terms, one for each state,

$$\Gamma_{N_M=1} = \sum_{i=0}^{n} \left[Q_i e^{\beta i \mu_L}\right] e^{\beta N_{L,free}\mu_L} e^{\beta N_S \mu_S}, \tag{10.21}$$

where Q_i is the canonical partition function for the system when it contains a single molecule ML_i, that is, one molecule of M bound to i molecules of L; i replaces $N_{L,bound}$.

We have shown earlier in this chapter (Sections 10.2 and 10.3) how the partition function Q_i of an ideal solution can be written as a product of terms, one for each molecule in the solution. We now apply this result, extending Eq. 10.3 to include also the free ligand molecules, and thereby obtain an expression for Q_i

$$Q_i = q_i \left(q_S\right)^{N_S} \frac{1}{N_{L,free}!} \left(q_L\right)^{N_{L,free}}, \tag{10.22}$$

where q_i is the (molecular) partition function q for molecules ML_i, q_L ditto for free ligand, and q_S replaces q_0. From Eq. 10.6 we know that $q_S = \exp\left(-\beta\mu_S\right)$, so that $\left(q_S\right)^{N_S}$ can be replaced with $\exp\left(-\beta N_S \mu_S\right)$ and, similarly, from Eq. 10.5 it follows that the terms in $N_{L,free}$ can be replaced with $\exp\left(-\beta N_{L,free}\mu_L\right)$. These factors exactly cancel the exponentials in Eq. 10.21, which simplifies to

$$\Gamma_{N_M=1} = \theta_M = q_M + \sum_{i=1}^{n} q_i e^{\beta i \mu_L}, \tag{10.23}$$

The function θ_M can be called the *molecular semigrand partition function* for macromolecule M in equilibrium with bound ligand L; θ_M gives the molecular partition function of macromolecule with *any* number of bound ligands as a sum of terms, each the product of the partition functions q_i of the macromolecule complexed with a different number of bound ligand and an exponential factor in the chemical potential of ligand. At equilibrium the macromolecules are distributed over these complexes according to

$$P_i = \frac{q_i \exp\left(\beta i \mu_L\right)}{\theta_M}, \tag{10.24}$$

(with $q_0 = q_M$) while the concentration c_i of the species having i bound ligands, relative to that of the ligand-free macromolecule c_0, is given by

$$\frac{c_i}{c_0} = \frac{P_i}{P_0} = \frac{q_i \exp\left(\beta i \mu_L\right)}{q_M}. \tag{10.25}$$

Because the solution is ideal and the macromolecules do not interact, we may apply the earlier stated *principle of multiplication of partition functions of independent components*, according to which the partition function of N_M independent macromolecules is found by raising the partition function for a single molecule to the power N_M (and dividing by $N_\mathrm{M}!$)

$$\Gamma\left(N_\mathrm{M}\right) = \frac{1}{N_\mathrm{M}!}\left(\theta_\mathrm{M}\right)^{N_\mathrm{M}}. \tag{10.26}$$

The semigrand partition function is related to the thermodynamic potential of the macromolecule μ_M according to

$$-k_\mathrm{B}T \ln \Gamma\left(N_\mathrm{M}\right) = N_\mathrm{M}\mu_\mathrm{M}. \tag{10.27}$$

(See discussion by Hill[4] and Appendix C. Except in experiments at elevated pressure, at which small volume differences become a factor influencing the equilibrium distribution, the contribution of the PV term in Eq. AC.21 can safely be neglected.)

The significance of the treatment here is that this ties the formulation in terms of stoichiometric binding constants and activity of ligand to the framework of statistical mechanics, which deals with molecular species and sets the stage for a detailed description of site-specific binding in Chapter 11. The results derived in the first part of this chapter and in Chapter 11 apply to material on coupled equilibria, allosteric function and helix-coil equilibria to be presented in subsequent chapters.

10.5.5 Generalization to Include Other Kinds of Molecular Interactions

So far, we have considered only the special case of strong specific interactions between macromolecule and ligand and excluded from our treatment cases of weak or repulsive interactions between macromolecules and other solutes, cosolvents or "crowders". This situation was treated thermodynamically in Chapter 9. The semigrand partition function can also be used to characterize weak interactions of "ligands" or cosolvents with M or even unfavorable interactions between "ligand" and macromolecule (exclusion or crowding). This was exactly the purpose of Stockmayer in his treatment of nonrandom partitioning of ions between the regions around polymers and bulk solution. While based on the same physical model and thus theoretically satisfying, the statistical mechanical treatment has no real advantage over the thermodynamic treatment given in Chapter 9 and will not be considered here.

10.6 STATISTICAL MECHANICAL MODELS OF NONIDEAL SOLUTIONS AND LIQUIDS

10.6.1 Mixtures

As should be evident from the discussion in Section 10.1, the development of a statistical thermodynamic theory of a nonideal liquid or solution involves the extremely difficult problem of *analytically* evaluating the potential energy of a many body system and integrating this to obtain z, or equivalently the configurational entropy (i.e., Ω). In this section, we discuss briefly approximate models of systems of interacting molecules that have proven effective. However, it should be stated at the outset that these models assume nonspecific and relatively weak interactions and thus have limited applicability.

The simplest approximation applies to solutions for which the AB interaction energies are not too different from those for AA or BB pairs. In this case, it is reasonable to approximate the entropy of mixing as that of an ideal solution, $\Delta S = k_{\mathrm{B}} \ln X_{\mathrm{A}} + k_{\mathrm{B}} \ln X_{\mathrm{B}}$, (Eq. 1.42) and to account for nonideality in terms of a nonzero enthalpy of mixing. This approach is termed *Regular Solution Theory*. If we assume that this nonzero enthalpy of mixing can be approximated as the nonzero internal energy of mixing, we can express the enthalpy of mixing in terms of the depths of potential well between molecules, ε, and the number of contacts between molecules, z,

$$\Delta H_{\mathrm{mix}} \approx \Delta E_{\mathrm{mix}} = -\xi \frac{N_{\mathrm{A}} N_{\mathrm{B}}}{N_{\mathrm{A}} + N_{\mathrm{B}}}; \quad \xi = zN \left(\frac{\varepsilon_{\mathrm{AB}} - \varepsilon_{\mathrm{AA}}}{2} - \frac{\varepsilon_{\mathrm{BB}}}{2} \right).$$

For the sake of illustration, we treat A as a solute in solvent B. Differentiating with respect to N_A, we obtain $\Delta\mu_A = \xi x_B^2$ for the change in chemical potential of A (relative to $\mu_{\mathrm{A}}^{\circ} + k_{\mathrm{B}} T \ln X_{\mathrm{A}}$) due to interactions. Adding this to the chemical potential of component A in an ideal mixture of A and B yields

$$\mu_A = \mu_A^{\circ} + k_{\mathrm{B}} T \ln x_A - k_{\mathrm{B}} T \xi x_A^2 = \mu_A^{\circ} + k_{\mathrm{B}} T \ln (\gamma x_A); \quad \gamma = \exp\left(\frac{-\xi}{k_B T} x_A^2 \right),$$

where γ is the activity coefficient of component A in the solution. When ξ is positive, the interaction between A and B is attractive, the enthalpy of the solution is less than that of the individual liquids, and the chemical potential of A in solution is reduced relative to an ideal mixture. As $\xi \to 0$, $\gamma \to 1$, and mixing becomes ideal. If ξ is negative, the enthalpy of solution is positive, the chemical potential of A dissolved in B becomes increasingly more positive until a point is reached at which it exceeds the chemical potential of pure A plus pure B, and then the two components become two separate phases. This behavior is reasonable, and Regular Solution Theory has been quite successful even in predicting phase boundaries, but only for liquids with weak intermolecular interactions, as might be expected.

10.6.2 Pure Liquids

A more sophisticated approach seeks to relate the partition function of a system of interacting molecules to the energy of pair interaction, $U(r)$, with r the intermolecular distance. It turns out (this is one difficult part of the theory) that the configurational part of the partition function (Eq. 6.17) can be found if we know the radial distribution function, $g(r)$; $g(r)$ describes the relative density of other molecules at some distance from any given central molecule. To make the problem tractable, any dependence of U on relative orientation of the molecules is ignored, and it is assumed that U is a two-body potential, that is, does not depend on the positions of other molecules.

For a given overall particle density, N/V the function $U(r)$ determines the distribution of particles, which is characterized in terms of the radial distribution function. Of course, in very dilute systems for which three-body correlations can be ignored, $g(r)$ is simply given by a Boltzmann distribution in $U(r)$. For condensed fluids, higher order correlations cannot be ignored, and repulsive parts of $U(r)$ become dominant.

Take as an example a hard-sphere potential, arguably the simplest intermolecular pair potential. At high dilution, the function $g(r)$ is zero for r less than twice the radius, R of the spheres, and uniform (equal to 1, as $g(r)$ is normalized) for any larger value of r. On the other hand, at sufficiently high concentration the spheres are crowded together and the distribution around a central particle (particle 1) drops from a highest value at contact distance $2R$ to a minimum, giving a first solvation shell containing at most 12 nearest neighbor particles (spherical close packing); other, gradually lower maxima and minima for second and higher solvation shells occur at larger separations. Note, however, that if the neighboring spheres did *not* interact with one another, then the distribution would again be uniform. Thus, it is essential to take into account the interaction between neighbor spheres (particles 2 and 3). This is another difficult problem, that is at the heart of condensed fluid theory, and we briefly summarize here the progress made toward solving it. (For details, see texts by McQuarrie and by Plischke and Bergersen cited at the end of this chapter.)

A first-order approach to treating correlations between three particles (1,2,3) is to assume that the true correlation function, $h(r_{12})$ between two particles, $h(r_{12}) = g(r_{12}) - 1$, is just the sum of a "direct" two-body correlation function, $c(r_{12})$, related directly to the two-body potential $U(r_{12})$, plus an "indirect" term reflecting the fact that particles 1 and 3 are directly correlated by $c(r_{13})$, while particle 3 is also correlated with particle 2 through $h(r_{23})$. Ornstein and Zernike proposed that the "indirect" correlation term was proportional to the average fluid density times an integral that accounts for the three-body correlation (the Ornstein–Zernike equation),

$$h\left(r_{12}\right) = c\left(r_{12}\right) + \rho \int d^3 r_3 \cdot c\left(r_{13}\right) \cdot h\left(r_{23}\right),$$

where ρ is the particle density and the integral is over all values of the position of particle 3, $\mathbf{r}_3$. However, the dependence of the direct correlation function $c(r)$ on $U(r)$ is not defined, so that, even for very simple forms of $U(r)$, this equation cannot provide $h(r)$ and therefore also $g(r)$ in terms of $U(r)$. The Ornstein–Zernike equation results from a first-order approach, and relates $g(r_{12})$ and the correlation between pairs of neighbor particles 2 and 3. Percus and Yevick were able to find an approximation for this correlation that depends on the potential, $U(r)$; the problem then became to solve the resulting nonlinear integral equation (the Percus–Yevick equation) for $g(r)$, given an expression for $U(r)$.

This last problem was solved in 1963 for a hard-sphere potential.[6] That solution is useful for dense systems; however, as this potential provides no attraction it cannot represent a separate liquid state. Solution of the Percus–Yevick equation was later extended to a square well potential.[7] While quite crude, this approximation captures the essence of an attractive potential even though it does not provide the subtlety of a soft potential that would add thermal energy associated with the potential.

NOTES

1. This is why the standard state of solute in dilute solution is different from that of pure solute (Henry's Law), and why the standard state for solvent is that of pure solvent (Raoult's Law).
2. The subscript "int" for "internal" seems no longer appropriate, and the factors Λ^3 of Eq. 6.9 are incorporated in the new functions.
3. Widom, B. Some topics in the theory of fluids. *J. Chem. Phys.* **39**: 2808–2812 (1963).
4. Hill, T. L., *Statistical Mechanics. Principles and Selected Applications*. New York: McGraw-Hill, 1956; Reprinted by Dover books, Chapter 3.
5. The semigrand ensemble and partition function were used by W.H. Stockmayer to treat light scattering in three-component systems (*J. Chem. Phys.* **18**: 58–61, 1950). He used the term *modified grand partition function*, as has one of us in past publications (e.g. *Biophys J.* **84**: 1238–51, 2003).
6. Lebowitz, J. Exact solution of generalized Percus-Yevick equation for a mixture of hard spheres. *Phys. Rev.* **133**: A895 (1964).
7. Martinez-Haya, B., A. Cuetos, A., Lago, S. Solution of the Percus-Yevick equation for square well spherocylinders. *Phys. Rev. E.* **67**: 051201 (2003).

SUGGESTED READING

1. Wyman, J. Linked functions and reciprocal effects in hemoglobin: a second look. *Adv. Protein Chem.* **19**: 223–286 (1964).
2. Wyman, J., Gill S. J., *Binding and Linkage: Functional Chemistry of Biological Macromolecules*. Mill Valley, CA: University Science Books, (1990).
3. Stockmayer, W.H. Light scattering in multi-component systems. *J. Chem. Phys.* **18**, 58–61, (1950).
4. Alberty, R.A. and Oppenheimer, I. Use of semigrand ensembles in chemical equilibrium calculation on complex organic systems. *J. Chem. Phys.* **91**, 1824–1828 (1989).
5. Hill, T.L., *Statistical Mechanics. Principles and Selected Applications*. New York: McGraw-Hill, 1956; reprinted by Dover books.
6. Plischke, M., Bergersen, B. *Equilibrium Statistical Physics*. 3rd ed., Singapore: World Scientific, 2006.

11

Analysis of Binding Equilibria in Terms of Partition Functions

11.1 ALTERNATE EQUIVALENT REPRESENTATIONS OF THE PARTITION FUNCTION

In the previous chapter we developed a statistical mechanical theory of specific binding of ligands to macromolecules at one or several binding sites, and in doing so derived two expressions for the partition function q_{ML_x} of macromolecule with n binding sites complexed with bound ligand L at equilibrium at a given concentration (or better, activity) of ligand. The first expression was derived following the seminal work of Wyman on the basis of stoichiometric binding theory, and thus ultimately on the basis of thermodynamic principles:

$$q_{ML_x} = q_M \sum_{i=0}^{n} \kappa_i a_L^i. \tag{11.1}$$

This equation gives the partition function of a macromolecule with n binding sites in terms of the partition function q_M of macromolecule with no ligand bound to it, the n (stoichiometric) binding constants κ_i, and the activity of ligand a_L (with $\kappa_0 = 1$).

A second expression was derived for this same system on the basis of principles of statistical mechanics (outlined in Chapters 5 and 6)

$$\theta_M \equiv q_{ML_x} = \sum_{i=0}^{n} q_i e^{\beta i \mu_L}. \tag{11.2}$$

Equilibria and Kinetics of Biological Macromolecules, First Edition. Jan Hermans and Barry Lentz.

This expresses the partition function as a sum of products of the partition functions q_i of macromolecular species ML_i with i bound ligands (i ranging from 0 to n; $q_0 \equiv q_M$) with the Boltzmann factors of the chemical potentials μ_L of the i bound ligand molecules.

This chapter is organized as follows. In the remainder of this section, we demonstrate the equivalence of the two expressions for q_{ML_x}. Next, we relate these to ligand binding and thermodynamic properties. We then investigate some of the complications that arise when the binding equilibria are treated in terms of site-specific binding and each stoichiometric species ML_i is explicitly treated as made up of a number of isomeric molecular species, depending on which particular sites are occupied by bound ligand. Finally, we indicate how these results apply to descriptions of coupled equilibria of biological macromolecules in following chapters.

The terms in the sums of the partition function Eq. 11.1, $q_M \kappa_i a_L^i$, and also those in Eq. 11.2, $q_i \exp(\beta i \mu_L)$, are proportional to the concentrations of species ML_i. Accordingly, for these two expressions to be equivalent, the following must be true

$$\frac{c_{ML_i}}{c_M} = \frac{P_i}{P_M} = \frac{q_i \exp(\beta i \mu_L)}{q_M} = \frac{q_M \kappa_i a_L^i}{q_M},$$

or,

$$\frac{q_i \exp(\beta i \mu_L)}{q_M a_L^i} = \kappa_i. \tag{11.3}$$

In order to show that this represents a true equality, we rewrite μ_L in the usual way in terms of μ_L^o and a_L, and replace q_M with an expression that relates q_M to the standard chemical potential μ_M^o of M (Eqs. 10.2 and 10.6),

$$q_M = V q_M^o = V \exp(\beta \mu_M^o), \tag{11.4}$$

and similarly for q_i, which gives

$$\begin{aligned}\frac{q_i \exp(\beta i \mu_L)}{q_M a_L} &= \frac{\exp\left(\beta \mu_i^o\right) \exp\left[\beta i \left(\mu_L^o + k_B T \ln a_L\right)\right]}{\exp(\beta \mu_M^o) a_L} \\ &= \exp[\beta(\mu_i^o + i\mu_L^o - \mu_M^o)].\end{aligned} \tag{11.5}$$

The expression on the right is exactly the thermodynamic equilibrium constant κ_i for the formation of ML_i from $M + iL$; according to Section 1.7, Eq. 1.67,

$$-k_B T \ln \kappa_i = \mu_{ML_i}^o - i\mu_L^o - \mu_M^o. \tag{11.6}$$

This confirms the equality postulated in Eq. 11.3.

11.2 GENERAL IMPLICATIONS

11.2.1 Relation to Chemical Potentials

The first term in θ_M (or q_{ML_x}) is the molecular partition function for the ligand-free macromolecule, $q_0 = q_M$. If we divide every term of θ_M by q_M, we obtain a *relative* molecular grand partition function, φ_M in which every term contains a ratio $q_i/q_M = \delta_i$ that depends only on intensive properties of each possible species relative to the unbound species:

$$\varphi_M = 1 + \sum_{i=1}^{n} \frac{q_i}{q_M} \exp(\beta i \mu_L) = 1 + \sum_{i=1}^{n} \delta_i \exp(\beta i \mu_L) \equiv 1 + \sum_{i=1}^{n} \kappa_i a_L^i. \quad (11.7)$$

(We note that $1/\varphi_M$ equals the fraction of molecules in the unliganded state M.) If we recall that a molecular partition function is related to the exponential of the molecular or unitary free energy (or the standard chemical potential) of a species, we see that

$$k_B T \ln \delta_i = k_B T \ln(q_i/q_M) = \mu_M^o - \mu_{ML_i}^o, \quad (11.8)$$

is the unitary free energy difference between species ML_i and M. Thus, $\delta_i > 1$ means that ML_i is stable and the partition function φ_M is large. We shall find this formulation most useful in making a connection with binding theory in the following sections. In Eq. 11.7, we can recognize the balance between the intrinsic unitary binding free energy of any given molecular species ($-k_B T \ln \delta_i$) and the free energy or "work" of transferring i ligand species between the compartment containing bulk solution and the compartment also containing the macromolecules ($i\mu_L$). This is precisely the relationship that we saw for the grand canonical partition function (Section 6.10). For any concentration of ligand, the species that optimizes this balance will dominate.

Finally, we note that φ_M, being the ratio of the two partition functions θ_M (or q_{ML_x}) and q_0, is related to a difference in standard chemical potentials for the equilibrium mixture of species and the ligand-free species M

$$-k_B T \ln \varphi_M = k_B T \ln(q_{ML_x}/q_M) = \mu_{ML_x}^o - \mu_M^o. \quad (11.9)$$

Accordingly $-k_B T \ln \varphi_M$ represents the difference in unitary free energy between the ligand-free macromolecule M and the equilibrium distribution of species ML_x.

11.2.2 Average Number of Ligands Bound per Macromolecule and Free Energy of Binding

The average number of ligand molecules bound per macromolecule is given by

$$\nu_{\mathrm{L}} = \frac{\sum_{i=0}^{n} i\delta_i \exp(\beta i \mu_{\mathrm{L}}^{\mathrm{o}}) a_{\mathrm{L}}^i}{\sum_{i=0}^{n} \delta_i \exp(\beta i \mu_{\mathrm{L}}^{\mathrm{o}}) a_{\mathrm{L}}^i} = \frac{\sum_{i=0}^{n} i\kappa_i a_{\mathrm{L}}^i}{1 + \sum_{i=1}^{n} \kappa_i a_{\mathrm{L}}^i} = \frac{\sum_{i=0}^{n} i\kappa_i a_{\mathrm{L}}^i}{\varphi_{\mathrm{M}}}$$

$$\nu_{\mathrm{L}} = \frac{\partial \ln \varphi_{\mathrm{M}}}{\partial \ln a_{\mathrm{L}}} = k_{\mathrm{B}} T \frac{\partial \ln \varphi_{\mathrm{M}}}{\partial \ln \mu_{\mathrm{L}}}. \tag{11.10}$$

In the second line of Eq. 11.10, we see that the molecular grand partition function defines the mean composition in the same way as we saw for the grand canonical partition function in Chapter 6.

We can use Eq. 11.9 and substitute for φ_{M} in Eq. 11.10

$$\nu_{\mathrm{L}} = \frac{\partial \ln \varphi_{\mathrm{M}}}{\partial \ln a_{\mathrm{L}}} = \frac{1}{k_{\mathrm{B}} T} \frac{\partial \ln(\mu_{\mathrm{ML}_x}^{\mathrm{o}} - \mu_{\mathrm{M}}^{\mathrm{o}})}{\partial \ln a_{\mathrm{L}}} = \frac{1}{k_{\mathrm{B}} T} \frac{\partial \ln \mu_{\mathrm{ML}_x}^{\mathrm{o}}}{\partial \ln a_{\mathrm{L}}}, \tag{11.11}$$

an important result that had earlier been derived strictly from thermodynamic principles (Chapter 9, Eq. 9.10). From a measured binding curve, that is, values of ν_{L} as a function of ligand activity a_{L} it is then possible by (numerical) integration to construct the dependence of the free energy as function of ligand activity

$$\mu_{\mathrm{ML}_x}^{\mathrm{o}}(a_{\mathrm{L}}) - \mu_{\mathrm{M}}^{\mathrm{o}} = k_{\mathrm{B}} T \int_{a_{\mathrm{L}}=0}^{a_{\mathrm{L}}} \nu_{\mathrm{L}} d(\ln a_{\mathrm{L}}). \tag{11.12}$$

By starting the integration at zero ligand concentration, one thereby obtains *from experimental data* an estimate of the change in standard chemical potential of the macromolecule as a result of binding ligand for a value of the activity of L corresponding to the upper integration limit (which we also named a_{L}; actually the upper limit of the integration is ln a_{L}).

11.3 SITE-SPECIFIC BINDING: GENERAL FORMULATION

Up to this point, we did not consider different distributions of ligands over specific binding sites, for which stoichiometric binding analysis cannot account. However, as has been pointed out in Chapter 8, a common goal of detailed studies of binding in macromolecular solutions is to reach a molecular description in terms of site-specific binding equilibria and site binding constants. In this section, we describe how a partition function may be used to considerable advantage as the basis for analyzing binding phenomena in terms of binding of ligands to specific sites on the macromolecule, a concept first developed in terms of thermodynamic partition functions by Wyman (see Suggested Reading at end of chapter).

Even in the simple case of a macromolecule M with n binding sites complexed with a single ligand L, each complex ML_i (except M and ML_n) is a mixture of isomers differing in the choice of site or sites at which ligand is bound. In the general asymmetrical case, each isomer makes a different, independent contribution to the partition function of species ML_i, the number of distinct isomeric substates of this species being equal to the combinatorial term $C_i^n = n!/[(n-i)!i!]$. The partition function of ML_i can then be written as

$$q_i = q_0 \exp(\beta i \mu_L)\delta_i = q_0 \exp(\beta i \mu_L)\sum_{j=1}^{C_i^n} \delta_{i,j} \equiv q_0 a_L^i \sum_{j=1}^{C_i^n} K_{i,j}, \tag{11.13}$$

by writing the (Boltzmann) factor δ_i as a sum of Boltzmann factors $\delta_{i,j}$ in the free energy difference between ligand-free macromolecule M and the jth isomer of macromolecule with a total of i bound ligand molecules. Alternatively, the ratio $q_{i,j}/q_0$ for each isomer can be expressed in terms of the activity of ligand and the equilibrium constant for converting the ligand-free species to the isomer.

The partition function for the system can be written as a nested sum over all stoichiometric species i and for each stoichiometric species over all distinct molecular species j consistent with that stoichiometry:

$$\theta_M = q_0 + q_0 \sum_{i=1}^{n} \exp(\beta i \mu_L) \sum_{j=1}^{C_i^n} \delta_{i,j}. \tag{11.14}$$

In case the n (here assumed distinguishable) sites are independent, each factor $\delta_{i,j}$ is simply the product of a series of Boltzmann factors in the free energy of adding a ligand to each of the individual sites occupied in the jth configuration, and proportional to the product of the i site binding constants for the occupied sites.

If the sites are not independent, but coupled, the expression for θ_M remains the same, but one or more of the factors $\delta_{i,j}$ are no longer equal to the product of the Boltzmann factors for the individual sites.

We can further generalize the expression to include binding of multiple types of ligand to multiple sites. For example, for binding of two types of ligands, A and B,

$$\theta_M = q_0 \sum_{i_A=0}^{n} \sum_{i_B=0}^{n-i_A} \exp(\beta i_A \mu_A) \exp(\beta i_B \mu_B) \sum_{j} \delta_{(i_A i_B),j}, \tag{11.15}$$

where the summation over j includes all distinct distributions of the i_A plus i_B ligands over the n binding sites (with $\delta = 1$ for the case $i_A = i_B = 0$).

We note briefly that when more than one type of ligand binds to a macromolecule, it is frequently observed that a change in the activity of type A, a_A

changes the extent of binding, ν_B of the other type. As was shown on the basis of thermodynamics in Chapter 9, this response is mutual (Eq. 9.12); an analogous result can be derived by double differentiation of θ_M in Eq. 11.15 and application of Eq. 11.10, remembering also that $\theta_M = q_0\varphi_M$, which gives

$$\frac{\partial \nu_B}{\partial \ln a_A} = \frac{\partial(\partial \ln \theta_M/\partial \ln a_B)}{\partial \ln a_A} = \frac{\partial(\partial \ln \theta_M/\partial \ln a_A)}{\partial \ln a_B} = \frac{\partial \nu_A}{\partial \ln a_B}. \tag{11.16}$$

Finally, one's model may also distinguish between conformational states of the molecule; for example, for a case with a single ligand type but two conformation states labeled T and R one has

$$\theta_{M,T} = q_{0,T} + q_{0,T}\sum_{i=1}^{n} \exp(\beta i \mu_L) \sum_j \delta_{(i,T),j}, \tag{11.17}$$

and similarly for R, with $q_{0,R}/q_{0,T}$ equal to the equilibrium constant K_{TR} for converting the ligand-free T state into the ligand-free R state. The sum of the partition functions for the two states $\theta_{M,T}$ and $\theta_{M,R}$ then equals the partition function of the macromolecule θ_M.

Although the notation adopted here is straightforward, the details require close attention.

11.4 USE OF SINGLE-SITE BINDING CONSTANTS

Ever since Monod, Wyman, and Changeux proposed the allosteric model for binding of oxygen to hemoglobin (to be discussed in Chapter 13), models to describe binding equilibria have been expressed in terms of single-site binding equilibria, using single-site binding constants. It is often possible to establish a "normal" value of the binding constant, in which case the model can advantageously focus on explaining deviations in terms of structure and of coupling to other equilibria.

However, a drawback of setting up one's model in this way is that the number of single-site binding constants grows rapidly with the number of binding sites. As we have seen, for each of the $0 < i < n$ complexes with different number of ligand molecules there are C_i^n unique distributions of i ligands over n sites. The total number of isomers is 2^n, each of which (except the ligand-free reference state) is characterized by a distinct value of $\delta_{i,j}$ or $K_{i,j}$ (Eq. 11.13). Unless this is assumed differently for a specific choice of model, the $2^n - 1$ constants are independent and together completely determine the equilibrium distribution.

The problem is that these 2^n possible molecular species are connected by $n \times 2^{n-1}$ possible single-site binding reactions, and the equilibrium constants for these single-site steps are clearly not independent. (See also Section 8.4.2.) The reason the number of possible site binding constants is so much larger than n is that each site binding constant depends on the distribution of ligands over all other sites. Even in the case of $n = 2$, of $2^n = 4$ unique species, only $2^n - 1 = 3$ independent site binding constants can be chosen (see Fig. 8.10). The case of $n = 3$ is illustrated in Fig. 11.1. In the left frame, we see that $7(2^n - 1)$ equilibria between ligand-free M and each other unique species can define the equilibrium (which was of

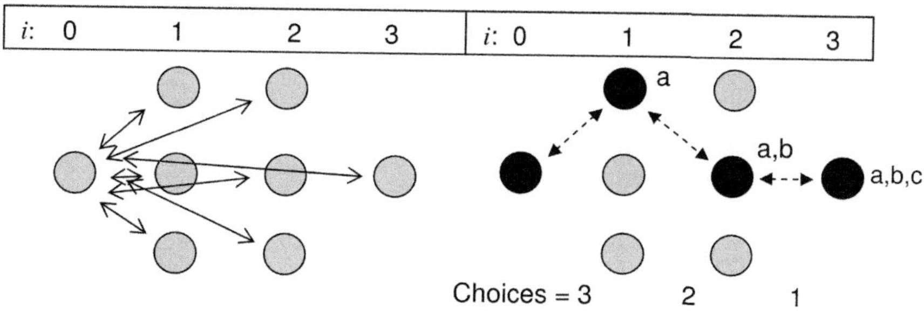

FIGURE 11.1 Two diagrams illustrating two valid ways of specifying $7(2n-1)$ independent equilibrium constants for the $8(2^n)$ different molecular species possible for the case of $n=3$ binding sites. The diagram on the left shows a set of equilibria that connect the unliganded macromolecule directly to the seven unique molecular species. The connections in the diagram on the right show one of six possible paths to obtain ML_3 by connecting molecular species using three steps involving site binding constants for sites a, *b*, or c in that order. The site occupancy is indicated beside each molecular species. In the network on the right, each species has just one connection to a species with a smaller value of i. But as ligands are added the number of choices for each step drops by 1, so that the number of possible paths to reach any species ML_i is the product $(n-i+1)\cdots 1$.

course the approach followed in Section 11.4.2). But in the right frame we see that as many as six different schemes consisting of successive one-step reactions, executed in different order, lead from M to ML_3, while a single step-wise path is adequate to account for each species (right frame of Fig. 11.1 shows one such path to get from M to ML_3). For $n=4$, there are a total of 16 possible species and a total number of 32 possible single-site binding constants, only 15 of which are independent. Table 11.1 summarizes key facts.

A sensible approach is then, first to define the equilibrium in terms of a partition function as in Eq. 11.14. This has the advantage that, for any choice of values of the 2^n-1 independent factors $\delta_{i,j}$ or $K_{i,j}$, any (measurable) mean equilibrium system property X can be readily evaluated from the (known) contribution $X_{i,j}$ that each species makes to the value of X. This is most simply expressed in terms of the probability distribution $P_{i,j}$ with

$$\langle X\rangle = P_0X_0 + \sum_{i=1}^{n}\sum_{j} P_{i,j}X_{i,j}$$

$$\langle X\rangle = \frac{1}{\varphi_{\mathrm{M}}}\left[X_0 + \sum_{i=1}^{n}\exp\left(\beta i\mu_{\mathrm{L}}\right)\sum_{j}\delta_{i,j}X_{i,j}\right]$$

$$\equiv \frac{1}{\varphi_{\mathrm{M}}}\left[X_0 + \sum_{i=1}^{n}a_{\mathrm{L}}^{i}\sum_{j}K_{i,j}X_{i,j}\right]. \tag{11.18}$$

TABLE 11.1 Summary of molecular species associated with adding a single type of ligand L, to a macromolecule with *n* sites

Total number of sites	n
Stoichiometric species: ML_i	i
Total number of stoichiometric species	n
Total number of full stoichiometric binding constants, κ_i	$n-1$
Number of molecular species associated with a stoichiometric species	C_i^n
Total number of molecular species $=\sum_{i=0}^{n} C_i^n$	2^n
Total number of possible paths to the molecular species ML_n	$\prod_{i=1}^{n}(n+1-i)$
Total number of single-site binding constants	$n \times 2^{n-1}$
Total number of unique site binding constants required to define all molecular species	$2^n - 1$

An important example of a measurable property is the average number of ligands bound, that is, the titration curve of macromolecule with ligand. We recall from Chapter 8 that even the most precise and complete titration can give only n independent constants (the stoichiometric binding constants κ_i), which for $n > 1$ is an insufficient number to determine the $n \times 2^{n-1}$ independent constants $\delta_{i,j}$ or $K_{i,j}$. We see that additional *types* of measurement are essential we are to obtain a complete description in terms of site binding constants. For example, changes of ultraviolet or visible absorption spectra or of nuclear magnetic resonance spectra are useful signals to quantitate binding to individual sites or to particular classes of sites.

Thus, we have been able to devise a clear path to formulate the independent constants for any given scheme, and evaluate equilibrium properties; the difficulty is to design and choose precise measurement techniques that can produce a sufficiently large set of independent data for determining the independent constants experimentally and deciding which model best fits these data.

11.5 PARTITION FUNCTION FOR SITE BINDING: ONE TYPE OF LIGAND, INDEPENDENT MULTIPLE SITES

In the simplest case, the ligand binding sites do not interact, and the affinity of any one site for ligand is unaffected by how many and which other sites have a ligand bound. In that simple case, each Boltzmann factor $\delta_{i,j}$ in which sites a, b, and so on, have ligand bound, equals the product $\delta_{1,a}\delta_{1,b}\cdots$ of the Boltzmann factors for the species in which *only* site a, site b, and so on, have bound ligand; similarly, $K_{i,j} = K_{1,a}K_{1,b}\cdots$.

As a further simplification of the model one may frequently assume that all sites of a given type have the same affinity (e.g., the four binding sites of hemoglobin for oxygen, or a protein's lysine side chains for hydrogen ion). In that case all (C_i^n) $\delta_{i,j}$ have the same value $\delta_{i,j} = \delta_1^i$ and $K_{i,j} = K_1^i$. Accordingly, the partition function Eq. 11.14 comes

$$\theta_{\mathrm{M}} = q_0 + q_0 \sum_{i=1}^{n} \exp(\beta i \mu_{\mathrm{L}}) C_i^n \delta_1^i] = q_0 \sum_{i=0}^{n} \exp(\beta i \mu_{\mathrm{L}}) C_i^n \delta_1^i]. \tag{11.19}$$

It is easy to verify that the sum is a binomial expansion; thus, the partition function for binding to independent and equivalent sites is

$$\theta_{\mathrm{M}} = q_0[1 + \exp(\beta \mu_{\mathrm{L}})\delta_1]^n \equiv q_0[1 + a_{\mathrm{L}} K_1]^n. \tag{11.20}$$

We note that the expression raised to the n-th power is, in fact, a (dimensionless) partition function for binding to a single site, and that the partition function for binding to multiple independent sites is the product of the partition functions for the n sites. This is another example of the principle of multiplication of partition functions of independent components that we have encountered several times. We can extend this to obtain an expression for the case of independent sites with different affinities, and write the partition function for binding to multiple independent nonequivalent sites as a product

$$\theta_{\mathrm{M}} = q_0 \prod_{x=1}^{n} [1 + \exp(\beta \mu_{\mathrm{L}})\delta_{1,x}] \equiv q_0 \prod_{x=1}^{n} (1 + a_{\mathrm{L}} K_{1,x}). \tag{11.21}$$

where the index x in turn indicates each individual site.

You may differentiate these expressions and obtain expressions for the average number of bound ligand molecules for these two models according to Eq. 11.10, and compare with results of Sections 8.4.3 and 8.4.4.

11.6 SITE BINDING TO INTERDEPENDENT OR COUPLED SITES

In the general case, the site binding constants depend on the state of the other sites, and when this is observed, we speak of coupled sites. Coupled equilibria are common in biological macromolecules, and often the coupling has important consequences for biological function.

We can formally describe the general case by referring the affinities of each state with ligand bound to multiple sites to the affinities of each individual occupied site when no other ligand is bound, as follows:

$$\begin{aligned} \delta_{i,j} &= \sigma_{i,j} \prod_x \delta_{1,x} \\ K_{i,j} &= \sigma_{i,j} \prod_x K_{1,x} \end{aligned}.$$

The products are over the particular i sites that have ligand bound in configuration j. This reformulation does not reduce the complexity of the problem, but it does separate the intrinsic affinity of the sites from the coupling factors $\sigma_{i,j}$ that reflect just interactions between sites. This separation is especially useful if one can ascribe the coupling to *direct site–site interaction*, such as that due to electrostatic repulsion of charged bound ligand molecules (ions).

However, in most, if not all, binding equilibria that play a role in biological function the coupling is indirect. This *indirect coupling* is due to a change in the conformation of the protein that is coupled to binding of multiple ligand molecules of the same or different type. In the 1960s, new measurements and new ideas about the nature of coupled equilibria in proteins ("linked functions"), including many regulated enzymes and, of course, hemoglobin, created the need for a new theoretical framework. Jeffries Wyman first developed such a framework, following ideas borrowed from statistical mechanics, and this proved enormously successful in explaining many observations. The parallel with statistical mechanics was explicitly recognized by Wyman and Gill; we have here recast this theory in terms of statistical mechanical partition functions.

Coupled equilibria are the principal topic of the following three chapters of this book. Chapter 12 discusses general principles and contains a discussion of self-association of monomeric proteins to linear and helical polymers and coupling to binding of ligand. Chapter 13 discusses allosteric proteins with main focus on hemoglobin as the example, and in Chapter 14 we discuss interactions of protein molecules with hydrogen ion and solvent.

SUGGESTED READING

1. Wyman, J. Linked functions and reciprocal effects in hemoglobin: A second look. *Adv. Protein Chem.* **19**: 223–286 (1964).
2. Wyman, J., Gill S. J., *Binding and Linkage: Functional Chemistry of Biological Macromolecules*. Mill Valley, CA: University Science Books, (1990).

12

Coupled Equilibria

12.1 INTRODUCTION

As mentioned, proteins are molecular machines that transduce free energy, in the sense that the free energy from one process involving the macromolecule is used to drive another process involving the same macromolecule. Indeed, the interactions between events at different sites of the same molecule give rise to unique properties of biological macromolecules that are exploited over and over again in the functioning of complex biological systems. Typically, a macromolecule is subject to multiple processes, which can be conformation change, binding to other small or large molecules, self-dimerization, and polymerization. These processes do not necessarily interact, but if they do, they are said to be coupled or linked.

The following are a few of the many cases where interaction between ligand binding and conformation change is at the core of a biological mechanism. Cooperative binding of oxygen, H^+, and glycerol diphosphate to hemoglobin are all coupled to conformation changes, and similar effects regulate the activity of many multimeric enzymes. In the bacterial photosystem, light absorption drives H^+ across a membrane against a concentration gradient, while their return drives synthesis of adenosine triphosphate (ATP). In active transport across membranes, hydrolysis of ATP drives molecules across a membrane against a concentration gradient. Hydrolysis of ATP in motor proteins in muscle and elsewhere causes conformation changes which are transduced into a mechanical force.

Our general understanding of the coupling underlying such biologically relevant systems has been provided with detailed studies of the three systems, which

Equilibria and Kinetics of Biological Macromolecules, First Edition. Jan Hermans and Barry Lentz.

are discussed in the following chapters, namely oxygen binding to hemoglobin, helix-coil transitions of peptides, and protein unfolding equilibria.

In this chapter, we discuss first the important case of coupling of conformation change and ligand binding; we introduce the allosteric or induced-fit model, and show how coupling to multiple sites leads to strong dependence of the conformation change on the concentration of ligand, and to so-called *cooperativity of the binding.* In the second part of the chapter we discuss self-association of macromolecules to form oligomers and polymers, and coupling of self-association and ligand binding, as well as coupling of successive self-association steps. Coupling of successive local conformation changes and the resulting cooperativity of conformational transitions of macromolecules are discussed in the context of helix-coil transitions of polypeptides in Chapter 16 and protein unfolding equilibria in Chapter 17.

12.2 SIMPLE CASE: COUPLING OF BINDING (ONE SITE) AND CONFORMATION CHANGE

A case of coupled binding of ligand to two sites on a macromolecule was discussed in the final section of Chapter 8. The mechanism of the coupling was left unspecified. As mentioned, most important cases of coupled ligand binding involve coupling to a conformation change. We consider first the simplest case, of a protein that can exist in two conformations, A and B, and that can bind a ligand molecule, L. We can describe this with the following scheme with two binding equilibria (one each for the A and B conformations) and two conformational equilibria (one for the protein without ligand, and the other for the protein with ligand bound)

$$\begin{array}{ccc} & \Delta G_A^\circ & \\ M_A + L & \leftrightarrow & M_AL \\ \Delta G_0^\circ \downarrow\uparrow & & \downarrow\uparrow \Delta G_L^\circ \\ M_B + L & \leftrightarrow & M_BL \\ & \Delta G_B^\circ & \end{array} \tag{12.1}$$

where ΔG_A° and ΔG_B° are the standard free energies of binding ligand to the two different conformations, and ΔG_0° is the standard free energy for the conformational equilibrium between states A and B in the absence of bound ligand, and ΔG_L° the same, but with a ligand molecule bound to the macromolecule.

In order for this to be a case of coupled equilibria, the affinity for ligand in these two conformations must be different, that is,

$$\Delta G_A^\circ \neq \Delta G_B^\circ. \tag{12.2}$$

One obtains an important result by stepping clockwise through the reaction cycle, starting at the upper left, converting 1 mole of M_A and 1 mole of L at standard conditions to 1 mole of M_BL at standard conditions and then returning to the starting state. As the free energy is a state function, the net free energy change in this cyclic process is equal to zero

$$\Delta G^\circ_A + \Delta G^\circ_L - \Delta G^\circ_B - \Delta G^\circ_0 = 0, \quad (12.3)$$

or,

$$\Delta G^\circ_A - \Delta G^\circ_B = \Delta G^\circ_0 - \Delta G^\circ_L. \quad (12.4)$$

This difference is the standard free energy characteristic of the coupling; if the difference is zero, then there is also none. Note the reciprocity: the standard free energy for the binding equilibrium depends on the state of equilibrium A, and the standard free energy for the conformation change depends on the state of equilibrium B, and the difference is the same for both.

We have here stated the coupling in terms of relations between thermodynamic functions, as this is particularly convenient. It is actually very easy to reformulate Eq. 12.4 in terms of equilibrium constants

$$\frac{K_B}{K_A} = \frac{K_L}{K_0}, \quad (12.5)$$

where K_A and K_B are binding constants for the two conformations, K_L is the equilibrium constant for the conformation change when ligand is bound, and K_0 is the equilibrium constant for the conformation change when no ligand is bound.

Expressing the equilibrium constants as ratios of partition functions, this last equation reduces to a simple equality

$$\frac{q^o_{MBL}}{q^o_{MB}} \bigg/ \frac{q^o_{MAL}}{q^o_{MA}} = \frac{q^o_{MBL}}{q^o_{MAL}} \bigg/ \frac{q^o_{MB}}{q^o_{MA}}.$$

The term *coupling* or *linkage* describes all cases in which the *state* of one equilibrium, A, affects the values of thermodynamic functions for another equilibrium, B, that is, *whenever* ΔG°_B (the *standard* free energy for equilibrium B) depends on the state of equilibrium A.

12.3 COUPLING OF BINDING TO MULTIPLE SITES AND CONFORMATION CHANGE

12.3.1 The Allosteric, Induced-fit, and Conformational Selection Models

The description of coupled binding sites given in Section 8.4.5 made no assumptions as to the reason why the sites were coupled. *Coupling between sites often implies a conformational change.* This might not be the case if the sites are closely juxtaposed and occupancy of one interferes with binding to the other. However, most cases of coupling are best discussed on the basis of specific models of binding accompanied by changes in a protein's conformation state.

A typical model assigns two (or more) alternative conformations to the protein, and a set of binding constants for one or more ligand species to each conformation. In the absence of long range interactions[1] between bound ligands, the binding sites are independent, and the apparent dependency that is observed in practice is attributed to conformation changes. In other words, binding to site 1 in conformation A does not change the affinity for site 2 in conformation A, but binding to site 1 changes the equilibrium in favor of conformation B, in which the affinity for site 2 is different (Fig. 12.1).

The model obviously implies that one conformation, A, is more stable when no ligand is bound and that the *other* conformation, B, is more stable when the ligand binding sites are filled. Hence,

$$K_{\mathrm{AB}} < 1$$
$$K_{\mathrm{AB}}\left(K_{\mathrm{B}}/K_{\mathrm{A}}\right)^2 > 1. \tag{12.6}$$

or in terms of partition functions

$$q^{\mathrm{o}}_{\mathrm{MB}}/q^{\mathrm{o}}_{\mathrm{MA}} < 1$$
$$q^{\mathrm{o}}_{\mathrm{MBL_2}}/q^{\mathrm{o}}_{\mathrm{MAL_2}} > 1. \tag{12.7}$$

A model that ascribes coupling of binding sites to conformation change is called an *allosteric model* (= with alternative shapes). This model was originally formulated by Jacques Monod, in studies of how a bacterial gene known as the *lac* operon allows for the effective digestion of lactose, but only when lactose is present as a nutrient. It achieves this with the *lac* repressor, which halts the production in the

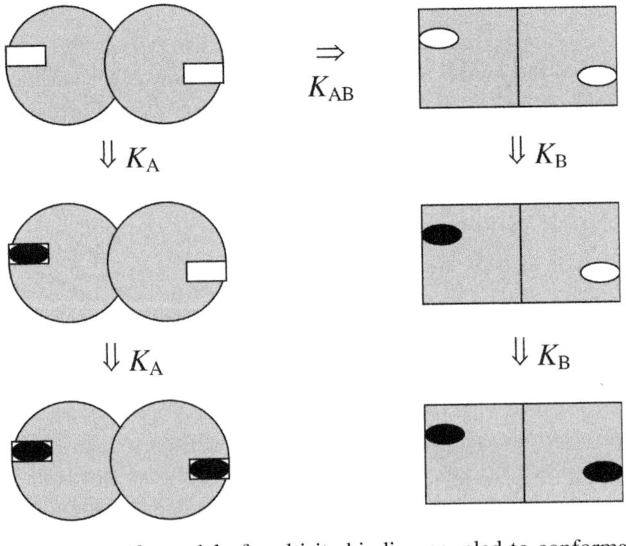

FIGURE 12.1 A *specific* model of multisite binding coupled to conformation change.

absence of lactose by repressing the synthesis of β-galactosidase, β-galactoside permease, and thiogalactoside transacetylase. Monod (working with François Jacob) determined that the *lac* repressor protein prevents transcription of the gene, and hypothesized that this protein can exist in two conformations, one with high affinity for the *lac* operon, and the other with high affinity for lactose, but insufficient affinity for the *lac* operon. (Monod's application of allostery to understand the unusual ligand-binding properties of hemoglobin will be discussed in the following chapter.)

Which conformation of an allosteric protein is observed, is determined by the presence (and concentration) of the ligand. It is appropriate to think of the conformation change as a response to the binding of ligand, something that is expressed well by the term *induced fit* (which was originally introduced by Dan Koshland). The term *induced fit* implies that a conformation change coincides with binding of ligand. More recently, the term *conformational selection* has been introduced into discussions of allosteric proteins. This term suggests that the ligand selects a particular, in the absence of ligand minor, conformation by strong binding.

It follows from a simple thermodynamic argument that, as long as one is concerned with equilibrium properties, the terms (ligand dependent) allostery, induced-fit, and conformational selection are interchangeable. Consider the complete scheme of two conformations, A and B of the same protein that is able to bind a ligand, L given in Eq. 12.1, in which the dominant species are M_A and M_BL. We introduce also a standard free energy difference for the dominant experimentally observed equilibrium $M_A + L \leftrightarrow M_BL$, $\Delta G^\circ_{A\to BL}$. We do not know its value, but we do know that the value is "reasonable", as the equilibrium is (by definition) measurable. A mathematician will say that $\Delta G^\circ_{A\to BL}$ is *bounded*. Also, the free energy of ligand binding to conformation B, $\Delta G^\circ_{B\to BL}$ is bounded, due to physical limitations on the molecular interactions between protein and ligand. Finally, we know from thermodynamics that $\Delta G^\circ_{A\to BL} = \Delta G^\circ_{A\to B} + \Delta G^\circ_{B\to BL}$. Consequently, also the standard free energy for the conformation of free protein from state A to state B, $\Delta G^\circ_{A\to B}$ is bounded, and hence, under equilibrium conditions, the proportion of state B is not zero, but finite. A similar argument will show that the proportion of state AL is not zero (except in the complete absence of ligand L). The concentrations of states B (ligand-free) and of state AL may be small and not measurable with current techniques; however, the advent of more sensitive measurement techniques will reveal the presence of one or the other of these states in many instances where these had thus far remained unresolved[2] (As has been the case thanks to recent advances in NMR spectroscopy.[3])

The term induced fit may seem appropriate to describe cases in which a disorganized macromolecule, p_u assumes a specifically ordered form, p_f when bound to a ligand molecule as p_fL. (See discussion of natively unfolded proteins in Section 17.12.) Even so, the ordered conformation, p_f must be relatively stable in the absence of ligand, as the need to form the ordered form, $p_u \to p_f$, decreases the net affinity of L for p_u, relative to the, ultimately limited (bounded), affinity of L for p_f.

In kinetic studies these different terms can make meaningful distinctions between different kinetic mechanisms: if in a first step the macromolecule changes conformation, after which the ligand binds, one should call this a case of conformational selection, whereas, if first the ligand binds weakly to the macromolecule and subsequently the conformation changes, this should be described as induced fit. Kinetic studies in which the concentration of ligand is changed can distinguish whether the conformation change from M_A to M_BL predominantly follows a path via state M_AL ("induced fit") or via state M_B ("conformational selection") as defined in Eq. 12.1.[4] The presence of measurable equilibrium concentrations of the intermediate states, i.e., of M_B in equilibrium with M_A in the absence of L, or of M_AL in equilibrium with M_BL in the presence of L, indicate that one or the other may be the preferred intermediate. However, this cannot be definitively concluded without additional measurement of the rate constants along the two paths that lead from M_A to M_BL. Analysis shows that, at least in some cases, the same system can predominantly adopt either path depending on the concentrations of protein and ligand.[5]

12.3.2 Evaluation of a Simple Allosteric Model

Given expressions for the partition functions of distinct molecular species, one can numerically evaluate the properties of such a model by expressing the concentrations of all species in terms of the partition function of a single species, usually the unliganded form in one particular conformation state and calculating results for a series of values of ligand activity, a_L. For example, for one ligand species and two protein conformations we have

$$c_{M*} = \sum_i c_{M_AL_i} + \sum_i c_{M_BL_i} = c_{M_A*} + c_{M_B*}. \tag{12.8}$$

or, in terms of partition functions,

$$q^o_{M*} = q^o_{M*_A} + q^o_{M*_B}. \tag{12.9}$$

The partition function of conformation A in all ligand-bound states is

$$q^o_{M_A*} = q^o_{M_A} \sum_i \kappa_{A,i} a^i_L. \tag{12.10}$$

where $\kappa_{A,i}$ is a stoichiometric-binding constant that can be expressed in terms of site-binding constants, and similarly for species $q^o_{M_B*}$, while the partition functions for M_B and M_A are related by an additional adjustable equilibrium constant

$$q^o_{M_B} = K_{AB} q^o_{M_A}. \tag{12.11}$$

The equilibrium concentrations of the two conformations are in proportion to these partition functions, so that the fraction of molecules in state A is

$$f_A = q_{M_A*} / (q_{M_A*} + q_{M_B*}). \tag{12.12}$$

The average number of ligands bound to the protein molecules in conformation state A is

$$\nu_{L,A} = \frac{\sum i q^{o}_{M_A L_i}}{q^{o}_{M_A *}}, \tag{12.13}$$

and the average number of ligands bound to all protein molecules is

$$\nu_L = f_A \nu_{L,A} + f_B \nu_{L,B}. \tag{12.14}$$

If the n sites in state A are equivalent, with binding constant K_A, then of course

$$\nu_{L,A} = n \frac{K_A a_L}{1 + K_A a_L}.$$

These equations provide the means to evaluate these models and find the best values of the equilibrium constant by comparing with experimental results.

If formation of oligomers or polymers does not occur (or, if it does, is not coupled to ligand binding), each partition function q is proportional to every other partition function (or, the concentration of each species is proportional to the concentration of every other species). In that case, the expressions for f_A and ν_L, etc. do not depend on the total macromolecule concentration. However, this is not true if the protein molecules form oligomers or polymers. In that case one may again express the concentrations of all species in terms of the concentration of monomer without ligand c_M, but the concentrations of dimers and higher oligomers are proportional to $c_M{}^2$ and higher powers. This then requires solution of a nonlinear equation relating c_M to the (known) total protein concentration. (See also Sections 12.5 and 12.6).

12.3.3 Cooperativity

According to the allosteric model, one conformation, A is more stable when no ligand is bound and the other conformation, B is more stable when the ligand-binding sites are filled. Thus, state M_A is always present in greater proportion than state M_B, and state M_BL_2 is always present in greater proportion than state M_AL_2. It is also the case, but less obviously, that the intermediates M_AL and M_BL are less prevalent than would be the case in the absence of the linkage. To show this, consider the midpoint of the binding, which is also the ligand concentration at which the concentrations of M_A and M_BL_2 are equal, for which

$$K_{AB}(a_L K_B)^2 = 1,$$
$$a_L K_B = K_{AB}{}^{-1/2} > 1, \tag{12.15}$$

(because $K_{AB} < 1$). Similar reasoning shows that $a_L K_A < 1$, while in the absence of linkage; that is, when $K_A = K_B = K$, the product $aK = 1$. That is, the species

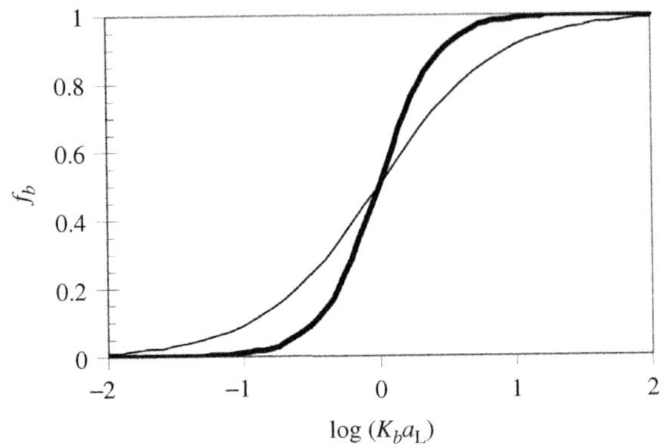

FIGURE 12.2 Binding curves, as fraction of sites with bound ligand (or saturation) *versus* log ($K_b a_L$), for (extremely) cooperative (steepest curve) and noncooperative two-site models.

M_AL is "disadvantaged" relative to M_A and the species M_BL is "disadvantaged" relative to M_BL_2.

In the extreme case, when $K_{AB} \ll 1$ and $K_B/K_A \gg 1$, the system becomes very simple, as only two species, M_A and M_BL_2 play a significant role at any value of a_L. In such a case, binding is said to be cooperative: any given macromolecule has either no ligand, or has two molecules of ligand bound to it. Also, in the extreme case, the equilibrium is adequately described with a one-step scheme

$$M_A + 2L \leftrightarrow M_BL_2, \tag{12.16}$$

for which the binding curve is twice as steep as it is in the absence of coupling (Fig. 12.2).

In general, the steepness of the binding curve for extremely cooperative n-site binding is n times the steepness in the non-cooperative case. As the conformation change is coupled to the binding of ligand, not only the extent of binding, but *also the conformation changes (in the extreme case) n times more rapidly with the ligand concentration.* Of course, in actually observed cases the ratio is less than in the extreme case, with values around three not atypical of proteins that show cooperative binding to four sites (such as hemoglobin). Note that the results obtained in this section are a special case of the general formulation given in Section 12.4, in particular Eq. 12.21.

12.4 FREE ENERGY OF BINDING CAN "DRIVE" CONFORMATION CHANGE

The description of coupling in terms of specific binding interactions given above is appropriate in many cases of biological significance. We discuss here a more general treatment based on thermodynamics, using the results of Chapter 9.

We again consider the case of two conformations of a macromolecule, A and B, each capable of binding ligand L, but with different affinities. Form A has the higher affinity for ligand, but, without bound ligand is less stable than form B, that is, has a larger standard chemical potential than form B,

$$\mu^{o}_{A,a_L=0} > \mu^{o}_{B,a_L=0}. \tag{12.17}$$

The binding of ligand to these two forms is schematically described in Fig. 12.3. At low values of a_L no ligand is bound to either form; at very high values of a_L both forms are saturated with ligand; while, in between, in the region where the number of ligand molecules bound, ν_L is different for the two conformations, with $\nu_{L,A} > \nu_{L,B}$. Consequently, at low $a_L \cdot \mu$ of neither form depends on a_L, while at very high values of $a_L \cdot \mu$ both forms change identically with a_L. Only where $\nu_{L,A} > \nu_{L,B}$, does $\mu_A - \mu_B$ depend on a_L; here μ_A decreases more rapidly than μ_B and the fraction f_A of molecules in state A increases. If the interval between the two binding curves is sufficiently large, then, at some value of a_L, μ_A will drop below μ_B and the predominant conformation will become A.

Given the dependence of chemical potential of a macromolecule M on ligand activity that results from binding of ligand, expressed in Eq. 9.11, we have

$$\Delta\mu^{o}_{M} = \mu^{o}_{M,a_L} - \mu^{o}_{M,a_L=0} = -k_B T \int_{a_L=0}^{a_L} \nu_L d(\ln a_L),$$

and applying this to both conformations, we obtain

$$\Delta\mu^{o}_{A} - \Delta\mu^{o}_{B} = -k_B T \int_{a_L=0}^{a_L} d(\ln a_L)(\nu_{L,A} - \nu_{L,B}). \tag{12.18}$$

The two binding curves show dashed in Fig. 12.3, and the solid curve indicates the transition from A to B.

The midpoint of the transition, where $f_A = 1/2$, occurs at that value of a_L, $a_L(1/2)$ at which the two conformations have the same chemical potential, that is, when

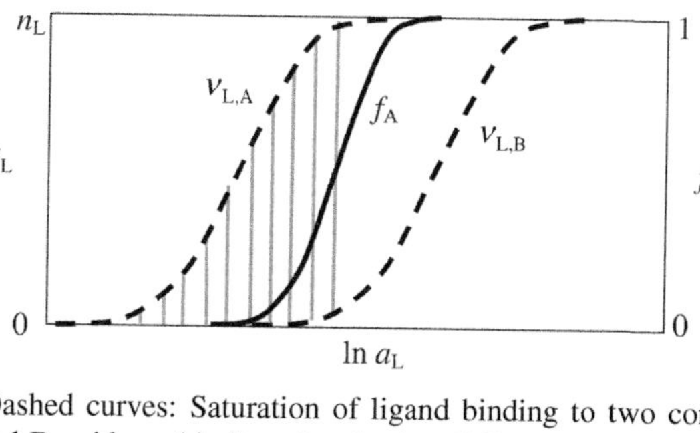

FIGURE 12.3 Dashed curves: Saturation of ligand binding to two conformations of the same protein, A and B, with n_L binding sites having different affinity for ligand in the two conformations. Solid curve: conformation changes from B to A with increase of a_L.

$$\mu^{\rm o}_{{\rm A},a_{\rm L}} - \mu^{\rm o}_{{\rm B},a_{\rm L}} = 0$$

$$\mu^{\rm o}_{{\rm A},a_{\rm L}=0} - \mu^{\rm o}_{{\rm B},a_{\rm L}=0} = k_{\rm B}T \int_{a_{\rm L}=0}^{a_{\rm L}(1/2)} d(\ln a_{\rm L})(\nu_{\rm L,A} - \nu_{\rm L,B}). \qquad (12.19)$$

The integral in the last equation corresponds to the area between the two binding curves that has been indicated with vertical hashing.

The ratio of the concentrations of all species B to all species A may be considered an equilibrium constant; however, this ratio does depend on the ligand activity. In the usual manner, one has

$$K_{{\rm AB},a_{\rm L}} = \frac{\sum[{\rm BL}_j]}{\sum[{\rm AL}_j]},$$

$$-k_{\rm B}T \ln K_{{\rm AB},a_{\rm L}} = \mu^{\rm o}_{{\rm B},a_{\rm L}} - \mu^{\rm o}_{{\rm A},a_{\rm L}}. \qquad (12.20)$$

and

$$\frac{\partial \ln K_{{\rm AB},a_{\rm L}}}{\partial \ln a_{\rm L}} = -\frac{1}{k_{\rm B}T}\frac{\partial(\mu^{\rm o}_{{\rm B},a_{\rm L}} - \mu^{\rm o}_{{\rm A},a_{\rm L}})}{\partial \ln a_{\rm L}} = \nu_{\rm L,B} - \nu_{\rm L,A}. \qquad (12.21)$$

This is sometimes called the *Wyman linkage function*. One sees that the ratio of the two conformations depends more strongly on the ligand activity, the greater the difference in the number of ligands bound in the two conformations. *Having several binding sites for the same ligand on the same protein (typically in a multimeric protein), all coupled to a conformation change, provides a mechanism for a sensitive response to changes in the concentration of the ligand.* The binding in such cases is said to be cooperative, but, in fact this cooperativity is only indirect. It is more insightful to say that the conformation change is cooperative, in the sense that all binding sites have the same conformation, which is determined by the state of the protein molecule. The binding then simply follows the conformation change.

One sees that the approach given above has produced a general treatment of coupled binding and conformation change without presupposing a specific model of the binding phenomenon. However, we hasten to add that all or nearly all cases of coupled binding and conformation change of importance in biology can be described in terms of binding of ligands to precisely defined binding sites, and that the equations describing binding of ligand to (multiple) binding sites should be used in these cases to predict the dependence of the saturation, $\nu_{\rm L}$ as a function of the ligand concentration or ligand activity.

12.5 FORMATION OF OLIGOMERS AND POLYMERS

Many proteins are made up of subunits, which are, in fact, smaller protein molecules that associate to form more or less stable complexes. The subunits may be identical or different. For example, human hemoglobin A (Hb A) contains two α-chains and two β-chains, and is a symmetrical dimer of $\alpha\beta$-units. If the complex is unstable, and dissociates under physiological conditions, this will have consequences if the

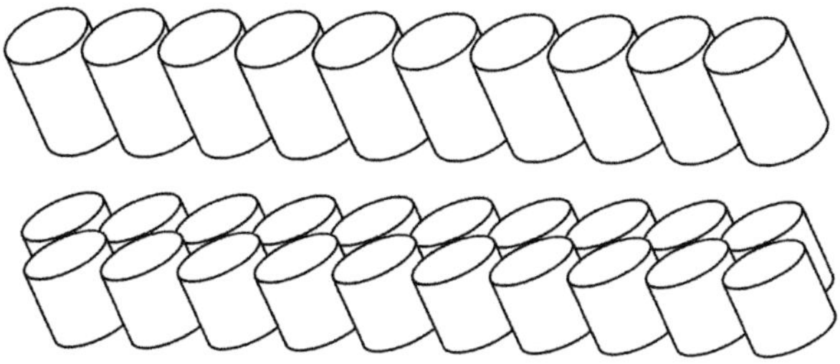

FIGURE 12.4 Examples of "linear" (top) and "helical" (bottom) polymers of subunits. In any linear polymer each subunit touches exactly two others; in this particular helical polymer, each subunit touches four others, and addition of one subunit at the end of the helix gives two new intersubunit contacts. (Actual polymers typically are twisted, rarely flat as shown here and in Fig. 12.5.)

protein's biological activity depends on the aggregation state, and if, in turn, this state is controlled by binding of ligands.

A number of proteins, including many of the proteins that form the cytoskeleton (such as actin and tubulin), self-associate to form polymers of indefinite length.[6] In general, the geometric structure of such a polymer of protein subunits approximates a helical arrangement. (See Section 15.3.) However, when the polymer is stabilized only by contacts between successive subunits (unit i interacting with units $i+1$ and $i-1$), and addition of one subunit at the end of the chain gives a single new intersubunit contact, we follow the convention introduced in note 6, and speak of a *linear polymer*. Only if each subunit forms stabilizing contacts with more than two other subunits, do we speak of a *helical polymer* (Fig. 12.4).

A polymer can be symmetric, or asymmetric, depending on the arrangement of the subunits. In most cases, the polymer is asymmetric, but in specific cases, for example, if the subunits are themselves arranged in symmetrical pairs, with their symmetry axes perpendicular to the helical axis, the polymer is also symmetric. This is shown in Fig. 12.5. Asymmetric polymers, such as actin fibers, can serve as tracks along which molecular motors (myosin) move cargos in one direction and not the other, or grow rapidly longer (and shorter) at one end, and not the other (microtubules).

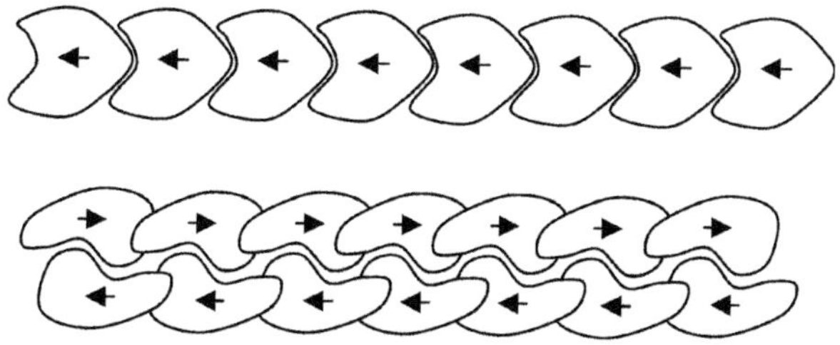

FIGURE 12.5 Top: Polymer with subunits aligned in the same direction has "plus" and "minus" ends. Bottom: Special arrangements of subunits can produce a polymer that does not have a "sense."

In order to describe the *polymerization equilibrium of linear polymers*, one uses a simple model with a single equilibrium constant for forming a dimer or for elongation of a polymer of any length by one subunit, according to

$$M_{i-1} + M \rightleftarrows M_i$$

$$\frac{c_{M_i}}{c_{M_{i-1}}} = K_p c_M$$

$$\frac{c_{M_i}}{c_M} = \{K_p c_M\}^{i-1}; \; i > 0. \tag{12.22}$$

In order to relate the total concentration of protein, c_{M^*} to the concentration of free monomer, c_M and the equilibrium constant for polymerization, K_p we can express the distribution, ϑ over monomer and all polymer species in terms of a sum of molar concentrations

$$\vartheta = c_{Mp*} = c_M + c_{M_2} + \cdots + c_{M_i} \cdots$$

$$\vartheta_l = c_M \sum_{i=0}^{\infty} \{K_p c_M\}^i = \frac{c_M}{1 - K_p c_M}, \tag{12.23}$$

where the subscript l denotes linear polymerization. If the function ϑ is differentiated with respect to the concentration of free *subunits* c_M, one obtains a sum of terms that correspond to the distribution of the *subunits* over the different species. This sum then equals the concentration of bound plus free subunits, c_{M^*}

$$\frac{\partial \vartheta_l}{\partial c_M} = \frac{1}{c_M}(c_M + 2c_{M_2} + \cdots + ic_{M_i} \cdots) = \frac{c_{M^*,l}}{c_M}$$

$$\frac{c_{M*,l}}{c_M} = \frac{1}{\{1 - K_p c_M\}^2}. \tag{12.24}$$

The result is a quadratic equation with root

$$Kc_M = 1 + \frac{1}{2y}\left[1 - \frac{1}{2}\sqrt{1 + 4y}\right]$$

$$y = K_p c_{M^*,l}. \tag{12.25}$$

(the negative square root giving the required value $c_M = c_{M^*}$ for $y = 0$.)

The mean polymer size is

$$\langle i \rangle_l = i_{n,l} = \frac{c_{M^*,l}}{\theta_l} = \frac{1}{1 - K_p c_M}. \tag{12.26}$$

The mean polymer size, $\langle i \rangle$ is also called the number average degree of polymerization, i_n. Of interest is also the larger "weight average" polymer size, i_w

$$i_{w,l} = \frac{\langle i^2 \rangle_l}{\langle i \rangle_l} = \frac{c_{\mathrm{M}}}{c_{\mathrm{M}^*,l}} \frac{\partial c_{\mathrm{M}^*,l}}{\partial c_{\mathrm{M}}} = \frac{1 - \{K_p c_{\mathrm{M}}\}^2}{\{1 - K_p c_{\mathrm{M}}\}^2}. \tag{12.27}$$

which can be determined by measurements of light scattering intensity. (See Section 6.8.).[7]

In the limit of large K_p,

$$i_{n,l} = \langle i \rangle_l \approx \sqrt{K_p c_{\mathrm{M}^*,l}}$$

$$i_{w,l} \approx 2\sqrt{K_p c_{\mathrm{M}^*,l}} = 2 i_{n,l}. \tag{12.28}$$

The ratio $i_w / i_n = 2$ is characteristic of an exponential length distribution.

A description of *formation of helical polymers* requires only a minor modification of the partition function, Eq. (12.23), which, however, leads to an important difference in behavior. A characteristic of a helical polymer is that a dimer of subunits does not have a full complement of intersubunit contacts, and is relatively unstable. After several subunits have been assembled into a nucleus, the addition of successive subunits is more favorable; one sees that the equilibrium for adding these later units is coupled to the equilibrium of forming a nucleus containing just a few monomers. In the simplest case the nucleus contains two subunits, and one then has

$$\frac{c_{\mathrm{M}_2}}{c_{\mathrm{M}}} = \sigma K_p c_{\mathrm{M}}$$

$$\frac{c_{\mathrm{M}_{i+1}}}{c_{\mathrm{M}_i}} = K_p c_{\mathrm{M}};\ i > 2$$

$$\frac{c_{\mathrm{M}_{i+1}}}{c_{\mathrm{M}}} = \sigma \{K_p c_{\mathrm{M}}\}^i;\ i > 0, \tag{12.29}$$

where the nucleation constant, σ is a small number, $\sigma =\ll 1$. Proceeding as before, one obtains for helical polymerization (subscript h)

$$\vartheta_h = c_{\mathrm{M}} \left[1 + \sigma \sum_{i=1}^{\infty} \{K_p c_{\mathrm{M}}\}^i \right]$$

$$\vartheta_h = c_{\mathrm{M}} \left\{ 1 - \sigma + \frac{\sigma}{1 - K_p c_{\mathrm{M}}} \right\}, \tag{12.30}$$

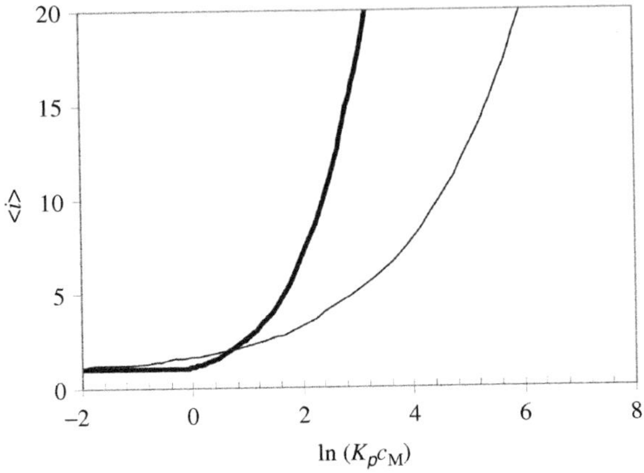

FIGURE 12.6 Degree of polymerization, as a function of total monomer concentration, c_{M*}. Thin curve: linear polymer. Heavy curve: helical polymer ($\sigma = 0.001$).

and differentiation gives

$$c_{M*,h} = c_M \frac{\partial \vartheta_{p,h}}{\partial c_M} = \sigma \sum_{i=0}^{\infty} (i+1) K_p^i c_M^i + (1-\sigma) c_M,$$

$$i_{n,h} = \frac{c_{M*,h}}{c_M} = \frac{1 + (1-\sigma)(K_p^2 c_M^2 - 2K_p c_M)}{(1 - K_p c_M)^2}. \tag{12.31}$$

Some results for formation of linear and helical polymers are shown in Figs. 12.6 and 12.7. One sees that, at all concentrations, helical polymerization is less when

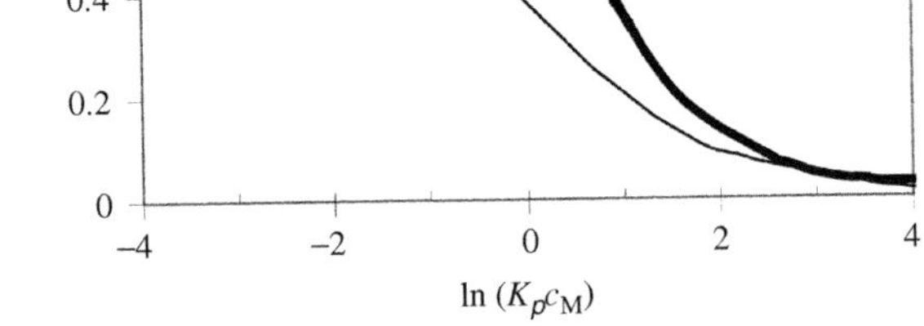

FIGURE 12.7 Fraction of protein remaining as unpolymerized monomer, as a function of total monomer concentration, c_M. Thin curve: linear polymer. Heavy curve: helical polymer ($\sigma = 0.001$).

measured by the disappearance of free monomer. However, the helical polymers contain on average more subunits. Both effects are understandable given that helical polymerization incurs a penalty for each oligomer; this penalty favors monomer and long oligomers. One sees that formation or dissolution of a polymer with a modest nucleation penalty can be controlled by smaller changes in environment, specifically concentration of a ligand with different affinity for monomer and polymer, which causes an effective change of K_p, as described in the following section.

12.6 COUPLED POLYMERIZATION AND LIGAND BINDING

The affinity for ligand of subunits in oligomers or polymers may be different from that of the free monomers for any one of several reasons, such as, occlusion of binding sites in the polymer, formation of binding sites between subunits in the polymer (Fig. 12.8), or conformation change with modification of binding sites on polymerization.

The corresponding equilibrium equations are as follows,

$$\begin{aligned} c_{\mathrm{M}_2} &= c_{\mathrm{M}}^2 K_p \\ c_{\mathrm{ML}} &= c_{\mathrm{M}} K_{b,1} a_{\mathrm{L}} \\ c_{\mathrm{M}_2\mathrm{L}} &= 2 c_{\mathrm{M}_2} K_{b,2} a_{\mathrm{L}} \\ c_{\mathrm{ML}_2} &= c_{\mathrm{M}} (K_{b,2} a_{\mathrm{L}})^2, \end{aligned} \tag{12.32}$$

and this can be rearranged to give

$$\begin{aligned} c_{\mathrm{M}^*} &= c_{\mathrm{M}} + c_{\mathrm{ML}} = c_{\mathrm{M}}(1 + K_{b,m} a_{\mathrm{L}}) \\ c_{\mathrm{M}_2^*} &= c_{\mathrm{M}_2} + c_{\mathrm{M}_2\mathrm{L}} + c_{\mathrm{M}_2\mathrm{L}_2} = c_{\mathrm{M}_2}(1 + K_{b,p} a_{\mathrm{L}})^2 \\ c_{\mathrm{M}_2^*} &= (c_{\mathrm{M}^*})^2 K_p \left(\frac{1 + K_{b,p} a_{\mathrm{L}}}{1 + K_{b,m} a_{\mathrm{L}}} \right)^2 = (c_{\mathrm{M}^*})^2 K'_p, \end{aligned} \tag{12.33}$$

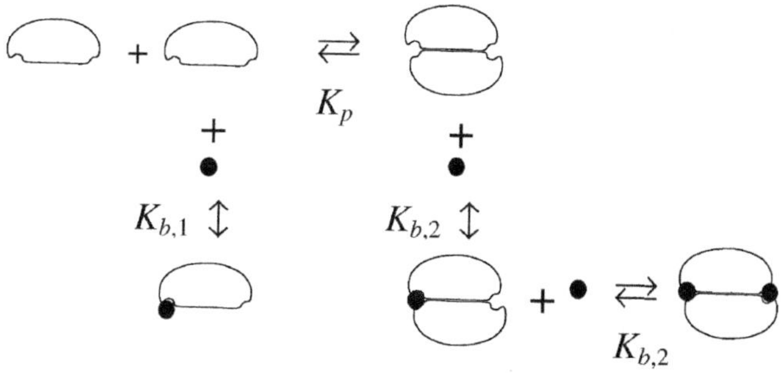

FIGURE 12.8 One possible scheme of dimerization of protein subunits enhanced by ligand binding ($K_{b,1} < K_{b,2}$).

the * indicating *any* species with or without bound ligand. This equation shows how dimerization in the presence of ligand proceeds with a simply modified binding constant K'_p that depends on the activity of ligand; the result is easily generalized to oligomers with more than two subunits and long polymers

$$K'_p = K_p\left(\frac{1 + K_{b,2}a_L}{1 + K_{b,1}a_L}\right)^n, \tag{12.34}$$

where the power n equals the number of (equivalent) binding sites in the whole oligomer, or in one subunit of the polymer.

Any study of polymerization promoted (or inhibited) by ligand binding is greatly facilitated by the ability to measure not only binding as a function of ligand activity a_L, but also binding *and* degree of polymerization as a function of both ligand activity and total protein concentration c_M. However, a complication is that a straightforward description in terms of stoichiometric binding constants, as in Eq. 8.24, is not possible for a monomer-to-oligomer or monomer-to-polymer equilibrium with changing affinity for ligand.

NOTES

1. If the ligands are ions, binding changes the net charge of the macromolecule, and the long-range effect of charge–charge interactions must be considered, successive binding of like-charged ions becoming progressively more difficult (lower affinity).
2. Cui, Q., Karplus, M. Allostery and cooperativity revisited. *Protein Sci* **17**: 1295–1307 (2008).
3. Volkman, B.F., Lipson, D., Wemmer, D.E., Kern, D. Two-state allosteric behavior in a single-domain signaling protein. *Science* **291**: 2429–2433 (2001).
4. Copeland, R.A. Conformational adaptation in drug–target interactions and residence time. *Future Med. Chem.* **3**: 1491–1501 (2011).
5. Hammes, G.C., Chang, Y.-C., Oas, T.G. Conformational selection or induced fit: A flux description of reaction mechanism. *Proc. Natl. Acad. Sci. USA* **106**: 13737–13741 (2009).
6. Oosawa, F., Kasai, M. A theory of linear and helical aggregation of macromolecules. *J. Mol. Biol.* **4**: 10–21 (1962).
7. Stain on a gel also reflects collective weight, rather than number, of molecules.

SUGGESTED READING

Wyman, J., Gill, S.J., *Binding and Linkage: Functional Chemistry of Biological Macromolecules*. Mill Valley, CA: University Science Books, 1990.

Oosawa, F., Asakura, S., *Thermodynamics of the Polymerization of Protein*. New York: Academic Press, 1975.

13

Allosteric Function

Red-blooded biophysicists study hemoglobin.

13.1 INTRODUCTION

At the time the allosteric model was developed, each protein was understood to have a unique structure which endowed that protein with a specific (often highly specific) biological function. The allosteric model hypothesized that some proteins could assume two well-defined structures, with different properties. An event, such as binding to one specific site, can then switch the conformation of such an allosteric molecule, and this perturbation will be "sensed" at a second, distant site; as a result, the activity of this second site will differ from what it is in the absence of a bound ligand at the first site. If, as a result of symmetry, the second site is equivalent to the first, the second site can then have a greater, or a lesser affinity for ligand than the first. Such indirect site–site interactions are distinctly different from the direct interactions between, for example, the two hydrogen ions that can bind to an oxalate ion, or cases in which the two ligand molecules are in direct contact because of proximity of the binding sites.[1]

The allosteric model was first introduced in order to explain the *cooperative* nature of the binding of oxygen to (vertebrate) hemoglobin, that is, the observed *increase* of affinity with the saturation, given the observation that the four binding sites for oxygen are a considerable distance apart, too far for a *direct* interaction between the binding sites. In this chapter, we first review well-known relevant material about structure and function of hemoglobin, next describe the two allosteric models, the one proposed by Monod, Wyman, and Changeux (MWC) and the other by Koshland, Némethy, and Filmer (KNF), and their application to oxygen

Equilibria and Kinetics of Biological Macromolecules, First Edition. Jan Hermans and Barry Lentz.

binding to hemoglobin. This is then followed by a summary of more recent studies of hemoglobin. The chapter concludes with sections in which we discuss other allosteric effects, including negative cooperativity, and the importance of allostery in protein function as it is currently understood.[2]

13.2 BACKGROUND ON HEMOGLOBIN

Hemoglobin is a tetramer, consisting of two α and two β chains. Each chain contains a heme group (Fe^{++} protoporphyrin IX complex). The three-dimensional structure of the α and β chains is similar, and similar to the structure of the single chain in myoglobin, which contains a single heme group. Each heme can reversibly bind one oxygen molecule. Myoglobin has greater affinity for oxygen than hemoglobin, which makes sense, as myoglobin, in the muscle, is a recipient of oxygen carried there by hemoglobin in the blood stream.

Hemoglobin's affinity for oxygen is lower at low pH; thus, the release of CO_2 into the blood as a result of metabolism (and its conversion to H_2CO_3, which lowers the pH) helps drive the oxygen off the hemoglobin. Hemoglobin's affinity is also controlled (tuned) by 2,3 diphosphoglycerate (DPG). DPG levels will change slowly in response to a change in altitude.

At low p_{O_2} the affinity is lower than expected, relative to the given midpoint, but at high p_{O_2} the affinity is raised. The result of this is a larger difference in saturation between the limits of physiological oxygen pressure, 1/5 bar = 150 Torr in the atmosphere, and 20 Torr or below in working muscle, or the 10 Torr needed to 90% saturate Mb (Fig. 13.1; The unit "Torr" is equivalent to the obsolete "mm of Hg".) This contrasts with myoglobin's oxygen binding curve, which is as expected for binding to a single site, with half saturation at a (partial) pressure of oxygen, p_{O_2} of 2 Torr.

(i) *The oxygen affinity of hemoglobin, as reflected in the* p_{O_2} *required for half saturation, is "tuned" for good performance, and,* (ii) *the "cooperative" effect increases the oxygen carrying capacity of the blood.*

The special ligand binding properties of hemoglobin are tied in with a conformation change, which occurs on binding of oxygen or of carbon monoxide. The state seen at low p_{O_2} is called T and the state seen at high p_{O_2} is called R. Here R stands for "relaxed", as in the R-state the affinity is high, as for monomeric myoglobin, and T stands for "tight", which expresses the idea that the structure is stressed and that this altered state has a lower affinity for oxygen.

Crystal structures of both states are known from the work of Max Perutz and his students; the major change in structure is a shift in packing of the interface between the two $\alpha\beta$ dimers; on closer inspection, this change is seen to be accompanied by subtle changes in the structure of the subunits. In particular, in the T-state the Fe atom is displaced out of the plane of the heme ring away from the side where the oxygen molecule binds, and this explains the lowered affinity. DPG is found to bind in a pocket between the $\alpha\beta$ dimers in the T-state, and participation in the interface of several ionizable groups (histidine side chains and α-amino groups) changes when the conformation changes.

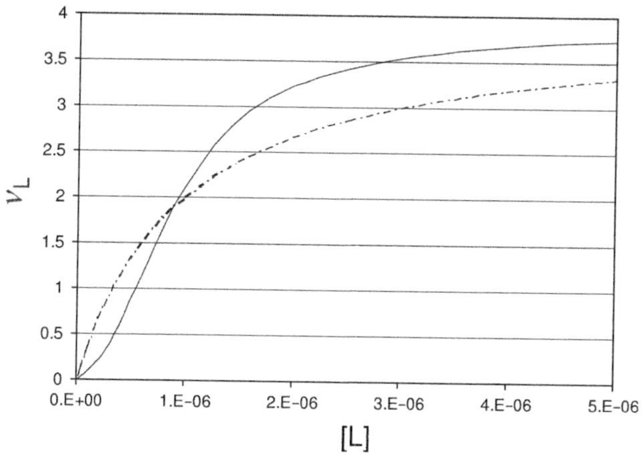

FIGURE 13.1 Curves with similar midpoints for binding to independent sites (dashed) and cooperative binding (continuous curve). The hyperbolic shape is found for binding of oxygen to myoglobin, and the sigmoid binding curve is observed for hemoglobin.

Thus, the basic model consists of one state (the T-state) with low affinity for oxygen and another state (the R-state) with high affinity. Furthermore, DPG binds more strongly to the T-state; also, a few binding sites for hydrogen ion have higher affinity in the T-state, which gives rise to a dependency of oxygen affinity on the pH, called *the Bohr effect.* At low concentration, the oxygenated form dissociates into $\alpha\beta$ dimers (that retain the R-state), but the *in vivo* concentration of hemoglobin is very high (50% of the mass of red blood cell content), and then dissociation into halves is negligible.

13.3 THE ALLOSTERIC OR INDUCED-FIT MODEL OF HEMOGLOBIN

These observations have been explained by invocation of an allosteric or induced fit model. Such a model supposes the existence of (at least) two conformations of the macromolecule, differing in their affinity for ligands in such a way that ligand binding induces a switch from one preferred conformation to another.

The hemoglobin molecule possesses four equivalent binding sites, each located on a different chain, and the allosteric model assumes two different conformations, called T and R, for each chain. The partition function of this allosteric model of oxygen binding to hemoglobin can then include five conformation states,

$$T_4, T_3R, T_2R_2, TR_3, \text{and } R_4,$$

and each state can bind from zero to four oxygen molecules, with a different binding constant for each site.

13.4 SIMPLIFIED ALLOSTERIC MODELS: CONCERTED AND SEQUENTIAL

One important assumption inherent in the allosteric model is that binding to *the four sites is independent if the conformation does not change*, that is, that coupling is *entirely* the result of a conformation change. In the case of hemoglobin, an obvious simplification is to make no distinction between α and β chains. We describe here these equilibria with partition functions of statistical mechanics, as introduced in Chapters 10 and 11.

Accordingly, we can then write the (molecular) partition function including all five conformation states of hemoglobin and for each conformation the equilibrium distribution of all combinations (as indicated by the *) of occupied or free ligand sites, q_{M^*} as

$$q_{M^*} = \sum_{k=0}^{4} q_{k^*} = \sum_{k=0}^{4} q_k \prod_{i=1}^{4} (1 + K_{k,i} p_{O_2}), \qquad (13.1)$$

with q_{k^*} the partition function of conformation k each with all combinations of bound ligand, q_k the molecular partition function of conformation k in the absence of bound ligand, and $K_{k,i}$ the binding constant to site i in conformation k. (In evaluating this model one should take into account also the number of ways k chains out of four can be assigned the R conformation, which equals 1 for $k = 0$ or 4, 4 for $k = 1$ or 3, and 6 for $k = 2$.)

This model has many more parameters than can be determined unequivocally from the experimental data, that is, from the degree of saturation of the binding sites as a function of the oxygen pressure, and simplifying assumptions have to be made. It therefore makes sense to also work with *only two binding constants*, one for an R subunit and the other for a T subunit, K_T and K_R. Even then, further simplification is needed.

Monod, Wyman, and Changeux proposed a *concerted allosteric model* in order to reproduce the essence of the observations with hemoglobin (MWC model). In this model, the four subunits of any hemoglobin molecule assume the same state. Thus, this model works with just two conformation states, T_4 and R_4, and either one can bind oxygen at four equivalent, independent sites, but with a different binding constant for T and for R,[3] as described by the following scheme

$$\begin{array}{ccc} & K_T & \\ T + 4O_2 & \rightleftarrows \cdots\cdots \rightleftarrows & T(O_2)_4 \\ K_{TR} \updownarrow & & \updownarrow K_{TRL} \\ R + 4O_2 & \rightleftarrows \cdots\cdots \rightleftarrows & R(O_2)_4 \\ & K_R & \end{array} \qquad (13.2)$$

The dots represent the states with 1, 2, and 3 bound oxygen molecules; K_{TR} and K_{TRL} are equilibrium constants for conformation changes and K_T and K_R are site binding constants.

The partition function is then

$$q_{MWC} = q_{T*} + q_{R*} = q_T(1 + K_T p_{O_2})^4 + q_R(1 + K_R p_{O_2})^4, \tag{13.3}$$

and in order to make the T-state more stable than the R-state in the absence of bound oxygen, and vice versa,

$$K_{TR} = \frac{q_R}{q_T} < 1$$

$$\frac{K_R}{K_T} > 1. \tag{13.4}$$

Subsequently, Koshland, Némethy, and Filmer proposed an alternative, *sequential allosteric model* (KNF model).[4] In this model, each subunit's conformation is determined by whether an oxygen molecule is bound to its heme group (R conformation), or not (T conformation). The conformation of a hemoglobin molecule with 1, 2, or 3 oxygen molecules bound is then a mixed one. The partition function has only five terms

$$q_{KN} = \sum_{i=0}^{4} C_i^4\, q_{T_{4-i}R_i}(K_b p_{O_2})^i = \sum_{i=0}^{4} \frac{4!}{i!(4-i)!}\, q_{T_{4-i}R_i}(K_b p_{O_2})^i. \tag{13.5}$$

The model favors the two extreme states, R_4 and $T_4(O_2)_4$ by assigning to each partly oxygenated state a penalty, σ corresponding to the occurrence of TR interfaces:

$$q_{T_{4-i}R_i} = \sigma_i\, q_{T_4}$$

$$\sigma_0 = \sigma_4 = 1 \tag{13.6}$$

$$\sigma_1, \sigma_2, \sigma_3 < 1,$$

that is,

$$q_{KN} = \sum_{i=0}^{4} C_i^4\, \sigma_i\, q_{T_4}(K_b p_{O_2})^i$$

$$q_{KN} = q_{T_4}\{1 + \sigma[4K_b p_{O_2} + 6(K_b p_{O_2})^2 + 4(K_b p_{O_2})^3] + (K_b p_{O_2})^4\}, \tag{13.7}$$

the latter if all σ_i have the same value.

The MWC and KNF models include the same two endpoints, but all intermediate, partly oxygenated states are different, as can be seen from the table of states (Table 13.1). The MWC model includes the states in boldface, and the KNF model includes the states in italics.[5]

TABLE 13.1 Possible conformations and oxygenation states, and their inclusion in the MWC (bold) and KNF models (italics)

Number O_2 Bound					
0	$\mathit{T_4}$	T_3R	T_2R_2	TR_3	$\mathbf{R_4}$
1	$\mathbf{T_4O_2}$	$\mathit{T_3RO_2}$	$T_2R_2O_2$	TR_3O_2	$\mathbf{R_4O_2}$
2	$\mathbf{T_4(O_2)_2}$	$T_3R(O_2)_2$	$\mathit{T_2R_2(O_2)_2}$	$TR_3(O_2)_2$	$\mathbf{R_4(O_2)_2}$
3	$\mathbf{T_4(O_2)_3}$	$T_3R(O_2)_3$	$T_2R_2(O_2)_3$	$\mathit{TR_3(O_2)_3}$	$\mathbf{R_4(O_2)_3}$
4	$\mathbf{T_4(O_2)_4}$	$T_3R(O_2)_4$	$T_2R_2(O_2)_4$	$TR_3(O_2)_4$	$\boldsymbol{R_4(O_2)_4}$

13.5 NUMERIC EXAMPLE

Figure 13.2 shows data calculated according to the two models to give approximate binding curves of the same *shape* as observed for hemoglobin, but lower cooperativity.[6] Important features are the midpoint, the slope at the midpoint, and the details at low and at high p_{O_2}.

The parameters for the KNF model were adjusted to optimize the fit to the data for the MWC model.[7] One sees that the difference between the two binding curves is very small, and that alone cannot serve to prefer one model over the other. On the other hand, as is also shown in Fig. 13.2, the extent of conformation change from R- to T-state is significantly different for the two models.

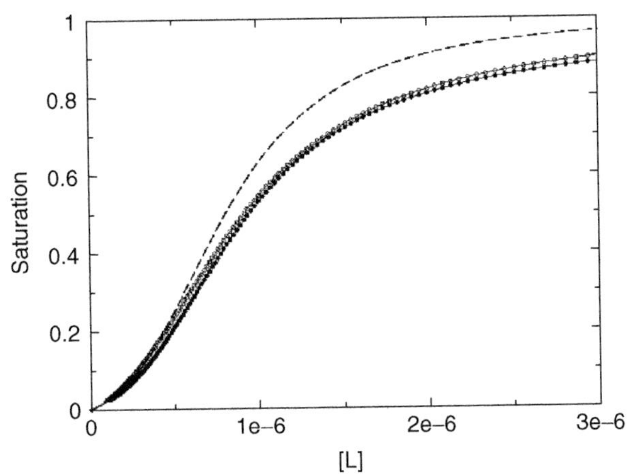

FIGURE 13.2 Oxygen binding to hemoglobin, as saturation, f_b modeled with the MWC and KNF models with relatively low cooperativity. (MWC: filled symbols; KNF: open symbols). The dashed line shows the fraction of molecules in the R state for the MWC model. For the KNF model the conformation change coincides with the oxygen binding curve.[7]

13.6 COMPARISON OF OXYGEN BINDING CURVES

Interpretation of actual measurements of binding of oxygen to hemoglobin is complicated by a number of factors. The measurements of oxygen affinity are not particularly difficult. Oxygen pressure can be determined manometrically. In concentrated solution (or in whole blood) the amount of oxygen bound to hemoglobin can be determined from the change of volume of a known initial amount of gaseous oxygen. In dilute solution, the fraction of oxy-hemoglobin may be determined from the large change in absorption in the visible ("blue to red"), in which case the assumption is made that the spectra of oxy-heme and deoxy-heme groups are not dependent on the conformation state and are the same for the α- and β-chains. It is important to collect measurements over a large range of saturation, f_b; the uncertainty in the measured saturation will tend to be more or less constant; interpretation in terms of models relies much on results at low and nearly complete saturation, where the relative error in f_b or $1-f_b$ is largest. It is important to avoid complications arising out of the dependence of oxygen affinity on pH, on hemoglobin concentration, and on effectors such as DPG. Our discussion *assumes* that all these complications have been avoided; if this is not the case, the models must be extended in order to take into account any changes in the environment influencing the binding of oxygen.

In any analysis of binding data for hemoglobin, it is useful to realize that at low p_{O_2} one species (that without bound oxygen) dominates. The small amount that is *not* in this form has *only one* oxygen molecule bound. Thus, the binding at low p_{O_2} can reveal the affinity in the first step. Similarly, the residual shortage of binding at high p_{O_2} can reveal the affinity in the last step. (This can be captured as the initial and final asymptotes of the Hill plot; see Section 13.12. It is nowadays simpler to fit the model's parameters to the measured binding curve using a computer fitting procedure.)

As we have seen in Chapter 8, the binding of a single type of ligand molecule to n sites on a macromolecule can be *fully* described with just n independent stoichiometric binding constants, κ_1 through κ_n. For hemoglobin, with four sites, the partition function is

$$q_4 = q_0(1 + \kappa_1 p_{O_2} + \kappa_2 (p_{O_2})^2 + \kappa_3 (p_{O_2})^3 + \kappa_4 (p_{O_2})^4). \qquad (13.8)$$

Adair in effect used this equation, but with single-step stoichiometric binding constants, $\mathcal{K}_1'$ through $\mathcal{K}_4'$ and combinatorial factors, making the following replacements

$$\begin{aligned}
\kappa_1 &= 4\mathcal{K}_1' \\
\kappa_2 &= 6\mathcal{K}_1'\mathcal{K}_2' \\
\kappa_3 &= 4\mathcal{K}_1'\mathcal{K}_2'\mathcal{K}_3' \\
\kappa_4 &= \mathcal{K}_1'\mathcal{K}_2'\mathcal{K}_3'\mathcal{K}_4'.
\end{aligned} \qquad (13.9)$$

Again, this analysis produces a perfect fit (barring experimental error), but does not provide insight why the affinity rises, and $\mathcal{K}_4'$ exceeds $\mathcal{K}_1'$.

The MWC model, with just three independent parameters, is only a modest simplification over the Adair model. The model's great strength lies in that it presents an interpretation of the observations in terms of a realistic physical model of binding coupled to conformation change; furthermore, the conformational difference between T and R states is well established experimentally, and one of the parameters in the model (K_{TR}) can, in principle, be determined independently. The KNF model (with all penalties equal) with only two independent parameters also relies on the observed difference in conformation of the two endpoints.[8]

The example given in the preceding section shows that MWC and KNF models give very similar behavior when used to model sigmoid binding curves not unlike those observed for hemoglobin. The oxygenation of hemoglobin can be more cooperative than the earlier example. For example, Imai gives the ratio K_T/K_R of the best fitting MWC model as 600 under one particular set of conditions.[9] For this *more* cooperative case, there is even less distinction between the binding curves predicted by the MWC and KNF models.

Hence, either of the two allosteric models can explain the binding data for hemoglobin reasonably well. A choice of which allosteric model is preferable for hemoglobin can in principle be made on the basis of measurements of binding and conformation change of the same solutions. It is disappointing that measurement of the extent of conformation change accompanying oxygenation apparently remain insufficiently precise to support one model over the other.

We may briefly consider here how to treat the effects of binding of ligands such as H^+ and DPG on the oxygen affinity in terms of the two models. For the MWC model this problem has a quite straightforward solution by simple alteration of the partition functions q_R and q_T. For example, one molecule of DPG is known to bind to deoxy-hemoglobin, which is in the T state, while binding to the R state can be ignored; consequently, in order to account for the presence of DPG, q_T is replaced by q_T', which is equal to the product of q_T and the partition function for binding DPG,

$$q_T' = q_T(1 + K_{b,\mathrm{DPG}} a_{\mathrm{DPG}}), \tag{13.10}$$

where $K_{b,\mathrm{DPG}}$ is the binding constant and a_{DPG} the activity of DPG. It is unnecessary to assume that the oxygen affinity of the T state changes as a result of the binding of DPG. In this way, coupling between oxygen binding and DPG binding is accounted for by a natural extension of the concerted allosteric model.

Given the observation that DPG binds at the interface across chains, one would then apply a similar equation to modify the partition function of the T_4 state of the KNF model.

13.7 SEPARATING OXYGEN BINDING AND CONFORMATION CHANGE OF HEMOGLOBIN

Owing to the high cooperativity of hemoglobin oxygenation, the population of molecules is, at any saturation, dominated by (at most) two species, T_4 and $R_4(O_2)_4$;

as a result, the composition of the small population of intermediates, which one would use to distinguish between MWC, KNF, and other models, has not been determined. This problem has been addressed by Eaton and coworkers by elegant new measurements, in which they were, in effect, able to separate oxygen binding and conformation change.

In this study it was found that crystals of T-state deoxy-hemoglobin can bind oxygen reversibly while retaining their crystal structure and the T-state tertiary structure. A remarkable feature of the experiment was that, by choosing a specific orientation of a single crystal relative to the direction of polarization of the light used to measure the absorption change accompanying oxygenation, Eaton and coworkers[10] could detect *separately* the oxygenation of α- and β-chains (Fig. 13.3). Overall, the binding of O_2 to these crystals is noncooperative. Apparently, the crystal packing forces are sufficiently strong to prevent a breakup of the crystal followed by transition of oxygenated molecules to the R-state. (A precise analysis of the binding curves shows that a slight cooperativity remains between α and β chains; this indicates a strong coupling within the $\alpha\beta$ dimer even in the absence of a change of the quaternary structure to the R state, something that is also indicated by the work reported in the following section.)

The affinity of the α-chains for oxygen was found to be between 2.5 and 5 times higher than that of the β-chains and somewhat lower than the affinity of hemoglobin at low saturation in solution. As mentioned, at low saturation the predominant molecular species are deoxy-hemoglobin and hemoglobin with a *single* bound oxygen molecule; the latter species is T_4O_2 according to the MWC model and T_3RO_2 according to the KNF model. The observation of a low affinity both in the T-state crystal (where the first oxygenated species is T_4O_2) and in solution at low saturation is evidence in favor of the MWC model.

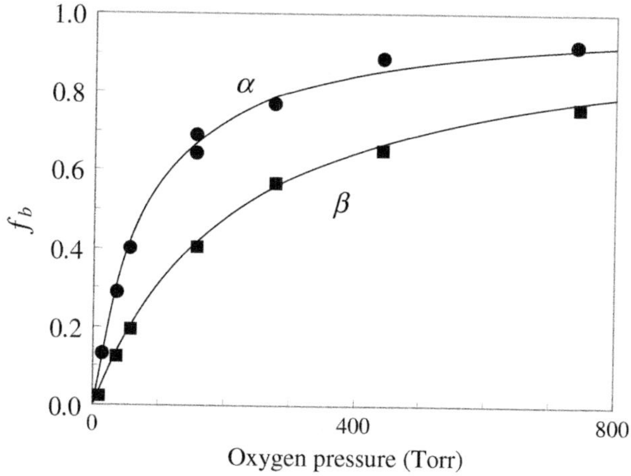

FIGURE 13.3 Saturation of oxygen binding sites of α- and β-chains in T-state crystals of hemoglobin as a function of oxygen pressure. Atmospheric pressure is 760 Torr. From Rivetti, C., Mozzarelli, A., Rossi, G.L., Henry, E.R., Eaton, W.A. Biochemistry **32**: 2888–2906 (1993).

A review by Eaton and coworkers[11] summarizes earlier results on oxygen binding to hemoglobin as well as these and other newer measurements.

13.8 EXPERIMENTS WITH HYBRID HEMOGLOBINS

As we have seen in the preceding section, the MWC model has become the favored description of the cooperativity of binding of oxygen to hemoglobin as a result of continued research and application of new methods. Additional new results are based on experiments with mutant hemoglobins, with modified heme groups that cannot be oxygenated, and with construction of mixed tetramers and of stable $\alpha\beta$ dimers.

Unavoidably, simple models such as MWC or KNF ultimately break down in the face of new experimental data. Ackers and Holt[12] have recently summarized the new understanding that has been reached on the basis of these new results. The assumption that α and β chains have the same oxygen affinity remains, and the relative concentrations of species with different numbers of bound oxygen are only modestly different from those required by the MWC model. In order to understand the radically new finding, one should realize that the packing of the four chains of hemoglobin involves four physically different interfaces, one $\alpha\alpha$, one $\beta\beta$, and two $\alpha\beta$. One of the latter is very stable and present in the $\alpha\beta$ dimers into which oxyhemoglobin dissociates at low concentration. The new finding is that this $\alpha\beta$ dimer interface plays its own role in enhancing cooperativity of intact, tetrameric hemoglobin; it was found that the binding of a second oxygen molecule is on the order of 10 times easier if this occurs on (the other chain of) the same $\alpha\beta$ dimer than when it occurs on a chain of the other $\alpha\beta$ dimer. This possibility is not envisaged by either of the simple MWC or KNF models, or even in the more comprehensive scheme of Table 13.1.

13.9 TWO-SITE PROTEINS, HALF-THE-SITES REACTIVITY, AND NEGATIVE COOPERATIVITY

The binding isotherm of an allosteric protein with just two binding sites is fully described by just two independent parameters (e.g., two stoichiometric binding constants). Consequently, the parameters of an MWC model of such a protein (frequently a dimer of identical subunits) are incompletely determined by the binding data, and an independent measurement of the conformational equilibrium constant K_{TR} will be required to establish the complete concerted model.

The KNF model with its two parameters should be able to provide a perfect fit to the binding data, and this fit also determines the model's parameters, K_b and penalty σ. The observation of a perfect fit obviously does not establish that an interpretation in terms of the sequential allosteric model is correct.

However, in principle, the sequential KNF model is the more versatile, because if the *TR interface(s) make(s) a stabilizing* contribution ($\sigma > 1$ in Eq. 13.7), this can account for *negative* cooperativity, while the MWC model cannot.

At least since the 1970s, it has been recognized that in many enzymes that are dimers of identical chains, one of the sites is much more reactive than the other. Such half-the-sites reactivity can arise if the two chains, although indistinguishable in a time average, in fact form at any one time an asymmetric structure with different geometry at the two active sites. Another possibility is that the binding of a first substrate molecule induces a conformation change of the whole molecule, which lowers the affinity of the second site; this then constitutes a case of induced fit according to the KNF model. Determining which of these situations is the case for a particular molecule must be based on the results of high resolution structural studies. Crystal structural studies of thymidylate synthase indicate that this enzyme is an example of this second behavior, as the enzyme can bind two substrate molecules, but binding to one monomer of a molecule of a cofactor required for reactivity changes the conformation so that another molecule of cofactor cannot bind to the other monomer.[13]

Fersht has proposed that a conformation change associated with binding ligands (tyrosine and/or ATP) to the second, low affinity site in another dimeric enzyme, tyrosyl tRNA synthase, would provide a mechanistic boost to catalysis.[14] Structural studies of this enzyme indicate a dual role for the subunits, in that the synthase contacts the final substrate, tyrosyl tRNA, which is an asymmetric molecule, with both subunits; the subunit loaded with the activated tyrosine contacts the 3′ end of the RNA where it transfers the tyrosine, while the other subunit contacts another part of the tRNA that includes the anticodon, and thereby ensures that tyrosine is transferred to *tyrosyl* tRNA.[15] This observation of different roles for the two subunits might provide an alternate rationale for why only one subunit at a time is charged with tyrosine.[16]

13.10 ALLOSTERIC EFFECTS IN PROTEIN FUNCTION

The allosteric model of hemoglobin cooperativity was at once accepted because it had just become known at the time that the four binding sites were not in close proximity. Not long thereafter x-ray crystallographic studies by Perutz and coworkers established clearly distinct structures for the oxygen-free T-state and the oxygenated R-state (deoxyhemoglobin and oxyhemoglobin). It makes sense not to restrict the application of this elegant model to this one protein (or the class of cooperative, multi-subunit proteins). In fact, as studies by x-ray crystallography and nuclear magnetic resonance (NMR) have accumulated structural information on hundreds of proteins, and as concepts of protein dynamics have been developed on the basis of both NMR measurements and simulations, it has become clear that allosteric effects are widespread and are intrinsic to the function of many proteins.[17]

The following is then a working definition of an allosteric effect: an event at one site of a protein molecule alters the conformation to the extent that this results in a change in function. In the case of hemoglobin, the "event" is the binding of one or more oxygen molecules, and the result is a change in binding affinity at other binding sites. In the case of the lac repressor protein, binding of lactose results in an allosteric change that decreases the affinity of the repressor for its cognate DNA, the lac operon, which is then freed to allow the synthesis of enzymes needed for metabolism based on lactose. The enzyme aspartate carbamoyltransferase contains six catalytic subunits (arranged as a complex of three dimers) and six regulatory subunits (similarly arranged). This enzyme occupies a key position in the synthetic pathway of cytidine triphosphate (CTP). Positive cooperativity with respect to the substrate aspartic acid enhances activity when this substrate is in good supply, while binding of CTP to the regulatory subunits decreases activity when the pathway's *end product* is plentiful.

A last refinement of this description of an allosteric mechanism recognizes that a protein's conformation is a state consisting of a *distribution* over many conformations that differ in details of side chain and backbone conformation, as a result of differences in (mostly, side chain) conformers corresponding to different local energy (or free energy) minima, as well of thermal motion about each energy minimum. This leads to a definition of allosteric effects, not in terms of a change of conformation, but in terms of a change of conformational ensemble. This redefinition allows for an allosteric effect (i.e., an effect at a distant site) that results from a change in the *size* of the conformational ensemble without a change in the *average* structure.[18]

Our final working definition of an allosteric effect will then read as follows: an event at one site of a protein molecule alters the *conformational ensemble* to the extent that this results in a change in function.

13.11 SICKLE CELL HEMOGLOBIN

Sickle cell hemoglobin (Hb S) is a genetic variant of human adult hemoglobin A (Hb A; one amino acid substitution in the β chains). The prevalence of Hb S among the natives of West Africa and their descendants is believed to have been caused by the act of protection against malaria by individuals with a mixture of Hb A and Hb S (AS heterozygotes). However, individuals whose adult hemoglobin is all Hb S suffer from sickle cell disease, in which the red blood cells may deform and then clog tiny arteries and veins. The red blood cells deform to the so-called *sickle shape* because the Hb S molecules form long aggregates that align themselves preferentially in one direction and stress the cell into a different shape. The actual occurrence of sickling is sensitive to very small physiological changes, as the aggregation is highly dependent on total Hb concentration, and is a property only of the deoxygenated form.[19] (See Section 19.10 for a kinetic model.)

13.12 HILL PLOT

Early studies of hemoglobin made much use of the *Hill plot*, which is a plot of

$$\log \frac{f_b}{1 - f_b} \text{ versus} \log p_{O_2}, \qquad (13.11)$$

where f_b is the saturation. It is easy to show that for binding to a single site with binding constant K_b, the Hill plot is a straight line with slope 1 and intercept log K_b.

When applied to a protein with cooperative binding to n sites, the plot has asymptotes at high and low p_{O_2} each with a slope of 1. For a cooperative system behaving according to MWC, KNF or a related model, the intercepts of these asymptotes (for log $p_{O_2} = 0$) equal the logarithms of the binding constants for the first and last binding steps, respectively. As in this case the last binding step has a higher binding constant than the first and the middle part of the Hill plot has a slope greater than 1. The slope is called *the Hill coefficient*, and is often given for the value at 50% saturation, although the maximal value may occur at 65–75% saturation.

Quite generally, the maximal Hill coefficient must be less than the number of binding sites. Thus, the Hill coefficient for human hemoglobin must be less than 4 because this value would imply perfect cooperativity; maximum values in a range near 3 are observed, with the precise value depending on conditions.

An interesting exercise is to calculate the binding curve (saturation as a function of p_{O_2}) for the MWC model that was used as an example (Fig. 13.2),[7] to use these data to construct a Hill plot, and to determine the Hill coefficient. It is important to compute enough points to draw reliable asymptotes each with a slope of precisely 1.

A recent example of use of a Hill plot is the study of chicken hemoglobin D (Fig. 13.4).[20] Sedimentation experiments show that, without bound oxygen, molecules of this four-chain protein (tetramers) associate to form dimers with eight globin chains (octamers). The asymptotes at low p_{O_2} are seen to shift down with increasing protein concentration, which indicates that the oxygen-free globin octamer has a significantly lower oxygen affinity than the oxygen-free tetramer.

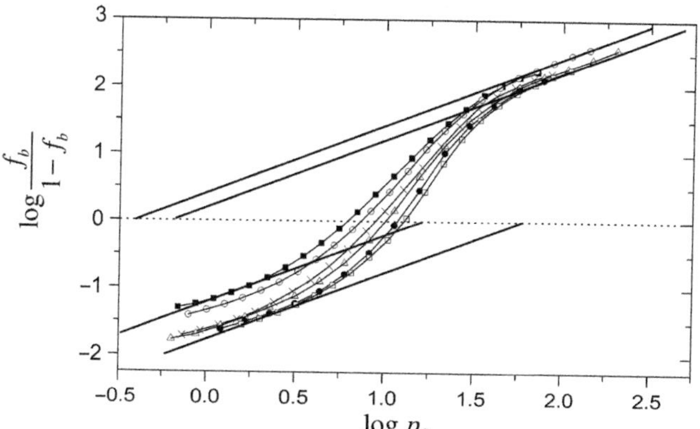

FIGURE 13.4 Hill plot for oxygen binding to chicken hemoglobin D. Hb concentrations in mM heme of curves sequentially from left to right are 0.084, 0.167, 0.334, 1.33, 2.67, and 4.0. The straight lines of slope 1 represent asymptotes at low and high p_{O_2}. From Rana, M.S., Knapp, J.E., Holland, R.A.B., Riggs, A.F. *Proteins:Struct. Funct. Genet.* **70**: 553–561 (2008).

(The shift is much less for the asymptotes at high p_{O_2}.) The maximal Hill coefficient changes from 2.5 for the most dilute, to slightly over 4.0 for the most concentrated solutions. This is a case where oligomerization is coupled to conformation change, and then to changes in ligand binding and results in an increase in cooperativity.

NOTES

1. In the former, interaction between the charged hydrogen atoms make the binding of the second more difficult than that of the first; in the latter, interactions between bound ligand molecules could be favorable or unfavorable, depending on the structural details.
2. Background reference: Smith, Hill, Lehman, Lefkowitz, Handler, White, *Principles of Biochemistry. Mammalian Biochemistry*. 7th ed., New York: McGraw Hill, 1983, chapter 4.
3. (a) Monod, J., Changeux, J.-P., Jacob, F. Allosteric proteins and cellular control systems. *J. Mol. Biol.* **6**: 306–329 (1963), (b) Monod, J., Wyman, J., Changeux, J.-P. On the nature of allosteric transitions: a plausible model. *J. Mol. Biol.* **12**: 88–118 (1965).
4. Koshland, D.E., Némethy, G., Filmer, D. Comparison of experimental binding data and theoretical models in proteins containing subunits. *Biochemistry*. **5**: 365–385 (1966).
5. Szabo, A., Karplus, M. A mathematical model for structure-function relations in hemoglobin. *J. Mol. Biol.* **72**: 163–197 (1972).
6. The Hill coefficient (see Section 13.12) of the model data is only 2.0, while it is typically greater than three for hemoglobin (depending on the exact conditions).
7. The parameters used for the *MWC model* are $K_{TR} = 0.01$, $K_T = 1.6 \times 10^5$, $K_R = 3.2 \times 10^6$; those used for the *KNF model* are penalty $\sigma_1 = \sigma_2 = \sigma_3 = 0.2$, $K_b = 1.1 \times 10^6$.
8. If the penalties σ_1 through σ_3 in Eq. 13.7 are unequal, then the KNF model becomes mathematically identical with the model with four independent stoichiometric binding constants, and hence trivially able to produce a perfect fit to the data.
9. Imai, K. Thermodynamic aspects of the cooperativity in four-step oxygenation equilibria of haemoglobin. *J. Mol. Biol.* **133**: 233–247 (1979). The conditions are 25 °C, pH 7.4, 10 mg/ml hemoglobin, 2 mM DPG, tris buffer, 0.1 M chloride.
10. (a). Mozzarelli, A., Rivetti, C., Rossi, G.L., Henry, E.R., Eaton, W.A. Crystals of haemoglobin with the T quaternary structure bind oxygen noncooperatively with no Bohr effect. *Nature (London)*. **351**: 416–419 (1991). (b). Rivetti, C., Mozzarelli, A., Rossi, G.L., Henry, E.R., Eaton, W.A. Oxygen binding by single crystals of hemoglobin. *Biochemistry* **32**: 2888–2906 (1993).
11. Eaton, W.A., Henry, E.R., Hofrichter, J., Mozzarelli, A. Is cooperative oxygen binding by hemoglobin really understood? *Nature Str. Biol.* **6**: 351–357 (1999).
12. Ackers, G., Holt, J. Asymmetric cooperativity in a symmetric tetramer: human hemoglobin. *J. Biol. Chem.* **281**: 11441–11443 (2006).
13. Anderson, A.C., O'Neil. R.H., DeLano, W.D., Stroud, R.M. The structural mechanism for half-the-sites reactivity in an enzyme, thymidylate synthase, involves a relay of changes between subunits. *Biochemistry* **38**: 13829–13836 (1999).
14. (a) Fersht, A.R. Demonstration of two active sites on a monomeric aminoacyl-tRNA synthetase. Possible roles of negative cooperativity and half-of-the-sites reactivity in oligomeric enzymes? *Biochemistry*. **14**: 5–12 (1975). (b) Fersht, A.R., Mulvey, R.S.,

Koch, G.L.E. Ligand binding and enzymic catalysis coupled through subunits in tyrosyl-tRNA synthetase. *Biochemistry* **14**: 13–18 (1975).

15. Yaremchuk, A., Kriklivyi, I., Tukalo, M., Cusack, S. Class I tyrosyl-tRNA synthetase has class II mode of cognate tRNA recognition. *EMBO J.* **21**: 3829–3840 (2002).
16. Yet another possible rationale: conformation change resulting from binding of substrate to one active site site effectively expels product bound to the other site in a see-saw effect.
17. Gunasekaran, K., Ma, B., Nussinov, R. Is allostery an intrinsic property of all dynamic proteins?. *Proteins: Struct. Funct. Genet.* **57**: 433–443 (2004).
18. Cooper, A., Dryden, D.T. Allostery without conformational change. A plausible model. *Eur. Biophys. J.* **11**: 103–109 (1984).
19. *For a series of biophysical studies, see* (a) Hofrichter J., Ross P.D., Eaton W.A. Supersaturation in sickle cell hemoglobin solutions. *Proc. Natl. Acad. Sci. USA* **73**: 3035–3039 (1976). (b) Ross P.D., Hofrichter J., Eaton W.A. Calorimetric and optical characterization of sickle cell hemoglobin gelation. *J. Mol. Biol.* **96**: 239–256 (1975). (c) Hofrichter J., Ross P.D., Eaton W.A. Kinetics and mechanism of deoxyhemoglobin S gelation: A new approach to understanding. *Proc. Natl. Acad. Sci. USA* **71**: 4864–4868 (1974).
20. Rana, M.S., Knapp, J.E., Holland, R.A.B., Riggs, A.F. Component D of chicken hemoglobin ... self-associate[s] upon oxygenation: Effect of oxygen binding. *Proteins: Struct. Funct. Genet.* **70**: 553–561 (2008).

SUGGESTED READING

Wyman, J., Gill, S.J. *Binding and Linkage: Functional Chemistry of Biological Macromolecules*. Mill Valley, CA: University Science Books, 1990.

Traut, T. *Allosteric Regulatory Enzymes*. New York: Springer, 2008.

14

Charged Groups: Binding of Hydrogen Ions, Solvation, and Charge–Charge Interactions

14.1 INTRODUCTION

Proteins contain a number of amino acids with side chains that carry a net charge of +1 or −1 elementary charges[1] under physiological conditions; nucleic acids contain one negative charge per phosphate group in the backbone. Many of these groups can be made to revert to their neutral form by a pH change, as a result of either binding or giving up a hydrogen ion. Some protein side chains and the side chains of nucleic acids are not charged at pH 7, but can bind or give up a hydrogen ion in the accessible pH range.

Binding of hydrogen ion is one of many biochemically important equilibria. For example, in several cases of *coupling* of hydrogen ion binding to other, functionally important, equilibria, and in many *enzyme-catalyzed reactions*, a hydrogen ion is shuttled between the protein and either substrate or product, sometimes both.

Hydrogen ion binding has been much studied, because it is easy to measure the activity of hydrogen ion very accurately and precisely with a pH meter and, *in vitro*, the activity of hydrogen ion can be varied over a wide range. Adjustment of pH is one of the simplest ways of creating nonnative conditions for studying biological macromolecules.

Interactions between polar groups with zero net charge (dipoles) are ubiquitous in compact, native structures of both proteins and nucleic acids, and are thought to make important contributions to stabilizing these structures. Remarkably, direct contacts between oppositely charged side chains (so-called *salt bridges*) are absent

Equilibria and Kinetics of Biological Macromolecules, First Edition. Jan Hermans and Barry Lentz.

in many proteins, and the implication is that these make only a moderate contribution to protein stability. Qualitatively, this is well understood: charged groups interact strongly with solvent water, and bringing two charged groups in contact requires stripping them of at least some of the tightly bound water molecules, which is energetically unfavorable.[2]

In this chapter we discuss interactions of charged groups, first, hydrogen ion binding equilibria and, second, solvation of charged groups in terms of the continuum dielectric model, including effects of ionic strength.

14.2 IONIZABLE GROUPS IN PEPTIDES

The side chains of four types of amino acids in proteins are normally in their charged form at neutral pH (positively charged for arginine and lysine; negatively charged for aspartate and glutamate). The side chains of two amino acids (cystine and tyrosine) convert into the negatively charged ion form at elevated pH, while the side chains of aspartate and glutamate residues convert into the neutral form at low pH and those of lysine residues do so at high pH. The charged (low pH) and neutral (high pH) forms of the side chain of histidine are in equilibrium in equal concentrations near pH 7.

As a reference for the hydrogen ion binding properties of an amino acid side chain (or nucleic acid base), one may take those of small molecule models having a related chemical structure. For example, for the dissociation of the carboxyl group of a glutamic acid side chain in a protein, one might use as models the dissociation of acetic or propionic acid, of glutamic acid itself, and of peptides containing a glutamic acid residue flanked by other residues. Such comparisons have yielded a set of "normal" dissociation constants (here expressed as $\mathrm{p}K_a = \log_{10} K_b = -\log_{10} K_d$; K_a refers to the acid dissociation constant) for this and other residues (plus the chain-terminal α-amino and α-carboxyl groups), summarized in Table 14.1.

These numbers are not very precise, because of small variations among the $\mathrm{p}K_a$ values of any reasonably chosen set of reference compounds.

It turns out that the side chain of a free amino acid actually is not a good model of a side chain of the same type of residue in a peptide, because in the free residue the nearby charged α-amino and α-carboxyl groups perturb the $\mathrm{p}K_a$ of the side chain. The best values of normal $\mathrm{p}K_a$s of side chains in proteins are based on measurements on series of peptides containing alanine and glycine and a single residue of an ionizable side chain, in which the effect of the terminal charges is found to rapidly diminish as the chain becomes longer.

The positively charged form of the histidine side chain has a hydrogen atom bonded to both nitrogen atoms. In the neutral form, there is a hydrogen on only one of the nitrogen atoms; in water, the two forms are in equilibrium, with the species with the hydrogen on N^{δ} favored by a factor of approximately 10, while for an uncharged histidine side chain buried in the interior of a protein, the placement of the hydrogen atom is determined by hydrogen bonding. In either case, the NH group acts as hydrogen donor and the "lone pair" nitrogen acts as an acceptor.

TABLE 14.1 Characteristics of ionizable groups in proteins

Amino Acid	Type of Group	Low pH Species	High pH Species	Approximate Normal pK_a
Glu, Asp, C-terminus	δ-, γ-carboxyl α-carboxyl	COOH	**COO^- + H^+**	4.0–4.5[a]
Lys N-terminus	ε-amino α-amino	**NH_3^+**	$NH_2 + H^+$	10 8
His	imidazole	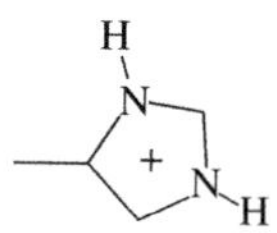	H^+ + 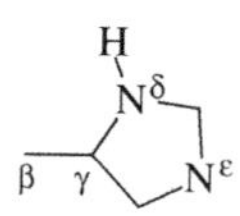 (or H on N^ε)	6.5
Tyr	phenol	C_6H_4—OH	**C_6H_4—O^- + H^+**	10
Cys	sulfhydryl or thiol	SH	**S^- + H^+**	9
Arg	guanidyl	**NH—$C(NH_2)_2^+$**	N—$C(NH_2)_2 + H^+$	14

Species with positive or negative net charge are shown in bold face.
[a]A slightly lower normal pK_a for aspartic than for glutamic acid side chains is generally accepted.

14.3 pH TITRATION OF A PROTEIN: RIBONUCLEASE—NORMAL AND ABNORMAL IONIZABLE GROUPS

The measurements required to obtain the acid–base titration curve of a protein, such as the one for bovine pancreatic ribonuclease A[3] of Fig. 14.1, are quite simple. A fairly concentrated solution of protein in salt solution (no buffer) is brought to higher and lower pH by the addition of small amounts of (usually) 1 M HCl and KOH and one determines the difference in titrant added to a solution of protein and to the same volume of solvent.

A first approximation to the acid–base titration curve of a protein is a set of dissociation curves for independently ionizing groups. Typically, the titration curve will show several smooth steps, one for each different category of side chains in the protein. These steps are indicated in the figure with boxes. Closer inspection shows that the midpoints of the three major steps do not lie at the "normal" pK_as of the side chains of Table 14.1. The titration is spread out over a larger range of pH than expected for normally titrating side chains, and the effect is larger at low ionic strength. This general effect is attributable to charge–charge interaction, the higher the (positive) charge, the more difficult it is to bind another hydrogen ion, and this results in a lower than normal effective pK_a; conversely, a negative net molecular charge raises pK_a. We shall come back to this effect later.

The titration can be expressed in terms of two scales. One, the number of hydrogen ions dissociated relative to the state at low pH is indicated with numbers from 0 to 30. The other, the mean net charge of the molecule, covers the same range, as dissociation of one hydrogen ion lowers the net charge by 1, but this

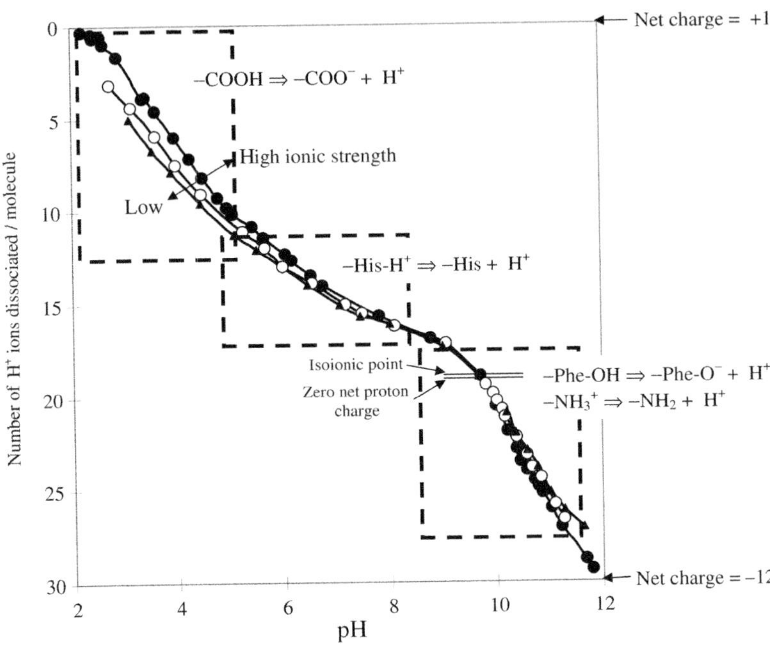

FIGURE 14.1 Titration curves of ribonuclease at three ionic strengths (0.01, 0.03, and 0.15 M). The boxes indicate titration (from left to right) of Asp and Glu, of His, and of Tyr and Lys. (See text for additional explanation.) Data from Tanford, C., Hauenstein, J.D. *J. Am. Chem. Soc.* **78**: 5287–5291 (1956).

scale starts at +18 at low pH, which corresponds to the sum of the numbers of histidine, lysine, and arginine side chains in the molecule plus 1 for the N-terminus. The isoionic or isoelectric point is the pH where the mean net charge is zero and the molecules have zero electrophoretic mobility; for this protein, it is found to coincide within experimental error with the pH at which the net proton charge is zero. Binding of ions from the solution can cause these to be different, but is not observed here.

The behavior of side chains of individual residues, as opposed to that of classes of residues, cannot be inferred from global (all-residue) titration data such as these. Even before the global titrations were performed by Tanford's group, the titration of the tyrosine side chains of ribonuclease was studied by following changes in the UV absorption spectrum that accompany ionization of tyrosine side chains in model compounds (including tyrosine itself; similar changes are observed for phenol). The neutral form of the side chain absorbs weakly above 290 nm, while the ionized form has an absorption maximum at 295 nm; calibration with tyrosine and peptides containing tyrosine allows the absorbance of a protein at 295 nm to be used as a measure of the number of tyrosine residues in the ionized (high pH) form. pH titration of ribonuclease[4] shows that of the six tyrosine side chains, three ionize with an (effective) pK_a between 9.5 and 10, while the other three ionize only at

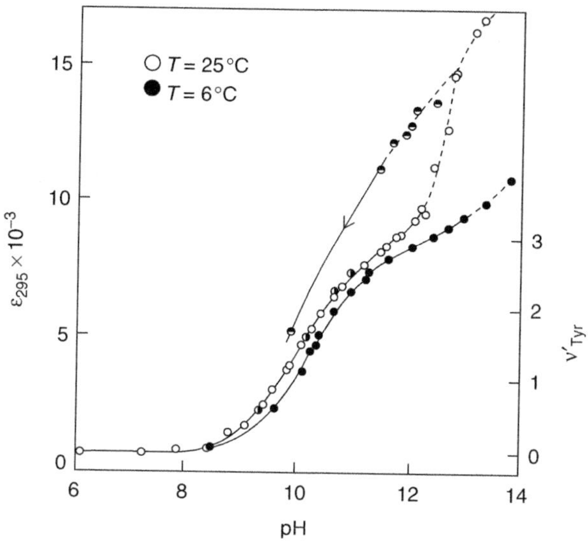

FIGURE 14.2 Titration of the tyrosine side chains of ribonuclease. The half-filled points are observed after exposure to pH 12.5 (upper) or 11.5 (lower). From Tanford C., Hauenstein J.D., Rands D.G. *J. Am. Chem. Soc.* **77**: 6409–6413 (1956).

much higher pH. Later work has shown that the latter occurs with concomitant unfolding of the protein. (Fig. 14.2)

Our model for the behavior of these three side chains is that in native ribonuclease these groups are an intrinsic part of the folded structure, which results in their having a very high pK_a, so that their ionization is not observed below the pH where unfolding occurs. The ionization of these three groups is then *coupled* to the unfolding equilibrium of ribonuclease (similar to the unfolding that at pH 7 is observed at elevated temperature or in the presence of denaturants such as urea or guanidine; cf. Chapter 17). Given these circumstances, the actual pK_as of the three abnormal tyrosine side chains in native ribonuclease have not been determined experimentally.

The three tyrosine side chains having a high effective pK_a are termed *abnormal*. Abnormal ionizable groups are not uncommon. For example, the unfolding of sperm whale myoglobin at low pH is coupled to binding of hydrogen ion to six ionizable groups, at least five of which are histidine side chains.

Global titration studies do not distinguish individual side chains; since these experiments were done, it has become possible to determine the pK_a of individual residues by measuring the pH dependence of NMR resonances of carbon-bound hydrogens. An early application to the C^{δ}-protons of the four histidine side chains of ribonuclease shows that two of these have normal pK_a of 6.6, while the other two (tentatively identified as the active-site histidines) have lower pK_a of 5.4 and 5.8, respectively.[5] In later studies, this technique is typically applied to H-resonances after these have been assigned to specific residues.[6]

14.4 LOCAL INTERACTIONS CAUSE pK_as TO BE ABNORMAL

The interior of a protein molecule is a less polar medium than water; however, buried polar side chain and backbone groups are nearly always stabilized by hydrogen bonds. Typically, the hydrogen bonds stabilize that particular form of the ionizable group that is present in the native structure at neutral pH; the other form of the side chain will interact more poorly with this environment, and, as a result, the pK_a of fully buried side chains will be abnormal (high if the native buried form is the low pH form, and low if this is the high pH form). The abnormality will vary depending on the degree of sequestration.

The uncharged forms of the side chains of Tyr, His, and Cys are frequently found in the core of the protein, with their polar, but neutral, ends forming hydrogen bonds with other parts of the protein. This is observed for the three tyrosine side chains of ribonuclease and the histidine side chains of myoglobin; these residues are usually identifiable on the basis of their interactions with other groups in high resolution three-dimensional protein structures. As mentioned, the pK_as of these groups in the native protein cannot be determined because the groups titrate only if the protein unfolds.

In less extreme cases, the surrounding structure of the native protein may shift a side chain's pK_a within the accessible pH range. Thus, a shift of one or even two units in the pK_a of a histidine side chain may yet allow this residue to be titrated without the protein unfolding; also, a pK_a shift of several units may move the titration of a carboxyl group (of glutamate or aspartate) into the physiological pH range.

Compared with what is the case for the side chains of histidine and tyrosine, the neutral forms of the side chains of aspartate, glutamate, and lysine are uncommon in the core of proteins. One can understand this in terms of the energetics of a two-step process: in step 1 the group is converted into its electrically neutral form, which for these side chains is unfavorable at pH 7, and in step 2 the electrically neutral form is transferred to the core (with formation of intramolecular hydrogen bonds). The first step then carries an extra cost in energy that must be overcome when the protein folds.

14.5 INTERNAL CHARGE–CHARGE INTERACTIONS: ION PAIRS OR SALT BRIDGES

The charged forms of aspartate, glutamate, histidine and lysine side chains, and the charged side chains of arginine are predominantly found on the protein surface, and may form hydrogen bonds with nearby charged surface groups or surface groups with dipoles. Hydrogen-bonded ion pairs (such as Lys and Glu; $-C^{\beta} - C^{\gamma} - C^{\delta} - NH_3{}^{+} \ldots CO_2{}^{-} - C^{\gamma} - C^{\beta}-$) are also called *salt bridges*. While in a vacuum the Coulomb energy of such interactions is very large and negative, in water the *net* free energy of the interaction is not, because to form the salt bridge, the favorable interactions of the fully solvated charged end groups with water molecules are greatly diminished.

Barlow and Thornton[7] analyzed the available three-dimensional structures of 38 proteins and, using accepted geometric criteria for what constitutes an ion pair, determined the presence of 224 such pairs, for an average of six per protein. They found that one-third of the charged residues in these proteins are involved in ion pairs and that three-fourths of those are concerned with stabilizing the tertiary (rather than the secondary) structure. Only 17% of ion pairs in these proteins were found buried in the interior. (In considering these statistics, it is useful to remember that the analyzed structures are of protein in crystals, where the close proximity of neighboring protein molecules and the use of peculiar crystallization-promoting solvents may favor arrangements in which surface side chains do not protrude from the protein as much as in aqueous solution, something that could favor formation of salt bridges.) Barlow and Thornton, citing also early estimates of the stabilization contributed by such ion pairs,[8,9] concluded that, overall, ion pairs make a significant contribution to the stability of globular proteins. Nevertheless, some proteins have no ion pairs at all while others contain a great many.

14.6 MEASURING STABILITY OF SALT BRIDGES FROM DOUBLE MUTANT CYCLES

Many studies have investigated the contributions of salt bridges by site-directed mutagenesis; most, but not all, of these studies found that the removal of a surface salt bridge destabilized the folded structure (or that the introduction of a new surface salt bridge stabilized it), to an extent that varied from case to case. The stabilization is context dependent: the stabilization varies from case to case because the local effect of replacing any particular residue with one of another type varies from case to case, and the local effects caused by replacing two residues with a pair that might form a salt bridge need to be accounted for before one can conclude that the introduced salt bridge has a stabilizing effect and, if so, of what magnitude.

In order to separate the contribution that arises from the *interaction* between the longer side chains (denoted by X and Y) when these are replaced by shorter, noninteracting side chains A and B, from the context dependence, one uses "double mutant cycle" analysis. This involves the preparation of four variants, each with a different combination of residues at the two sites: XY, XB, AY, and AB, and measuring the stability (as the standard free energy of unfolding ΔG^{o}) of each of these four structures. The overall change in stability caused by replacing the AB pair with the XY pair is

$$\Delta\Delta G^{o}_{\mathrm{overall}} = \Delta G^{o}_{\mathrm{XY}} - \Delta G^{o}_{\mathrm{AB}}. \tag{14.1}$$

The context-dependent part is found as

$$\Delta\Delta G^{o}_{\mathrm{context}} = \Delta G^{o}_{\mathrm{XB}} - \Delta G^{o}_{\mathrm{AB}} + \Delta G^{o}_{\mathrm{AY}} - \Delta G^{o}_{\mathrm{AB}}, \tag{14.2}$$

and the contribution due specifically to interactions between the side chains X and Y is then found by subtracting these two quantities, which gives

$$\Delta G^{\circ}(\mathrm{DMC}) = \Delta G^{\circ}_{\mathrm{XY}} - (\Delta G^{\circ}_{\mathrm{XB}} + \Delta G^{\circ}_{\mathrm{AY}}) + \Delta G^{\circ}_{\mathrm{AB}}. \qquad (14.3)$$

In a recent example, Makhatadze et al.[10] applied this to the small globular protein ubiquitin, which contains a pair of oppositely charged residues, Lys 11 and Glu 34, located properly on the protein surface so as to be able to form a salt bridge, and analyzed the effects of replacing each of these residues with an uncharged residue in three double-mutant cycles (mutating K11/E34 to A11/S34, to A11/L34, and to A11/K34; here Ala = A, Ser = S, Leu = L, Glu = E, and Lys = K) and also investigated a double mutant cycle in which the two charged side chains have changed places (mutating E11/K34 to A11/S34). The values of ΔG°(DMC) for these four cycles varied little; the salt bridges were found to stabilize the folded protein by an amount that varied from 3.5 to 3.8 kJ/mol (circa 0.85 kcal/mol). These experiments also demonstrated a very marked context dependence, as the K11/E34 and E11/K34 variants are not equally stable, and the K11/E34 variant is slightly less stable than the A11/S34 variant and is considerably less stable than the A11/L34 variant.

Makhatadze et al.[10] analyzed the salt bridges described in 11 referenced publications by other authors in terms of the Tanford-Kirkwood electrostatic model[11] (a variant of the continuum-dielectric model discussed later in this chapter). Here, they also used a double mutant cycle, in which they compared the Kirkwood–Tanford electrostatic free energies $\Delta G^{\circ}_{\text{q-q}}$ of the four variants XY, AX, XB, and AB by evaluating

$$\Delta G^{\circ}_{\text{q-q}}(\mathrm{DMC}) = \Delta G^{\circ}_{\text{q-q,XY}} - (\Delta G^{\circ}_{\text{q-q,AY}} + \Delta G^{\circ}_{\text{q-q,XB}}) + \Delta G^{\circ}_{\text{q-q,AB}}. \qquad (14.4)$$

The observed correlation between $\Delta G^{\circ}_{\text{q-q}}$(DMC) and ΔG°(DMC) indicates that charge–charge interactions dominate the energetics of salt bridge formation.

14.7 SALT BRIDGES STABILIZE PROTEINS FROM THERMOPHILIC ORGANISMS

Proteins from thermophilic organisms usually show high intrinsic thermal stability, but their structures are very similar to their mesophilic homologs. With only a few available sequences of highly thermostable proteins, Perutz proposed that the greater stability was achieved by the presence of a large proportion of salt bridges,[8] but when additional structures became available, this did not appear to be a general rule.

As an example of recent work, we cite a study by Ge et al.,[12] who used a double mutant cycle approach in order to estimate the effect of temperature on the contribution made by two highly conserved salt bridges to the stability of the hyperthermophilic protein Ssh10b at two different temperatures. The coupling free energies were found to be 2.4 ± 0.4 kJ/mol at 298 K and 2.2 ± 0.4 kJ/mol at 353 K for Glu-54/Arg-57, and 6.0 ± 0.2 kJ/mol at 298 K and 5.9 ± 0.6 kJ/mol at 353 K for Glu-36/Lys-68. One sees that the contribution of these two salt bridges to the stability of this protein changes little with temperature; these salt bridges are resilient to temperature increases and thus are indeed suited for stabilizing this protein at the high temperatures encountered by hyperthermophilic organisms, as had been proposed by Perutz.[8]

Szilágyi and Závodszky[13] have compiled and analyzed a nonredundant data set comprising all high quality structures of thermophilic proteins and their mesophilic homologs from the Protein Data Bank (64 mesophilic and 29 thermophilic protein subunits, representing 25 protein families). From the atomic coordinates, 13 structural parameters were calculated, compared, and evaluated using statistical methods. The authors conclude that different protein families adapt to higher temperatures by different sets of structural devices. The only generally observed rule is an increase in the number of ion pairs with increasing growth temperature. The number of hydrogen bonds and the polarity of buried surfaces exhibit no clear-cut tendency to change with growth temperature.

14.8 CHARGED SIDE CHAINS IN ENZYME CATALYSIS AND PROTEIN SOLUBILITY

Those side chains having normal pK_a close to 7 are obvious candidates for the role of successively taking up and giving off a hydrogen ion (or the reverse) in a biological mechanism, as the free energy changes for protonation and deprotonation will be small. However, other groups may assume these functions, which is then usually facilitated by a change in (= abnormality of) the groups' pK_a, because of peculiarities of the local protein environment.

In their electrically neutral form, ionizable small molecules, such as acetic acid and methyl amine, are reasonably soluble in nonpolar liquids, but the ionic, charged species, that is, acetate ion and methyl ammonium ion, are not. The ionic forms do interact favorably with water. It is therefore understandable why the ionic ends of charged ionizable side chains (at neutral pH: Glu, Asp, Lys, and Arg; cf. Table 14.1) are predominantly found on the surface of globular proteins: the charged species are not "soluble" in the nonpolar interior. The ionic surfaces serve to keep globular protein molecules from forming nonspecific contacts and precipitating.

14.9 ACCOUNTING FOR CHARGE–CHARGE AND CHARGE–SOLVENT INTERACTIONS

We have noted a variety of effects involving charged side chains. These include the mild interaction between charged side chains on the surface of a protein that spreads out the titration curve, the stronger interactions between titratable groups and the protein environment, and the propensity of charged side chains (and other polar groups) to remain solvated (hydrated). In principle, one should seek to understand these effects with a common theory of charge–charge and charge–solvent interactions based on statistical mechanics. A problem with this approach is that one would have to consider a multitude of solvent configurations for *each* conformation of the solute; while these solvent configurations might be accessible via dynamics or Monte Carlo simulations, this would require a lengthy process.

Instead, excellent results are obtained with a much easier-to-use method in which the solvent is represented as a continuum with a high dielectric constant. To understand why this is a good model, one should remember that the interaction of a charged group with solvent results in an energetically favorable reorientation and reordering of the nearby solvent water molecules, whose dipole moments tend to point toward negative and away from positive charges. From studies by Debye it is known that orientation of solvent dipoles makes a major contribution to the dielectric behavior; the dielectric constant of water is around 80 times that of vacuum, thanks to the ability of water molecules to orient themselves in the described manner.

Thus, the continuum dielectric model treats the solvent as a continuum with a high dielectric constant; this solvent continuum is polarized by the charges in the solute molecules, and the energetics are described with equations of classical electrostatics in terms of the electrostatic potential inside the protein and in the surrounding solvent. This model allows one to estimate the electrostatic contribution to the solvation free energy and overall produces a highly successful quantitative representation of solvation of biological macromolecules and its effect on the interaction of charged groups.

14.10 THE CONTINUUM DIELECTRIC MODEL

The continuum dielectric model represents a folded macromolecule in solvent in terms of two volumes having very different dielectric constant, ε, low in the macromolecule's interior and high in the surrounding volume that represents the solvent. The shape of the macromolecule is based on available information about the molecule's structure, and a surface is assumed that separates the two volumes with different values of ε. Finally, a charge distribution is chosen for the protein (Fig. 14.3). More specifically, this model makes the following four assumptions:

The volume is divided into clearly separated "inside" and "outside" spaces. The boundary between inside and outside is usually chosen to be the *molecular surface*, as defined in Appendix H.

Inside the macromolecule, the charge distribution is known (we have a reasonable idea, as we have atomic coordinates from x-ray or NMR studies and estimates of charge distribution from models such as molecular mechanics).

Inside the macromolecule, the medium is a vacuum, except for the known charges ($\varepsilon = 1$; some prefer to use a higher dielectric constant of 2 or 4 for the interior).

Outside the macromolecule, the medium has a very high dielectric constant (ε of water = 78 at 25 °C).

This medium is polarized by the charges inside, and the polarization tends to mitigate the direct Coulomb interactions. The charge distribution resulting from polarization and the polarization energy can be found by solving the Poisson equation

$$\nabla \cdot \varepsilon \nabla \Phi = -\frac{\rho}{\varepsilon_0}. \tag{14.5}$$

Here ε_0 is the permittivity of vacuum and ε the dielectric constant of the medium (relative to that of vacuum). This equation relates the electrostatic potential, Φ to the (macromolecule's) charge density, ρ in a medium of nonuniform dielectric constant. (For more details, see Appendix G.) The energy of the entire system is given in terms of the electrostatic potential and the charge distribution by

$$\Delta G_{\mathrm{el}} = \frac{1}{2}\int_V dV\,\rho(V)\Phi(V) = \frac{1}{2}\sum_i q(r_i)\Phi(r_i), \tag{14.6}$$

where alternate forms are given for a continuous charge distribution and a distribution of point charges, q. The energy can be thought of as consisting of two

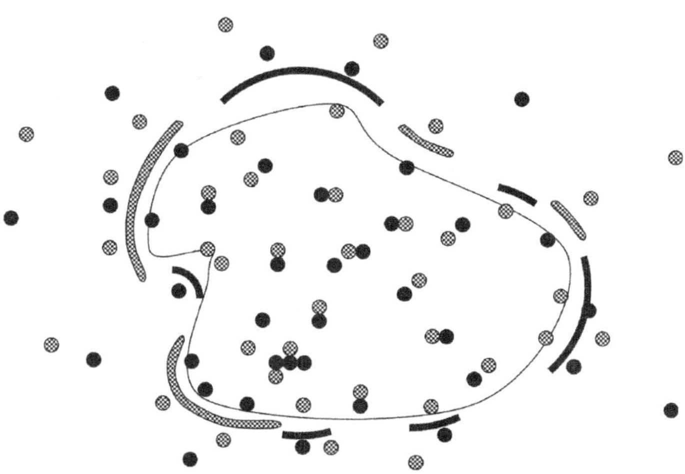

FIGURE 14.3 Schematic representation of the continuum dielectric model. The protein is enclosed by a boundary (molecular surface), inside which the dielectric constant is set to 1 and the atoms are represented with explicit charges (dark and hashed represent opposite signs). Outside the protein, the solvent (with the dielectric constant of water) is polarized in response to the charge distribution inside the protein (dark and hashed bars). Counter ions of different sign (dark and hashed) are in excess near protein charges of opposite sign.

parts: the energy for the same charge distribution in vacuo, and a contribution to the solvation free energy

$$\Delta G_{\mathrm{el}} = \Delta U_{\mathrm{coul},0} + \Delta G_{\mathrm{sol}} = \frac{1}{2}\sum_i \sum_j q(r_i)q(r_j) + \frac{1}{2}\sum_i q(r_i)\Phi_{\mathrm{pol}}(r_i), \quad (14.7)$$

where the potential Φ_{pol} reflects the polarization of the solvent. Free energy: because polarization of solvent requires an increased ordering of solvent molecules, the lower energy that results is in part offset by an entropy loss.

14.11 APPLICATION TO A CHARGED SPHERICAL PARTICLE

The continuum dielectric model has analytical solutions only in simple cases. The following discussion covers two such systems, one a sphere with a uniform charge distribution throughout its volume and the other a sphere with a uniform charge distribution on its surface, with different dielectric constants inside and outside the sphere. It turns out that both the electrostatic potential in the solvent surrounding the particles and the electrostatic contribution to the solvation free energy are identical for these two slightly different models.

The so-called *Born model* represents a solute ion as a spherical particle with radius r and a charge q inside. Solution of the Poisson equation (see Appendix G) gives the following expression for the electrostatic contribution to the free energy of solvation (transfer of the ion from vacuum to a solvent)

$$\Delta G_{\mathrm{el,sol}} = -\frac{q^2}{2r}\left(1 - \frac{1}{\varepsilon}\right)\frac{1}{4\pi\varepsilon_0}. \quad (14.8)$$

This free energy is inversely proportional to the radius of the ion, and for water (which has $\varepsilon = 80$) the magnitude of this free energy is almost equal to the Coulomb energy of a pair of charges in vacuum at a distance equal to twice the radius of the sphere.

Precisely the same electrostatic potential results for an alternative system with uniform dielectric constant equal to 1, in which an additional charge, q_{pol} is distributed over the surface of the sphere, with

$$q_{\mathrm{pol}} = -q\left(1 - \frac{1}{\varepsilon_{\mathrm{a}}}\right). \quad (14.9)$$

Solvent polarization is seen to result in an induced negative charge near the surface of a positively charged particle; this was to be expected from a model based on preferential orientation of dipoles. Note that if the result of polarization is formulated in terms of an additional induced charge distribution, then both the original charges and the induced charges contribute to the electrostatic potential, but the energy is found by integrating the product of *only the original charge distribution* and this electrostatic potential.

14.12 ACCOUNTING FOR IONIC STRENGTH: THE POISSON–BOLTZMANN EQUATION AND DEBYE–HÜCKEL THEORY

The Poisson equation is an excellent starting point to take into account the distribution of dissolved ions and also to assess the influence of ionic strength on the solvation free energy of a macromolecule. Dissolved ions experience the electrostatic potential of one another and of a macromolecular solute. For simplicity, consider a solution containing equal concentrations, c_{ion} of oppositely charged ions, such as Na^+ and Cl^-. The electrostatic energy of a positive ion is

$$U_+ = e\Phi(\mathbf{r}), \tag{14.10}$$

and similarly for U_- with a minus sign. The ions will distribute themselves according to a Boltzmann distribution, with excess positive ions where the potential is negative, and vice versa, so that

$$c_+ = c_{\text{ion}} \exp(-\beta U_+) = c_{\text{ion}} \exp[-\beta e \Phi(\mathbf{r})]. \tag{14.11}$$

One now modifies the Poisson equation to take into account the additional charge distribution due to the dissolved ions

$$\nabla \cdot \varepsilon \nabla \Phi = -\frac{[\rho_0 + e c_{\text{ion}}(e^{-\beta e\Phi} - e^{\beta e\Phi})]}{\varepsilon_0}. \tag{14.12}$$

This is the *Poisson–Boltzmann equation.* (Even for very simple models this equation does not have any analytical solution.)

The exponentials in this equation can be expanded, and terms in higher powers of $e\Phi$ ignored (if $e\Phi$ is small) to give the *linearized* form of the Poisson–Boltzmann equation

$$\nabla \cdot \varepsilon \nabla \Phi = -\frac{(\rho_0 - 2\beta e^2 \Phi c_{\text{ion}})}{\varepsilon_0}. \tag{14.13}$$

This equation is the basis of the *Debye–Hückel theory* of ionic solutions. It can be solved analytically for a few simple cases. In particular, the electrostatic potential around a spherical charged particle in a (dilute) electrolyte solution becomes (Appendix G)

$$\Phi_{\text{a}}(r) = \frac{q e^{-\kappa r}}{4\pi \varepsilon_0 \varepsilon_{\text{a}} r} \frac{e^{\kappa R}}{1 + \kappa R}. \tag{14.14}$$

Here, κ is the Debye–Hückel parameter, the inverse of a distance that decreases proportionally with the square root of the ionic strength; for 0.01 M 1–1 electrolyte $1/\kappa$ is around 30 Å.

This equation indicates that at the surface of a charged spherical solute the electrostatic potential is lower by a factor $(1 + \kappa R)$ in the presence of the electrolyte, and the potential, and thus also the excess concentration of ions of opposite sign,

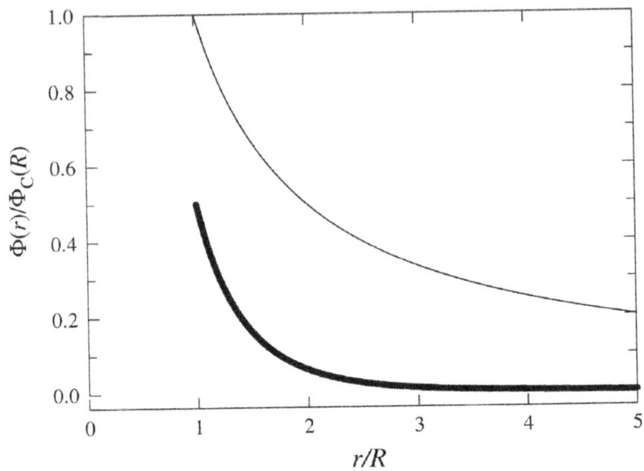

FIGURE 14.4 Electrostatic potential around a spherical charged particle, for Coulomb potential (thin curve) and Debye–Hückel potential with $\kappa R = 1$ (heavy curve; $\Phi_C(R)$ is the Coulomb potential at the surface of the sphere).

drops off more rapidly with distance away from the sphere's surface, by a factor $e^{-\kappa(r-R)}/r$, instead of just $1/r$ (Fig. 14.4).

In a more concentrated electrolyte solution, the linear approximation of the Debye–Hückel theory becomes progressively worse. Also, the value of $1/\kappa$ for 1 M ionic strength, at 3 Å is comparable to the radius of small ions. Tightly bound water molecules having a relatively fixed orientation should not be treated as part of a continuum solvent model.

The free energy of solvation of a spherical charged particle in a (dilute) electrolyte solution changes by

$$\Delta\Delta G_{\mathrm{el,sol}} = -\frac{q}{4\pi\varepsilon_0\varepsilon R}\left[1 - \frac{1}{1+\kappa R}\right], \tag{14.15}$$

as a result of the introduction of electrolyte.

14.13 NUMERICAL TREATMENT VIA FINITE DIFFERENCES

The continuum dielectric model for the *general case of a protein molecule surrounded by a medium of high dielectric coefficient* can be dealt with only numerically. A much used approach divides the system containing protein and a large (finite) volume of solvent into a large number of discrete volume elements, each with a given value of ε, ρ and Φ; the values of Φ are iteratively adjusted to obey the Poisson or Poisson–Boltzmann equation. The finite difference method has been implemented in several readily available computer programs, such as DELPHI,[14] UHBD and the more recently developed, faster APBS.[15]

As was already mentioned, the dielectric coefficient of the protein, ε_P is frequently set to 1, sometimes 2 or 4; for the values of the atomic charges of the protein, q_i one may use the charges from a molecular mechanics force field. The calculated polarization energy is very sensitive to the atomic radii used to determine the molecular surface, and these radii are parameters that are usually optimized by a fit to experimental data for small molecules, such as free energies of transfer from vacuum to water. (See Section 14.5.2.)

14.14 STRENGTHS AND LIMITATIONS OF THE CONTINUUM DIELECTRIC MODEL

Many approaches to molecular thermodynamics are based on molecular energies, and deal with one conformation at a time. For these, estimating the corresponding *free* energies requires an additional integration over conformation space that can be extremely difficult, and sometimes impossible, to accomplish. The continuum dielectric approach is an exception, as it, in fact, integrates over all solvent configurations, if only for a single conformation of the macromolecule, and gives *the electrostatic contribution to the solvation free energy*. It is also uniquely able to account for the effect of ionic strength. It should, however, be kept in mind that numerical evaluation of the continuum dielectric model for a protein molecule takes a considerable amount of computation, and that with the same effort a large number of different configurations of solvent and macromolecule can be evaluated for a molecular mechanics model, even with explicit representation of solvent.

The Poisson and Poisson–Boltzmann equations assume a *linear polarization*, that is, a polarization of the medium that is proportional to the local electric field (the gradient of the electrostatic potential). In very strong electric fields, polarization is typically less than that expected by extrapolating a linear model of the polarization observed in weaker fields; such saturation easily occurs if the dielectric constant has a strong component because of orientation of molecular dipoles, which in strong fields are essentially perfectly oriented, so that further increase of the field may affect only the smaller component, that results from the redistribution of electrons and nuclei and increases the molecular dipole moment.

The continuum treatment will be least appropriate at short distances from the molecular surface, where the details of the solvent's molecular structure can be expected to be important. A linear treatment will break down in situations where water molecules are strongly ordered, such as water molecules adjacent to small multivalent ions, or multiply hydrogen-bonded water molecules in pockets on the surface of protein molecules. (One may, of course, explicitly represent the bound water molecule as a part of the molecular space, rather than of the solvent space.)

The reader should be conscious of these limitations when considering the following applications.

14.15 APPLICATIONS OF THE CONTINUUM DIELECTRIC MODEL TO MACROMOLECULES

14.15.1 Lang's Simple Model of the Effect of Charge–Charge Interactions on pK_a

An early model based on Debye–Hückel theory was proposed by Linderstrøm-Lang, who sought to account for interactions between charged groups on a protein's surface and how these influence binding of hydrogen ion. This model represents the protein as a sphere in a medium with a dielectric constant ε_a, and distributes over the surface of the sphere a charge Ze, where Z is the average net charge (number of positively minus negatively charged groups) and e is the elementary charge. The electrostatic potential at the surface of this model is

$$\Phi_\mathrm{a}(R) = \frac{Ze}{4\pi\varepsilon_0\varepsilon_\mathrm{a}R}\frac{1}{1+\kappa R}. \qquad (14.16)$$

This changes the standard free energy of binding an additional hydrogen ion by an amount

$$\delta G_\mathrm{el} = \frac{Ze^2}{4\pi\varepsilon_0\varepsilon_\mathrm{a}R}\frac{1}{1+\kappa R}. \qquad (14.17)$$

This again changes the binding constant from its value in the absence of charge–charge interaction ($Z = 0$)

$$\frac{K_a}{K_{a,0}} = \exp\frac{\delta G_\mathrm{el}}{k_\mathrm{B}T}, \qquad (14.18)$$

and the corresponding difference in pK_a will be given by

$$\Delta \mathrm{p}K_a = \frac{-\delta G_\mathrm{el}}{\ln 10\, k_\mathrm{B}T}. \qquad (14.19)$$

(The factor $\ln 10 = 2.303\ldots$ is needed to convert to 10-based logarithms.) For the simple spherical model with a uniform surface charge density, this then gives

$$\Delta \mathrm{p}K_a \approx \frac{-1}{\ln 10\, k_\mathrm{B}T}\frac{Ze^2}{4\pi\varepsilon_0\varepsilon_\mathrm{a}R}\frac{1}{1+\kappa R}. \qquad (14.20)$$

This model assumes that the side chains are solvated as they would be in a peptide, and thus have normal pK_a, *except* for this repulsion.

According to this model, pK_a is lowered when Z is positive and increased when Z is negative, as observed. This model produces reasonable agreement with the observed dependence of the hydrogen ion titration curve on net charge and ionic strength (Fig. 14.1).

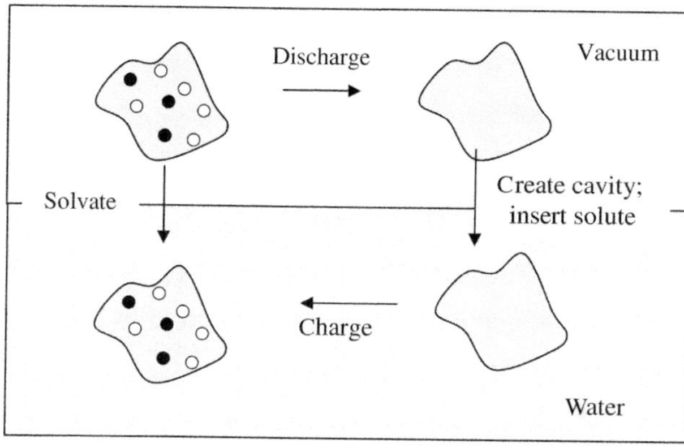

FIGURE 14.5 Decomposition of the solvation process into three steps.[16]

14.15.2 Solvation Free Energies Provide Model Parameters

One can simplify the solvation of a (macro)molecule in terms of a three-step process if one assumes that molecular shape and charge distribution are not affected by the solvation process (Fig. 14.5). In the first step, the charges are set to zero, in the second step a cavity is made in the solvent in which to fit the protein and the uncharged protein inserted into this cavity, and in the third step the charges are reintroduced. The free energy for the first step can be estimated by Coulomb's law; the free energy for the second step is in first approximation equal to the product of a cavity constant and the area of the protein–water interface (either the solvent-accessible or the molecular surface, see Appendix H); the free energy for the third step is the electrostatic free energy and can be estimated using the continuum dielectric model with the Poisson–Boltzmann equation (Eq. 14.7).

Parameter sets consisting of atomic radii, partial charges, and the cavity constant have been developed using on the one hand parameters of molecular mechanics force fields and on the other hand sets of experimentally determined transfer free energies of small molecules. As an example, we cite a study in which such parameters were adjusted to reproduce the solvation free energies of 65 small molecules, ranging from -0.3 to -10 kcal/mol with an average error of 0.44 kcal/mol.[16] For two exceptional molecules, 1,2-ethanediamine and 1,2-ethanediol, the transfer from vacuum to water was much more favorable for the model than found experimentally (by 3.4 and 4.3 kcal/mol, respectively).

14.15.3 Macromolecular Charge Distribution and pK_a

By numerically solving the electrostatic potential, Φ of the continuum dielectric model, this can be used to estimate electrostatic contributions to ion binding, and therefore also the deviations of pK_as from normal values, for any specific model of a protein's shape and internal charge distribution. The procedure is simple; the free energy difference for changing the charge by $+e$ by attaching a hydrogen ion to a side chain at position **r** is evaluated as

$$\delta G_{\text{el}} = e\Phi(\mathbf{r}). \tag{14.21}$$

This must be calculated twice: once for the side chain on the protein, and once for the same side chain in a short peptide in a conformation in which the side chain is unencumbered by short-range contacts. The pK_a shift then follows by applying Eq. 14.19 to the difference between the two values of δG_{el}.

A complication arises if more than one group can bind a hydrogen ion in the same pH-range. In that case, the electrostatic contribution to the solvation free energy of all alternatively charged species (2^n for n such groups) must be evaluated with the Poisson–Boltzmann equation; the extent of saturation with bound ions is then found by considering all these species. Tests of this method performed on a number of small proteins show limited success; the RMS deviation of pK_as of four small proteins from measured values varies between 0.7 and 1.4.[17]

There are two reasons why these calculations do not agree better with the experimental results. *The model assumes a single conformation* for the protein in both ionization states, for which one then uses the experimental structure, here the x-ray coordinates. This is a shortcoming not only because the conformation is a dynamic average but also because the average conformations of the two forms, with and without the added hydrogen ion, need not be the same. This will be the case especially if addition or extraction of this hydrogen ion is unusually difficult because the side chain interacts with the rest of the protein structure and if this either requires or excludes the hydrogen ion because of the presence of one or more hydrogen bonds. A second difficulty is related to the presence of *bound water molecules*. An example is a salt bridge between the (protonated) His-31 and Asp-70 side chains in T4-lysozyme, which, in the crystal structure, is connected to four ordered water molecules. If these water molecules are treated as part of the solvent, the pK_a shifts of the side chains are too small (to 8.0 and 3.3, respectively), but if these are explicitly included in the protein structure, then the pK_a shifts are too large (to 9.9 and circa 0), compared with the experimental values of 9.0 and 1.4.[17] In the former case, the four water molecules are underpolarized and contribute too little to the electric field, and in the latter case, they are overpolarized and contribute too much.

The continuum dielectric model has been found to perform well in modeling the electrostatic interactions between side chains and side chains and solvent that affect the titration of ionizable polypeptides and determine the pH-dependence of the helix-coil equilibrium.[18]

14.15.4 Molecular Interactions

Charge–charge interaction between enzyme and substrate can have a large effect on the catalytic rate. A favorable interaction enhances the rate of diffusion, which is important when the reaction rate is diffusion-limited, and, if this is not the case, raises the concentration of substrate near the enzyme. Section 19.14 discusses a simple model, proposed by Smoluchowski, of the diffusion-limited reaction in

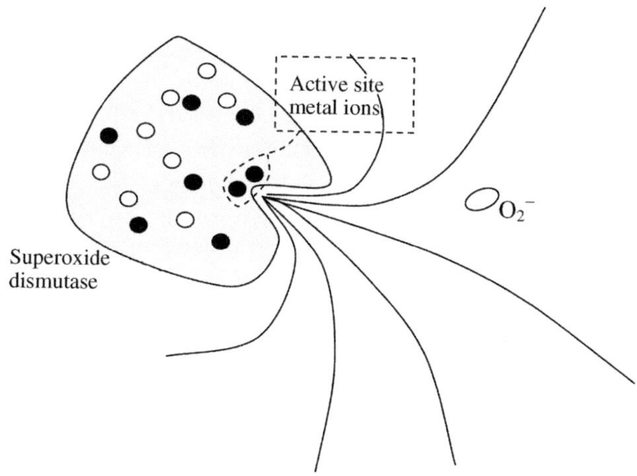

FIGURE 14.6 Electrostatic lines of force "directing" the diffusion of negatively charged superoxide ion O_{2-} to the metal ions at the active site of superoxide dismutase.

which enzyme and substrate are represented as featureless spheres. This model can be extended to incorporate an electrostatic force between enzyme and substrate, and slower than diffusion-limited catalysis (see Section 21.10). However, the spherical symmetry of the model limits its application, and an attractive alternative is to treat both molecules in atomic detail, represent the electrostatic interactions using the continuum dielectric method, the dynamics with a stochastic model using frictional and random forces according to the Langevin equation (Section 21.5; this simulation algorithm is also called *Brownian dynamics*), and the catalysis with an appropriate model such as the Michaelis–Menten scheme (Section 19.9).

This approach has been applied with considerable success in reproducing the rate enhancement of several enzymes with their charged substrates: triose phosphate isomerase with glyceraldehyde phosphate, super oxide dismutase with super oxide, and acetylcholine esterase with acetyl choline, that operate near or at the diffusion-controlled limit.[19] These studies reproduce the favorable long-range interaction of the charged substrate with the charge distribution of the enzyme, which guides the substrate's diffusive motion toward the site with largest opposite charge (Fig. 14.6). The rate enhancement caused by the electrostatic force compensates for the fact that the active site is only a quite small fraction of the enzyme surface, and is effective even if the enzyme molecule has zero net charge, but the active site contains metal ions or side chains having charge opposite that of the substrate.

NOTES

1. The electronic charge (1.6022×10^{-19} C).
2. DNA forms a stable double helical structure in spite of the close proximity of negatively charged phosphate groups that this requires; in this case, the repulsion is mitigated by solvent and counter ions.

3. Tanford, C., Hauenstein, J.D. Hydrogen ion equilibria of ribonuclease. *J. Am. Chem. Soc.* **78**: 5287–5291 (1956).
4. (a) Shugar, D. The ultraviolet absorption spectrum of ribonuclease. *Biochem. J.* **52**: 142–149 (1952). (b) Tanford C., Hauenstein J.D., Rands D.G. Phenolic hydroxyl ionization in proteins. II. Ribonuclease. *J. Am. Chem. Soc.* **77**: 6409–6413 (1956).
5. Bradbury, J.H., Scheraga, H.A. Structural studies of ribonuclease. XXIV. The application of nuclear magnetic resonance spectroscopy to distinguish between the histidine residues of ribonuclease. *J. Am. Chem. Soc.* **88**: 4240–4246 (1966).
6. (a) Wüthrich, K., Wagner, G., Richarz, R., Perkins, S.J. Individual assignments of the methyl resonances in the proton nuclear magnetic resonance spectrum of the basic pancreatic trypsin inhibitor. *Biochemistry* **17**: 2253–2263 (1978). (b) Richarz, R., Wüthrich, K. High-field carbon-13 nuclear magnetic resonance studies at 90.5 MHz of the basic pancreatic trypsin inhibitor. *Biochemistry* **17**: 2263–2269 (1978).
7. Barlow, D.J., Thornton, J.M. Ion-pairs in proteins. *J. Mol. Biol.* **168**: 867–885 (1983).
8. Perutz, M. F., Raidt, H. Stereochemical basis of heat stability in bacterial ferredoxins and in haemoglobin A2. *Nature (London)* **255**: 256–259 (1975).
9. Fersht, A.R. Conformational equilibria in α- and δ-chymotrypsin : The energetics and importance of the salt bridge. *J. Mol. Biol.* **64**: 497–509 (1972).
10. Makhatadze, G.I., Loladze, V.V., Ermolenko, D.N., Chen, X.F., Thomas, S.T. Contribution of surface salt bridges to protein stability: Guidelines for protein engineering. *J. Mol. Biol.* **327**: 1135–1148 (2003).
11. (a) Tanford, C., Kirkwood, J.G. Theory of protein titration curves. I. General equations for impenetrable spheres. *J. Am. Chem. Soc.* **79**: 5333–5339 (1957). (b) Tanford, C. *Physical Chemistry of Macromolecules*. New York: Wiley & Sons, Inc., 1967.
12. Ge, M., Xia, X.-Y., Xian-Ming Pan, X.-M. Salt bridges in the hyperthermophilic protein Ssh10b are resilient to temperature increases. *J. Biol. Chem.* **283**: 31690–31696 (2008).
13. Szilágyi, A., Závodszky, P. Structural differences between mesophilic, moderately thermophilic and extremely thermophilic protein subunits: results of a comprehensive survey. *Structure* **8**: 493–504 (2000).
14. (a) Sharp, K.A., Honig, B. Calculating total electrostatic energies with the nonlinear Poisson-Boltzmann equation. *J. Phys. Chem.* **94**: 7684–7692 (1990). (b) Sitkoff, D., Sharp, K.A., Honig, B. Accurate calculation of hydration free energies using macroscopic solvent models. *J. Phys. Chem.* **98**: 1978–1988 (1994).
15. (a) Luty, B., Davis, M., McCammon, J.A. Electrostatic energy calculations by a finite-difference method: rapid calculation of charge-solvent interaction energies. *J. Comput. Chem.* **13**: 768–771 (1992). (b) Baker, N.A., Sept, D., Holst, M.J., McCammon, J.A. The adaptive multilevel finite element solution of the Poisson–Boltzmann equation on massively parallel computers. *IBM J. Res. & Dev.* **45**: 427–438 (2001).
16. Sitkoff, D., Sharp, K.A., Honig, B.H. Accurate calculation of hydration free energies using macroscopic solvent models. *J. Phys. Chem.* **98**: 1978–1988 (1994).
17. (a) Yang, A.-S., Gunner, M.R., Sampogna, R., Sharp, K., Honig, B. On the calculation of pK_as of proteins *Proteins: Struct. Funct. Genet.* **15**: 252–265 (1993). (b) Antosiewicz, J., McCammon, J.A., Gilson, M.K. The determinants of pK_as in proteins. *Biochemistry* **24**: 7819–7783 (1996).

18. Vila, J.A., Ripoll, D.R., Villegas, M.E., Vorobjev, Y.N., Scheraga, H.A. Role of hydrophobicity and solvent-mediated charge–charge interactions in stabilizing alpha-helices. *Biophys. J.* **75**: 2637–2646 (1998).

19. (a) Madura, J.D., McCammon, J.A. Brownian dynamics simulation of diffusional encounters between triose phosphate isomerase and D-glyceraldehyde phosphate. *J. Phys. Chem.* **93**: 7285–7287 (1989). (b) Sines, J., Allison, S., Wierzbicki, A., McCammon, J.A. Brownian dynamics simulation of the superoxide-superoxide dismutase reaction: Iron and manganese enzymes. *J. Phys. Chem.* **94**: 959–961 (1990). (c) Antosiewicz, J., S.T. Wlodek, J.A. McCammon. Acetylcholinesterase: Role of the enzyme's charge distribution in steering charged ligands toward the active site. *Biopolymers* **39**: 85–94 (1996).

SUGGESTED READING

Tanford, C., *Physical Chemistry of Macromolecules*. New York: John Wiley & Sons, Inc., 1961.

Part 4

Conformational Stability and Conformation Change

In this fourth part of the book we discuss conformational equilibria of macromolecules. The principles of conformation change in small molecules, chiefly a consequence of internal rotation about single bonds, apply equally to macromolecules. But to this are added two levels of complexity: the very large number of internal degrees of freedom in a single macromolecule creates an enormous number of possible different conformations, and the possibility to form close contacts between parts of a macromolecule that are separated by many chemical bonds sometimes produces uniquely stable structures. Each of these complexities has led to a large body of knowledge.

We begin this part of the book with Chapter 15, which covers some very basic principles of polymer physics, mainly the statistical description of the conformation of "random coil" polymer chains in solution. Next, we describe the formation of ordered helical structures in biopolymers and helix-coil equilibria in Chapter 16, and this is followed by Chapter 17 on folding–unfolding equilibria of globular proteins. Finally, Chapter 18 is devoted to two mechanisms that provide elasticity to biological materials, one based on the extension of random coil polymer chains, the other based on unfolding-refolding equilibria of globular protein domains.

Equilibria and Kinetics of Biological Macromolecules, First Edition. Jan Hermans and Barry Lentz.

15

Some Elements of Polymer Physics

15.1 INTRODUCTION

A polymer is a large molecule made up of monomers linked together by chemical bonds. In many synthetic polymers the monomer units are all the same or the polymer is a random sequence of a few types of monomer, in either case of ill determined length; with few exceptions (such as cellulose and starch), biological macromolecules are made up of *unique* sequences, having a fixed length, of nonidentical monomers of a similar type. The class of biological polymers is very distinct from its synthetic relatives in that a large proportion of its members assume well defined three-dimensional structures in solution or in a solid phase. Few synthetic polymers assume ordered structures in solution, while local crystalline order may be present in the solid, but in that case the ordered regions are irregular because the polymer molecules vary in length.

The greater part of polymer physics deals in terms of statistical mechanics with the physical properties of solutions and melts containing structurally disordered polymers, while the greater part of molecular biophysics deals with the ordered states of biological macromolecules. However, as biological macromolecules form ordered structures only subsequent to their biological synthesis, questions arise on how the formation of order is accomplished, and why a molecule containing a particular unique sequence of monomers assumes a corresponding particular unique structure. Knowledge of both fields is needed if one is to address these questions; hence, the inclusion in this book of the following very brief introduction to the physics of polymer chains in solution.

Equilibria and Kinetics of Biological Macromolecules, First Edition. Jan Hermans and Barry Lentz.

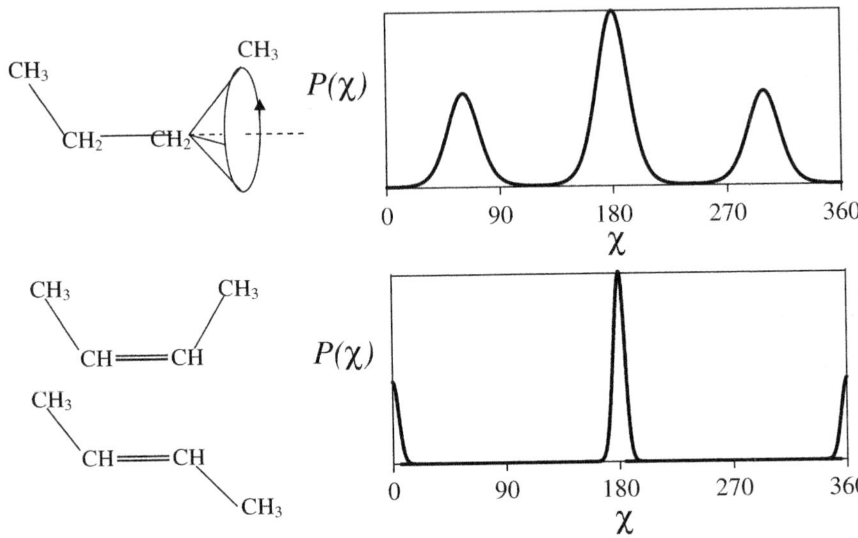

FIGURE 15.1 Models and (schematic) probability distributions of torsion angle, χ. butane has three stable conformers, gauche$^+$, trans and gauche$^-$ (top). butene has two stable conformers, cis and trans (bottom).

15.2 CONFORMATIONAL VARIATION IN SMALL MOLECULES

Small molecules in solution or liquid assume different conformations, but the extent to which the conformation of a small molecule varies is limited and can, in first instance, be catalogued in terms of a small number of intramolecular torsions. In a sample at thermal equilibrium, the value of a torsion angle of any one molecule cannot be exactly predicted; the values of the torsion angle are not equally probable, but statistically follow a Boltzmann distribution in the energy.

As a simple example, the conformation of a molecule of butane in vapor or solution can be defined in first instance by the C_2—C_3 torsion angle. Three distinct conformations are found to be common (trans, gauche$^+$ and gauche$^-$), and for each of these states the value of the torsion angle varies from the mean value for that state (180°, 60°, or −60°) to an extent determined by the curvature of the energy profile and the temperature. As another example, butene assumes two conformations (cis and trans), but these conformations are more rigid than those found in butane, and deviations from the mean torsion angle of 0° or 180° are small. (Fig. 15.1).

15.3 CONFORMATIONAL VARIATION IN CHAIN MOLECULES

A chain molecule contains bonds that are also found in small molecules, with the distinction that the same pattern is repeated many times. Such a linear polymer is different not because the chemistry is different, but because there are so *many* bonds and each can assume a range of torsion angles, in many cases quite independently of each other.

Theorem If the values of all successive torsion angles in a chain molecule are the same (or if they follow a repeating pattern), then the atoms of that molecule lie at constant equal intervals on *helical* curves, one curve per atom in a repeating unit. The molecule is said to be in a helical conformation.

The points on a helical curve obey the following equations ($-\infty < t < +\infty$)

$$x = r\cos(2\pi t)$$
$$y = r\sin(2\pi t)$$
$$z = at, \tag{15.1}$$

in a local Cartesian coordinate system in which the helix axis is the z-axis. The helix has radius r and makes one revolution over a distance a along the z-axis, called *the helix pitch*. Sometimes the result is a closed structure, that is, a ring. (We omit a proof.)

However, given *any* degree of uncertainty in the choice of each torsion angle, the molecule's shape will be irregular, and each molecule will have its own shape. Thus, a long polymer chain can have a great many different conformations and the *ensemble* of these conformations is called *the random coil state*, or also *the statistical coil state*.

Nature appears to have favored molecules that assume stable helical conformations in solution and in solid states. Frequently, a change of conditions can be found that destabilize the helical state of a biological macromolecule and cause a transition to the random coil state, and this is also the case for the folded structure of globular proteins. These transitions are the subject of Chapters 16 and 17, while the rest of this chapter deals with the properties of the random coil state.

15.4 THE IDEAL RANDOM COIL AND THE CHARACTERISTIC RATIO

Similar to the concept of ideal gas and ideal solution, we introduce that of the ideal random coil, in which interactions between distant parts of the chain are ignored.

Theorem One property of ideal random coils is that the distribution, $P(R)$ of the distance, R that separates beginning and end of the chain, approaches a Gaussian function when the chain is sufficiently long. Each *component* of the vector connecting the ends will obey a distribution of the form (written for the x-component)

$$P(R_x)dR_x \sim \exp\left(-\frac{\frac{1}{2}R_x{}^2}{\langle R_x{}^2\rangle}\right)dR_x, \tag{15.2}$$

the distribution of the end-to-end vector, $\boldsymbol{R} = (R_x, R_y, R_z)$ will be

$$P(\boldsymbol{R})d\boldsymbol{R} \sim \exp\left(-\frac{\frac{3}{2}\boldsymbol{R}^2}{\langle R^2\rangle}\right) d\boldsymbol{R}, \tag{15.3}$$

and the distribution of the end-to-end *distance*, $R = (R_x^2 + R_y^2 + R_z^2)^{1/2}$ will be

$$P(R)dR \sim \exp\left(-\frac{\frac{3}{2}R^2}{\langle R^2\rangle}\right) R^2 dR. \tag{15.4}$$

Furthermore, the mean *square* value of the end-to-end distance is proportional to the length of the chain (again, for long ideal random coil chains)

$$\langle R^2\rangle = C_r n\, l^2, \tag{15.5}$$

as expressed in the number of monomers, n and the square of the chain length, l per monomer. The characteristic ratio, C_r varies from polymer to polymer and depends on conditions such as solvent and temperature. For a freely jointed chain (the unrestricted random walk model),

$$\langle R^2\rangle = n\, l^2, \tag{15.6}$$

and the characteristic ratio is equal to 1.

This theorem is proved by straightforward application of the random walk model and the associated central limit theorem, according to which the addition of contributions of varying size, with as only condition that positive and negative contributions of the same size are equally probable, produces as answer a sum that obeys a Gaussian distribution if the summation is repeated independently many times. (See Appendix B.)

For any real polymer, the random walk's perfect Gaussian distribution is perturbed by the fact that two chain segments cannot occupy the same volume; how one may correct for self-avoidance is discussed below. Ignoring the correction for the moment, one sees that an individual polymer molecule is coiled within a volume having radius on the order of the root-mean-square end-to-end distance, which is proportional to $n^{1/2}$. This volume is, therefore, proportional to $n^{3/2}$; it is typically much larger than the volume of solvent that is displaced by the macromolecule, and grows more rapidly than the molecules' mass. Thus, a relatively dilute solution of high molecular-weight polymer can effectively be "full" of solute molecules (and as a consequence be gel-like or very viscous).

15.5 THE PERSISTENCE LENGTH AS A MEASURE OF CHAIN FLEXIBILITY

In order to achieve Gaussian chain statistics, the nature of the chain's flexibility need not be specified. A model of a polymer containing C—C bonds with rotational isomerism, as for butane in Fig. 15.1, obeys Gaussian statistics, but so does a model of a gradually curved elastic rod, the amount and sense of local curvature being randomly distributed, and its extent determined by thermal energy. In other words, the observation of Gaussian chain statistics, for example, in terms of the dependence of molecular dimensions on chain length, allows no conclusions as to the cause of the chain's flexibility.

In order to capture a greater variety of chain models, the relation between end-to-end distance and length of the chain can be expressed in terms of the so-called persistence length. This concept expresses the chain's inherent flexibility (more precisely, its inherent stiffness) by noting that as one moves along the chain, the chain's direction changes in an unpredictable manner, more rapidly for a flexible chain, more slowly for a stiff chain, and the direction steadily becomes less and less correlated with that of the starting segment. While a rigid rod extends in one direction over its full length, a flexible randomly coiled molecule will, on an average, extend only to a limited extent in the direction in which the chain starts, and this distance is called *the persistence length.* For very long ideal random coil molecules, the mean square of the end-to-end distance is related to the persistence length, a by

$$\langle R^2 \rangle = 2aL, \tag{15.7}$$

where L is the contour length of the chain.

This result is important because it allows us to apply the random coil concept to molecules and supra-molecular structures that are not strictly linear chain molecules, but are (almost) gradually curved elastic rods. Specifically, (long) DNA molecules in solution have properties of random coils. (Because this polymer is negatively charged, and charge–charge interactions carry over large distances, DNA and other polyelectrolytes are far from *ideal* random coils). The limiting model for gradually curved elastic molecules is called *the wormlike chain.* (Fig. 15.2).

15.6 CONFORMATION OF SELF-AVOIDING CHAINS

The statistics of self-avoiding random walks can be approximated by a model in which each monomer is assigned a volume v_c, which is not accessible to any other

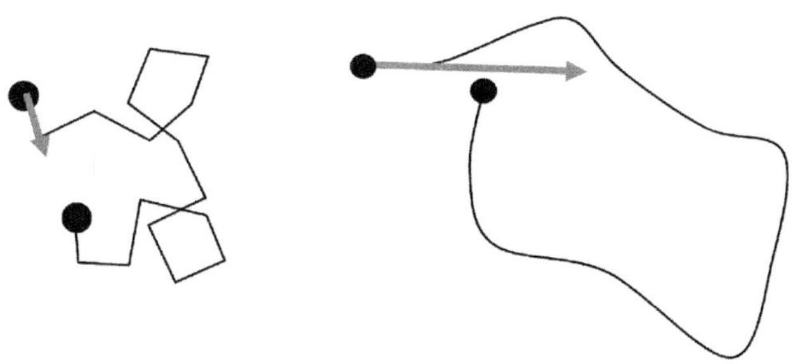

FIGURE 15.2 Left: Model of a molecular chain, that is, a chain of chemical bonds. Right: Model of a wormlike chain. Both are shown with (hypothetical) persistence vectors.

monomer. (The volume v_c is an unknown proportionality constant.) If the volume occupied by the coil is modeled as a sphere, with volume proportional to R^3, then a fraction nv_c of this volume is not accessible to any particular monomer; this reduces the number of possible locations of this monomer by a fraction equal to nv_c/R^3, leaving a fraction $1 - nv_c/R^3 \approx \exp(nv_c/R^3)$ of the volume available for placing this monomer. This applies to all n monomers, and one gets

$$P_{\mathrm{SA}}(R)dR \sim \exp\left[-\frac{3R^2}{2\langle R^2\rangle} - \frac{n^2 v_c}{2R^3}\right] R^2 dR. \tag{15.8}$$

The most probable end-to-end distance, R^* is then found by maximizing this probability, which gives

$$R^* \sim n^{3/5}, \tag{15.9}$$

in the limit of large N. The exponent is significantly larger than the value of ½ obtained without self-avoidance, and is remarkably close to the value of 0.588 that is found by successively approximating the self-avoiding walk's distribution with series expansion. Experiments and simulations show that the variation of the radius of gyration, R_g of polymer chains with chain length closely follows this prediction, that is, it is found that

$$R_g \sim n^{0.6} l. \tag{15.10}$$

15.7 DEPENDENCE OF CHAIN CONFORMATION ON SOLVENT CONDITIONS; "THETA" CONDITIONS

The interaction of polymer molecules with solvent molecules affects the conformation and solubility of the polymer. The probability distribution for the end-to-end distance can now be approximated with

$$P_{\mathrm{SA},E}(R)dR \sim \exp\left[-\frac{3R^2}{2\langle R^2\rangle} - \frac{n^2 v_c}{2R^3}\right] \exp\left[-\frac{E(R)}{k_B T}\right] R^2 dR. \tag{15.11}$$

where the "contact energy", $E(R)$ consists of the (free) energy contributions from interactions between polymer segments, which are more frequent when the conformation is compact and R is small. It is of course an approximation to integrate over all values of R using the average energy, $E(R)$ instead of summing over all conformations each with its own energy.

A simple theory to describe solvent–polymer interactions was developed by Flory and Huggins. Flory–Huggins theory uses a lattice model; a lattice site can be filled by either a solvent molecule or a segment of the polymer, and the model's thermodynamics are limited to the effects of direct contacts between lattice sites, long-range interactions being ignored. The (free) energy of a given conformation contains $nz/2$ contact terms, where n is the number of lattice sites and z the number of contacts between a site and its neighbors. The contact (free) energy for the i-th conformation is then

$$E_i = -n_{\mathrm{pp}}^{(i)}\varepsilon_{\mathrm{pp}} - n_{\mathrm{ps}}^{(i)}\varepsilon_{\mathrm{ps}} - n_{\mathrm{ss}}^{(i)}\varepsilon_{\mathrm{ss}}, \tag{15.12}$$

where $-\varepsilon_{\mathrm{pp}}$ represents the energy of forming a contact between two polymer segments and $n_{\mathrm{pp}}^{(i)}$ the number of these contacts in conformation number i, and similarly for polymer–solvent and solvent–solvent contacts. To form one new pp contact, two ps contacts must be broken, and one new ss contact formed. Thus, relative to a conformation without any pp contacts, the energy difference is

$$\Delta E_i = -n_{\mathrm{pp}}^{(i)}\varepsilon_{\mathrm{pp}} + 2n_{\mathrm{pp}}^{(i)}\varepsilon_{\mathrm{ps}} - n_{\mathrm{pp}}^{(i)}\varepsilon_{\mathrm{ss}} = -2n_{\mathrm{pp}}^{(i)}\Delta\varepsilon. \tag{15.13}$$

where the energy $\Delta\varepsilon$ reflects the quality of solvation. As was done in order to describe the excluded volume effect, the number of pp contacts per contact site is set proportional to the density of chain segments in the volume equal to R^3, that is,

$$\langle n_{\mathrm{pp}}^{(i)}\rangle \approx \frac{1}{2}zn\frac{nv_{\mathrm{c}}}{R^3}. \tag{15.14}$$

One then has

$$P_{\mathrm{SA},E}(R)dR \sim \exp\left[-\frac{3R^2}{2\langle R^2\rangle} - \frac{n^2 v_{\mathrm{c}}}{2R^3}\right]\exp\left[\frac{n^2 v_{\mathrm{c}}}{2R^3}2\chi\right]R^2 dR$$

$$\chi = \frac{z\Delta\varepsilon}{k_{\mathrm{B}}T}. \tag{15.15}$$

Large end-to-end distances are favored when χ is negative; negative values of χ and hence $\Delta\varepsilon$ denote what is called *a good solvent*. Because of the factor n^2 multiplying χ, the transition from an open chain for $\chi < 0$ to a compact chain when $\chi > 0$ is quite abrupt for large n, as is shown by experimental results for the dimensions of long polymer molecules in an organic solvent (Fig. 15.3).[1] The collapse is reminiscent of the abrupt folding of a globular protein as the temperature is lowered; however, the disordered state of the compact polymer is, rather, suggestive of the molten globule conformation. (See Chapter 17.)

Conditions under which $\chi = 0$ are called *theta conditions*. Theta conditions also quite sharply separate regions in which a polymer is soluble and in which it is not. The effect is so pronounced (i.e., solubility changes so rapidly with conditions) because transfer of a polymer molecule from solution to a precipitate involves formation of a large number of pp contacts with other polymer molecules in the precipitate, a number that scales as n, the number of segments in a polymer molecule.

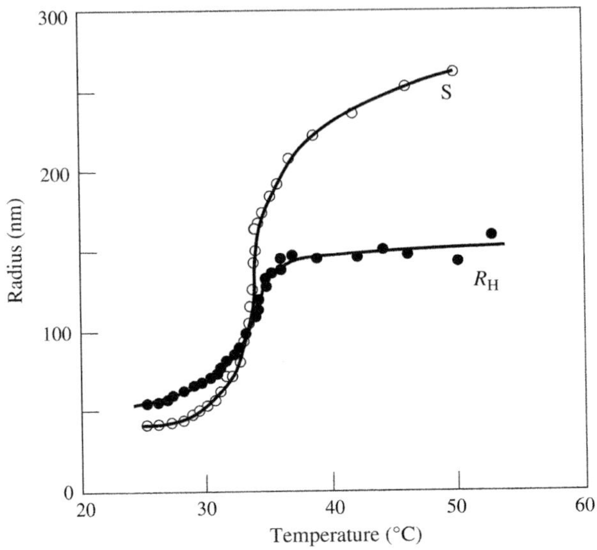

FIGURE 15.3 Collapse of high molecular-weight polystyrene chains in cyclohexane in dilute solution at the theta temperature (35 °C). S is the radius of gyration and R_H the hydrodynamic radius (which governs such properties as sedimentation velocity and solution viscosity. From Sun, S.T., Nishio, I., Swislow, G., Tanaka, T. *J. Chem. Phys.* **73**: 5971–5975 (1980).

15.8 RELATING CHAIN STATISTICS TO MOLECULAR STRUCTURE

Flory and his students[2] have shown how intrinsic chain properties such as characteristic ratio and persistence length can be related to the geometry of the chain, expressed in terms of bond lengths, bond angles, and a *distribution* of dihedral angles, and developed the requisite mathematical expressions.

The math is relatively simple for the special case in which the distribution of the torsion angles in each residue is independent of the values of the torsion angles of nearby residues. For a macromolecule made up of independent residues, such properties as persistence length can be calculated directly given the geometry and the conformational distribution in terms of backbone dihedral angles. The independent-residue model is well applicable if the most common conformations are relatively extended, for then situations that are excluded because atoms of residues two or three units apart clash with one another, are very rare, and ignoring their contribution to ensemble averages will not be problematic. It was argued that the model should be applicable to denatured proteins because successive C^{α} atoms are separated by a distance of around 3.74 Å by a relatively rigid spacer, the C^{α}—C—N—C^{α} group (see Fig. 3.4), while an independent-residue model is inappropriate for a repeating polymer not having such spacers, such as poly(ethylene), $(CH_2)_n$.

In Chapter 17 we shall discuss evidence that shows that denatured proteins in 6 M guanidine (with any disulfide crosslinks removed by reduction) behave as random coils. Viscosity measurements by Tanford and coworkers give a value of 5.0 for the characteristic ratio under these conditions, based on the dependence of the end-to-end distance on the number of residues, with a value $l = 3.74\,\text{Å}$. This value of the characteristic ratio is in agreement with an independent-residue model, with a simple common joint probability distribution of the variable backbone torsions angles φ (about N—C$^{\alpha}$) and ψ (about C$^{\alpha}$—C) based on energy considerations, in which most residues are relatively extended (β and P_{II} conformations)[3].

If successive residues interact more strongly, and the conformational distributions of successive residues are not independent, the problem is much more difficult. Simulations of molecular mechanics models can be helpful by eliminating conformations with chain overlap and also by including energetic terms favoring or disfavoring polymer–polymer interactions. One may, for example, use a (relatively compute-intensive) Monte Carlo process to generate chain conformations in numbers proportional to their Boltzmann factor, $\exp(-E^{\circ}/k_BT)$ or $\exp(-G^{\circ}/k_BT)$.

15.9 POLYELECTROLYTES

Polyelectrolytes are polymers containing ionic monomer units; synthetic polyelectrolytes are usually highly soluble in water. Proteins are polyelectrolytes, but the charge density is low; at neutral pH a protein like ribonuclease contains both negative and positive charges, and only a fraction of the side chains is ionic. In contrast, nucleic acids (both RNA and DNA) are highly charged and true polyelectrolytes. A number of nonbiological polyelectrolytes have been prepared by synthetic methods (such as, poly-L-glutamic acid and poly-L-lysine) or as derivatives of biological polymers (e.g., carboxymethyl cellulose).

Because of the long range of electrostatic forces, the Flory–Huggins model, which considers only nearest-neighbor interactions is not an effective model for polyelectrolytes. Both the statistics of individual polyelectrolyte molecules and the interactions between molecules are strongly dependent on electrostatic forces between chain elements; as these forces depend strongly on the ionic strength, this is also the case for the average chain dimensions of polyelectrolyte molecules in solution.

This is illustrated by the data of Fig. 15.4, which shows results of viscosity measurements of carboxymethyl cellulose, a water-soluble derivative of cellulose in which significant numbers of —OH groups have been replaced with —O—CH_2—COO^-. The relative specific viscosity increment, $(\eta - \eta_0)/(\eta_0 c)$ is plotted as a function of the polymer concentration, c at a series of ionic strengths (where η is solution viscosity and η_0 solvent viscosity), the polymer's contribution to the ionic strength also being accounted for.

The significance of this ratio is based on the Einstein equation for the relative specific viscosity increment of *noninteracting spheres* (the intrinsic viscosity, $[\eta]$; in practice this means extrapolation to zero concentration). The equation is

$$[\eta] = \lim_{c\to 0} \frac{\eta - \eta_0}{\eta_0 c} = 2.5\frac{V}{M} \qquad (15.16)$$

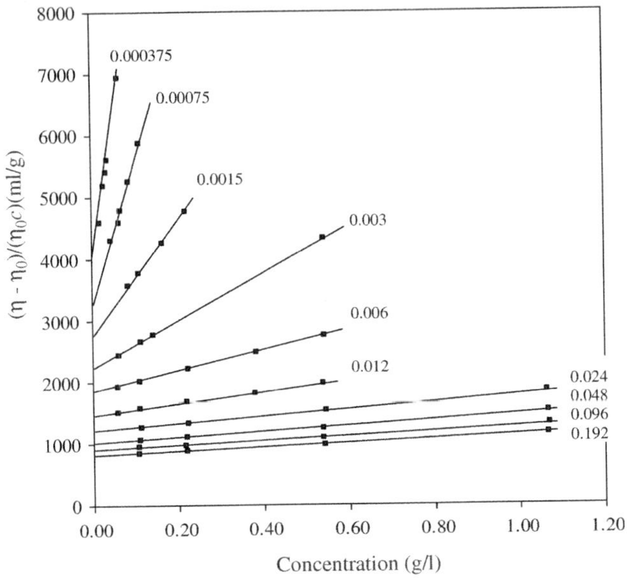

FIGURE 15.4 Viscosity of carboxymethyl cellulose solutions at different ionic strength (in M). The average molecular weight is 150,000, for circa 725 glucose monomer units per molecule; the number of substituted —CH_2—COO^- groups is 0.56 per monomer. From Pals, D.T.F, Hermans, J.J. *Rec. Trav. Chim.* **71**: 433 (1952).

where V/M is the ratio of volume to mass of a solute molecule. Using this as a model, one may define the effective (hydrodynamic) volume of random coil polymer molecules.

A polymer molecule's hydrodynamic volume should scale approximately as the cube of the root-mean-square end to end distance R, so that $[\eta]$ should scale as R^3. One sees that $[\eta]$ increases more than 10-fold over this range of ionic strength, and, therefore, that R *of these carboxymethyl cellulose molecules more than doubles when the ionic strength is reduced from moderate to very low*. Also, the interaction between molecules, as evidenced by the concentration dependence of $(\eta - \eta_0)/(\eta_0 c)$, is more pronounced at low ionic strength (even relative to the increase of $[\eta]$).

NOTES

1. Sun, S.T., Nishio, I., Swislow, G., Tanaka, T. The coil–globule transition: Radius of gyration of polystyrene in cyclohexane. *J. Chem. Phys.* **73**: 5971–5975 (1980).
2. Flory, P.J., *Statistical Mechanics of Chain Molecules*. New York: Interscience (1969).
3. Brant, D.A., Flory,P.J. The configuration of random polypeptide chains. I. Experimental results; II. Theory. *J. Am. Chem. Soc.* **87**: 2788–2791; 2791–2800 (1965).

SUGGESTED READING

Much of our understanding of polymers is based on the work of Paul Flory. A classic reference is Flory, P.J., *Principles of Polymer Chemistry*. Ithaca, NY: Cornell University Press, 1953.

Tanford, C., *Physical Chemistry of Macromolecules*. John Wiley & Sons, Inc.: New York, 1967.

The material in this chapter relies also on a comprehensive short treatise by Doi, M., *Introduction to Polymer Physics*. Oxford: Clarendon Press, 1996.

16

Helix-Coil Equilibria

A scientific paper is . . . a cunningly contrived piece of rhetoric. It has only one purpose; it must persuade the reader of the veracity of the observer, his disinterestedness, his logical infallibility, and the complete necessity of his conclusions.

—J. M. Ziman, loc. cit.

16.1 INTRODUCTION: MULTISTATE TRANSITIONS OF HELICAL POLYMERS

A variety of polypeptides and nucleic acids form helical structures.[1] These extend far in one direction, but are quite limited in size in the other two directions. Inherently, the equilibria between the *helical* (= folded) state and the *random coil* (= unfolded) state *cannot* be described in terms of just two states. However, in most cases, the transitions are *cooperative* as a result of linkage of the conformational equilibria of successive residues; a suitable model was found to be one that had been developed by Ising to describe cooperative magnetization, which requires a relatively complicated mathematical treatment.

16.2 SINGLE-STRANDED POLY (A): A COMPLETELY NON-COOPERATIVE TRANSITION

At low temperature, poly (ribo adenine), or poly (A) forms a helical structure in which the adenine bases of successive residues stack on top of each other. A transition to a disordered structure is observed when the temperature is raised, and can be followed by the change of the UV absorption. (Fig. 16.1).

A key observation is that this curve (for poly A) is *independent of the length of the polymer*, all the way down to the dimer.[2,3] (Also independent of the concentration.) This means that the stacking equilibrium of any pair of adjacent bases is

Equilibria and Kinetics of Biological Macromolecules, First Edition. Jan Hermans and Barry Lentz.

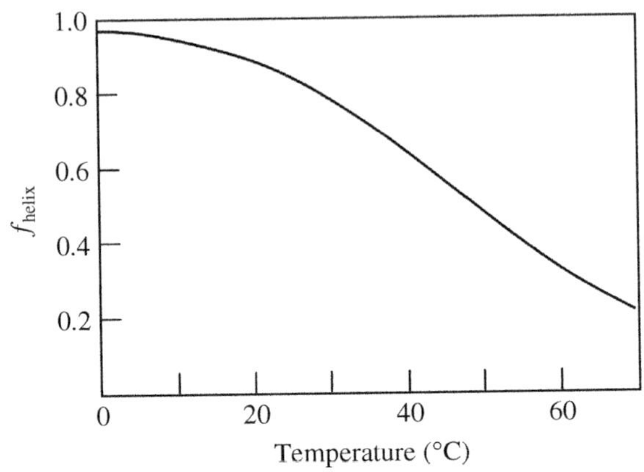

FIGURE 16.1 F_{helix}, the fractional amount of stacking interaction of Poly A, as a function of temperature. (From Eisenberg, H., Felsenfeld, *J. Mol. Biol.* **30**: 17–37 (1967); calculated from the data of Leng and Felsenfeld.[2])

independent of what is happening elsewhere, that is, of the extent of stacking, or of whether adjacent bases are stacked or free. Consequently, the ratio f_{helix}/f_{coil} can be equated with the equilibrium constant for stacking a *single* pair of adjacent bases. The corresponding enthalpy

$$\Delta H^{o} = -k_B \frac{\partial \ln K}{\partial (1/T)} = -k_B \frac{\partial \ln(f_{helix}/f_{coil})}{\partial (1/T)}, \qquad (16.1)$$

is not insignificant, around -12 kcal/mol.

The transition of poly (A) is said to be multistate, in that a partly denatured state is an ensemble of many different species, each with different distribution of helical residues over the length of the molecules. This contrasts with a two-state transition, in which only fully helical and fully coiled molecules are observable under any conditions. Note that if the helix-coil transition of poly (A) followed a two-state model, then the enthalpy change should be proportional to the number of residues, and, therefore, the steepness of the temperature transition also should be proportional to the number of residues.

16.3 SYNTHETIC POLYPEPTIDES

The early observations on formation of helices in solutions of synthetic polypeptides of different lengths showed a more complicated situation. In this case, longer molecules form more stable helices and the temperature transitions become steeper with increasing molecular size (Fig. 16.2).[4] While this is (qualitatively) what one expects for an all-or-none model, it is also observed that as the molecules are made even longer, the transition becomes independent of molecular size; therefore, an all-or-none (i.e., two-state) model is not applicable.

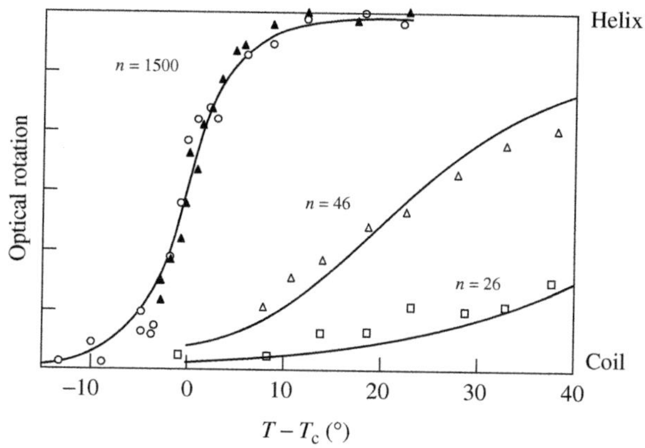

FIGURE 16.2 Helix-coil transition of poly-γ-benzyl-L-glutamate chains with different number of residues, n. T_c is the midpoint for the transition of very long molecules. Curves are theoretical. Under the conditions of these experiments, the helix is stable at higher temperature. From Zimm, B.H., Doty, P., Iso, K. *Proc. Natl. Acad. Sci.* USA **45**: 1601–1607 (1959).

John Schellman formulated a multistate model to describe the unfolding of helices.[5,6] This model recognizes that it is difficult to put one residue in a helical conformation, as this will entail a decrease of conformational entropy, and perhaps also an increase of the energy. The same loss of entropy results when additional residues are added to this incipient helix. However, after the helix completes its first turn, newly added residues interact specifically with residues already in the helix; such new intramolecular contacts, in particular hydrogen bonds, can now exert a stabilizing influence. Ultimately, the free energy contributions from these favorable interactions must offset the unfavorable contributions, otherwise no helix would be observed (Fig. 16.3.) This overall favorable multistep process, in which early steps are unfavorable, is a typical example of a process in which one or more unfavorable *nucleation* steps are followed by favorable *growth* steps. If enough detail is provided, such a model can be used for a statistical thermodynamic analysis; this will be discussed for the helix-coil transition in this chapter. A related kinetic model is briefly referenced in Chapter 20.

We define K_j as the equilibrium constant for a residue to assume a helical conformation. The subscript j indicates how long a helical stretch is formed as a result. Thus, K_1 is for a single residue to assume a helical conformation in an otherwise coiled molecule, K_2 pertains to either of the two residues adjacent to a single helical residue, and so on. The equilibrium constant for forming this stretch of two adjacent helical residues in a nonhelical molecule is then K_1K_2, and the equilibrium constant for forming a helix with k residues is $K_1K_2 \ldots K_k$. For this to describe a nucleation-growth model, the first steps must be unfavorable and later steps favorable. In the Schellman model for the helix-coil equilibrium, the first three steps are unfavorable, so that K_1, K_2, and K_3 are all less than 1, but

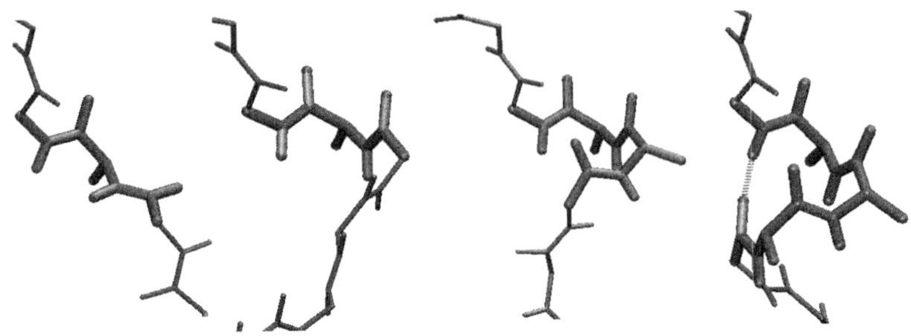

FIGURE 16.3 Model of helix initiation in a coiled polypeptide molecule according to Schellman. In each successive structure one residue assumes an α-helical conformation. The first two steps are unfavorable (confinement without stabilizing contacts). A stabilizing hydrogen bond is formed only in the third step (vmd graphic).

steps beyond the third are favorable, so that K_4 and higher are greater than 1 if measurable helix forms.

In the Schellman model, the three initial sequences ...chc..., ...chhc..., and ...chhhc... are successively less stable. The model may be *simplified* by letting

$$\begin{aligned} K_1 &= s\sigma \\ K_1K_2 &= s^2\sigma \\ K_1K_2K_3 &= s^3\sigma \\ &\mathit{etc}. \end{aligned} \tag{16.2}$$

The simplification introduced in Eq. 16.2 treats formation of a single helical residue as the helix initiation step, while in the Schellman model helix initiation consists of formation of a full turn of the helix by four adjacent residues. The simplified model contains just two independent parameters, namely, the helix initiation parameter, σ which is small

$$\sigma \ll 1, \tag{16.3}$$

and the helix-growth parameter, s

$$s = \frac{K_1K_2 \ldots K_{k+1}}{K_1K_2 \ldots K_k}. \tag{16.4}$$

For a short peptide, little or no helix formation is possible, unless s is greater than 1; furthermore, when this is the case, and helix formation is observed, long helices are more stable than short ones.

Is the simplification of Eq. 16.2 justified? It turns out that this depends on the details of the system that the model seeks to describe. The relative stability of very short helical segments is not well represented, as a single growth parameter, independent of the length of the helical segment, j describes all cases. On quite general grounds, the growth parameter should approach a constant value for large j, but it is a poor assumption that s has the same value for elongation of short helices. Therefore, if the helix content is the result of a mixture of short helical segments, then the approximation is a poor one; on the other hand, if the helical segments are nearly all long, then the approximation is acceptable. As we shall see, application of the model to the helix-coil equilibrium of very long polypeptide molecules shows that for these the helical segments are quite long, which for these very long molecules then justifies the approximation made in formulating the model in terms of only two equilibrium constants, s and σ.

16.4 ZIMM–BRAGG, GIBBS–DIMARZIO, AND LIFSON–ROIG ANALYSES

Schellman stated the model, but did not give a complete description that agrees with the observations, because he did not consider the formation of multiple helical segments in very long polypeptide molecules.[7] With this added sophistication, Schellman's model is generally accepted to account for the cooperativity of helix-coil transitions of long polypeptides containing a single residue (see comment elsewhere so-called homopolymers), and with modifications to reflect the results of experiments on short polypeptides with specific sequences that have become possible because of the use of automated peptide synthesis.

The model enumerates the possible conformations by assigning one of two possible states, helix (h) and coil (c), to each residue. A particular conformation is then described by a sequence of h–s and c–s. For example, for an oligopeptide containing five residues, one has the $2^5 = 32$ states of column 1 of Table 16.1. These states have been grouped in categories with same number of helical residues, i and same number of helices, j. The *number of states* in each category, $m(i, j)$ is also given in the table.

The last column lists the "statistical weight" (the official term), which is the equilibrium constant for forming one of these states from the all-coil state. According to the (simplified) Schellman model, the *statistical weight*, w is a function of the number of helical residues and of the number of separate helices in the state and is given by

$$w_{ij} = \sigma^j s^i. \tag{16.5}$$

The two parameters σ and s reflect the physics of helix initiation and helix growth, respectively, and depend on the type of amino acid, and conditions such as temperature and solvent composition. Table 16.2 shows the factors m_{ij} for a poly amino acid with $n = 20$ residues.

The above model contains everything that is needed to formulate a partition function containing all helical and coiled states in the appropriate proportions. The following equation expresses this in terms of the ratio of the partition functions for the mixture of all part helical, part coil states to that of the all-coil state, as a

TABLE 16.1 Helix/coil states of a pentapeptide

States	Number of Helical Residues (i)	Number of Helices (j)	Number of Instances ($m(i, j)$)	Statistical Weight w_{ij}
ccccc	0	0	1	1
hcccc ... cccch	1	1	5	σs
hhccc ... ccchh	2	1	4	σs^2
hchcc ..... chchc	2	2	6	$\sigma^2 s^2$
hhhcc	3	1	3	σs^3
hhchc.....	3	2	6	$\sigma^2 s^3$
hchch	3	3	1	$\sigma^3 s^3$
hhhhc	4	1	2	σs^4
hhhch	4	2	3	$\sigma^2 s^4$
hhhhh	5	1	1	σs^5

The Total Number of Distinct Conformation States is $2^5 = 32$

TABLE 16.2 Factors m_{ij} for a 20-residue polypeptide

i or j	0	1	2	3	4	5	6	7	8	9	10
0	1	0	0	0	0	0	0	0	0	0	0
1	0	20	0	0	0	0	0	0	0	0	0
2	0	19	171	0	0	0	0	0	0	0	0
3	0	18	306	816	0	0	0	0	0	0	0
4	0	17	408	2040	2380	0	0	0	0	0	0
5	0	16	480	3360	7280	4368	0	0	0	0	0
6	0	15	525	4550	13,650	15,015	5005	0	0	0	0
7	0	14	546	5460	20,020	30,030	18,018	3432	0	0	0
8	0	13	546	6006	25,025	45,045	36,036	12,012	1287	0	0
9	0	12	528	6160	27,720	55,440	51,744	22,176	3960	220	0
10	0	11	495	5940	27,720	58,212	58,212	27,720	5940	495	11
11	0	10	450	5400	25,200	52,920	52,920	25,200	5400	450	10
12	0	9	396	4620	20,790	41,580	38,808	16,632	2970	165	0
13	0	8	336	3696	15,400	27,720	22,176	7392	792	0	0
14	0	7	273	2730	10,010	15,015	9009	1716	0	0	0
15	0	6	210	1820	5460	6006	2002	0	0	0	0
16	0	5	150	1050	2275	1365	0	0	0	0	0
17	0	4	96	480	560	0	0	0	0	0	0
18	0	3	51	136	0	0	0	0	0	0	0
19	0	2	18	0	0	0	0	0	0	0	0
20	0	1	0	0	0	0	0	0	0	0	0

The number of helical residues, i increases down the table, and the number of helices, j increases across the page. The total number of states is $2^{20} = 1048576$.

sum over categories (i, j), and takes into account both the number of states in a category and the statistical weight for that category:

$$\frac{q_{\text{hc}}}{q_{\text{coil}}} = \sum_{i=0}^{n} \sum_{j} m_{ij} w_{ij} = \sum_{i=0}^{n} \sum_{j} m_{ij} \sigma^{j} s^{i}. \tag{16.6}$$

Note that the partition function not only gives us free energy, and therefore also the entropy and enthalpy in the usual way, but that also the mean number of helical residues, $\langle i\rangle$ and the mean number of helices, $\langle j\rangle$ per molecule can be obtained from q

$$\frac{\partial \ln q_{\text{hc}}}{\partial \ln s} = s\frac{\sum_i\sum_j i\, m_{ij}\sigma^j s^{i-1}}{\sum_i\sum_j m_{ij}\sigma^j s^i} = \frac{\sum_i\sum_j i\, m_{ij}\sigma^j s^i}{\sum_i\sum_j m_{ij}\sigma^j s^i} = \langle i\rangle, \quad (16.7)$$

$$\frac{\partial \ln q_{\text{hc}}}{\partial \ln \sigma} = \langle j\rangle. \quad (16.8)$$

16.5 SOLUTION OF THE PARTITION FUNCTION

We must still find a closed expression for the partition function, q_{hc} of Eq. 16.6, that is, an expression that does not contain a sum of terms whose number becomes extremely large for large n. This is not straightforward, as the factors m_{ij} are products of combinatorial terms. The three named sets of authors made their contribution by providing this analysis. Gibbs and DiMarzio[8] solved the partition function in the limit of large n, by finding the maximum term in the sum, while Zimm and Bragg[9] derived algebraic expressions for the partition function of polymers of any length.

The mathematics that was used in order to solve Eq. 16.6 are reproduced in Appendix E. The partition function is known for all values of the length of the molecule (number of residues, n.) Here we give the result for large n, following the notation of Zimm and Bragg,

$$q_{\text{hc}}/q_{\text{coil}} = \left[\frac{1}{2}\left\{1 + s + \sqrt{(1-s)^2 + 4\sigma s}\right\}\right]^n. \quad (16.9)$$

Given a closed expression for q, it is also possible to obtain closed expressions for $< i >$ and $< j >$ by differentiating q with respect to s and σ (Eqs. 16.7 and 16.8).

Interesting results are for large n,

$$\langle i\rangle = n/2 \quad \text{for } s = 1, \text{any value of } \sigma$$
$$\langle j\rangle = (n/2)\sqrt{\sigma}/(1+\sqrt{\sigma}) \quad \text{for } s = 1. \quad (16.10)$$

This gives the fraction helix and the average length of a helical segment (still for large n)

$$f_{\text{helix}} = \frac{\langle i\rangle}{n} = \frac{1}{2} \quad \text{for } s = 1, \textit{any} \text{ value of } \sigma,$$
$$\langle\text{helix length}\rangle = \frac{\langle i\rangle}{\langle j\rangle} = \frac{1+\sqrt{\sigma}}{\sqrt{\sigma}} \quad \text{for } s = 1. \quad (16.11)$$

The slope of the transition curve is determined by the dependence of s and σ on the external conditions, such as temperature or solvent composition, for the former

$$\frac{df_{\text{helix}}}{dT} = \frac{\partial f_{\text{helix}}}{\partial s}\frac{ds}{dT} + \frac{\partial f_{\text{helix}}}{\partial \sigma}\frac{d\sigma}{dT}. \tag{16.12}$$

Expressions for $\partial f_{\text{helix}}/\partial s$ and $\partial f_{\text{helix}}/\partial \sigma$ can be obtained by twice differentiating the partition function. For large n and for $f_{\text{helix}} = 1/2$ (i.e., $s = 1$), one has

$$\begin{aligned} \frac{\partial f_{\text{helix}}}{\partial \ln s} &= \frac{1}{4\sqrt{\sigma}} \\ \frac{\partial f_{\text{helix}}}{\partial \sigma} &= 0. \end{aligned} \tag{16.13}$$

One sees that for large n both the midpoint and the slope are independent of n, as is observed experimentally.

Why does the length of the helices at the midpoint of the transition not depend on n when n is very large? This is because the combinatorial terms w_{ij} become so large that the dominant species (i, j) have several helices, and thus, when n is large, the state of the molecule's ends no longer makes a significant difference. One may consider the average cooperative unit to consist of $1/\sqrt{\sigma}$ residues.

Lifson–Roig model. Lifson and Roig[10] modified the Zimm–Bragg approach in order to more realistically represent the statistical weights of a greater variety of states and to account for the fact that a stable helix requires *three* successive helical residues and thereby to use realistic values for the statistical weights of sequences containing very short stretches of helical residues, ...chc..., ...chhc..., and ...chhhc.... (In contrast, in the formulation used by Zimm and Bragg these have almost identical (low) stability, while in the original Schellman model these three sequences are *successively* less stable, something that is also represented in the Lifson–Roig model.)

At the time of publication, this refinement was more than was necessary to account for the available experimental data. However, some years later, when studies with short synthetic oligopeptides strongly revived interest in the study of helix-coil equilibria, the multiple new results led to a renewed interest in the Lifson–Roig approach. These studies included the effects of see comment elsewhere so-called helix capping, that is, the stabilization of helical segments by interaction with different preceding and successive nonhelical residues (or chain termini). In order to analyze these results, Doig et al.[11] (see also Doig and Baldwin[12]) extended the Lifson–Roig approach to separate the contributions from helix initiation (i.e., capping on the "left" of a group of helical residues) and helix termination (i.e., capping on the "right").[13]

16.6 EXPERIMENTS ON SYNTHETIC HOMO-POLYPEPTIDES AND PROTEIN FRAGMENTS

The theory has been found to describe the available experiments with long synthetic polypeptides (homopolymers). Homopolymers of many amino acids are not well soluble in water. Polymers of Glu (and of Lys) are very soluble, but at neutral pH these are in the random coil form because of the repulsion between the charged side chains. However, at extreme values of pH where the side chains are not charged (low pH for Glu and high pH for Lys), the polymers do form stable helices. For these polymers, binding of hydrogen ion and conformation change are coupled processes; Zimm and Rice[14] have shown how this can be exploited, and how to determine the stability of the helix of uncharged poly (Glu) relative to the coil state by careful measurements of binding of hydrogen ion in a potentiometric titration experiment. The same method has been applied to measure the stability of uncharged poly (Lys). For both, the stability corresponds to a value of equilibrium constant for helix elongation s of around 1.15 at 25 °C in 0.1 M KCl.[15] Values of σ varying in a range from below 10^{-4} to 4×10^{-3} have been found depending on composition of polymer and solvent and on temperature.

Some early experiments used oligopeptide fragments of proteins prepared by specific proteolysis and fractionation of the fragments by column chromatography[16–18]. Some peptides were found to form a measurable amount of α-helix in solution and undergo a temperature-dependent transition from helix to coil (with highest observed helix content at 0 °C), while others had little or no tendency to do so. This tendency correlated positively with whether the peptides were in a helical conformation in the folded protein.[18] However, many peptides prepared in this manner were poorly soluble in water, and ultimately the number of peptides studied was insufficient to establish useful relations between amino acid composition and tendency to form α-helix, unlike what had been possible for natively folded proteins.[19]

16.7 EXPERIMENTAL DETERMINATION OF HELIX PROPENSITIES IN SYNTHETIC PEPTIDES

The field languished until chemical methods of peptide synthesis became available in the 1980s, and it became possible to study oligopeptides of medium length (up to 50 residues long) each having a specific different amino acid sequence. These experiments have enabled the establishment of "canonical" values of the helix propensities, s for different amino acid types, and of the contribution each amino acid type makes to helix initiation in aqueous solutions. (Work by Baldwin, DeGrado, Kallenbach, Stellwagen, and their students.[20])

Peptides containing alanine were found to have higher helix propensity than peptides containing other naturally occurring amino acids. Measurable helix formation in water-soluble oligopeptides was achieved by sequences of alanine containing also a small number of hydrophilic residues, such as glutamic acid and lysine.

TABLE 16.3 Values of s for some amino acid types

Amino Acid	$-k_B T \ln s$(kcal/mol)
Alanine	−0.3 to −0.2
Leucine	−0.1 to 0.0
Valine, Isoleucine	0.5
Glycine	1.0
Proline	3

Thus, Scholtz et al. [21] studied Ala-based peptides up to 50 amino acid residues long, containing repeating sequences (−Ala–Glu–Ala–Ala–Lys–Ala–). In these polymers helix content increases with chain length; by applying the theory of the helix-coil equilibrium to their results, it was then possible to establish effective values of both parameters of the helix-coil model, $s = 1.35$ and $s = 3 \times 10^{-3}$ at 0 °C.

These peptides were found to be only partly helical (even at 0 °C). Higher helicity could be achieved by incorporating special helix-stabilizing features in the design:

- salt bridges between side chains four residues apart in the chain (see example discussed in Section 16.8);
- negative charge near the N-terminus and positive charge near the C-terminus to give favorable interaction with the helix dipole (see example discussed in Section 16.9);
- "capping" of the helix ends by interaction with side chains, in particular Asp or Asn;[11,12]
- start or end the molecule with a helix-initiating template molecule.[22]

Different laboratories, using different experimental models, in particular using different strategies to stabilize the helix, have (of course) obtained slightly different estimates. (Helix content is typically determined from the circular dichroism (CD) spectrum. An article by Chin et al. reports the CD signal of short helices.[23])

Table 16.3 shows estimates of s. As mentioned, alanine, with no substituents on C^β, forms particularly stable helices ($s > 1$). The majority of amino acid residues has a single substituent on C^β and these residues have rather similar s-values near 1. So-called beta branching (as in valine, isoleucine, and threonine) again lowers s. The inability to form a helical hydrogen bond to the residue three steps back along the chain explains the very low value of s for proline. (Helices with proline in the middle found in proteins typically are kinked by 15° or more.) Finally, the low value of s of around 0.2 for glycine may be explained by the greater freedom of the glycine backbone in the random coil state, rather than by an absence of stabilizing interactions in the helix state.

Muñoz and Serrano[24] used an empirical analysis of these and many other results and estimated a set of energy contributions that together can account for the stability of the α-helical conformation of peptides of different amino acid sequence

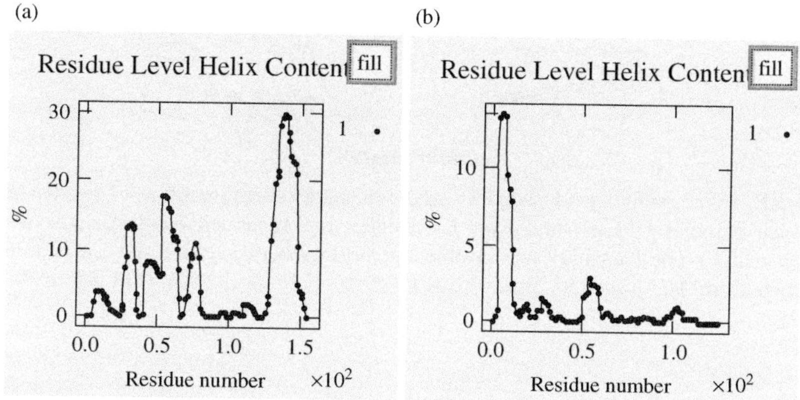

FIGURE 16.4 Residue helicity of the unfolded chains of sperm whale myoglobin (a) and bovine pancreatic ribonuclease A (b) predicted on the basis of the helix propensity of the amino acid sequences of these two proteins with Muñoz and Serrano's Agadir program. *From Nature: Struct. Biol.* **1**: 399–409 (1994).

in aqueous solution, at different temperatures, and pH and ionic strength. In the original article they analyzed the average helicity of 323 peptides and the helicity per residue derived from NMR.

The Agadir program developed by Luis Serrano[24] combines current best estimates of s and contributions to σ with a mathematical model based on the theory of the helix-coil transition, in which all states (from all-coil through all-helix) are considered, and produces estimates of helix fraction of each residue in peptides of any specified amino acid sequence. Serrano has established and maintains a web server which makes this algorithm freely accessible to users.[25]

Figure 16.4 shows the helicity of the peptide chains of two proteins in water at 278 K, predicted by this algorithm on the basis of helix propensities, in the absence of long-range interactions. One sees that the unfolded chain of sperm whale myoglobin, a highly α-helical protein, has high helix propensity, while the unfolded chain of ribonuclease A has low helix propensity, except in the N-terminal peptide, which forms an α-helix in the folded structure (as do residues 24–33 and 50–57). Residues 80–120 of myoglobin are partly α-helical in the native structure, but, as can be seen, these residues have but scant tendency to form helix in the unfolded state. These results illustrate an important rule: *helix propensity of the unfolded chain correlates with helicity of folded proteins, however, with significant exceptions.*

16.8 HELIX STABILIZATION BY SALT BRIDGES IN OLIGOMERS CONTAINING Glu AND Lys

As mentioned, short alanine-containing peptides, made soluble by incorporating a few residues of either lysine or of glutamic acid show only partial helicity.

FIGURE 16.5 Amino acid sequence and schematic structure of a 17-residue peptide in which pairs of Glu (E) and Lys (K) residues are separated by 3 Ala (A) residues. (Ac = acetyl.) The bars indicate possible salt bridges between Glu and Lys side chains. Redrawn from Marqusee, S., Baldwin, R.L. *Proc. Natl. Acad. Sci. USA* **84**: 8898–8902 (1987).

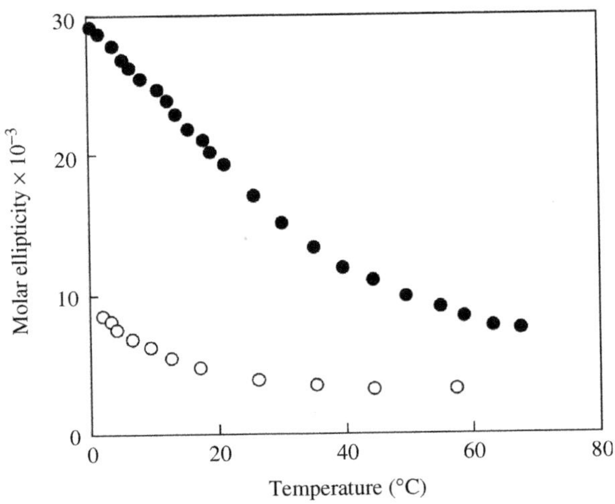

FIGURE 16.6 Thermal unfolding profiles of two peptides each containing 3 Lys and 3 Glu residues, and the remainder Ala. Filled symbols: peptide described in Fig. 16.5, which is capable of forming three salt bridges. Open symbols: analogous 16-residue peptide incapable of forming salt bridges, as its pairs of Glu and Lys residues are separated by 2 Ala residues. Neutral pH, 0.01 M NaCl. Redrawn from Marqusee, S., Baldwin, R.L. *Proc. Natl. Acad. Sci. USA* **84**: 8898–8902 (1987).

Marqusee et al.[26] studied a series of peptides containing besides the alanine residues an equal number of lysine and glutamic acid residues, the placement of which they varied.

These molecules are water-soluble and have measurable helix propensity at neutral pH. It was then found that by placing Glu and Lys residues four residues apart (in positions i and $i + 4$ in an oligomer containing three such pairs and otherwise only alanine residues, as shown in Fig. 16.5, the helicity of a peptide only 17 residues long was as high as 75% at 0 °C. However, a similarly designed 16-residue peptide in which the Glu and Lys residues were three residues apart had much lower helicity (Fig. 16.6).

Model building indicated that, with a spacing of four, but not with other spacings, the side chains of each pair of Lys and Glu residues are located favorably for the formation of a salt bridge in the helical state, as schematically indicated in Fig.16.5. We may account for such a helix-stabilizing local interaction by introducing a factor $\tau(\tau > 1)$ in any term in the partition function q_{hc} of Eq. 16.6 that corresponds to a helix containing the stabilizing interaction, in this case a Glu-Lys pair separated by three Ala residues (in fact, a factor τ for each such Glu-Lys pair). As mentioned, values of the parameters s and σ are known for the reference polymers not containing salt bridges. Using such a modified expression for the partition function, it is then relatively easy to make a rough estimate of $\tau = 4$ for the factor that expresses the increase in helix stability by a single salt bridge in these oligomers.

16.9 HELIX STABILIZATION BY CHARGED GROUPS INTERACTING WITH THE HELIX DIPOLE

The concept of the helix dipole was developed by Hol et al., [27] in first instance in order to rationalize the high frequency with which phosphate groups are found to bind to proteins near the N-termini of α-helices. Hol et al. estimated that the local dipoles of peptide NH and CO groups (which point approximately along the helix axis in the direction of the N-terminus) together form a global line dipole, called *the helix dipole*, corresponding to circa one half an elementary charge at each helix end, positive charge at the N-terminus and negative charge at the C-terminus. The former should interact favorably with a negatively charged bound phosphate group.

Experiments with helix-forming oligopeptides[28,29] tested this model by demonstrating considerable stabilization of the helix as a result of changing the charged state of groups near the chain termini. Thus, an oligomer with an electrically neutral acetylated N-terminus (CH_3–CO–NH–) formed helix more readily than the same oligomer with a positively charged free α-amino group (NH_3^+–). Succinylation (giving end group COO^-–CH_2–CO–NH–) was found to lead to an even more stable helix, while substitution of the N-terminal residue with lysine, which results in having two positive charges near the N-terminus, gave the lowest tendency to helix formation. Similarly, replacement of the negatively charged free α-carboxyl group (–COO^-) at the C-terminus with an electrically neutral group (CO–NH_2) increased helix-forming tendency. Some helix stabilization could be observed also by bringing the charge of the α-amino or α-carboxyl group to zero by either lowering or else raising the pH, and all these effects were found to depend strongly on the ionic strength.

16.10 HELIX-COIL EQUILIBRIA OF NUCLEIC ACIDS

The following overview reviews background that can be found in more detail in advanced biochemistry textbooks.

The transition from double helix to coil of double-stranded DNA or RNA is readily followed by measuring the change of UV absorption spectrum of dilute solutions. This "melting" transition can be obtained by raising the temperature, and similar transitions are observed by changing the pH to low or high values; these pH changes alter the ionization state of the bases and thereby disrupt base pairing. Such changes of the UV spectrum are observed as well in temperature transitions of *single*-stranded NA polymers (see Section 16.2); apparently, a change in the spectrum is associated with the parallel stacking of nucleic acid bases that occurs when the helical conformation is formed.

The complementary strands of DNA or RNA in a double-helical molecule contain only two kinds of base pairs: guanine and cytosine form GC base pairs in both DNA and RNA, adenine and thymine form AT base pairs in DNA and adenine and uracil form AU base pairs in RNA. (All three types of base pairs occur in DNA–RNA hybrids.)

GC base pairs have three hydrogen bonds each and are more helix-stabilizing than AT or AU base pairs, which have only two hydrogen bonds each. The melting temperatures, T_m of DNA from different sources vary linearly with fraction of GC base pairs, f_{GC} over a range of 41 °C, and the enthalpy accompanying the transition has been measured calorimetrically. In 0.2 M salt these are given by

$$\begin{aligned} T_m &= [69.3 + 41 f_{GC}]^{\circ}\mathrm{C}, \\ \Delta H^{\circ} &= [4460 + 69(T - 293)]\ \mathrm{cal/base\ pair}. \end{aligned} \tag{16.14}$$

The treatment of helix-coil equilibria of double-stranded nucleic acids is more complex than that of polypeptides, and the inherent difficulties have not entirely been solved. The difficulties encountered in developing a physical model of these transitions have a variety of origins (see Fig. 16.7).

As in the case of the helix-coil transition of polypeptides, partly helical molecules must be considered when modeling the transition quantitatively. The base sequence of genomic DNA is set by whatever information it encodes; any given short sequence of bases may occur multiple times, but long sequences are increasingly rare, and the longest are unique. Thus, the sequences of the strands of a largely but incompletely helical molecule are aligned, and the origin of a single double-helical stretch of length i can lie at only $n-i$ positions in a molecule containing two strands each having n base pairs. [However, in a partly helical pair of synthetic *homo*polymers of length n nucleotides, for example, poly (dG) and poly (dC), the origin of a single double-helical stretch of length i can lie at $n-i$ positions on *each* strand, so that there are $(n-i)^2$ molecular states having one helical stretch containing i base pairs.]

Nucleic acid molecules are highly charged, and the electrostatic interactions between phosphates make a large contribution to the relative stability of the helical state, which depends strongly on ionic strength. Especially at low ionic strength, these electrostatic interactions are long-range and a model with few equilibrium constants, such as was effective for polypeptides, may not be adequate for nucleic acids at low ionic strength.

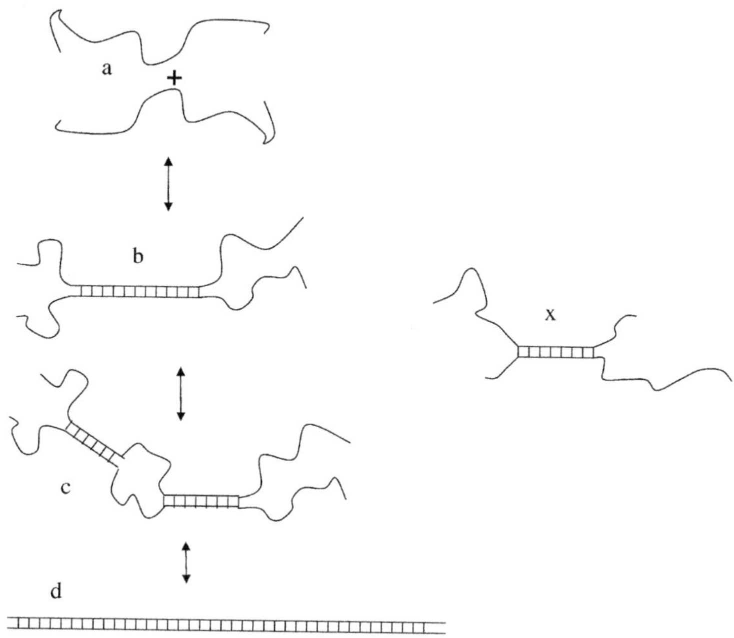

FIGURE 16.7 Helix formation in double-stranded DNA. The diagram shows on the left pairs of complementary DNA strands: (a) separated randomly coiled strands; (b) partly helical with a single helical segment; (c) partly helical with with an internal loop; (d) the complete double-helical structure. (x) On the right, two non-complementary chains associate thanks to the incidental presence of a short complementary stretch.

When fully unfolded, the molecule dissociates into two parts; hence, equilibrium and kinetics of folding and unfolding are dependent on concentration.

The double helix can fray at the ends as well as open up an unfolded loop in the middle. The loop structure restricts the conformations of the residues in the loop, and the residues in the internal coiled loop have higher free energy than residues in coil segments that terminate in the molecule's ends. This primarily entropic contribution in favor of the helical state depends on the size of the loop, and more strongly disadvantages formation of large loops.

The rate of renaturation of double-stranded DNA from strand-separated single-stranded pairs depends inversely on the concentration and on the size of the unique genome of the DNA, as a larger genome corresponds in effect to a more dilute solution of complementary molecules. In all cases, when fully unfolded and dissociated chains reassociate, short complementary sequences may not correspond to the correct molecular alignment and/or longer double-helical stretches may contain one or two mispaired bases. These abortive pairs of chains have to disassociate before a correct pairing occurs; this effect can be minimized by choosing conditions under which the double helix is barely stable, and where imperfectly formed double helical structure can more readily dissociate for the strands to then associate with proper partners (so-called annealing).

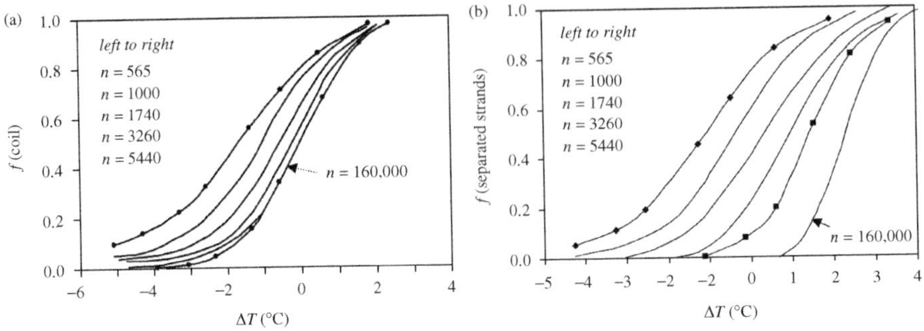

FIGURE 16.8 Melting transitions of fragments of T2 DNA of different length (expressed as average number of base pairs, n), and of the intact molecule ($n = 160,000$). (a) Total fraction of coil. (b) Fraction present as single-stranded molecules. The reference temperature is the midpoint of the transition of intact T2 DNA. Data from Crothers, D., Kallenbach, N.R., Zimm, B.H. *J. Mol. Biol.* **11**: 802–820 (1965).

16.11 MELTING TRANSITION OF DNA

As an application, we discuss a combined experimental and theoretical study of DNA melting transitions by Crothers, Kallenbach, and Zimm.[30]

Materials. The article reports the melting profiles of samples of DNA each containing fragments of a different average size, varying in a range from circa 300 to 4000 base pairs; the fragments were prepared by breaking DNA obtained from T2 bacteriophage (circular DNA containing 160,000 base pairs) by subjecting this in solution to varying high shear stress for different lengths of time. Use of DNA of low molecular weight and an ionic strength of 0.2 M assures reversibility of the observed melting; this is of course essential if one wishes to compare measurements with the theory of the helix-coil equilibrium.

Separation of melting transitions with and without chain separation. An equilibrated, partly melted sample at temperature T contains a mixture of partly "unzippered" double helices and dissociated coiled single strands. As zippering and un-zippering are faster than dissociation and reassociation of strands, the authors were able to separately determine the extent of strand dissociation at temperature T by quickly returning to a temperature well below the melting point, and observing the slow component requiring reassociation of strands prior to rapid zippering. Results are shown in Fig. 16.8.

Comparison with theory. As for helix-forming polypeptides, the melting transition of shorter fragments occurs at lower temperatures, and as the length of the fragments increases, a limit is reached. For short fragments, strand separation accompanies the melting transition. On the other hand, no appreciable strand separation of intact T2 DNA occurs below the midpoint of the melting transition. According to the theory of the helix-coil transition, in long partly melted molecules, *multiple* helical segments alternate with coiled segments (as in Fig. 16.7, diagram c); just as for long polypeptides, very long molecules can have

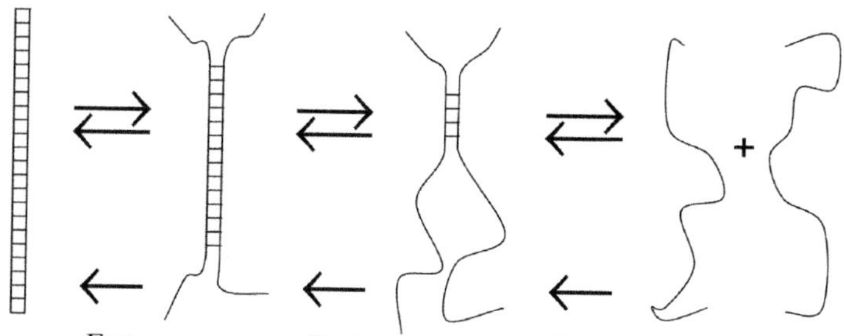

FIGURE 16.9 Transition of a T2 DNA fragment from double helix to partly helical to separated coiled strands. Top row of arrows indicate equilibration at temperatures in the transition range. Bottom row of arrows indicate return to the native state when the temperature is reduced to well below the transition range. A kinetic study of the return to the helical state allows a determination of the extent of strand separation from the ratio of fast and slow recovery of signal.

a great many alternating segments, and these double-stranded species predominate because they are greatly favored by the combinatorial factor m_{ij} of Eq. 16.6.

For a more accurate analysis we must write a partition function for this system. In order to do this, we shall consider the melting process in two steps.

In the first step partly helical double-stranded molecules form in which coiled regions alternate with double-helical regions. A problem is that *internal* coiled stretches (as in Fig. 16.7, diagram c) are constrained to looped conformations that are less favorable than unconstrained conformations, and that the unfavorable effect increases with increasing size of the coiled stretch. However, by limiting application of the helix-coil transition to short chain fragments, Crothers and coworkers could in first approximation ignore forms with internal coiled segments, limiting the model to chains melted only at the ends, with a single double-helical segment (Fig. 16.9). When limited to these species, the partition function, q_b becomes (compare the partition function for the polypeptide case, Eq. 16.6).

$$\frac{q_b}{q_1} = \sum_{i=1}^{n} m_i w_i = \sum_{i=1}^{n} (n - i + 1)s^{i-1}. \tag{16.15}$$

Here q_1 is the partition function for the two chains held together by a single bond; m_i is the number of ways a stretch of i complementary residue pairs can be chosen in a molecule of length n residues, which equals $(n - i + 1)$. Furthermore, w_i is the statistical weight of a stretch of i helical residue pairs, which contains a factor s for each base pair except the first.

Estimated values of s for GC and AT base pairs follow by setting $s = 1$ at the (extrapolated) transition midpoints for all-GC and all-AT DNA (110 and 69 °C) and adjusting to the experimental temperature using the measured enthalpy. (See Eq. 16.14; the values of s for GC and AT are approximately in the ratio $s_{GC}/s_{AT} = 5$.) If the base sequence of the DNA is unknown, a simple first approximation is to assume a random sequence of bases. For such a random distribution, DNA fragments containing hundreds of base pairs vary statistically in base *composition* with a root-mean-square deviation on the order of 5%, which corresponds to a deviation in melting temperature of 2 °C. This contribution cannot be neglected given the narrow range of the temperature transitions of Fig. 16.8.

In the second step each no-longer helical pair of chains dissociates into two separate complementary coiled molecules. This is a simple equilibrium for which we can write

$$\frac{q_1}{q_{\text{coil}}} = \kappa \frac{N_{\text{coil}}}{V}$$

$$q_{\text{hc}} = q_{\text{coil}} + q_{\text{b}}. \tag{16.16}$$

where q_{coil} is the partition function of dissociated coil, κ is an equilibrium constant, N_{coil}/V is the concentration of coiled complementary molecules, q_{b} is given by Eq. 16.15 and q_{hc} is the partition function for the mixture of (partly) helical duplexes and dissociated coiled molecules. The fraction of dissociated molecules is given by the ratio q_1/q_{hc}, and the mean number of helical base pairs per molecule is obtained with Eq. 16.7.

When the solution is returned to low temperature, the reformation of near 100% double helix in partly helical molecules is rapid, much faster than the reassociation of separated strands. Consequently, the change in absorbance accompanying renaturation occurs in two separable steps (Fig. 16.9), and this allows the authors to determine the fraction of separated strands at each denaturing temperature shown in Fig. 16.8.

Transition curves were calculated numerically according to this simple model for a great number of possible random sequences (at a constant overall fraction GC) and averaged, for fragments containing from 400 to 4000 base pairs. The value of κ could then be chosen to reproduce the measured fraction of separated strands for the smallest fragment. This model accurately reproduced the transition midpoints, while the calculated slopes were somewhat too steep, more so for the larger fragments, which can increasingly form coiled loops, as in Fig. 16.7, diagram c, which is something that the model ignores. The model was found to reproduce the transition to dissociated complementary chains reasonably well for the shortest fragments, but predicts too strong a temperature dependence for the longer ones. Overall this simple implementation of the theory of the helix-coil transition produces a remarkably good representation of a complex phenomenon.

The design of this particular study avoided the difficulties of having to deal with the problem of modeling the contribution of partly double-helical conformations with internal coiled stretches, discussion of which is outside the scope of this book. Interested readers are referred to the cited book by Poland and Scheraga.[6]

NOTES

1. As discussed in Section 15.3, if every element of a chain of identical elements is given the same conformation, a helix results.
2. Leng, M., Felsenfeld, G. A study of polyadenylic acid at neutral pH. *J. Mol. Biol.* **15**: 455–466 (1966).
3. Eisenberg, H., Felsenfeld, G. Studies of the temperature-dependent conformation and phase separation of polyriboadenylic acid solutions at neutral pH. *J. Mol. Biol.* **30**: 17–37 (1967).
4. Zimm, B.H., Doty, P., Iso, K. Determination of the parameters for helix formation in poly-γ-benzyl-L-glutamate. *Proc. Natl. Acad. Sci. USA* **45**: 1601–1607 (1959).
5. (a) Schellman, J.A. The stability of hydrogen-bonded peptide structures in aqueous solution. *Compt. rend. trav. lab. Carlsberg, Sér. chim.* **29**: 230–259 (1955) (reproduced in note 6), (b) Qian, H., Schellman, J.A. Helix-coil theories: A comparative study for finite length polypeptides. *J. Phys. Chem.* **96**: 3987–3994 (1992).
6. Many of the early papers on helix-coil transitions have been reprinted in: Poland, D., Scheraga, H.A. *Theory of Helix-Coil Transitions in Biopolymers; Statistical Mechanical Theory of Order–Disorder Transitions in Biological Macromolecules.* New York: Academic Press, 1970.
7. The transition of a model with a single helical segment becomes indefinitely steeper with increasing number of residues.[5] The steepness of the transition of a model with multiple helical segments reaches a *finite* limit for very long molecules, as is experimentally observed.
8. Gibbs, J.H., DiMarzio, E.A. Statistical mechanics of helix-coil transitions in biological macromolecules. *J. Chem. Phys.* **30**: 271–282 (1959).
9. Zimm, B.H., Bragg, J.K. Theory of the phase transition between helix and random coil in polypeptide chains. *J. Chem. Phys.* **31**: 526–535 (1959).
10. Lifson, S., Roig, A. On the theory of helix-coil transition in polypeptides. *J. Chem. Phys.* **34**: 1963–1974 (1961).
11. Doig, A.J., Chakrabartty, A., Klingler, T.M., Baldwin, R.L. Determination of free energies of N-capping in α-helices by modification of the Lifson-Roig helix-coil theory to include N- and C-capping. *Biochemistry* **33**: 3396–3403 (1994).
12. Doig A.J. and Baldwin R.L., N- and C-capping preferences for all 20 amino acids in alpha-helical peptides. *Protein Sci.* **4**: 1325–1336 (1995).
13. The original publication by Lifson and Roig used a 4×4 matrix and still only two adjustable parameters. Doig et al. used a 4×4 matrix with four adjustable parameters, two of which represent the contributions from helix initiation. (See Appendix E for a description of the matrix method.)
14. Zimm, B.H., Rice, S.A. The helix-coil transition in charged macromolecules. *Mol. Phys.* **3**: 391–407 (1960).
15. Hermans, J. Experimental free energy and enthalpy of formation of the α-helix. *J. Phys. Chem.* **70**: 510–415 (1966).

16. Klee, W.A. Conformation of ribonuclease S-peptide. *Biochemistry* **7**: 2731–2736 (1968).
17. Epand, R.M., Scheraga, H.A. The influence of long-range interactions on the structure of myoglobin. *Biochemistry* **7**: 2864–2872 (1968).
18. Hermans, J., Puett, D. Relative effects of primary and tertiary structure on helix formation in myoglobin and α-lactalbumin. *Biopolymers* **10**: 895–914 (1971).
19. Chou, P.Y., Fasman, G.D. Prediction of the secondary structure of proteins from their amino acid sequence. *Adv. Enzymol.* **47**: 45–148 (1978).
20. Reviews of generation-two studies of helix-coil equilibria in short peptides: (a) Scholtz, J.M., Baldwin, R.L. The mechanism of α-helix formation of peptides. *Ann. Rev. Biophys. Biomol. Struct.* **21**: 95–118 (1992), (b) Chakrabartty, A., Baldwin, R.L. Stability of alpha-helices. *Adv. Protein Chem.* **46**: 141–176 (1995). Similar studies of beta structures: Smith, C.K., Regan, L. Construction and design of β-sheets. *Acc. Chem. Res.* **30**: 153–161 (1997). Earlier work on helix-coil equilibria by Scheraga and students employed random copolymers containing a nonnatural amino acid as a "host" residue; subsequent work has shown that helix propensities derived from this work are not transferable.
21. Scholtz, J.M., Qian, H., York, E.J., Stewart, J.M., Baldwin, R.L. Parameters of helix–coil transition theory for alanine-based peptides of varying chain lengths in water. *Biopolymers* **31**: 1463–1470 (1991).
22. For example: Kemp, D.S., Rothman, J.H. Efficient helix nucleation by a macrocyclic triproline-derived template. *Tetrah. Lett.* **36**: 4023–4026 (1995).
23. Chin, D.-H., Woody, R.W., Rohl, C.A., Baldwin, R.L. Circular dichroism spectra of short, fixed-nucleus alanine helices. *Proc. Natl. Acad. Sci. USA* **99**: 15416–15421 (2002).
24. Muñoz, V., Serrano, L. Elucidating the folding problem of helical peptides using empirical parameters. *Nature: Struct. Biol.* **1**: 399–409 (1994).
25. Serrano, L. Agadir web server: an algorithm to predict the helical content of peptides. http://agadir.crg.es, 1994.
26. Marqusee, S., Baldwin, R.L. Helix stabilization by $Glu^{-} \ldots Lys^{+}$ salt bridges in short peptides of de novo design. *Proc. Natl. Acad. Sci. USA* **84**: 8898–8902 (1987).
27. Hol, W.G., van Duijnen, P.T., Berendsen, H.J. The alpha-helix dipole and the properties of proteins. *Nature (London)* **273**: 443–446. (1978).
28. Shoemaker, K.R., Kim, P.S., York, E.J., Stewart, J.M., Baldwin, R.L. Tests of the helix dipole model for stabilization of α-helices. *Nature (London)* **326**: 563–567 (1987).
29. Fairman, R., Shoemaker, K.R., York, E.J., Stewart, J.M., Baldwin, R.L. Further studies of the helix dipole model: Effects of a free α-$NH3^{+}$ or α-COO^{-} group on helix stability. *Proteins: Struct. Funct. Genet.* **5**: 1–7 (1989).
30. Crothers, D., Kallenbach, N.R., Zimm, B.H. The melting transition of low molecular-weight DNA: Theory and experiment. *J. Mol. Biol.* **11**: 802–820 (1965).

17

Protein Unfolding Equilibria

17.1 INTRODUCTION

Many small proteins can be unfolded (or denatured) by changes in solvent composition, pH and/or temperature, and will fold back into the native conformation when the original conditions are restored. Under extreme conditions the protein may undergo irreversible chemical changes, and in that case refolding may no longer be possible. (e.g., as a result of deamidation of Asn and Gln side chains or scrambling of disulfide bridges by sulfhydryl exchange.) Also, in concentrated solution an unfolded protein may form a precipitate, rather than refold, when denaturing conditions (some of which improve solvation) are removed.

The implication of this observation is that these proteins fold spontaneously following biosynthesis. Apparent exceptions are proteins that fold immediately after biosynthesis, and form a stable native structure, but are destabilized by posttranslational modifications such as cleavage of the polypeptide chain at one or several specific sites, but remain in a folded state because the rate of unfolding is very low. As the equilibrium constant favors the unfolded state, these modified proteins do not spontaneously refold once they have been unfolded. An example is the two-chain protein insulin; this folds into a native conformation as a single polypeptide chain called *proinsulin*, which is converted by cleavage of peptide bonds to the active and *still folded* hormone when this is needed.

A remarkable finding made by Anfinsen is that proteins fold spontaneously with concomitant formation of covalent bonds, that is, disulfide bridges linking Cys residues, according to

Equilibria and Kinetics of Biological Macromolecules, First Edition. Jan Hermans and Barry Lentz.

$$2\text{–SH} \Rightarrow \text{–SS–} + 2\text{H}.$$

The oxidation can be accomplished by molecular oxygen, or by exchange with oxidized dimeric sulfhydryl compounds. Conversely, addition of a reduced sulfhydryl compound to a solution of protein containing disulfide bridges can lead to a slow reversible unfolding with concomitant reduction of disulfide bridges.

The discussion in this chapter is based on studies of reversible unfolding of individual globular proteins for which the equilibrium is established relatively rapidly, and side effects that lead to irreversible changes are absent or can be controlled.

17.2 THE TWO-STATE APPROXIMATION

Another remarkable aspect of reversible folding and unfolding of many proteins is its two-state character, which was first proposed by Rufus Lumry and his students Biltonen and Brandts.[1] The experiments show that for many proteins only two states are significantly populated in an equilibrium mixture containing folded and unfolded molecules, and this is also true in many cases during the kinetic processes of protein folding and unfolding. *Strong* evidence in favor of two-state behavior is obtained when all measures of a protein's state change proportionally as conditions are changed and the equilibrium shifts from native to denatured or the reverse. For example, the *far circular dichroism (CD) spectrum* can be used to measure secondary structure, changes in the *near UV CD spectrum* and *fluorescence* reflect burial of aromatic groups, and *biological activity* reports the presence of a binding or a catalytic site, all of which is absent in the unfolded state, but which would be present to a different extent in partly folded states. Results for ribonuclease,[2] shown in Fig. 17.1, demonstrate that this protein meets such a requirement for two-state unfolding. The coincidence of transition curves measured by following different properties is *necessary* as a condition for establishing the two-state nature of a transition, but not by itself *sufficient* evidence. (In fact, as we shall see subsequently, the unfolding of ribonuclease is not two-state at pH 6.18.) *Conclusive* evidence for two-state behavior of temperature-induced transitions can be obtained from calorimetric measurements of the transition enthalpy, as discussed in Section 17.4.

It is clear that during the *kinetic* process of folding or unfolding a protein molecule will pass through a series of stages of increasing or decreasing partial folding. However, whenever folding is two-state, these intermediate states do not accumulate to the extent that they are observable and can be studied.

How to reconcile the two-state behavior with our understanding that a protein's structure should be considered as an ensemble containing a great number of possible conformations? To do this, we introduce the concept of a *reaction coordinate* that tracks the ordering of the protein into its native structure. As such a coordinate one could choose, for example, the number of native close atomic contacts, which would be high in the folded, native state, and low in a fully unfolded state, and intermediate between these extremes in partly folded states. Other definitions are possible, and in reality there is no accepted uniquely appropriate definition of a

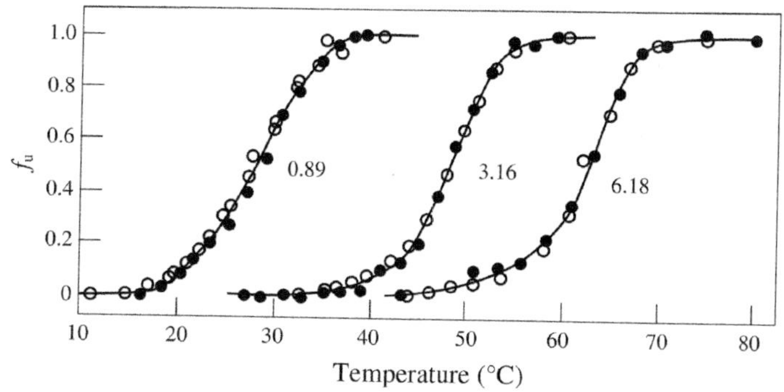

FIGURE 17.1 Different measures of unfolding of ribonuclease A (changes in optical rotation and extinction coefficient) give close to the same extent of unfolding, f_u, as a function of temperature at three pH values. From Hermans, J., Scheraga, H.A. *J. Am. Chem. Soc.* **83**: 3283–3292 (1961).

reaction coordinate for protein folding. The problem is that the difference between the folded state and the unfolded state involves a great many degrees of freedom that can change independently, and that many different intermediate partly folded states are possible for any given intermediate value of a reaction coordinate. The multitude of possible folding/unfolding pathways makes this an unusually complex case; there are many other reactions for which appropriate reaction coordinates can be found without great difficulty.

In spite of this problem, the concept of a reaction coordinate remains useful, and is helpful in reaching a qualitative understanding. In Figs. 17.2 and 17.3 we have drawn a series of schematic diagrams showing how the free energy of a protein varies as a function of the reaction coordinate in the case of a two-state folding equilibrium, for different overall conditions (temperatures or concentrations of denaturant).

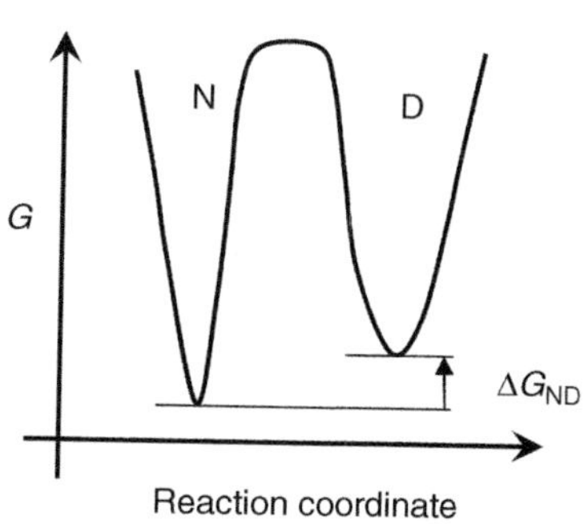

FIGURE 17.2 Free energy as a function of a reaction coordinate for a two-state equilibrium between native and denatured states (N and D) of a protein. Under native conditions ($G_N^\circ < G_D^\circ$).

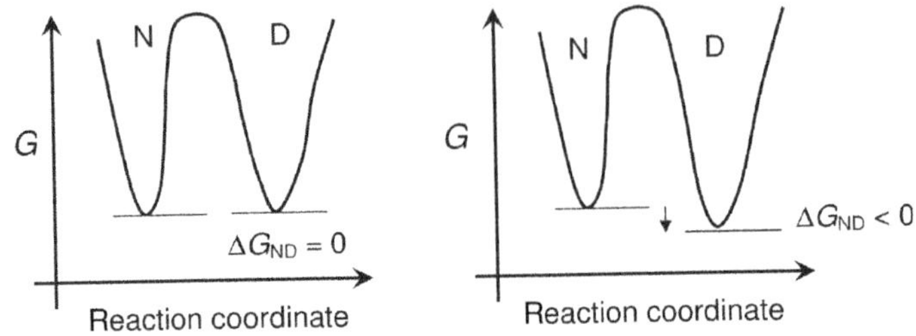

FIGURE 17.3 Diagrams for the same system, but, respectively, at the transition temperature ($G^\circ_N < G^\circ_D$), and above the transition temperature ($G^\circ_N < G^\circ_D$).

In the first diagram, the free energy difference ΔG°_{ND} is positive, that is, this diagram is drawn for a low temperature at which the native (= folded) state is more stable than the denatured (= unfolded state). The denatured state is the very large ensemble of states in which the values of the torsion angles vary widely from molecule to molecule. The native state is a much smaller ensemble in which the residues in the interior of the protein have fixed conformations, although even there torsion angles can vary as a result of vibrational motion; side chains on the surface preserve freedom to assume different conformations, so that molecules in the native state still do not possess a completely unique conformation.

The native state has a lower energy and the denatured state has a higher entropy; raising the temperature will therefore lower the value of ΔG°_{ND} until at the transition temperature it becomes zero, and at still higher temperatures, negative. The key element that makes this diagram typical of a two-state transition is the presence of a large *free energy barrier separating the two states*. This free energy barrier corresponds to the transition state, which is itself an ensemble containing a particular set of partly folded molecules, about which we know the least.

17.3 WORKING WITH THE TWO-STATE MODEL

The two-state model is very convenient because it allows a ready conversion of experimental data to thermodynamic functions. The needed data consist of a series of measurements of a *molecular property* that depends on the conformation, at different values of a *condition* that induces unfolding. Examples of a suitable measured property, R include CD spectrum, fluorescence, biological activity, and NMR spectrum, and possible conditions, X include temperature and denaturant concentration. The first step is to convert the measurements of R to a common form, namely, the fraction of molecules in the unfolded state, f_u. One here equates the *fraction* f_u with the *fractional* change in the measured property,

$$f_u(X) = \frac{R(X) - R_{folded}(X)}{R_{unfolded}(X) - R_{folded}(X)}, \qquad (17.1)$$

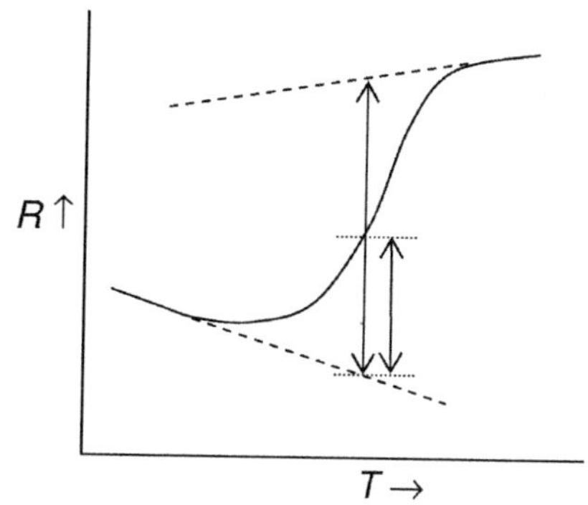

FIGURE 17.4 Use of (typically) spectroscopic data, R to measure fraction of protein that is unfolded, f_u as the ratio $(R - R_{\text{folded}})/(R_{\text{unfolded}} - R_{\text{folded}})$, with linear extrapolations of baselines.

where R_{folded} and R_{unfolded} represent the extreme values of the signal R. As the notation indicates these may themselves depend on the unfolding condition, X. One determines them by extrapolating measurements outside the transition region, that is, at low values of X, to establish a *baseline* of R_{folded} and at high values of X, to establish a *baseline* of R_{unfolded} (assuming high values of X promote unfolding. Frequently, the baselines depend linearly on X. See Fig. 17.4).

So far, our treatment has been strictly phenomenological. But if the two-state model applies, then one can write simple equations for a true equilibrium constant and the corresponding standard free energy change

$$\begin{aligned} K_u &= f_u/(1 - f_u) \\ \Delta G_u^\circ &= -k_B T \ln K_u, \end{aligned} \tag{17.2}$$

and at the midpoint

$$f_u = \frac{1}{2}; \quad K_u = 1; \quad \Delta G_u^\circ = 0. \tag{17.3}$$

The temperature dependence of K_u can be used to give the corresponding standard enthalpy with the van't Hoff equation

$$\Delta H^\circ_{\text{van't H}} = \frac{\partial(\Delta G^\circ/T)}{\partial(1/T)} = -k_B \frac{\partial \ln K}{\partial(1/T)}, \tag{17.4}$$

and the temperature dependence of ΔH° can be used to calculate the specific heat change, ΔC_P°

$$\Delta C_P^\circ = \frac{d\Delta H^\circ}{dT}. \tag{17.5}$$

An important *limitation* is a consequence of the logarithmic dependence of ΔG° on f_u in Eq. 17.1. Given an unavoidable statistical error in the measured values of

R, δR, and a corresponding statistical error f_u, δf_u the statistical error in ΔG_u° is approximately equal to $[d(\Delta G^\circ)/df_u]\delta f_u$

$$\delta(\Delta G_u^\circ) \approx -\frac{k_B T}{K_u}\frac{\delta f_u}{(1-f_u)^2} = -\frac{k_B T}{f_u(1-f_u)}\delta f_u, \tag{17.6}$$

which is smallest when $f_u = 1/2$, and becomes larger as f_u approaches 0 or 1. In practice, precise values of the free energy can be obtained from direct spectroscopic measurements only in the range $0.02 < f_u < 0.98$, which translates into an accessible range of circa $-2 < \Delta G_u^\circ < +2\,\mathrm{kcal/mol}$. Thus, if the standard free energy of unfolding of a protein exceeds 2 kcal/mol, its magnitude will have to be determined indirectly; how this is done in practice is the subject of Sections 17.4 and 17.5.

17.4 CALORIMETRIC MEASUREMENTS OF THE THERMODYNAMICS OF PROTEIN UNFOLDING

As was discussed in Chapter 1, if any one of the thermodynamic functions G, H, or C_P is accurately known as a *function of temperature*, the other two can be calculated. Therefore, thermodynamic functions can be determined from measurements of H or C_P with a sensitive calorimeter.

In order to study protein unfolding, one uses a *differential scanning calorimeter* (see Section 1.9), in which sample and reference are both heated at a slow constant rate, dT/dt, and the (excess) heat absorbed by the sample, dH/dt is measured and converted to give the (excess) specific heat, C_p according to

$$C_P = dH/dT = \frac{dH/dt}{dT/dt}. \tag{17.7}$$

At low temperature the protein is folded, and the measurement produces C_P of the folded protein. At high temperature the protein is unfolded, and the measurement produces C_P of the unfolded protein. In the transition region, the heat required to unfold the protein is measured in addition, and this produces a peak in the curve of C_P as a function of temperature, T (Fig. 17.5).

Integration of the excess above the baseline gives the standard enthalpy of unfolding.

$$\Delta H_u^\circ(T_m) = \int_{\text{low } T}^{\text{high } T} dT(C_P - C_{P,0}), \tag{17.8}$$

where $c_{P,0}$ is the baseline value.[3] The standard entropy difference is found with

$$\Delta S_u^\circ(T_m) = \int_{\text{low } T}^{\text{high } T} \frac{dT}{T}(C_P - C_{P,0}). \tag{17.9}$$

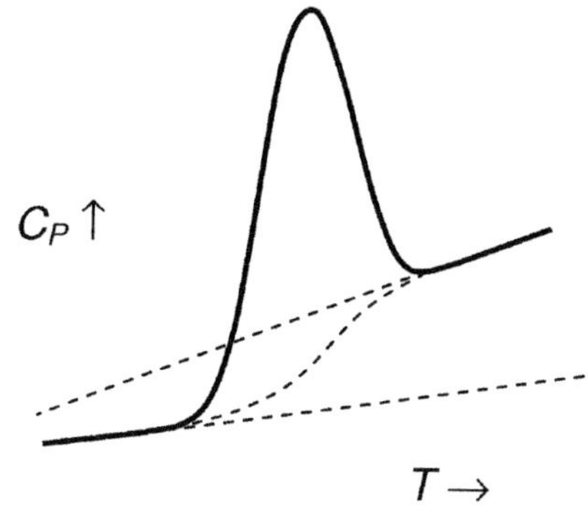

FIGURE 17.5 Excess specific heat in the temperature range in which the protein unfolds (solid curve). Low and high temperature baselines are extrapolated into the transition range; the dashed curve is a baseline constructed taking into account the fraction of unfolded protein at each temperature.

As indicated, these are the values for the transition temperature, T_{m} (the temperature at the midpoint of the transition).

The value of the enthalpy and entropy differences at other temperatures are found by integrating the difference in specific heat between folded and unfolded states

$$\Delta H^{\circ}(T) = \Delta H^{\circ}(T_{\mathrm{m}}) + \int_{T_{\mathrm{m}}}^{T} dT (C_{P,U} - C_{P,F})$$

$$\Delta S^{\circ}(T) = \Delta S^{\circ}(T_{\mathrm{m}}) + \int_{T_{\mathrm{m}}}^{T} \frac{dT}{T} (C_{P,U} - C_{P,F}). \tag{17.10}$$

Thus, a careful scanning calorimetric measurement can establish if the transition is two-state, and, if so, provide ΔG°, ΔH°, and Δc_p of the unfolding reaction.

The calorimetric measurement gives the specific heat per *gram* of protein in the sample, and this is then multiplied by the protein's molecular weight to give specific heat (and subsequently other thermodynamic functions) of the protein in units of energy per *mole*. In contrast, the van't Hoff enthalpy is based on the temperature dependence of the equilibrium constant (Eq. 1.73), and is obtained directly in units of energy per mole, without a knowledge of the molecular weight being needed. Equality of the van't Hoff enthalpy and the calorimetric enthalpy *establishes unequivocally if the transition obeys a two-state model.*

To illustrate this, consider a model protein consisting of two similar domains that are connected only by a disordered segment of polypeptide chain. The folding of the two domains (subscripts 1 and 2) is then independent, the equilibrium constant is

$$K_{1/2} = \frac{f_{u,1}}{(1 - f_{u,1})} = \frac{f_{u,2}}{(1 - f_{u,2})}, \tag{17.11}$$

and the temperature dependence of $K_{1/2}$ gives a van't Hoff enthalpy corresponding to the heat of unfolding one domain; in contrast, the calorimetric enthalpy will be twice as large (if the molecular weight used is that of the dimer).

The original definition of the van't Hoff enthalpy in terms of the temperature dependence of the equilibrium constant is unsatisfactory when a transition is not two-state, and consequently involves more than one equilibrium constant. Privalov obtained a more generally applicable definition of the van't Hoff enthalpy based on the relation between fluctuations of the energy and the specific heat that was derived on the basis of the statistical mechanics of the canonical ensemble in Chapter 6, Eq. 6.35, or, rather, the analogous expression

$$\mathrm{var}(H) = k_B T^2 C_P. \tag{17.12}$$

Considering *only the contributions from conformation change*, one sees that in a two-state equilibrium with reaction enthalpy H_0, individual molecules have $H = 0$ if they are folded (fraction $1 - f_u$), and $H = H_0$ if they are unfolded (fraction f_u). The mean value of H is then $f_u H_0$, and the variance of H is given by

$$\mathrm{var}(H) = f_u(H_0 - f_u H_0)^2 + (1 - f_u)(f_u H_0)^2 = f_u(1 - f_u)H_0^2. \tag{17.13}$$

For the two-state equilibrium the van't Hoff enthalpy equals the reaction enthalpy, H_0 so that the following expression for the van't Hoff enthalpy[4]

$$(\Delta H_{\mathrm{van't\,H}})^2 = \frac{1}{f_u(1 - f_u)} k_B T^2 C_P, \tag{17.14}$$

is correct for the two-state case and provides a general expression that is useful also for multistate transitions.

17.5 UNFOLDING THERMODYNAMICS OF RIBONUCLEASE

Bovine pancreatic ribonuclease has been a subject of much research in biophysics ever since Armour and Co. in the 1950s instituted its preparation in a highly purified form, together with chymotrypsinogen and trypsinogen, as a by-product of the preparation of bologna and other meat products. Ribonuclease was one of the first proteins to have its amino acid sequence and crystal structure determined, and it was the subject of Anfinsen's successful refolding of the reduced form, alluded to in the introduction.

The two-state model was first applied to the unfolding of ribonuclease in the low pH range by Brandts and Hunt.[5] The UV absorption was measured as a function of temperature for several pH values, with transition midpoints between 20 and 50 °C (Fig. 17.6).

These data were converted first to values of f_u, then to standard free energies, ΔG°, according to the two-state model. Ideally, one would use a computed least-squares fit to obtain the best values of thermodynamic parameters underlying these

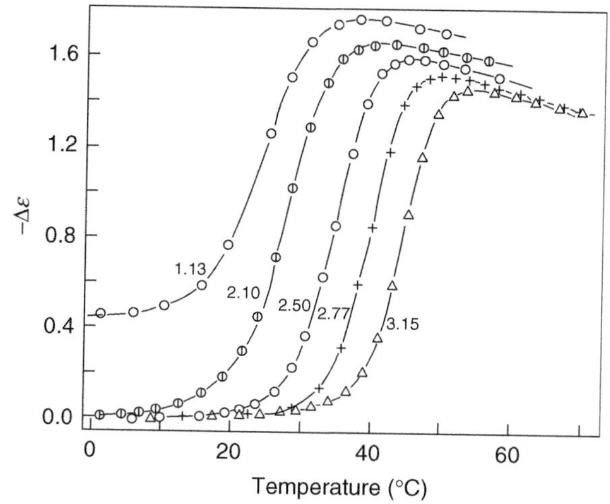

FIGURE 17.6 The change in extinction coefficient at 287 nm for ribonuclease A at five pH values. The samples at pH 2.77 and 3.15 were buffered with 0.04 M glycine, while the other samples contain only HCl. Protein concentrations are 0.03–0.04 g/100 ml. From Brandts, J.F., Hunt, L. *J. Am. Chem. Soc.* **89**: 4826–4838 (1967).

transitions. However, the precision with which curvature can be determined from these data already is not very high, and parameters of a still higher order cannot be estimated. Thus, one may hope to obtain values of ΔG°, ΔH°, and ΔC_P° for each of these transitions, with no information available about the temperature dependence of ΔC_P°. Therefore, the data were used to find the best fitting theoretical free energy expression with independent values of ΔH°, ΔS°, and (constant) ΔC_P° at one temperature (30 °C; $T_0 = 303$ K) at each pH, which, taking into account the temperature dependence of ΔH° and ΔS° (see Section 1.3), is

$$\begin{aligned}\Delta G^\circ(T) &= \Delta H^\circ(T) - T\Delta S^\circ(T)\\ &= \Delta H^\circ(T_0) + \Delta C_P(T - T_0) - T\Delta S^\circ(T_0) - T\Delta C_P \ln\frac{T}{T_0}.\end{aligned} \qquad (17.15)$$

If this fit is done for each of the five pH values, the resulting values are quite scattered; ΔH° ranges from 51 to 63 kcal/mol and ΔC_P° ranges from 1.3 to 2.4. However, very good fits can be obtained if the values of ΔC_P° are restricted to all be the same, at 1.8 kcal/mol/K; then $\Delta H^\circ(T_0)$ varies between 58 and 62 kcal/mol and $\Delta S^\circ(T_0)$ lies between 180 and 210 cal/(mol K).[6] In summary,

$$\begin{aligned}\Delta H_u^\circ(303) &= 60\,\text{kcal/mol}\\ \Delta S_u^\circ(303) &= 195\,\text{cal/(mol}\cdot\text{K)}\\ \Delta C_{P,u}^\circ(303) &= 1.8\,\text{kcal/(mol}\cdot\text{K)}.\end{aligned} \qquad (17.16)$$

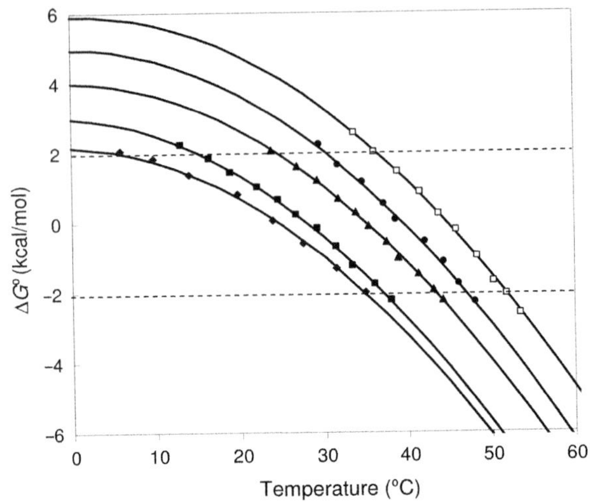

FIGURE 17.7 Temperature dependence of the standard free energy for the ribonuclease transition at different pH values. The points represent values of ΔG° calculated from measurements of the extinction coefficient, and the solid curves correspond to analytical values of ΔG° obtained as described in the text. Dashed lines indicate the range of relatively high experimental accuracy. Data from Brandts, J.F., Hunt, L. *J. Am. Chem. Soc.* **89**: 4826–4838 (1967).

The small differences in these thermodynamic functions as a function of pH are not discernable in the noise. The result of the fit is shown in Fig. 17.7.[7]

These and similar results for chymotrypsinogen provided the first evidence for *a large positive* ΔC_P°, which has been found to be generally true of the thermodynamics of unfolding of globular proteins.

The validity of the two-state model was confirmed by comparison with a calorimetric study of unfolding at 45 °C made in Julian Sturtevant's laboratory, which gave $\Delta H_{\text{cal}}^\circ = 86 \pm 5\,\text{kcal/mol}$ and $\Delta C_{P,\text{cal}}^\circ = 2.3 \pm 0.4\,\text{kcal/(mol} \cdot \text{K)}$ and corresponding $\Delta H_{\text{van'tH}} = 88\,\text{kcal/mol}$ and $\Delta C_{P,\text{van'tH}} = 2.5\,\text{kcal/(mol} \cdot \text{K)}$. (Using the thermodynamic data from the fit (Eq. 17.16), one gets a value of ΔH° at 45 °C of $60 + 15 \times 1.8 = 87$.)

Two very interesting *predictions* follow if one extrapolates the thermodynamic functions for the unfolding equilibrium of ribonuclease A to temperatures outside the range of these measurements: a large increase in ΔH° at higher temperatures, and a decrease of ΔH° to zero at temperatures around 0 °C, and negative values at temperatures lower than that. Both predictions have been corroborated experimentally, as discussed next and in the following section.

Extrapolation to higher temperature. At neutral pH, the temperature transition has a midpoint at circa 65 °C (Fig. 17.1; this is the pH range of maximum stability.) Extrapolation of the thermodynamic data of Eq. 17.16 to 65 °C predicts a value of ΔH° at this temperature of 121 kcal/mol.

TABLE 17.1 Thermodynamic parameters for the thermal unfolding of ribonuclease a at pH 6.8 from careful calorimetric measurements[a]

	ΔH°_u (kcal/mol)	ΔS°_u (cal/mol/K)	T_m (°C)	$\Delta H_{van'tH}$ (kcal/mol)[b]
Pretransition	37.7	118	47.4	—
Main transition	113.7	338	63.8	—
Overall	149.8	444	—	115

[a] Stelea, S.D., Pancoska, P., Benight, A.S., Keiderling, T.A. *Protein Sci.* **10**: 970–978 (2001).
[b] Calculated with Eq. 17.14.

The transition of ribonuclease at neutral pH has been reinvestigated with spectroscopic and calorimetric measurements, and this work has given a value for the transition enthalpy of 150 kcal/mol. The transition was found to be divided into two parts, a pretransition between 30 and 60 °C and a main transition centered at 63.8 °C.[8] (Table 17.1.) Analysis of far-UV CD and Fourier transform infrared (FTIR) spectra indicates modifications in the secondary structure at these lower pretransition temperatures; the authors attribute these initial changes to the fraying of the helical segments as well as changes in the β-strands. These results confirm the validity of the extrapolation of the thermodynamic values of Eq. 17.16 to higher temperatures, with a (relatively) constant, large ΔC_P.

We can now estimate the standard free energy for unfolding of ribonuclease at neutral pH by extrapolation, using a constant ΔC°_P, according to Eq. 17.15.

$$\Delta G^\circ_u(298, \text{pH} \sim 7) \approx 9\,\text{kcal/mol}, \tag{17.17}$$

which corresponds to a value of K_u at room temperature around 3×10^{-7}, that is, 1 in 3×10^6 molecules of ribonuclease is fully unfolded in solution at room temperature. This is a quite typical result. Measured values of ΔG°_u for different globular proteins range from 0 to 20 kcal/mol at room temperature. (e.g., for hen egg white lysozyme the value is estimated to be 6 kcal/mol, and for sperm whale myoglobin circa 15 kcal/mol.)

As said earlier, use of the two-state model greatly simplifies the treatment of protein unfolding equilibria. The case of ribonuclease is instructive because here the two-state model applies under some, but not all conditions. We have seen also that careful calorimetric measurements are required to draw firm conclusions. It is wise to undertake extensive equilibrium and kinetic studies of protein folding under different conditions, and the effect of mutations, only with a protein exhibiting two-state unfolding over a wide range of conditions. Fersht and coworkers have undertaken extensive studies of folding of two small proteins and carefully determined how closely these meet the two-state condition: the CI2 protease inhibitor consists of a single compact domain protein (64 amino acid residues), and the ribonuclease barnase has two domains (110 residues); the barnase fold resembles that of bovine pancreatic ribonuclease. Neither protein has disulfide bridges; results of kinetic studies of folding of these two proteins are discussed in Chapter 20.

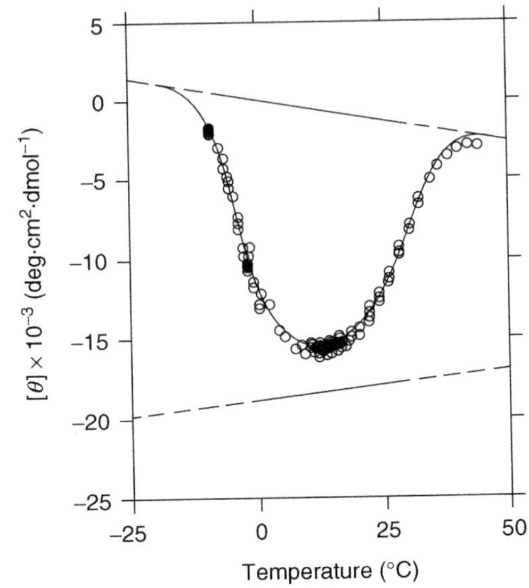

FIGURE 17.8 Thermal melting followed by circular dichroism at 223 nm of T4 lysozyme mutant I3C-C97/C54T, which is free of sulfhydryl groups and has a genetically engineered disulfide bridge between positions 3 and 97. The solutions contain 3 M GuHCl. The straight lines are baselines for folded (bottom) and unfolded (top) states, and the curve is a fit to the data assuming a constant value of ΔC_P°. From Chen, B., Schellman, J. A. *Biochemistry* **28**: 685–691 (1989).

17.6 COLD DENATURATION

According to a second prediction following from application of the two-state model to ribonuclease, the extrapolated values of K_u and ΔG_u° both have extrema (a minimum and a maximum, respectively) at low temperature. The prediction is therefore that if cooled to, say, $-20\,^\circ$C, ribonuclease will be less stable than at $0\,^\circ$C and would show some unfolding. Of course, such low temperatures are not achievable in aqueous solution. It has turned out that for other proteins also, the predicted temperature of maximum stability lies just out of range, that is, below 273 K, and cold denaturation was for a number of years an unverified prediction.

Ultimately, a mutant of T4 lysozyme was found with maximum stability at $12\,^\circ$C[9] (Fig. 17.8). More recently, partial unfolding of cytochrome c was found to proceed farther at both low and at high temperatures.[10] These results confirm the prediction of cold denaturation that followed from Brandt's studies of chymotrypsinogen and ribonuclease using a two-state model.

17.7 SOLVENT-INDUCED UNFOLDING: GUANIDINE HYDROCHLORIDE AND UREA

Some compounds when added at high concentration to a protein solution cause the protein to become less stable, and such compounds are called denaturants.

The most studied of these have been inorganic salts, alcohols (ethanol), urea [$O{=}C(-NH_2)_2$], and guanidinium chloride [$C(-NH_2)_3{}^+Cl^-$]. The much-studied effects of salts are complex and there is currently little interest in study or applications of salt–protein interactions. Ethanol is a good denaturant, but its effects are mixed: it increases the strength of hydrogen bonds, to the point that denatured proteins in ethanol can have considerable α-helix content, and/or become poorly soluble. The denaturing effect of ethanol is thus based on the fact that it interacts more favorably than water with exposed hydrophobic groups.

Urea and guanidinium chloride (GuCl)[11] are universal protein denaturants because they interact more favorably than water not only with exposed hydrophobic groups but also with the polypeptide backbone, and thus also unfold α-helices. The concentration of GuCl can be raised to around 8 M, and at this concentration the majority of globular proteins unfolds; the relation between radius of gyration and molecular weight of proteins in 6 M GuCl is consistent with all proteins being in a similar statistical coil state,[12] only with different chain length (see Chapter 15).

By application of a two-state model, one can readily analyze the effect of urea or guanidine on the thermodynamics of unfolding. The already cited study by Brandts and Hunt[5] shows a linear dependence of ΔG° of ribonuclease (at pH 3.15) on the urea concentration, with a constant ΔC_P° (of circa 1.8 kcal/mol/K, within experimental uncertainty). ΔG° is lower by slightly over 6 kcal/mol in 4 M urea.

Subsequently, unfolding of proteins by guanidine was systematically measured by Tanford and coworkers, who confirmed the linear dependence of ΔG° on denaturant concentration, c_U. Furthermore, by taking advantage of the pH dependence of the stability they varied the concentration of guanidine at which the unfolding was half complete, and they were thereby able to establish that in first approximation the slope, or *m-value*,

$$m = -\frac{d\Delta G_u^\circ}{dc_U}, \tag{17.18}$$

is for any protein a constant over a wide range of guanidine concentration.[13] Because of the constancy of the slope, a single set of values of ΔG_u° *measured as a function of denaturant concentration (at room temperature and neutral pH) can be extrapolated to give the standard free energy in the absence of denaturant.*

Results for unfolding of lysozyme by guanidine and by urea are given in Fig. 17.9. Linear extrapolation of these results gives identical values of $\Delta G^\circ(298, \text{pH7})$ of 5.8 kcal/mol. For ribonuclease, the same approach gives $\Delta G^\circ(298, \text{pH7}) = 9.3\,\text{kcal/mol}$ for measurements with guanidine, and 7.7 kcal/mol for measurements with urea.[14] The result for ribonuclease results agrees well with the value obtained by extrapolation of temperature-induced unfolding discussed in Section 17.5.

Measurement of unfolding by guanidine and linear extrapolation to zero has come into wide use as a procedure for determining the standard free energy of unfolding of globular proteins at room temperature and neutral pH (roughly, physiological conditions).

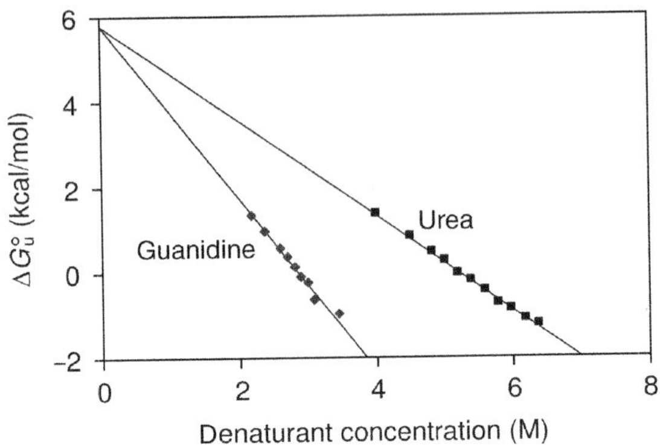

FIGURE 17.9 Free energy of unfolding of hen egg white lysozyme ΔG_u°, as a function of denaturant concentration. Data from Greene, R.F., Pace, C.N. *J. Biol. Chem.* **249**: 5388–5393 (1974).

The m-value is generally taken as a measure of the extent of burial of hydrophobic groups, that is, the difference of this extent between native and denatured. It is useful in interpretation of changes in stability on mutation; a related kinetic m-value is important in interpretation of unfolding and folding kinetics. More about this later.

17.8 MIXED SOLVENTS: DENATURANTS AND STABILIZERS

Many compounds are soluble in water to very high concentrations, and a few liquids are completely miscible with water. Many of these *cosolutes* do not react specifically with dissolved biological macromolecules. However, nearly always, the addition of such a cosolute to an aqueous solution of a macromolecule changes that molecule's chemical potential, with consequences for other equilibria in which the macromolecule participates. Indeed, interest in cosolutes is due precisely to the effect that these have on the stability of the native, folded conformation of biological macromolecules.

In Chapter 9, we presented a strictly thermodynamic analysis of molecular interactions that presupposes nothing about the mechanism of these interactions or the identity and structure of entities responsible for it. This analysis can deal with both preferential association and exclusion, and provides a framework for analyzing the weak, relatively nonspecific interactions between macromolecules and cosolutes.[15]

The connection between, on the one hand, stabilization of one conformation relative to the other, and, on the other hand, the difference in preferential interaction with the cosolute, can be treated with Eq. 12.21. Applying this to the two conformations, native, f, and unfolded, u, in the presence of a cosolute (species 2) we obtain

$$\frac{1}{k_B T}\frac{\partial(\mu_u - \mu_f)}{\partial \ln a_2} = -\nu_{2,u} + \nu_{2,f}, \tag{17.19}$$

or

$$\frac{\partial \ln K_{fu}}{\partial \ln a_2} = \nu_{2,u} - \nu_{2,f} = \Delta\nu_2. \tag{17.20}$$

The observed connection is rigorous: if addition of a cosolute destabilizes the native form of a protein (i.e., results in an increase in K_{fu}), then this *implies* that the cosolute associates preferentially with the unfolded form, and vice versa. (See additional discussion in Section 9.7 "Accounting for interactions of macromolecule and water".)

The *observed linear dependence* of $\ln K_{fu}$ on the concentration of denaturants guanidine or urea observed by Tanford implies that, for these, $\Delta\nu_2$ is proportional to a_2, which *is typical of binding to sites with a low affinity*, the unfolded state having a greater number of sites and/or sites with a (slightly) higher affinity than the native conformation. A linear free energy dependence is also observed for *stabilization* of the folded state by increasing the concentrations of trimethylamine N-oxide (TMAO); in experiments with ribonuclease T1, the effects of urea and TMAO are additive, with stabilization of the folded state by TMAO compensating for the destabilization by urea when present in a molar ratio of roughly 1:1.6.[16]

A number of *physical models* have been proposed of how the interactions with different cosolutes result in changes in stability. For example, the effectiveness of urea and guanidine as protein denaturants is thought to be a result of two properties of these molecules: (i) in mixtures with water they are slightly better at solvating nonpolar groups than is water alone and (ii) they are good hydrogen bond formers and generally interact well with polar solutes, and may even favor the disrupted hydrogen bonds of the denatured state. According to this model, urea and guanidine associate favorably with protein molecules, but more strongly with the unfolded form.

Another example: various sugars ("polyols") as cosolutes stabilize the native form, and one model has volume exclusion (which is more pronounced for the more voluminous unfolded form) as the responsible mechanism. Volume exclusion by cosolute is an unfavorable interaction; in contrast, preferential association with cosolute is a favorable interaction.

It is, therefore, of some interest to determine $\partial\mu_u/\partial \ln a_2$ or $\partial\mu_f/\partial \ln a_2$ individually. For example, negative values will indicate *favorable* interaction with the cosolute and suggest a binding model. The more common methods for determining these require equilibration with either an ideal gas or a pure crystalline form, which are not realizable for most macromolecules; accurate measurement of small amounts of cosolute bound or excluded in an equilibrium dialysis type experiment, which can, in principle, be used to give $\partial\mu/\partial \ln a_2$ via Eq. 9.9, is quite difficult against a high background concentration of cosolute. Timasheff and coworkers have reported measurements of preferential interactions of proteins with cosolutes, including chloroethanol, guanidine hydrochloride, and urea.[17]

In the absence of such information, one may *assume* that interactions between solvent and the unfolded state dominate (and that $\partial\mu_U/\partial \ln a_2$ determines the sign of $\partial(\mu_U - \mu_N)/\partial \ln a_2$). This assumption is reasonable because solvent–protein interactions are (obviously) much fewer for the native conformation than for the unfolded state. This is exactly what the cited experiments[17] with guanidine hydrochloride show: guanidine, a denaturant, binds preferentially to the protein, in both native and unfolded forms.

As a substitute, one can use values of $\partial\mu/\partial \ln a_2$ of small molecules that are chemically related to proteins, and that form pure crystals or are measurably volatile. Nozaki and Tanford measured in this manner the free energy of transfer of amino acids from crystalline form to various mixed solvents, and reduced the results to estimate the differences in free energy of transfer of amino acid side chains and of the peptide group. (Reported in a series of four papers between 1963 and 1971.[18] Some of these results have been reproduced in Table 17.2.) An additional assumption is that the effects on different parts are additive, which enables one to estimate the free energy of transfer of a fully unfolded and solvated protein from vacuum to one of these mixed solvents. A valuable addition to these reference data was made by Wolfenden, who reported the free energies of transfer of amino acid side chains from water to vacuum (Table 17.3).[19] The advantage of studying transfer from vacuum is clear, as no consideration need be given to intermolecular interactions in that phase.

It is noteworthy that the transfer free energy contributions from water to *guanidine hydrochloride* solutions shown in Table 17.2 are all negative. This denaturant interacts favorably with *all* parts of an unfolded protein, hydrophobic side chains,

TABLE 17.2 Contributions of amino acid side chains and peptide group to free energy of transfer at 25.1° in aqeous guanidine hydrochloride and ethanol (in cal/mol)[a]

Side Chain	4 M Guanidine Hydrochloride	40% Ethanol
Alanine	−30	−60
Leucine	−355	−375
Phenylalanine	−580	−545
Tryptophan	−980	−880
Methionine	−400	—
Threonine	−120	—
Tyrosine	−605	−615
Histidine	−385	−100
Asparagine	−490	90
Glutamine	−315	55
Valine	—	−195
Peptide group		
diglycine[b]	−175	270
triglycine[b]	−395	15
tetraglycine[c]	−205	

[a] From Nozaki, Y., Tanford, C. *J. Biol. Chem.* **238**: 4074–4081 (1963); *J. Biol. Chem.* **245**: 1648–1652 (1970); *J. Biol. Chem.* **246**: 2211–2217 (1971); *J. Biol. Chem.* **240**: 3568–3573 (1965).

[b] Diglycine less glycine; triglycine less diglycine.

[c] *N*-Acetyltetraglycine ethyl ester less ethyl acetate, divided by 4.

TABLE 17.3 Affinities of amino acid side chains for water[a,b]

Amino acid	RH	ΔG°_{tr}	Amino acid	RH	ΔG°_{tr}
Proline[b]		2.48	Serine	CH_3-OH	−5.06
Glycine	H_2	2.39	Tryptophan	3-CH_3-indole	−5.88
Leucine	CH-$(CH_3)_3$	2.28	Tyrosine	4-CH_3-C_6H_4-OH	−6.11
Isoleucine	C_4H_{10}	2.15	Glutamine	C_2H_5-$CONH_2$	−9.38
Valine	CH_3-CH_2-CH_3	1.99	Lysine	C_4H_9-NH_2	−9.52
Alanine	CH_4	1.94	Asparagine	CH_3-$CONH_2$	−9.68
Phenylalanine	C_6H_5-CH_3	−0.76	Glutamic acid	C_2H_5-COOH	−10.20
Cystine	CH_3-SH	−1.24	Histidine	4-CH_3-imidazole	−10.27
Methionine	CH_3-*s*-C_2H_5	−1.48	Asparagine	CH_3-COOH	−10.95
Threonine	C_2H_5-OH	−4.88	Arginine	*n*-Propylguanidine	−19.92

Affinities are measured as the free energies of transfer of side-chain analog molecules, RH from the vapor phase into water, where R is the amino acid's side-chain, at pH 7. Units are kcal/mol.

[a] For ionizable molecules, the free energy of transfer depends on pH because only the neutral form is transferred from water to the vapor phase.

[b] From Wolfenden, R., Andersson, L., Cullis, P.M., Southgate, C.C.B. *Biochemistry* **20**: 849–855 (1981).

hydrophilic side chains, and backbones. Every intramolecular contact in a native protein that is broken in the unfolded chain contributes to the denaturing effect of guanidine hydrochloride (and similarly for urea). The native structures of some proteins may be stable even in 6 M guanidine hydrochloride at room temperature (perhaps a protein from a hyperthermophilic organism—see subsequent text); otherwise, these numbers suggest that all remaining proteins assume a disordered structure with highly solvated backbone and side chains in this mixed solvent, although the (undoubtedly statistical) nature of this structure is still being debated.

The transfer free energies for *ethanol–water mixtures* (Table 17.2) show that this cosolute favors the transfer of hydrophobic side chains, but has the opposite effect on the transfer of hydrophilic side chains and the polypeptide backbone. Ethanol also acts as a protein denaturant; it has been less studied, because at high concentrations it is also a protein precipitant. Apparently, the stabilization of the unfolded form due to the exposure of hydrophobic side chains on unfolding exceeds ethanol's unfavorable effect because of the exposure of polypeptide backbone. The situation is different for the α-helical structure of a synthetic polypeptide (poly (L-glutamate)), which is *stabilized* by the presence of ethanol as a cosolute, as one might expect from the positive free energy of transfer of the peptide group from water to ethanol, and the obvious lack of a hydrophobic core in this structure. Also, as expected from the Nozaki–Tanford transfer data, the α-helix is *de*stabilized in the presence of urea.[20]

Recent revived interest in effects of mixed solvents has focused on the so-called osmolytes, a class of small hydrophilic organic molecules used by cells of numerous water-stressed organisms to help maintain cell volume; many have been found to stabilize the native structure of proteins. Transfer measurements for sucrose and sarcosine[21] show that, unlike ethanol, these cosolutes have a mixed effect on hydrophobic side chains, but, similar to ethanol, make transfer of the polypeptide backbone more difficult; the net effect is stabilization of the native structure with its many internally hydrogen-bonded peptide groups. A reasonable model is to assume that exposed peptide groups and osmolyte molecules are strongly hydrated, but interact less well with each other. As a result, the osmolyte molecules avoid the immediate vicinity of peptide groups; this makes $\langle n_2 \rangle$ negative and gives a positive contribution to $\partial \mu / \partial \ln a_2$ by each exposed peptide group. Such a model combines effects of preferential hydration and excluded volume. Note how water plays a critical role in this model, and, nevertheless, is not a factor in the formal thermodynamic treatment that we have employed.

17.9 UNFOLDING IS NOT TWO-STATE UNDER NATIVE CONDITIONS: HYDROGEN EXCHANGE

17.9.1 Detection of Localized Unfolding

We have already indicated that, with a spectroscopic measurement, observations of folding below a few percent, or close to 100% (i.e., $f_u < 0.02$ or $f_u > 0.98$) cannot be used to compute thermodynamic functions because, given a constant statistical error in the measurement of f_u the relative statistical error in either f_u or in $1-f_u$ becomes very large.

In order to detect and accurately measure small concentrations one has to resort to amplification, which can be achieved by coupling the measurement of f_u to that of a chemical reaction whose products accumulate. For example, small amounts of enzyme can be detected accurately by changes in concentration of substrate and product, and thus a small fraction of folded, enzymatically active protein molecules can be detected and accurately measured in the presence of a large excess of unfolded protein molecules. And, in order to detect a small amount of unfolded protein, one can exploit chemical changes in the protein itself to detect and quantitate a small fraction of reactive protein. The experiment can, for example, measure the ease with which buried side chains are chemically altered in a reaction that requires these to be exposed to solvent, or the enzymatic proteolysis of a protein at bonds that are buried in the interior of the native conformation.

A difficulty with methods based on chemical alteration of the protein is that most such changes destabilize the molecule, and that yet other groups (side chains or backbone) become more exposed and react also, so that substitution at different sites is not independent (except at very low degree of substitution). However, isotope exchange leaves the stability of a folded protein nearly unaltered. In the 1950s, Linderstrøm-Lang developed methods to measure *hydrogen isotope exchange*, that is, the exchange of hydrogen atoms of a protein for deuterium atoms in heavy water as solvent (or with tritium with use of tritiated water), for example,

$$> \mathrm{NH} + \mathrm{D_2O} \xrightarrow{k_x} > \mathrm{ND} + \mathrm{DOH}. \qquad (17.21)$$

Hydrogen atoms bonded to carbon do not exchange in these experiments, and hydrogen atoms bonded to ionizable groups exchange rapidly. Thus, the measurements give information about the accessibility of amide hydrogens, in the first

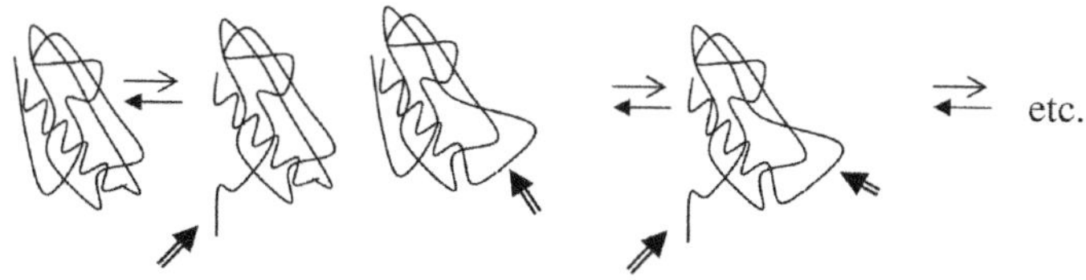

FIGURE 17.10 Under native conditions, *small* fractions of globular protein molecules are locally unfolded, in different parts of the structure. (Equilibrium is far to the left.) More completely unfolded states are progressively *less* common. Hydrogen atoms that are protected in the folded state can exchange locally in the unfolded sections (arrows).

instance those bonded to the backbone nitrogen atoms. The original technique required the accurate measurement of density of water droplets extracted from protein solutions in which exchange had taken place, and there was no information from what part of the protein the exchanged atoms had come. Nevertheless, the results were informative, as some hydrogen atoms exchanged as rapidly as in model peptides, while many exchanged more slowly, some at rates many orders of magnitude smaller. Because the side chains of most asparagine and glutamine residues are exposed, these do not contribute significantly to the slowly exchanging set; this set consists uniquely of hydrogen atoms of backbone peptide groups that are hydrogen-bonded in the native state, such as in α-helices and β-sheets.

With nuclear magnetic resonance measurements, it has since become possible to measure the exchange rate of *individual* hydrogen atoms. Given the full assignment of proton chemical shifts, the experiment consists simply of observing the decay of proton intensities as a function of time of the protein in a D_2O solution. (Because of the wide spread of rates, time points are collected at exponentially spaced intervals.) With the assumption that amide hydrogens participating in secondary structure do not exchange at a measurable rate, the exchange rates can be used to compute the fraction of molecules in which each individual hydrogen atom is exposed.

The hydrogen exchange experiments show that some hydrogens exchange only in fully unfolded molecules, but the variety of rates is clear indication that some amide hydrogens are more exposed than others, and that *partial unfolding* to various extent in different parts of the molecule is responsible for the observations (Fig. 17.10).

17.9.2 Mechanism and Calibration of Hydrogen Exchange

We should consider not only the extent of local unfolding but also the rates, k_1 and k_{-1} at which the native form, N converts into partly or wholly unfolded forms, U and the reverse, and the exchange rate of an exposed hydrogen, $k_{x,0}$ to give the labeled protein (U*, which reverts to N*) according to the following scheme

$$\mathrm{N} \underset{k_{-1}}{\overset{k_1}{\rightleftharpoons}} \mathrm{U} \xrightarrow{k_{x,0}} \mathrm{U}^* \xrightarrow{k_{-1}} \mathrm{N}^*. \tag{17.22}$$

The rate at which the protein is labeled is then[22]

$$k_{\mathrm{x}} = \frac{k_1}{k_{x,0} + k_{-1}} k_{\mathrm{x},0}. \tag{17.23}$$

When refolding is slow relative to exchange, the observed exchange rate constant, k_{x}, measures the rate of unfolding, that is,

$$k_{\mathrm{x}} \approx k_1. \tag{17.24}$$

This is called the EX1 mechanism.

On the other hand, when exchange is slow relative to refolding, the local extent of unfolding, f_{u}', can be obtained from k_{x} with

$$k_{\mathrm{x}} \approx \frac{k_1}{k_{-1}} k_{\mathrm{x},0} \approx f_{\mathrm{u}}' k_{\mathrm{x},0}, \tag{17.25}$$

where $k_{\mathrm{x},0}$ is the exchange rate constant of a fully exposed hydrogen. When $k_{\mathrm{x},0} \ll k_{-1}$, the exchange is said to occur by the EX2 mechanism. This is by far the most common circumstance.

Experiments with model peptides established also that the exchange is both acid- and base-catalyzed. Catalysis by base is much more efficient, and the exchange rate is minimal at a relatively low pH. An ongoing exchange reaction can, therefore, be quenched and saved for later analysis by acidifying the solution. Also, the pH dependence of the exchange measured in a pH range where the protein is maximally stable can establish if one is dealing with EX1 or EX2 conditions; for EX1, the exchange rate does not depend on $k_{\mathrm{x},0}$, so the measured rate will be independent of pH, while for EX2 the rate is proportional to $k_{\mathrm{x},0}$ and hence will vary as $[\mathrm{OH}^-]$, that is, increase by a factor of 10 for an increase of pH by one unit.

Degrees of (local or global) unfolding, f_{u}', too small to be measured directly by, for example, spectroscopy, correspond also to very small rates of hydrogen exchange, k_{x}. Still, measurable exchange will occur if this is tracked over a sufficiently long time (many hours), and thereby these small values of f_{u}' can be determined.

The exchange rate, $k_{\mathrm{x},0}$, of a hydrogen atom in the exposed state depends on the local amino acid sequence and, in first instance, on the nature of the residue to which it belongs. Using calibration data giving values of $k_{\mathrm{x},0}$ for short peptides,[23] one can then convert the observed exchange rate of each hydrogen atom, k_{x}, to the local extent of unfolding, f_{u}', and from it calculate the standard free energy for the local unfolding equilibrium required for the exchange to occur

$$\Delta G_{\mathrm{HX}}^{\circ} = -k_{\mathrm{B}}T \ln[f_{\mathrm{u}}'/(1 - f_{\mathrm{u}}')] \approx -k_{\mathrm{B}}T \ln f_{\mathrm{u}}'. \tag{17.26}$$

17.9.3 Case Study: Hydrogen Exchange of Cytochrome c

For some proteins, it has been found that the slowest exchange rates correspond to a free energy for local unfolding, $\Delta G^{\circ}_{\mathrm{HX}}$ equal to that for two-state unfolding under the same conditions obtained from measurements of temperature or solvent denaturation, and also with the same m-value (i.e., the same dependence of $\Delta G^{\circ}_{\mathrm{HX}}$ on denaturant concentration). This is, for example, true of the slowest exchanging hydrogens (among which the NH of leucine 98, indicated as L98 in Fig. 17.11) of cytochrome c. Other residues exchange more rapidly, albeit not freely, and these also have smaller m-values, *which shows that the extent of unfolding required for their exchange is less than global.* The maximally slowly exchanging hydrogen atoms were found to belong to residues that cluster in one part of the three-dimensional structure (labeled 1 in the structure diagram of Fig. 17.12).

As shown by the data of Fig. 17.11, at higher denaturant concentration the exchange rates of more rapidly exchanging hydrogens approach that for the slowest hydrogen atoms (as they must, given their smaller m-values; global unfolding sets an upper limit on the exchange rate of all hydrogens). Analysis of all measured exchange rates in cytochrome c and their dependence on denaturant concentration led to a grouping of the more rapidly exchanging hydrogen atoms into clusters on the basis of these unfolding isotherms. A subtlety is that some residues "join" a

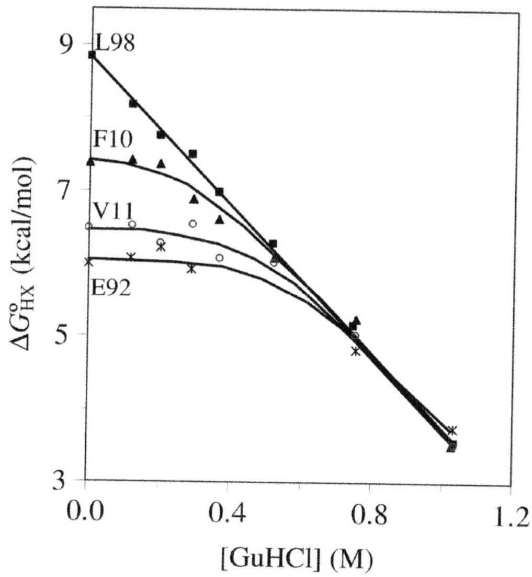

FIGURE 17.11 Hydrogen exchange results for some hydrogens in cytochrome c, as a function of concentration of guanidine hydrochloride. The exchange of leucine 98 proceeds only in a globally unfolded molecule, and displays the largest m-value. At low denaturant, some other hydrogens exchange through local fluctuations with smaller m-values. As the molecule is destabilized at higher denaturant concentration, their exchange becomes dominated by global unfolding. From Bai, Y., Milne, J.S., Mayne, L., Englander, S.W. *Proteins: Struct. Funct. Genet.* **17**: 75–86 (1993).

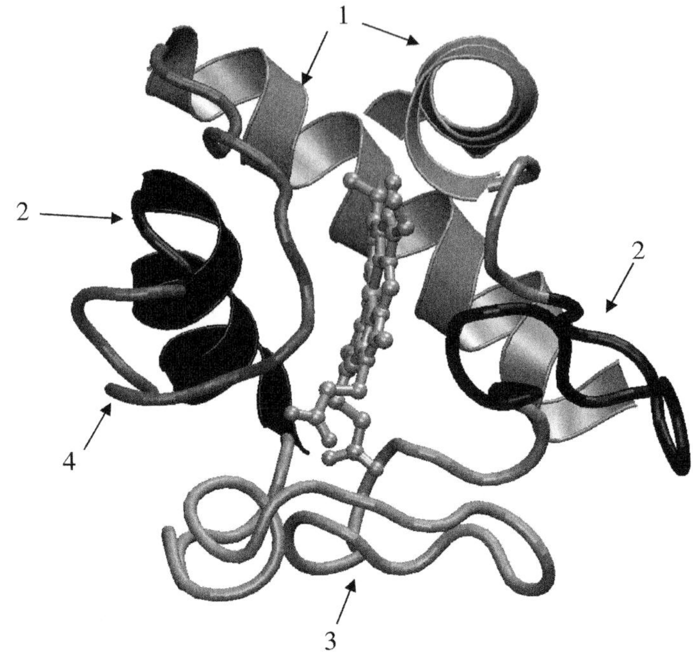

FIGURE 17.12 Four clusters in cytochrome c having different unfolding isotherms (in order of decreasing stability, numbered 1 through 4). The protein segments defined by the four clusters account for the whole cytochrome c protein. From Bai, Y., Milne, J.S., Mayne, L., Englander, S.W. *Proteins: Struct. Funct. Genet.* **17**: 75–86 (1993); vmd graphic. (*See insert for color representation of the figure.*)

particular cluster at some denaturant concentration. Fig. 17.12 shows where these clusters are located the three-dimensional structure of this protein.

To summarize, when not far from the transition midpoint, the dominant mode of unfolding is global, and the equilibrium distribution can be described with a two-state model. On the other hand, when the native structure is very stable, that is, well below the transition midpoint, where the proportion of any partly or wholly unfolded molecules is very small, the exchange process in many parts of the protein is dominated by *local* unfolding, while exchange in a very stable core may happen only if the entire molecule unfolds.

In the chapter on kinetics of protein folding (Chapter 20) we shall discuss how measurements of hydrogen exchange can be used to identify those parts of a protein that fold first.

17.10 NATURE OF THE TWO STATES

17.10.1 The Native State

All the evidence indicates that the native solution state of a globular protein such as ribonuclease is indeed folded into a well-defined three-dimensional structure required for the protein's biological activity, in the case of ribonuclease, as a catalyst. An appropriate reference for this folded state is a high resolution crystal structure or a structure determined with NMR spectroscopy. The unfolded or denatured state no longer possesses this highly ordered structure. Crystal structures of very high resolution show evidence of alternative side-chain conformations, while at lower resolution these side chains do not give a sufficiently high electron density and are simply marked as disordered; this sets a lower limit to the extent of disorder, as flexibility would be more pronounced in solution than within the confinement of a crystal. Also, the biological function of many proteins requires structural flexibility; this is obviously the case for allosteric proteins (see Chapter 13) but surely not limited to these, as many proteins must be flexible in order to admit a substrate to an active site. Hence, the question naturally arises as to what one can say about the fluctuations of the solution structure of native proteins. For the unfolded state, the parallel question is then what (remnant) structure is imposed on the molecule by steric and other energetic requirements, and how this varies with denaturing conditions. We address these in turn.

Fluctuation of the native structure. Protein crystallography has provided additional evidence of flexibility of at least some proteins, by revealing differences in structure in crystals prepared under different crystallization conditions that produce alternative crystal packing. These differences can be much more significant than those produced by disorder of a few side chains. While the *fact* of conformational fluctuation is well documented by experimental observations, its *extent* for any given protein molecule in aqueous solution is not readily accessible experimentally.

It is here that simulations, even with the inherently inaccurate molecular mechanics model, have provided the much needed insight, with the aid of an analysis called *principal component decomposition*. The procedure (cf. Section AF.6) consists of doing a long simulation of one protein molecule in solution, and from the trajectory calculating the deviation of each atomic position from the mean (the conformations are first superimposed by rigid-body rotation and translation), and accumulating mean square deviations and cross terms

$$\begin{aligned} c_{ii} &= (X_i - \langle X_i \rangle)^2 \\ c_{ij} &= (X_i - \langle X_i \rangle)(X_j - \langle X_j \rangle), \end{aligned} \tag{17.27}$$

where X_i is any one of the $3N$ coordinates of the N atoms. (However, not all atoms need be used in the analysis.) These coefficients constitute a square symmetric correlation matrix, **C**. The $3N$ eigenvectors, $\boldsymbol{\xi}_i$ and corresponding eigenvalues, λ_i of this matrix can be obtained by solving the system of linear equations represented by

$$\mathbf{C}\xi = \lambda \boldsymbol{\xi}. \tag{17.28}$$

The eigenvectors describe orthogonal internal displacements of the molecule that represent an average of the actual motion produced by the simulation, and are called the *principal components*. For each principal component, its eigenvalue determines the amplitude, and principal components are ordered according to decreasing eigenvalue. It must be stressed that the principal components are only an approximation of the actual anharmonic motions, and that repeated independent simulations will not produce exactly the same principal components.[24]

Studies by Berendsen and coworkers have shown that a few principal components with the largest eigenvalues suffice to account for the major part of the structural variability of the simulated molecule.[25] Application to dynamics of T4 lysozyme shows the dominant motions involve relative displacements of the two domains that form the sides of the active site cleft. Each domain internally is much less flexible. This protein is a depolymerase, and the domains must move apart to accept the polymeric carbohydrate substrate, then close to form the active site.

Principal component analysis of a set of crystallographic structures of T4 lysozyme and mutants showed very similar fluctuations as were observed in the dynamics simulation.[25] A much longer simulation of T4 lysozyme has shown a spontaneous opening followed by a reclosing of the active site cleft.[26]

17.10.2 The Unfolded State

The unfolded state is an ensemble spread over a very large number of different conformations separated by energy barriers; a logical reference for this state is the random-coil solution state of synthetic polymer molecules. Internal rotations about single bonds are the principal degrees of freedom that produce this diversity. There are two of these in the backbone of each amino acid residue (about N—C$^\alpha$ and C$^\alpha$—C), and from 0 to 5 about single bonds in the side chain. A good idea of the accessibility of different backbone conformations is provided by the data of Fig. 17.13, which shows a distribution of backbone conformation of amino acid residues in high resolution crystal structures.[27] (The conformational freedom of proline residues is more, and of glycine residues is less restricted than of alanine.)

Highly populated regions have lowest energy, by inference, also in solution. The most highly populated conformation in the upper left quadrant is also found in various β-structures, and correspond to relatively extended structures. Conformations with values of ψ near 0° introduce bends in the chain, and are a key to compact folding. (Successive residues with conformations with φ and ψ both near –60° form a right-handed α-helix.)

Optical rotation and CD spectra report on a protein's secondary structure (in the first approximation, in terms of three types: α-helices, beta sheet, and "unstructured"). Measurements of denatured proteins show varying content of secondary structure depending on conditions; furthermore, proteins with high α-helical content in their native structures also show a higher than average helical content if denatured at high temperature or at low pH. This indicates that in an appreciable number among the ensemble of denatured molecules under these conditions contain parts that are structured roughly as they are in the native protein. However, in 6 M guanidine, the spectra of different proteins are similar, and also the hydrodynamic radii of denatured (and reduced) proteins scale with molecular weight as expected for random-coil polymers.[28]

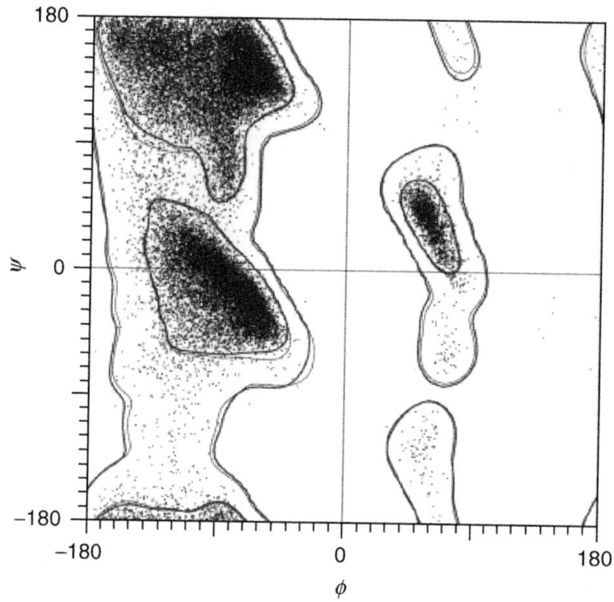

FIGURE 17.13 Backbone dihedral angle distributions for well-defined non-Gly, non-Pro, non-prePro residues from a database of 500 high resolution x-ray structures, with validation contours for favored and allowed regions. Residues in helix or sheet secondary structure have been excluded. From Lovell, S.C., Davis, I.W., Arendall, W.B., de Bakker, P.I.W., Word, J.M., Prisant, M.G., Richardson, J.S., Richardson, D.C. *Proteins: Struct. Funct. Genet.* **50**: 437–450 (2003).

It appears then that heat- or acid-denaturation imposes less stringent denaturing conditions than 6 M guanidine, and allows local and long-range interactions to form in at least some of the ensemble of denatured molecules. This raises the question of the structure of a denatured molecule under *native* conditions. Under these conditions, the protein is of course in a metastable state, in which it remains only briefly after denaturing conditions have been removed (by cooling, changing the pH, or diluting the guanidine). Accordingly, this state is relevant to an interpretation of kinetic studies of protein folding, and also to the process of protein folding following or accompanying biosynthesis of the protein. Evidence suggests that the state of the unfolded chain of at least some proteins under native conditions belongs in a third class, neither native nor fully denatured, which will be discussed in the following section.

Finally, a few words are in order about the random-coil state of proteins in 6 M guanidine. As we have seen in Chapter 15, the random-coil structure of polymer chains is based on the random walk model, the properties of which are obtained by application of the central limit theorem. According to this theorem, the Gaussian statistics characteristic of the random walk model result from an accumulation of the results of successive random choices; for a polymer molecule, these are choices in the values of successive torsion angles of chemical bonds in the polymer's backbone. The observation of random-coil behavior says only that the distribution of possible choices (i.e., torsion angles) is similar in the various proteins used in the analysis of hydrodynamic radii. It does not even say that the available choices are the same for every residue, only that any differences average out over the entire sequence of residues. The actual conformational distribution of the residues does determine the average dimensions of the macromolecules, as expressed by the characteristic ratio; in Section 15.4, we have discussed interpretations of this ratio in terms of the variability of the structure of an individual residue and application to measurements of denatured proteins in 6 M guanidine.

17.11 A THIRD STATE: THE MOLTEN GLOBULE[29]

The molten globule is an elusive third structural form of globular proteins; its existence was recognized in the 1980s as a result of the work of groups headed by Oleg Ptitsyn[30] in Russia and by Akiyoshi Wada[31] in Japan.

The molten globule state has characteristics intermediate between those of the folded, native structure and the unfolded state. Some of these characteristics are typical of native structure: a low intrinsic viscosity indicates relative compactness, and the CD spectrum in the far ultraviolet (peptide region) and the amide I infrared spectrum indicate organized secondary structure. Other physical properties are consonant with a disordered, denatured structure: CD in the near ultraviolet (aromatic side-chain region) is typical of a structureless environment for these side chains, and hydrogen–deuterium exchange is much faster than in the native form.

An *essential* finding in the experiments reported by the groups of Ptitsyn and Wada is the observation with increasing denaturing conditions of *two* successive, separate cooperative transitions. The evidence is most compelling if separate transitions can be followed with different signals, as this immediately excludes a two-state model. (See Section 17.2) Thus, with increasing concentration of guanidine HCl, α-lactalbumin shows a first transition centered near 1.3 M GuHCl, in which the ellipticity (CD signal) at 270 nm changes, and then in a second transition, centered at around 4.5 M GuHCl, the ellipticity at 224 nm changes to that typical of a fully denatured molecule; one estimates that as much as 90% of the molecules is in the intermediate state at 2 M GuHCl. Similarly, cytochrome c at pH 3.2 shows two transitions with increasing temperature, the first, centered at 50 °C, is accompanied by a change in CD at 282 nm, and in the second, centered at 70 °C, the CD signal changes at 222 nm; in this case, as much as 50% of the molecules is in the intermediate state at 60 °C.

Accordingly, Ptitsyn and Wada gave the name molten globule to this state. In short, a molten globule state is a disordered state, although more compact than the unfolded state, but possessing appreciable secondary structure.

Other observable characteristics of a molten globule state are an absence of fine structure in the H^1 NMR spectrum consistent with a disordered state. A useful test to establish if a protein has a well-ordered folded structure, and is, therefore, *not* a molten globule, is to establish, via NMR spectroscopy, the presence of a sufficient number of nuclear overhauser effect (NOE) cross peaks. Such cross peaks are observed only for pairs of nuclei that *remain* at a relatively short distance, and are largely absent from the spectrum of a molten globule state (and, *a fortiori*, from the spectrum of unfolded states).[32] It has also been found that molecules in molten globule states preferentially bind probe molecules of ANS (amino naphthalene sulfonate), with an increase in fluorescence by the probe molecule typical of transfer to a less polar environment.

These observations lead to an analogy between, on the one hand, three states of matter, crystalline solid, liquid, and vapor, and, on the other hand, three states of a protein, namely, native, molten globule, and unfolded. In both cases, the order decreases from first to last, that is, entropy (and enthalpy) increase from first to last.[33] Another possible analog of the collapsed molten globule state of a polypeptide chain is the collapsed state of synthetic homopolymer molecules in solution at and below the theta temperature. (This has been discussed in Section 15.7.)

In a somewhat speculative vein, we might follow this analogy of molten globule and liquid state a step further by considering the typical simple solid–liquid vapor phase diagram of Chapter 1 (Fig. 1.7). At low temperature and elevated pressure, the solid state is the most stable one, while at low temperature and low pressure, the vapor is observed. The liquid state first appears at temperatures and pressures above the triple point, where solid, gas, and liquid are in equilibrium. Liquid then lies between solid and vapor in the diagram. The liquid/vapor line terminates at the critical point; beyond this point liquid and vapor are indistinguishable, and the uniqueness of the liquid state disappears. If one converts this phase diagram into a diagram separating the native, unfolded and molten globule states in Fig. 17.14, one sees then that the *molten globule state occupies a limited part of a typical phase diagram.* For a single protein molecule in solution, a state variable equivalent to pressure is lacking, and one is limited to an

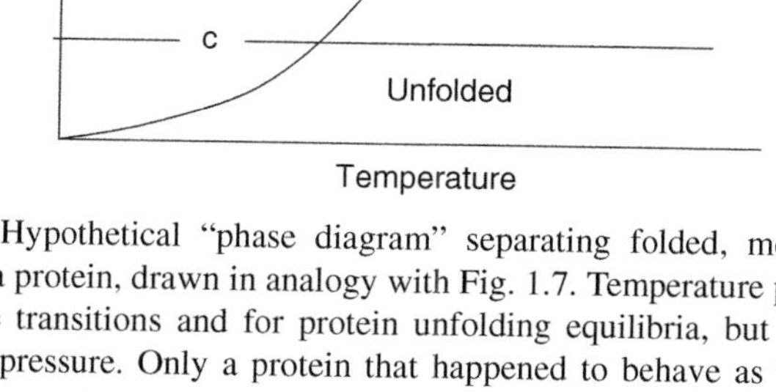

FIGURE 17.14 Hypothetical "phase diagram" separating folded, molten globule, and unfolded states of a protein, drawn in analogy with Fig. 1.7. Temperature plays an equivalent role for real phase transitions and for protein unfolding equilibria, but the latter have no equivalent for the pressure. Only a protein that happened to behave as "b" would show a molten globule in some temperature range.

(unknown) particular value of the variable "?" of the diagram; the variable marked "?" might correspond to a change in denaturant concentration, pH, or salt concentration. Accordingly, *only* if conditions place us near level "b" of this variable, a molten globule state will be observed over a short temperature interval. (In the case of α-lactalbumin and cytochrome c changes in pH or concentration of added salts are needed to shift the conditions to allow observation of the molten globule state.)

From the fact that molten globule states are rarely encountered with native proteins, one would conclude that most globular proteins that have a native, ordered state fall outside region b. On the other hand, engineered *designed* proteins are frequently found to behave as molten globules. It appears that it may be relatively easy to engineer theta conditions, simply by introducing a multitude of hydrophobic side chains, and much more difficult to design the close packing required to form a stable folded state typical of naturally occurring proteins. A surprise has been the relatively high stability of the molten globule state of some of these engineered constructs.

17.12 RANGE OF STABILITY

The stability of single-chain proteins, as measured by the standard free energy of unfolding at room temperature and neutral pH, covers a range. Highly stable proteins have ΔG° as high as +20 kcal/mol, while the least stable proteins assume their folded structure only in the presence of their biological target, that is, these natively unfolded proteins have $\Delta G^\circ < 0$;[34] and, presumably, some polypeptides do not fold at all. It would be tempting to look for rules about protein stability, such as that unfolding at neutral pH takes place somewhere below 100 °C, and, at room temperature, below 6 M guanidine, and, perhaps, that small proteins are less stable than larger ones. However, to show how differently a small protein *can* behave, we cite a study of a small protein (54 amino acid residues) from a hyperthermophilic organism, that is, able to live near and even above 100 °C, rubredoxin from *Pyrococcus furiosus*. At 100 °C. this protein's unfolding is characterized by a value of ΔG° of over +15 kcal/mol, and at 60 °C in 6 M guanidine ΔG° is estimated to be larger than +5 kcal/mol. The extrapolated stabilities in 0 M guanidine at 60 and 100 °C do not differ much, and it is not clear where the temperature of maximum stability lies. The native protein contains an FeS_4 cluster (in the cited study, Fe was replaced by Zn), and it is possible that this affects the unfolding thermodynamics in an unexpected manner. However, the exceptional stability is not shared by rubredoxin from a mesophilic organism, *Clostridium pasteurianum*; comparison of crystal structures has not shown an obvious, qualitative difference between the rubredoxins from these two organisms.[35]

This is not a unique case, as *many* components of *P. furiosus* must possess a stability that appears exceptional when compared with similar components from mesophilic organisms. Protein engineering studies have produced numerous exceptionally stable mutant proteins, by very limited number of amino acid replacements;

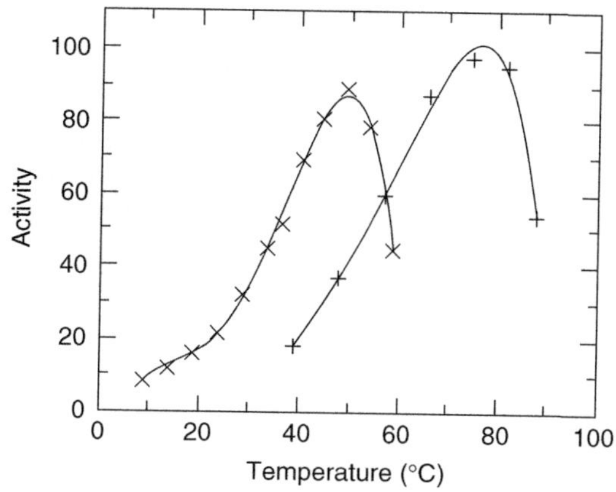

FIGURE 17.15 Relative activity–temperature profiles of *Escherichia coli* (×) and *Thermus thermophilus* (+) 3-isopropylmalate dehydrogenase (IPMDH). From Zavodszky, P., Kardos, J., Svingor, S., Petsko, P.A. *Proc. Natl. Acad. Sci. USA* **95**: 7406–7411 (1998).

these include hydrolytic enzymes that are stable under "laundry conditions," that is, in hot water in the presence of detergents.[36]

Thus, one concludes that many proteins, at least of mesophilic organisms, do not have maximal stability, and even the native state of some proteins is to be *not* folded. It is instructive to speculate about "reasons," of which one can consider several, all of which may be a factor.

- Proteins should have only limited stability because they need to be destroyed as part of normal cell function.
- Protein function (enzyme activity) is optimal when the protein is not hyperstable.

Activity generally requires some flexibility or other conformation change of the enzyme, and this can be facilitated by the presence of small voids, that is, packing defects, which would give lowered stability. This is likely to be the case for mesophilic and thermophilic variants of an enzyme (3-isopropylmalate dehydrogenase) of which the activity–temperature profiles are similar, but the entire profile of the thermophile is shifted to a higher temperature range (Fig. 17.15).[37] The protein from the thermophile is more stable than that of the mesophile, and the maximum activities are comparable, but the thermophile protein has much lower activity at moderate temperatures.

A more subtle argument for why moderate stability is preferable is related to what has been learned about the kinetics of protein folding (see Chapter 20).

Ideally, protein folding proceeds rapidly to a unique structure. Formation of metastable kinetic intermediates is a major obstacle to rapid folding. These intermediate folds (obviously) have free energies above that of the native fold. Thus, if the native fold is only barely stable relative to the unfolded state, then fewer (or no) stable alternative folds exist.

Finally, a simple argument says that, in many situations, modest stability is *adequate* for biological function, in which case a requirement of high stability bears a "cost" of not adopting any of the large number of amino acid sequences that achieve moderately stable structures, and thereby limiting the available choices.

Against this background, the existence of natively unfolded proteins[34] should not present any great puzzle. These fold into an active conformation only in complex with their targets, so we can write the following reaction

$$\begin{aligned} p_u &\rightleftarrows p_f \\ p_f + P &\rightleftarrows p_f \equiv P' \end{aligned} \tag{17.29}$$

where the natively unfolded protein, p_u, is in equilibrium with a low proportion of the folded form, p_f, which binds very strongly to its biological target, P. The affinity of p_u for P is of course not as strong, as the binding free energy is increased (becomes less negative) by an amount equal to the standard free energy of folding p. If the equilibrium constant for the folding of p were on the order of 10^{-3}, the binding constant, K_b, would decrease by a factor of 10^3, so that a K_b of 10^9 for p_f would change to 10^6 for p_u. (The factor 10^{-3} is somewhat larger than a typical α-helix initiation constant, σ.) For the binding constant to be this large, the target, P, would then have to possess a well-organized binding site.

One sees that natively unfolded proteins can serve as ligands of medium strength for binding to other, structured macromolecules. Being unfolded when not bound to the target, they are easily removed from the system by proteolysis. Lacking the extensive scaffolding that would be needed to give them the kind of well-defined tertiary structure that supports the active site of enzymes, biosynthesis of these proteins is less "costly." As if that was not enough of an advantage, it has been suggested that recognition of a portion of yet unfolded p by the target P and formation of an initial weak complex would speed up the association, as the rest of p would then quickly fold onto the binding site.

17.13 DECOMPOSITION OF THE THERMODYNAMIC PARAMETERS FOR UNFOLDING

As soon as thermodynamic values for protein unfolding became available thanks to application of the two-state model, attempts were made to rationalize these values as sums of component terms attributable to different kinds of intra- and intermolecular interactions that contribute favorably or unfavorably to the stability of the folded state. There is a reasonable consensus as to what these contributing factors are. However, our understanding of the relation between protein structure and thermodynamics of unfolding is at best semiquantitative. The standard free energy change, ΔG_u° is a difference between large opposing contributions, and cannot be predicted with any accuracy. Neither are we very successful at predicting the effect of perturbations, that is, how ΔG_u° will change as a result of site-directed mutagenesis, nor what differences in structure determine the differences in stability between proteins from thermophiles and mesophiles.

Factors that contribute to conformational stability: The two main factors that contribute to conformational stability are intramolecular hydrogen bonds and hydrophobic interactions.

Formation of the type of *hydrogen bond* that predominates in proteins (between electrically neutral NH or OH and O or N) in a vacuum or an apolar medium is accompanied by a large release of energy ($\Delta E \sim -5\,\text{kcal/mol}$); however, the energy released by formation of such a hydrogen bond in aqueous solution is much less because hydrogen bonds to water molecules must be broken for the intramolecular hydrogen bond to form. A variety of measurements (stability of α-helices, changes in stability of proteins due to mutations) has led to an estimate of $\Delta E \sim -1\,\text{kcal/mol}$. This is a relatively small contribution, but when multiplied by the large number of intramolecular hydrogen bonds in typical globular proteins, the result is a large favorable energy contribution.

The other major contribution to stability comes from *hydrophobic interactions* between apolar amino acid side chains (see Chapter 4). Again, the free energy contribution per contact between, say, two methyl groups is small (on the order of 70 cal/(mol Å^2) of buried surface), but the very large number of these contacts makes for a large favorable contribution.

In addition to these very common interactions, hydrogen bonds between pairs of positively/negatively charged groups (say, an interaction between side chains of arginine and glutamic acid) occur less frequently. These interactions are called *salt bridges*. As for hydrogen bond formation, the *net* energy, counting the need to strip some water molecules from the charged groups before a salt bridge is formed, make these not a major contributor to protein stability. (However, there is indication that (some) proteins from hyperthermophiles owe their enhanced stability to favorable interactions between charged groups, especially salt bridges, not present in the corresponding mesophilic proteins. See also Section 16.8.)

A factor that tends sometimes to be overlooked is the attractive intramolecular *van der Waals energy* (see Eq. 3.24 and surrounding text). If the hydrophobic interactions between side chains are modeled on the basis of free energy of transfer of side chains from water to a nonpolar solvent (à la Nozaki and Tanford; see Section 17.8), then the van der Waals interactions in the folded state are subsumed in the free energy of transfer. However, as an alternative, one can decompose the transfer of a side chain from water to the interior of a protein into two steps, transfer from water to vacuum, followed by transfer from vacuum into the protein interior. The first step can then be modeled with the values obtained by Wolfenden and coworkers (Table 17.3), and the second step by using the Lennard-Jones energy of Eq. 3.24 and the Coulomb energy. The Coulomb energy term ensures that nearly all buried polar side chains are hydrogen-bonded. The Lennard-Jones term produces a favorable energy component (unless the molecule is "over-packed"), which varies very rapidly (as r^{-6}) with interatomic distance, r; again, the individual contributions are small (Fig. 3.7), but the number of terms is large.

The advantages of "good packing" apply not only to van der Waals interactions but also to intramolecular hydrogen bonds, whose energy depends on distances and angles between the hydrogen bond forming groups. The more successful protein design programs try to account accurately for the effect of such details on the conformational free energy.

Factors that contribute to protein conformational instability. The main contribution opposing folding is the *conformational entropy*. The folded state of a protein is a much smaller ensemble of alternative conformations (separated by energy barriers) than the unfolded state. Also, the flexibility of each conformation in the folded ensemble, as expressed by the range of deformation permitted within a given change of the energy, is much less than that of a typical unfolded conformation. It is difficult to estimate the contribution quantitatively, in part because of uncertainty about the distribution of conformations in the unfolded state.

A rough estimate of the average conformational entropy contribution to the free energy of folding, Δs_{C} is estimated to be circa -5.5 cal/(mol K) per amino acid residue.[5] This includes backbone and side-chain entropy; obviously, the contribution of an individual residue depends on amino acid type and on the residual freedom of the side chain in the folded state. In α-helical polypeptides, the side chains are relatively free and for these Δs_{C} can be attributed largely to differences in backbone entropy; with a stabilizing contribution from one hydrogen bond, Δg_{H} per residue of $-1{,}000$ cal/mol, and given that the equilibrium constant for helix growth is very close to 1, so that $\Delta g_{\mathrm{H}} - T\Delta s_{\mathrm{H}} \approx 0$, one obtains $\Delta s_{\mathrm{H}} \approx -3\,\mathrm{cal}/(\mathrm{mol}\cdot\mathrm{K})$, a reasonable estimate equal to a little over half that for an entire residue in a folded protein.

Contributions to ΔH°, ΔS°, and ΔC_P°. Our best estimate is that hydrogen bonds make, on average, a contribution of -1 kcal/mol to the enthalpy of folding, and small contributions to entropy and specific heat. The favorable contribution from hydrophobic interactions is decomposed in a favorable entropy and smaller unfavorable enthalpy contribution. Loss of conformational freedom makes a large unfavorable contribution to the entropy. Finally, hydrophobic interactions make the dominant contribution to the specific heat, ΔC_P° and thus to the correspondingly strong temperature dependence of ΔH° and ΔS°.

NOTES

1. Lumry, R., Biltonen, R., Brandts, J.F. Validity of the "two-state" hypothesis for conformational transitions of proteins. *Biopolymers* **4**: 917–944 (1966).
2. Hermans, J., Scheraga, H.A. Structural studies of ribonuclease. V. Reversible change of configuration. *J. Am. Chem. Soc.* **83**: 3283–3292 (1961).
3. The baseline value, $c_{P,0}$ must take into account the extrapolated values of c_P for folded and unfolded protein and the proportion of each that is present, that is, $c_{P,0} = f_{\mathrm{f}} c_{P,f} + f_{\mathrm{u}} c_{P,u}$, where f_{u} and estimated with $f_{\mathrm{u}}(T) = 1 - f_{\mathrm{f}}(T) \approx \int_{\mathrm{low}\,T}^{T} dT (C_P - C_{P,0})/\Delta H_{\mathrm{u}}(T_{\mathrm{m}})$.
4. Privalov, P.L., Khechinashvili, N.N. A thermodynamic approach to the problem of stabilization of globular protein structure. *J. Mol. Biol.* **86**: 665–684 (1974).

5. Brandts, J.F., Hunt, L. The thermodynamics of protein denaturation. 3. The denaturation of ribonuclease in water and in aqueous urea and aqueous ethanol mixtures. *J. Am. Chem. Soc.* **89**: 4826–4838 (1967).
6. The unit "cal/mol/K" is sometimes called the *entropy unit*, abbreviated as "eu."
7. The analysis given here differs slightly from that in the original paper; the curves in Fig. 17.7 are not those of Fig. 2 of that paper.
8. Stelea, S.D., Pancoska, P., Benight, A.S., Keiderling, T.A. Thermal unfolding of ribonuclease A in phosphate at neutral pH: deviations from the two-state model. *Protein Sci.* **10**: 970–978 (2001).
9. Chen, B., Schellman, J.A. Low temperature unfolding of a mutant of phage T4 lysozyme. 1. Equilibrium studies. *Biochemistry* **28**: 685–691 (1989).
10. Nelson, C.J., LaConte, M.J., Bowler, B.E. Direct detection of heat and cold denaturation for partial unfolding of a protein. *J. Am. Chem. Soc.* **123**: 7453–7454 (2001).
11. Urea was used much in early studies, but has been replaced by GuCl for technical reasons, as in the presence of water urea slowly forms products that react chemically with proteins.
12. Tanford, C., Kawahara, K., Lapanje, S. Proteins in 6 M guanidine hydrochloride: demonstration of random coil behavior. *J. Biol. Chem.* **241**: 1921–1923 (1966).
13. Tanford, C. *Advan. Protein Chem.* **24**: 1–95 (1970).
14. Greene, R.F., Pace, C.N. Urea and guanidine hydrochloride denaturation of ribonuclease, lysozyme, α-chymotrypsin, and β-lactoglobulin. *J. Biol. Chem.* **249**: 5388–5393 (1974).
15. Davis-Searles, P.R., Saunders, A.J., Erie, D.A., Winzor, D.J., Pielak, G.J. Interpreting the effects of small uncharged solutes on protein-folding equilbria. *Annu. Rev. Biophys. Biomol. Struct.* **30**: 271–306 (2001).
16. Lin, T.Y., Timasheff, S.N. Why do some organisms use a urea-methylamine mixture as osmolyte? Thermodynamic compensation of urea and trimethylamine N-oxide interactions with protein. *Biochemistry* **33**: 12695–12701 (1994).
17. (a) Timasheff, S.N. Protein–solvent interactions and protein conformation. *Acc. Chem. Res.* **3**: 62–68 (1970). (b) Lee, J.C., Timasheff, S.N. Partial specific volumes and interactions with solvent components of proteins in guanidine hydrochloride. *Biochemistry* **13**: 26–49 (1979).
18. Nozaki, Y., Tanford, C. The solubility of amino acids and related compounds in aqueous urea solutions. *J. Biol. Chem.* **238**: 4074–4081 (1963); The solubility of amino acids, diglycine, and triglycine in aqueous guanidine hydrochloride solutions. *J. Biol. Chem.* **245**: 1648–1652 (1970); The solubility of amino acids and two glycine peptides in aqueous ethanol and dioxane Solutions. *J. Biol. Chem.* **246**: 2211–2217 (1971); The solubility of amino acids and related compounds in aqueous ethylene glycol solutions. *J. Biol. Chem.* **240**: 3568–3573 (1965).
19. Wolfenden, R., Andersson, L., Cullis, P.M., Southgate, C.C.B. Affinities of amino acid side chains for solvent water. *Biochemistry* **20**: 849–855 (1981).
20. Hermans, J. The effect of protein denaturants on the stability of the alpha helix. *J. Am. Chem. Soc.* **88**: 2418–2422 (1966).
21. Liu, Y., Bolen, D.W. The peptide backbone plays a dominant role in protein stabilization by naturally occurring osmolytes. *Biochemistry* **34**: 12884–12891 (1995).
22. According to the steady-state approximation, cf. Chapter 19.

23. Bai, Y., Milne, J.S., Mayne, L., Englander, S.W. Primary structure effects on peptide group hydrogen exchange. *Proteins: Struct. Funct. Genet.* **17**: 75–86 (1993).
24. If the actual motion is harmonic (the potential is quadratic and deviations oscillate as $\cos \omega t$), the principal components are identical to the normal mode vectors.
25. (a) Amadei, A., Linssen, A.B.M., Berendsen, H.J.C. Essential dynamics of proteins. *Proteins: Struct. Funct. Genet.* **17**: 412–425 (1993). (b) de Groot, B.L., Hayward, S., van Aalten, D.M.F., Amadei, A., Berendsen, H.J.C. Domain motions in bacteriophage T4 lysozyme: a comparison between molecular dynamics and crystallographic data. *Proteins: Struct. Funct. Genet.* **31**: 116–127 (1998).
26. Schlick, T., Skeel, R.D., Brünger, A.T., Kalé, L.V., Board, J.A., Hermans, J., Schulten, K. Algorithmic challenges in computational molecular biophysics. *J. Comput. Phys.* **151**: 9–48 (1999).
27. Lovell, S.C., Davis, I.W., Arendall, W.B., de Bakker, P.I.W., Word, J.M., Prisant, M.G., Richardson, J.S., Richardson, D.C. Structure validation by C^{α} geometry: φ, ψ and C^{β} deviation. *Proteins: Struct. Funct. Genet.* **50**: 437–450 (2003).
28. Tanford, C., Kawahara, K., Lapanje, S. Proteins in 6 M guanidine hydrochloride: Demonstration of random coil behavior. *J. Biol. Chem.* **241**: 1921–1923 (1966). The hydrodynamic radius scales as $N^{0.67}$; compare this with the expected behavior of the end-to-end distance in a self-avoiding random walk model, which scales as $N^{0.59}$ (Eq. 15.10).
29. At temperatures below 200 K, proteins enter a "glassy" state, in which simple harmonic vibrations predominate, and large-scale collective motions involving transitions over energy barriers are absent. The low temperature dynamic behavior of individual protein molecules is reminiscent of the changes in viscosity and other properties of liquids when they form a glass. No biological relevance has been attributed to this state. See Ringe, D., Petsko, G. A. The "glass transition" in protein dynamics: what it is, why it occurs, and how to exploit it. *Biophys. Chem*: **105**: 667–680 (2003).
30. (a) Dolgikh, D.A, Gilmanshin, R.I., Brazhnikov, E.V., Bychkova, V.E., Semisotnov, G.V., Venyaminov, S.Y., Ptitsyn, O.B. α-Lactalbumin: Compact state with fluctuating tertiary structure? *FEBS Letters* **136**: 311–315 (1981). (b) Pfeil, W., Bychkova, G.V., Ptitsyn, O.B. Physical nature of the phase transtion in globular proteins. Calorimetric study of α-lactalbumin. *FEBS Letters* **198**: 287–291 (1986). (c) Ptitsyn, O.B. Molten globule and protein folding. *Advances in Protein Chemistry*: 83–229 (1995). (d) Finkelstein, A.V., Ptitsyn, O.B. *Protein Physics*. Amsterdam: Academic Press - Elsevier Science, 2002.
31. (a) Ohgushi, M., Wada, A. 'Molten-globule state': a compact form of gobular proteins with mobile side chains. *FEBS Letters* **164**: 21–24 (1983). (b) Kuroda, Y., Ohgushi, M., Wada, A. Thermodynamic characterization of cytochrome c at low pH. Observation of the molten globule state and of the cold denaturation process. *J. Mol. Biol.* **223**: 1139–1153 (1992).
32. The intensity of NOE cross peaks varies as $1/r^6$, and NOE peaks are observed only for pairs of nuclei at quite short distances, r. In a well-ordered structure, a relatively small set of pairs of nuclei is close enough to give measurable NOE peaks. In a disordered structure many more pairs will be occasionally close enough, but only for a short time, and the resulting many weak NOE peaks will not exceed the noise level.
33. See also Pande, V.J., Rokhsar, D.S. Is the molten globule a third phase of proteins? *Proc. Natl. Acad. Sci. USA* **95**: 1490–1494 (1998).

34. Radivojac, P., Iakoucheva, L.M., Oldfield, C.J., Obradovic, Z., Uversky, V.N, Dunker, A.K. Intrinsic disorder and functional proteomics. *Biophys. J.* **92**: 1439–1456 (2007).

35. Day, M.W., Hsu, B.T., Joshua-Tor, L., Park, J.-B., Zhou, Z.H., Adams, M.W.W., Rees, D.C. X-ray crystal structures of the oxidized and reduced forms of the rubredoxin from the marine hyperthermophilic archaebacterium *Pyrococcus furiosus*. *Protein Sci.* **1**: 1494–1507 (1992).

36. An arbitrary example: *Detergent composition comprising an alkyl sulfate and a subtilisin variant*, US patent 5389307, "This invention relates to a detergent composition comprising an alkyl sulfate and one or more subtilisins and/or subtilisin variants, wherein said subtilisins and subtilisin variants have a glutamic acid residue in position 195 and/or an alanine residue in position 222."

37. Zavodszky, P., Kardos, J., Svingor, S., Petsko, P.A. Adjustment of conformational flexibility is a key event in the thermal adaptation of proteins. *Proc. Natl. Acad. Sci. USA* **95**: 7406–7411 (1998).

SUGGESTED READING

Fersht, A.R. *Structure and Mechanism in Protein Science*. New York: Freeman; 1999.

Englander, S.W. Hydrogen exchange: the modern legacy of Linderstrøm-Lang. *Protein Sci.* **6**: 1101–1109 (1997).

Finkelstein, A.V., Ptitsyn, O.B.*Protein Physics*. Amsterdam: Academic Press - Elsevier Science; 2002.

Timasheff, S.N. Control of protein stability and reactions by weakly interacting cosolvents: the simplicity of the complicated. *Adv. Prot. Chem.* **51**: 355–432 (1998).

Radivojac, P., Iakoucheva, L.M., Oldfield, C.J., Obradovic, Z., Uversky, V.N, Dunker, A.K. Intrinsic disorder and functional proteomics. *Biophys. J.* **92**: 1439–1456 (2007).

Uversky V.N., Oldfield C.J., Dunker A.K. Showing your ID: intrinsic disorder as an ID for recognition, regulation and cell signaling. *J. Mol. Recognit.* **18**: 343–384 (2005).

Schellman J.A. Fifty years of solvent denaturation. *Biophys. Chem.* **96**: 91–101 (2002).

Schellman J.A. Protein stability in mixed solvents: a balance of contact interactions and excluded volume. *Biophys. J.* **85**: 108–125 (2003).

18

Elasticity of Biological Materials

18.1 BACKGROUND

Numerous components of cells, organs, and tissues possess important material properties that cover a range of hardness and of elasticity. Widely found materials include cellulose and chitin, both polymeric carbohydrates. Here again proteins demonstrate their versatility, by providing a great number of different components providing strength and flexibility. However, we shall limit our discussion to a small number of examples.

An elastic material returns to its original shape when stress is applied and then released. A hard material deforms a little when stressed. Confusingly, a hard material is considered to have a large elastic stretch modulus, M_L the (normalized) ratio between strain and stress

$$M_L = \frac{\tau_L L}{\delta L} = \frac{\tau_L}{\sigma_L} \tag{18.1}$$

where $\delta L/L = \sigma_L$ is the relative extension on application of a stress τ_L (= force per unit cross section). Elasticity can also be measured as the response (shear σ) to an applied shear stress τ which deforms a rectangular cross section, measuring X by Y, into a parallelogram; this gives the shear modulus M_S

$$M_S = \frac{\tau_{S,X} Y}{\delta X} = \frac{\tau_{S,X}}{\sigma_S} \tag{18.2}$$

where the stress is applied along X, as illustrated in Fig. 18.1.

Equilibria and Kinetics of Biological Macromolecules, First Edition. Jan Hermans and Barry Lentz.

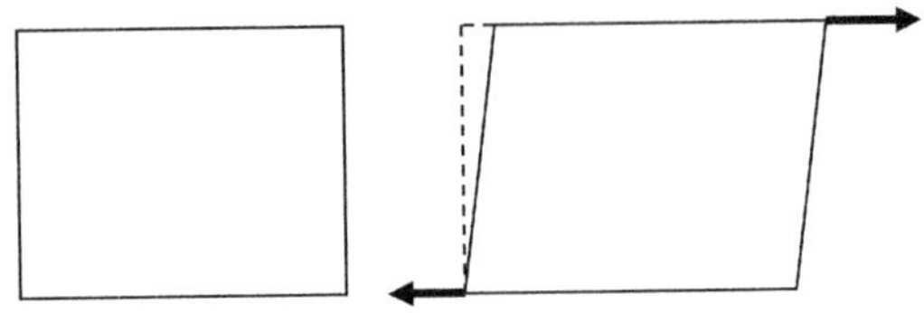

FIGURE 18.1 Deformation by an applied shear stress.

By equating the work required to achieve a given (reversible) deformation with the *elastic (free) energy*, this is found to be

$$A_{\text{shear}} = \frac{1}{2M}{\tau_S}^2 = \frac{1}{2}M\sigma^2, \tag{18.3}$$

per unit of volume.

Metals and inorganic crystals tend to be hard; deformation of organic materials is difficult if this requires the stretching of chemical bonds and easier if the stretched contacts are noncovalent. However, noncovalent contacts also break more readily, and successful elastic structures combine a high breaking strength with a low elastic modulus. Structural elements consisting of proteins, whether easily deformed or not, owe high breaking strength when fully stretched to the presence of uninterrupted chains of covalent bonds (peptide bonds and side-chain crosslinks). These biological materials are elastic if in the absence of stress the chains are *not* fully extended; furthermore, elastic behavior requires that the deformation resulting from application of stress is reversed on removal of the stress.

Biophysical studies have identified two mechanisms by which biological systems achieve significant elasticity, based, respectively, on the elastic properties of polymer networks, and on transitions between folded (structured) and unfolded (unstructured) molecular states. These will be discussed in turn.

18.2 RUBBER-LIKE ELASTICITY OF POLYMER NETWORKS

Polymeric rubber-like materials contain coiled polymer molecules, and in a useful rubber these molecules are linked together into an infinite network as a result of the presence of crosslinks (Fig. 18.2). If the sample has been crosslinked in the absence of strain, the chains will obey random chain statistics. If such a material is stressed, it is easily deformed because the chains locally slip past each other, and are restrained only by the occasional crosslinks.

One of the triumphs of the random-coil model of polymer chains introduced in Chapter 15 is that it allows an understanding of the phenomenon of rubber elasticity. Rubber-like materials are invariably made up of chain molecules in the random-coil state. (Whenever a polymeric material forms a crystalline state, as evidenced by the complexity of its x-ray diffraction pattern, rubber elasticity is absent. Such is the case for a polymer-like cellulose, which is naturally produced in crystalline form,

FIGURE 18.2 Network of crosslinked polymer chains; crosslinks are indicated by stars.

but also rubber itself can assume a crystalline state at low temperature, at which the typical high elasticity disappears.) Vulcanization introduces covalent crosslinks between rubber molecules; this produces a stronger material because it prevents relative slippage of the chains. The number of crosslinks must be sufficiently high that the molecules are linked into a continuous network; the material becomes harder the more crosslinks are introduced. Thus, if the polymer chains are very long, then the number of crosslinks per unit of volume needed to link the chains to a network can be quite small, and this will then produce an easily deformed elastic material, that is, a material with a low modulus of elasticity. (See Fig. 18.2.)

A striking property of rubber-like materials is that the stress for a given strain rises, and, conversely, the strain obtained for a given stress decreases, *in proportion to the absolute temperature*, a dependence that is readily observed for real rubber samples. Kuhn proposed that the elastic behavior could be understood by assuming that the equilibrium distribution of chain dimensions is perturbed by the deformation and that this perturbation is accompanied by a decrease of the chain entropy (and thus an increase of the free energy), which is in contrast to the elastic behavior of ordinary solids, where the deformation is accompanied by an increase of the energy. The assumption qualitatively explains the observation that rubber-like materials become more rigid when the temperature is raised, while ordinary materials become softer.

18.3 THEORY OF RUBBER ELASTICITY

Theories of elasticity suppose that every point in a sheared sample moves (to the right in Fig. 18.3) in proportion to its distance from the reference surface (here, bottom plane). The theory of rubber elasticity supposes that in a rubber-like material this remains true for the crosslinks between chains, but that the chain segments between crosslinks can move freely. This means that a chain with end-to-end vector $\mathbf{R} = (R_x, R_y, R_z)$ before stress is applied has end-to-end vector $\mathbf{R}' = (R_x + \sigma R_y, R_y, R_z)$ after the deformation of the material.

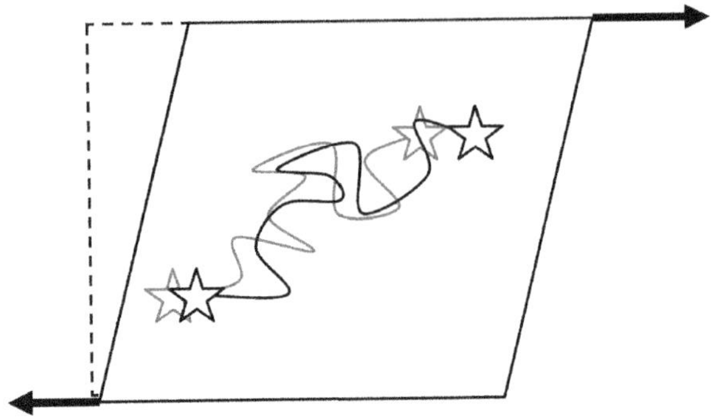

FIGURE 18.3 Deformation of a polymer chain segment between two crosslinks in a sheared sample. The crosslinks follow the deformation of the sample, but the chain conformation remains relatively free.

The free energy change needed for deformation of this chain is then based on the change in chain entropy as the probability of the end-to-end distance changes from $P(\mathbf{R})$ to $P'(\mathbf{R})$,

$$\begin{aligned} P'(\mathbf{R})/P(\mathbf{R}) &= \exp(-\alpha R'^2)/\exp(-\alpha R^2) \\ &= \exp[-\alpha(R'^2 - R^2)] = \exp[-\alpha(2R_x\sigma R_y + \sigma^2 R_y{}^2)] \\ &= \exp(-2\alpha R_x \sigma R_y)\exp(-\alpha\sigma^2 R_y{}^2). \end{aligned} \tag{18.4}$$

Here we have used the probability distribution of Eq. 15.3 and abbreviated $\alpha = 1/(2{<}R^2{>})$. We note that equal number of chains have a given y-coordinate R_y, as have the opposite value $-R_y$. The joint probability of changing the end-to-end distance of one of each is then

$$\frac{P_1'(\mathbf{R})P_2'(\mathbf{R})}{P_1(\mathbf{R})P_2(\mathbf{R})} = \left[\exp\left(-\frac{1}{2}\sigma^2 R_y{}^2/\langle R^2\rangle\right)\right]^2. \tag{18.5}$$

The joint probability is decreased, as the exponent is negative. The relative change in the joint probability distribution computed over all n_c chains in a sample becomes

$$\frac{P'_{\text{all}}}{P_{\text{all}}} = \prod_{i=1}^{n_c} \frac{P'(\mathbf{R}_i)}{P(\mathbf{R}_i)} = \prod_{i=1}^{n_c}\left[\exp\left(-\frac{3\sigma^2 R_{y,i}{}^2}{2\langle R^2\rangle}\right)\right]^2 = \exp\left(-\frac{3\sigma^2\sum_{i=1}^{nc} R_{y,i}{}^2}{2\langle R^2\rangle}\right),$$

$$\frac{P'_{\text{all}}}{P_{\text{all}}} = \exp\left(-\frac{3\sigma^2 n_c\langle R_y{}^2\rangle}{2\langle R^2\rangle}\right). \tag{18.6}$$

The mean square of the y-component of the end-to-end vector is simply one-third of the mean square of the vector itself,[1] so that

$$\frac{P'_{\text{all}}}{P_{\text{all}}} = \exp\left(-\frac{\sigma^2 n_c \langle R^2 \rangle}{2\langle R^2 \rangle}\right) = \exp\left(-\frac{1}{2}\sigma^2 n_c\right). \quad (18.7)$$

Finally, the corresponding free energy change is

$$\Delta A_{\text{shear}} = k_B T \ln(P/P') = \frac{1}{2} k_B T \sigma^2 n_c, \quad (18.8)$$

and the modulus M follows with Eq. 18.3

$$M = n_c k_B T. \quad (18.9)$$

Here n_c represents the number of "active" cross links per unit of volume. (Roughly, links over and above the number required to simply link all the molecules together without forming a network.)

18.4 RUBBER-LIKE ELASTICITY OF ELASTIN

Elastin is a nearly perfectly elastic material, responsible for, amongst others, the elastic properties of skin. In a classic article Hoeve and Flory demonstrated that elastin provides tissues with rubber-like elasticity; these tissues are "reinforced" with a network of collagen fibers, and once this is pulled taut, further deformation is very difficult.[2]

A great deal has been learned about elastin that was unknown at the time of that initial study.[3] Elastin consists of a protein that is predominantly a series of many pentapeptide repeats, with sequence (Gly–Val–Gly–Val–Pro–). Recent studies have been done with pure elastin and with a polymer, $(\text{GVGVP})_n$ consisting entirely of these pentapeptide sequences. Because each pentapeptide repeat contains two glycine residues, the backbone does not have a strong tendency to form a regular secondary structure. Glycine residues are quite soluble in water; on the other hand, the presence of three very hydrophobic side chains suggests a tendency to insolubility.

Individual elastin molecules have been studied with force extension measurements. (For technical details of this measurement see Section 18.6.) Measurements by Urry et al. show linear force-extension of $(\text{GVGVP})_{251}$ up to a length of 500 nm, after which the force rises more rapidly as full extension is reached. The strain is quickly recovered when the stress is removed. The persistence length is estimated to be around 0.5 nm. (See Fig. 18.4)

A remarkable observation is that solutions of $(\text{GVGVP})_n$ are stable at low temperature, but form a gel-like precipitate containing 60% water when the temperature is raised above around 30 °C. This apparently inverse transition is explicable on

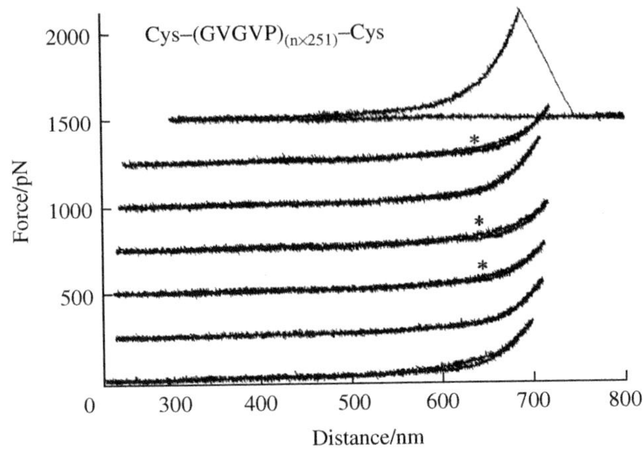

FIGURE 18.4 Single-chain force-extension curves for Cys–$(GVGVP)_{n\times 251}$–Cys at a temperature below the onset temperature for hydrophobic folding and assembly for this composition. The initial trace appears at the lowest position on the graph and subsequent traces of the same chain, without intervening detachment, are displaced 250 pN vertically. The second and fifth curves exhibit perfect reversibility. The curves marked by an asterisk, and the initial curve, followed a period of at least 30 s in the relaxed state, where one possibility is that some hydrophobic back folding occurs. From Urry, D.W., Hugel, T., Seitz, M., Gaub, H.E., Sheiba, L., Dea, J., Xu, J., Pa, T. *Phil. Trans. R. Soc. Lond. B* **357**: 169–184 (2002).

the basis of thermodynamics of interactions between hydrophobic groups discussed earlier in connection with unfolding equilibria of proteins. Specifically, for globular proteins there is a temperature of maximum stability, because of the large negative ΔC_P of hydrophobic interactions, and at low temperatures cold denaturation may be observed (or is implied by extrapolation; cf. Section 17.6); the process of formation of an elastin gel is analogous because the stability of the gel is due to hydrophobic contacts. (The higher transition temperature for elastin gelation may be due to the absence of stabilization by hydrogen bonds.) The elastin molecules in the gel phase can be crosslinked, and the resulting network is an ideal elastomer, which will recover its original length following deformation under stress.

Thus, Hoeve and Flory's original conclusion that elastin displays rubber-like elasticity, is fully confirmed by more recent studies. Conformational preferences of the relaxed elastin chain and their relation to the elastic properties, if any, remain a matter of conjecture where one faces the general problem of trying to establish conformational preferences of unfolded protein molecules. (See Section 17.10.)

18.5 TITIN AND TENASCIN: ELASTICITY BASED ON TRANSITIONS BETWEEN CONFORMATION STATES

Another possible elastic mechanism relies on conversion of a folded protein to an unfolded chain, by pulling in opposite directions on the two chain ends. The modulus of the folded structure is relatively high because of the presence of intramolecular hydrogen bonds and other noncovalent interactions. The free energy of the folded molecule does not change much as a result of the application of a stretching force, F (as the amount of stretch is small); however, the (root-mean-square) end-to-end distance of the unfolded chain is greater than the (fixed) distance, d_{0N} in the folded state and thus in the presence of the applied force the free energy of the unfolded chain is lower

$$\Delta A_{FU}(F) = \Delta A_{FU}(0) - F\left(\sqrt{\langle R_{0U}^2 \rangle} - d_{0N}\right). \tag{18.10}$$

When this becomes negative as the stress exceeds a threshold, the molecule unfolds. Any *further* extension will then be as for a rubber.

When a material containing polymer molecules is stretched, one observes the superposition of many molecular stress–strain curves; thus, models of the response of individual molecules to stress tended to remain speculative until the advent of single-molecule force extension experiments (see the following section). Experiments with single molecules of elastic components of muscle (tenascin) and extracellular matrix (titin) have established that when stressed these molecules unfold and stretch, and when the stress is removed, they refold and become short again.

The folded tenascin molecule is composed of a series of small domains connected by short segments of relatively extended chain. When the molecule is slowly stretched, the force rises slowly, then more rapidly as the slack in the connecting segments is taken up; finally, when the force exceeds a first limit, one domain unfolds, the force drops abruptly and then rises again as the slack in the newly unfolded chain segment is taken up, until a second domain unfolds, etc. until all domains have become unfolded, after which further extension leads to rupture. The unfolding force is greater if the chain is pulled faster; when the strain is released, the molecule *slowly* resumes its original length.[4,5] (See Fig. 18.5.)

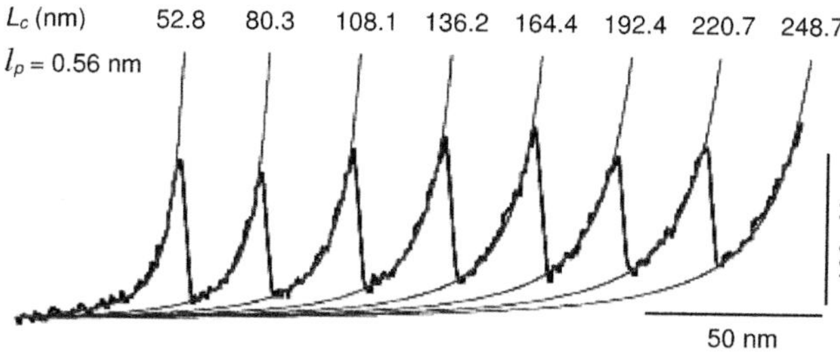

FIGURE 18.5 Force *versus* length during extension of a single tenascin molecule.[4,5] Each curve is a fit to a different ascending portion according to Eq. 18.11, with a different value of the contour length L_c but with the same value of the persistence length l_p.[5] From Fisher, T.E., Oberhauser, A.F., Carrion-Vazquez, M., Marszalek, P.M., Fernandez, J.M. *TIBS* **24**: 379–384 (1999).

The behavior observed in these experiments has been modeled on the basis of a statistical worm-like chain model of the unfolded chain, for which the force for a given end-to-end distance, R_{0x}, is given by the following empirical relationship

$$F = \frac{k_B T}{l_p}\left\{\frac{R_{0x}}{L_c} + \frac{1}{4}\left(1 - \frac{R_{0x}}{L_c}\right)^{-2} - \frac{1}{4}\right\}. \tag{18.11}$$

(This model can be applied also to the stress–strain curve for molecules of elastin in Fig. 18.4.) Here, L_c is the contour length (see Chapter 15), which grows by an increment equal to 28 nm for each additional unfolded domain and l_p is the persistence length for which a value of 0.56 nm was used. The model gives an essentially perfect description of the segments of stress–strain curves observed between unfolding events.

Being made up of a string of small folded domains that unfold reversibly under applied stress, tenascin molecules exert a force of *medium* magnitude when stretched to many times their resting length (in which the domains are folded). The molecules are less stiff than a molecule that remains folded, and stiffer than an unfolded chain.

Furthermore, apart from the sawtooth behavior as successive domains unfold, which averages out in a sample containing many molecules, the *force is remarkably constant over a wide range of extension.*

The behavior is further affected by the rate of refolding and unfolding of the domains, but this aspect will not be discussed here.

18.6 SINGLE-MOLECULE FORCE-EXTENSION MEASUREMENT

To measure the force-extension curve of a molecule, one end must be attached to a fixed surface, and the other to a surface that can be moved relative to the first. In addition, the force on the moving surface must be measurable. Titin and tenascin molecules were found to attach themselves spontaneously to surfaces as different as gold and latex.

One technique uses the atomic force microscope (AFM). AFM experiments use as a probe a very fine tip attached to an arm that also functions as a light spring. When the tip touches a surface, the spring is slightly bent, and the curvature is measured (from a change in the direction of reflected light; Fig. 18.6). In a typical AFM experiment a surface is scanned with some force applied to the spring, so that the tip always lightly touches the surface, the experiment records the vertical position of the tip and the results are presented as a map of hills and valleys. In order to measure force extension curves, the experiment is modified by having the tip at some distance from the surface "search" for the free end of a molecule that is attached to the surface at its other end.[5] Once such a free end attaches itself spontaneously, the AFM tip can be moved away from the surface, and the force it experiences can be recorded from the curvature of the springy arm.

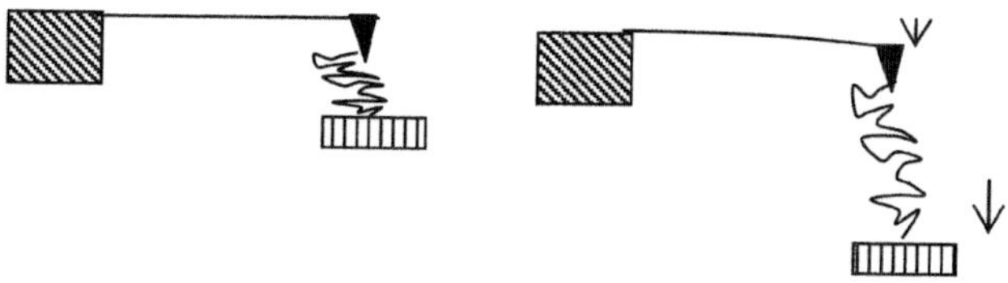

FIGURE 18.6 Use of AFM apparatus to measure a molecular force-extension curve. On the right, the chain has been stretched and the cantilever arm is curved because of the stretching force.

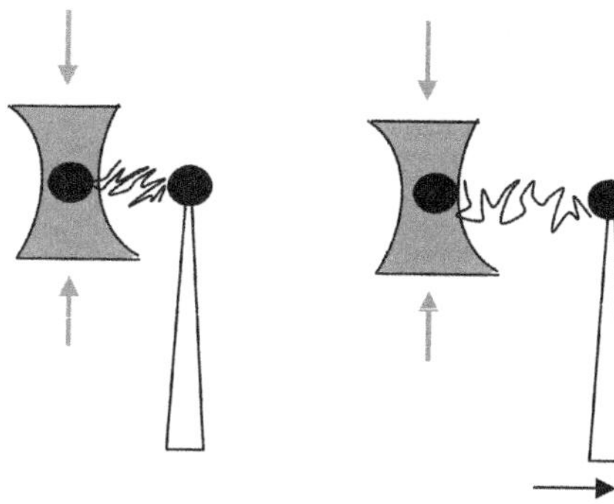

FIGURE 18.7 Use of optical tweezers to measure a molecular force-extension curve. On the right, the chain has been stretched and the trapped bead has moved to the right of center of the optical trap because of the stretching force.

The other technique focuses light from a laser to create an optical trap or optical tweezers for a latex bead (Fig. 18.7).[6,7] The trap controls the position of the bead, but not completely, as when the bead experiences a force, it moves away from the original equilibrium position in the trap and the relative displacement is a measure of the force. One end of a titin or tenascin molecule is attached to the trapped bead, and the other end is attached to another bead, which is held in place with the point of a very fine pipet.

NOTES

1. $\langle R_x{}^2 \rangle = \langle R_y{}^2 \rangle = \langle R_z{}^2 \rangle$ and $R^2 = R_x{}^2 + R_y{}^2 + R_z{}^2$.

2. Hoeve, C.A., Flory, P.J. The elastic properties of elastin. *J. Am. Chem. Soc.* **78**: 6523–6526 (1956).
3. Reviewed by Urry, D.W., Hugel, T., Seitz, M., Gaub, H.E., Sheiba, L., Dea, J., Xu, J., Pa, T. Elastin: A representative ideal protein elastomer. *Phil. Trans. R. Soc. Lond. B* **357**: 169–184 (2002).
4. Oberhauser, A.F., Marszalek, P.E., Erickson, H.P., Fernandez, J.M. The molecular elasticity of the extracellular matrix protein tenascin. *Nature* **393**: 181–185 (1998).

5. Fisher, T.E., Oberhauser, A.F., Carrion-Vazquez, M., Marszalek, P.M., Fernandez, J.M. The study of protein mechanics with the atomic force microscope. *TIBS* **24**: 379–384 (1999).
6. Simmons, R.M., Finer, J.T., Chu, S., Spudich, J.A. Quantitative measurements of force and displacement using an optical trap. *Biophys. J.* **70**: 1813–1822 (1996).
7. Optical tweezers are capable of manipulating nanometer and micrometer-sized dielectric particles by exerting extremely small forces via a highly focused laser beam. The beam is typically focused by sending it through a microscope objective. The narrowest point of the focused beam, known as the beam waist, contains a very strong electric field gradient. It turns out that dielectric particles are attracted along the gradient to the region of strongest electric field, which is the center of the beam. (from Wikipedia)

Part 5

Kinetics and Irreversible Processes

We have devoted the greater part of this book to a discussion of equilibria because we believe that material to be an essential part of understanding biology in terms of physics; however, the understanding so reached is also completely inadequate! Any living organism is made up of components that are continuously undergoing chemical and physical transformations, and hence is not nearly in equilibrium. By virtue of the Second Law, these transformations all *decrease* the free energy and most are irreversible; (with a few exceptions) the entire biological circus is kept going by the conversion of the sun's radiant energy to chemical free energy by photosynthetic organisms as the one input of free energy.

In biology, time is of the essence. The rates at which the transformations occur determine each organism's viability; thus, a discussion of kinetics of irreversible processes is integral to reaching an understanding of the physics of biological macromolecules. But it remains also the smaller part, as it typically makes little sense to study the kinetics of a process if the equilibrium has not first been measured and analyzed.

We begin the discussion of kinetic theory in Chapter 19 with a variety of approaches to measurement of reaction kinetics and analysis in terms of traditional chemical kinetics models. This basic chapter is followed by Chapter 20 in which these ideas are applied to kinetic measurement and theory of protein folding. Then, in Chapter 21 we introduce the theory of kinetic, irreversible processes in terms of stochastic models.

Equilibria and Kinetics of Biological Macromolecules, First Edition. Jan Hermans and Barry Lentz.

19

Kinetics

19.1 INTRODUCTION

In this chapter, we address a variety of practical and theoretical aspects of kinetics, basing this entirely on the concept of the *chemical reaction equation*. In its simplest form this is written as

$$A \xrightarrow{k_{AB}} B,$$

and implies a *reaction rate*, which is the rate of change of the concentration of reactant A, c_A, or [A]. This simple case corresponds to a rate given by the following equation

$$d[A]/dt = -k_{AB}[A].$$

Here, k_{AB} (or $k_{A\rightarrow B}$) is the *rate constant.*

We first discuss several methods of measuring fast reaction rates. Next, we discuss the behavior of a number of specific model reactions, and introduce the concepts of relaxation time, steady state model, and nucleation and growth mechanism. Following this, we introduce both the concept of the transition state, which is then applied to enzymatic catalysis and the design of enzyme inhibitors, and the contrasting model of the diffusion-controlled reaction. Finally, we show how the transition state concept can be applied to estimate reaction rates (including enzyme catalytic rates) from simulations.

Equilibria and Kinetics of Biological Macromolecules, First Edition. Jan Hermans and Barry Lentz.

19.2 MEASURING FAST KINETICS BY RAPID PERTURBATION

In principle, kinetic experiments are possible using the same (often spectroscopic) signals that are used in the measurements that are used to study the equilibria. In practice, a limitation is imposed by the time required for the selected measurement, with both the time constant of the spectrometer (or other measure of concentrations of reactants and products) and the time required to prepare the reactants as important factors, as the studied reaction should not make significant progress during either of these two times. Light intensity in light absorption and fluorescence experiments can be measured with photomultiplier tubes with instrument response times on the order of a nanosecond, and when these are used with an adequate stream of photons, the limiting factor becomes the delay encountered in preparing the sample in the reactive, nonequilibrium state. The following three methods have been developed in order to *rapidly* perturb the equilibrium of solutions containing biological macromolecules.

Rapid mixing experiments are widely used in kinetic studies. Here, "rapid" means that the mixing must be complete in a time shorter than the reaction time. Reactions with rates of more than 1/s require special *stopped flow equipment.* In a stopped flow experiment, two reactants, each in its own syringe, are mixed by a force exerted simultaneously on both syringe plungers (Fig. 19.1). The mixing takes place in a specially designed mixing jet, and well-designed instruments have mixing times that allow studies of reactions on the *millisecond timescale*. The mixing stops abruptly when a preset volume has been exactly dispensed into the observation chamber (e.g., spectrophotometer cell) and out into the "stop" syringe. The system can initially be quite far from equilibrium, and large changes in concentrations of different species can be produced. Stopped flow experiments can be used effectively to determine rates of ligand binding and pH or solvent-induced protein folding and unfolding.

Temperature-jump experiments can be used to study equilibria that are sensitive to temperature change, in which case a *sudden change in temperature* can trigger an observable change in concentrations of reactants and products. A simple temperature-jump setup can be achieved with a spectrophotometer cell in a jacket of circulating water, which is suddenly switched to a different temperature.

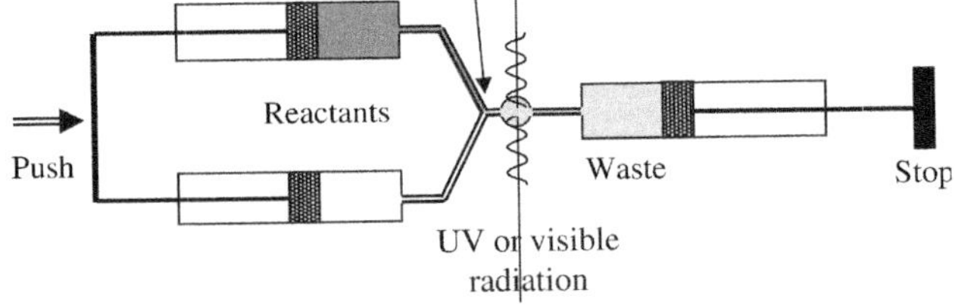

FIGURE 19.1 Scheme of a stopped flow apparatus. Outgoing radiation is measured to give absorption or fluorescence intensity.

This setup permits a large temperature change, but the time required to change the temperature of the sample is on the order of seconds.

In the 1960s, Manfred Eigen developed a fast temperature-jump experiment in which the temperature was raised on the microsecond timescale by discharge of a capacitor to pass a current through the solution. Currently, such temperature jumps can be achieved on the *nanosecond timescale* by heating a small volume of the solution by illumination with a (pulsed) infrared laser; in the cited paper, a single 10 ns, 2 mJ IR pulse at 2 μm wavelength resulted in a temperature change of 18 K in a volume of 900 nl in 20 ns.[1] The generated heat dissipates within a few milliseconds. The contents of the heated volume can be studied by standard spectroscopy, such as UV or IR absorption and fluorescence. The T-jump method represents an enormous improvement in speed over the stopped flow method, but at the price of a relatively small perturbation, always toward higher temperature; such a small perturbation will produce measurable changes in the concentrations of reactants and products *only* if the system is in an equilibrium state not far from the reaction midpoint.

Photoreactive molecules can generate nonequilibrium states from equilibrium ensembles by absorption of light. In most systems, absorption of a photon creates an excited state that returns to the ground state by radiative energy transfer, but in some compounds the excited state decays into a reactive state. The advantage of photoactivation is that (with a powerful laser) this can be accomplished *very rapidly*, so that very fast reactions of the photoactivated system can be observed (*picosecond timescale*). A disadvantage is that it is difficult to design the experiments because of the limited number of usable photoreactive molecules.

Some molecules are rendered highly reactive as a result of photoactivation and can react chemically with groups on other molecules. Application in kinetic studies is limited by the possibility of undesirable side reactions. More generally useful for kinetic studies is a class of compounds called *caged molecules*, which release less reactive species that are typical of ligands (substrates, inhibitors, cofactors) of biological macromolecules.[2] Two examples are given in Fig. 19.2. Uncaged carboxylic acids can of course serve as specific ligands, but will also cause a rapid shift to lower pH, and for either reason can be useful in bulk kinetic experiments.[3]

An example of an intrinsically photoreactive system is the *heme–CO complex* (in CO–myoglobin and CO–hemoglobin), which dissociates when illuminated. A laser can supply a flash sufficient to dissociate all molecules in the solution in

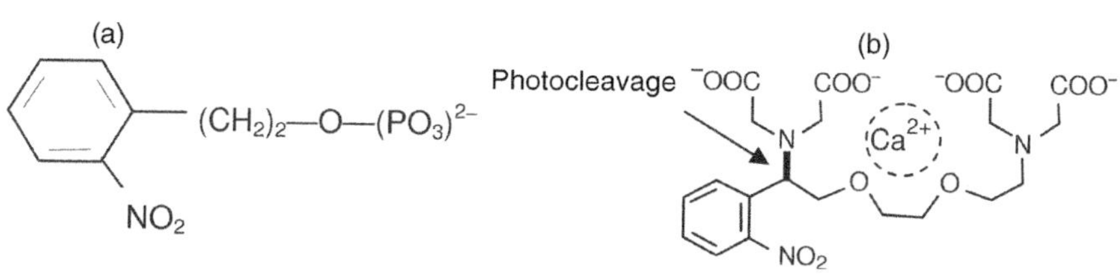

FIGURE 19.2 (a) A caged form of inorganic phosphate. (b) Caged calcium ion. From Invitrogen Co.

a matter of a picosecond. The subsequent rebinding of CO to myoglobin can be followed by measuring changes of the heme spectrum on a picosecond to nanosecond timescale, and reflects (i) differences in the state of the myoglobin and (ii) variation in the temporary disposition of the CO molecule inside the myoglobin molecule. Discussion of these results falls outside the scope of this book. We cite a seminal paper from the laboratory of Hans Frauenfelder, which investigated events subsequent to photodissociation of CO myoglobin at cryogenic temperatures.[4]

19.3 FAST RATES FROM SPECTROSCOPIC LINE SHAPE AND RELAXATION TIMES

Line widths of many types of spectra depend on the dynamics (motion and change of environment), and, as a result, accurate spectroscopy can be an important source of kinetic information. The equations are slightly complicated, and the theory is often presented in a nonintuitive manner.

Although any problem involving absorption and emission of radiation can be dealt with conclusively only by considering this in terms of quantum mechanics, the following classical approach is useful to clarify underlying principles. In its simplest form, the incoming radiation is a sine or cosine wave (fixed frequency ν_0, angular velocity $\omega_0 = 2\pi\nu_0$), which we represent with complex notation as $\exp(i\omega_0 t)$. For a variety of causes, the incoming radiation is not perfectly coherent (these causes include *energy transfer*, *molecular motion*, and *conformation change*) and this is taken into account by representing the emitted radiation as a wave of which the amplitude decays exponentially with relaxation time τ,

$$I_0 \propto \exp(-t/\tau)\exp(i\omega_0 t). \tag{19.1}$$

Each emitter is then considered to emit an irregularly spaced series of these waves.

The signal's correlation time, τ, is related to how rapidly and to what extent the incoming signal is perturbed by time-dependent changes in the emitting molecules; the slower these are, the longer the correlation time.

The detector used to measure the spectrum of incoming radiation in effect reports a Fourier transform. This means that the signal measured at a particular frequency ν (angular velocity $\omega = 2\pi\nu$; $\omega = \omega_0 + \delta\omega$) is given by

$$\begin{aligned} I(\omega) &\propto \int \exp(-t/\tau)\exp(i\omega_0 t)\exp(-i\omega t)dt \\ &= \int \exp(-t/\tau)\exp(-i\delta\omega t)dt. \end{aligned} \tag{19.2}$$

Integration (over the range $0 < t < \infty$) gives

$$I(\omega) \propto \mathrm{Re}\left(\frac{1}{1/\tau + i\delta\omega}\right) = \frac{1/\tau}{1/\tau^2 + (\delta\omega)^2}. \tag{19.3}$$

This equation determines the line width, which depends on how rapidly $I(\omega)$ drops when $\delta\omega$ changes away from 0. Substitution of

$$\delta\omega = 1/\tau, \tag{19.4}$$

gives a value of $I(\omega)$, which is lower by a factor 2 than the maximum value (for $\delta\omega = 0$); thus, the line width at half height is equal to $2/\tau$. The line shape described by Eq. 19.3 is called *Lorentzian.*

Thus, *the line width of the received signal varies as the inverse of the correlation time* τ. *Dynamic changes of the system break the continuity of the emitted radiation and thus determine the value of the correlation time.* We note that the relation between line width and dynamics does not depend on the value of ω_0, or, more generally, on the type of spectroscopy.[5] The practicality of extracting specific kinetic or dynamic information from measured line widths depends only on the frequency resolution of the detection; also, background noise can become an obstacle to accurate measurement, especially if the lines are broad.

Of interest here is perturbation of the emitted wave as a result of transitions between the two conformation states that produce signals of slightly different frequencies, ω_1 and $\omega_2 = \omega_1 + \Delta\omega$. If the correlation times are long and the lines consequently sharp, the spectrum will show a line at each of these two frequencies, in proportion to the concentrations of the two states. As the correlation time becomes shorter, the lines broaden and the signal observed at frequencies ω lying *between* ω_1 and ω_2 will contain contributions from both incoming frequencies. For still shorter correlation times, the two lines merge into a single line at intermediate ω (as is the case for the example given subsequently), and eventually the line again becomes sharp, when $1/\tau \gg \delta\omega_0$, something that is not correctly described by the simplistic model of Eq. 19.1.

It turns out that ^{1}H NMR spectroscopy is well suited to detecting the kinetics of protein conformation changes in the millisecond to microsecond range. The equations are additionally complicated by the fact that the line shapes depend also on the so-called transverse relaxation times, T_2. However, their exact form need not be reproduced here.

Oas and coworkers[6] have made excellent use of the inherent possibilities in studies of the folding of a small protein, the N-terminal domain of phage lambda repressor protein, denoted as λ_{6-85}. They showed that with a knowledge of line shapes (resonances of just two protons) of fully native and fully denatured protein (denatured in a high concentration of urea), it was possible to deduce both forward and reverse rate constants, k_f and k_u from line shapes measured at intermediate denaturant concentrations, where a mixture of folded and unfolded molecules is present (Fig. 19.3). The measurements closely reproduce the equilibrium degree of unfolding determined independently from circular dichroism spectra.

The second of the two cited papers reports use of *T_2 NMR relaxation times* to determine forward and reverse rate constants of λ_{6-85} protein at low urea concentrations (down to 0 M).[6b] These confirm the linear extrapolations of $\ln k$

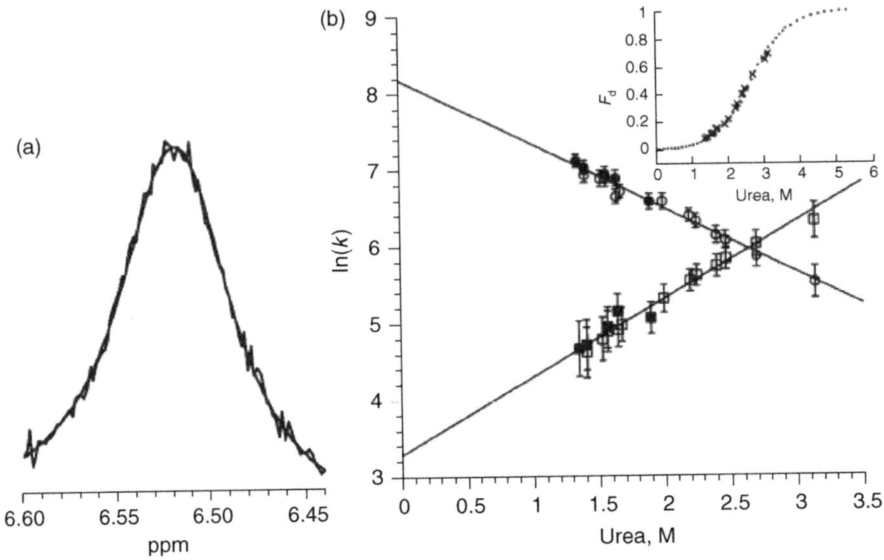

FIGURE 19.3 (a) Experimental and modeled NMR line shape for Tyr 22–δH atom of λ_{6-85} protein in 1.41 M urea. The model curve uses $k_f = 1000$ and $k_u = 136$/s; here 1 ppm corresponds to 500/s on a scale of frequency. (b) Circles: Forward and reverse rate constants for folding of λ_{6-85} protein as a function of urea concentration, determined by analysis of NMR line shapes of Tyr 22 – δH and Tyr – 22 εH. Inset shows fraction unfolded from NMR line shape (crosses) and from circular dichroism (curve). From Huang, G.S., Oas, T.G. *Proc. Natl. Acad. Sci. USA* **92**: 6878–6882 (1995).

made in Fig. 19.3, and give an extrapolated value at 0 M urea of $k_f = 4900/\text{s}$ at 37 °C. These measurements of T_2 also gave values of the rate constant for folding, $k_f = 30/\text{s}$ and the equilibrium constant for unfolding $K_{fu} = 1/163$. Both rate constants agree with the linear extrapolations of Fig. 19.3. It is impressive that the NMR-based measurement gives values of k_f and K_{fu} under conditions where these could not have been determined directly with traditional methods, and without the sudden change in conditions needed for a classical kinetic measurement (as with stopped flow). These results are discussed further in Section 20.6.

19.4 RELAXATION TIME IN A UNIMOLECULAR REACTION

We begin our discussion of various individual kinetic mechanisms by considering, as a simple system, a unimolecular reaction

$$\mathrm{A} \underset{k_-}{\overset{k_+}{\rightleftharpoons}} \mathrm{B}$$

with forward and reverse rate constants k_+ and k_-. The time dependence of the concentrations of the reactants is given by the following *master equation*

$$\frac{d[\mathrm{A}]}{dt} = -k_+[\mathrm{A}] + k_-[\mathrm{B}] = -\frac{d[\mathrm{B}]}{dt}$$

or

$$\frac{dc_\mathrm{A}}{dt} = -k_+c_\mathrm{A} + k_-c_\mathrm{B} = -\frac{dc_\mathrm{B}}{dt}$$

or

$$\frac{d\mathrm{A}}{dt} = -k_+\mathrm{A} + k_-\mathrm{B} = -\frac{d\mathrm{B}}{dt}$$

or

$$\dot{\mathrm{A}} = -k_+\mathrm{A} + k_-\mathrm{B} = -\dot{\mathrm{B}}, \tag{19.5}$$

according to a variety of commonly used notations. At equilibrium

$$\mathrm{A}_\mathrm{eq} = \frac{k_-}{k_+}\mathrm{B}_\mathrm{eq}. \tag{19.6}$$

Under other conditions, the actual concentrations will be

$$\begin{aligned} \mathrm{A} &= \mathrm{A}_\mathrm{eq} + \delta\mathrm{A} \\ \mathrm{B} &= \mathrm{B}_\mathrm{eq} + \delta\mathrm{B} \\ \delta\mathrm{A} &= -\delta\mathrm{B}, \end{aligned} \tag{19.7}$$

and substitution gives

$$\begin{aligned} \frac{d\mathrm{A}}{dt} = \frac{d\delta\mathrm{A}}{dt} &= -k_+\mathrm{A}_\mathrm{eq} + k_-\mathrm{B}_\mathrm{eq} - k_+\delta\mathrm{A} + k_-\delta\mathrm{B} = -(k_+ + k_-)\delta\mathrm{A} \\ \frac{1}{\delta\mathrm{A}}\frac{d(\delta\mathrm{A})}{dt} &= -(k_+ + k_-) \\ \delta\mathrm{A} &= \delta\mathrm{A}_{t=0} \times \exp[-(k_+ + k_-)t], \end{aligned} \tag{19.8}$$

This is called *exponential relaxation* of $\delta\mathrm{A}$ with *relaxation time* $\tau = 1/(k_+ + k_-)$. Note that when the system is far from equilibrium, k_- is negligible relative to k_+, or vice versa.

19.5 RELAXATION TIME IN A BIMOLECULAR REACTION

Next, we consider a simple bimolecular reaction

$$2\mathrm{A} \underset{k_-}{\overset{k_+}{\rightleftharpoons}} \mathrm{B}$$

with rate constants k_+ and k_-. The master equation is

$$\frac{d\mathrm{A}}{dt} = -k_+\mathrm{A}^2 + \frac{1}{2}k_-\mathrm{B} = -\frac{1}{2}\frac{d\mathrm{B}}{dt}$$

or

$$\dot{\mathrm{A}} = -k_+\mathrm{A}^2 + \frac{1}{2}k_-\mathrm{B} = -\frac{1}{2}\dot{\mathrm{B}}. \tag{19.9}$$

The equation for this simple case can be solved analytically, but more complex cases that involve multiple consecutive or parallel second-order reactions, become very difficult to solve.

Hence, we consider only a *perturbation from equilibrium*. At equilibrium

$$\mathrm{A}_{\mathrm{eq}}^2 = \frac{k_-}{2k_+}\mathrm{B}_{\mathrm{eq}}. \tag{19.10}$$

The actual concentrations will be

$$\begin{aligned} \mathrm{A} &= \mathrm{A}_{\mathrm{eq}} + \delta\mathrm{A} \\ \mathrm{B} &= \mathrm{B}_{\mathrm{eq}} + \delta\mathrm{B} \\ 2\delta\mathrm{A} &= -\delta\mathrm{B}, \end{aligned} \tag{19.11}$$

and substitution gives

$$\frac{d\mathrm{A}}{dt} = -k_+\mathrm{A}_{\mathrm{eq}}^2 + \frac{1}{2}k_-\mathrm{B}_{\mathrm{eq}} - 2k_+\mathrm{A}_{\mathrm{eq}}\delta\mathrm{A} - k_+(\delta\mathrm{A})^2 + \frac{1}{2}k_-\delta\mathrm{B}. \tag{19.12}$$

If the system is not far from equilibrium, then the term in $(\delta\mathrm{A})^2$ can be ignored

$$\frac{d\delta\mathrm{A}}{dt} = -2k_+\delta\mathrm{A} + \frac{1}{2}k_-\delta\mathrm{B} = -(2k_+\mathrm{A} + k_-)\delta\mathrm{A}, \tag{19.13}$$

$$\frac{1}{\delta\mathrm{A}}\frac{d(\delta\mathrm{A})}{dt} = -(2k_+\mathrm{A} + k_-)$$

$$\delta\mathrm{A} = \delta\mathrm{A}_{t=0} \times \exp[-(2k_+\mathrm{A} + k_-)t]. \tag{19.14}$$

This is an exponential relaxation of $\delta\mathrm{A}$ with relaxation time $\tau = 1/(2k_+\mathrm{A} + k_-)$. Such a perturbation treatment can be applied to any reaction, or to any individual step in a complex reaction scheme, and is useful as long as the concentrations are not far from equilibrium.

19.6 MULTIPLE REACTIONS

If there is more than one reaction, either in parallel, or consecutive, or both, connecting n species, one begins with writing all corresponding differential equations of the type of Eq. 19.13. This will give, as the master equations, n linear differential equations in the n unknowns, which are the perturbations from equilibrium of the concentrations, δc_i

$$\frac{d\delta c_i}{dt} = \sum_j k_{ij}\delta c_j, \quad i = 1, 2, \ldots n. \tag{19.15}$$

Such a system of simultaneous linear differential equations can be solved analytically by standard methods. The solution contains $n - 1$ relaxation times, and gives each δc_i as a sum of exponentials

$$\delta c_i = \sum_j a_{ij} \exp[-\lambda_j t], \quad i = 1, 2, \ldots n. \tag{19.16}$$

The coefficients in the exponents, $\lambda_j = 1/\tau_j$ are a function of the rate constants (and the overall kinetic scheme), while the coefficients a_{ij} depend (also) on the initial conditions. [The coefficients λ_j also are $(-1\times)$ the eigenvalues of the original coefficient matrix, $\mathbf{k}$.] One of the values of λ is zero, which corresponds to an infinite relaxation time, and therefore to the equilibrium state attained in the limit of a very long time, t.

(See Appendix F. When two eigenvalues coincide exactly because of regularities of the kinetic scheme, the solution differs from that given in Eq. 19.16. An example of such a special case is given in Section 19.8.)

19.7 NUMERIC INTEGRATION OF THE MASTER EQUATION

An alternative to linearizing a complex set of nonlinear kinetic equations is to use numeric integration (using of course a computer). The equations contain second and higher degree terms, and instead of Eq. 19.15 we have the following set of equations

$$\frac{dc_i}{dt} = \sum_j k_{ij}c_j + \sum_{j,l} k_{ijl}c_j c_l + \cdots, \quad i = 1, 2, \ldots n. \tag{19.17}$$

Given values of all c_i, the first derivatives can all be evaluated. The numeric integration then proceeds by extrapolating over a small time interval, δt, with

$$c_i(t_0 + \delta t) = c_i(t_0) + \delta t \frac{dc_i}{dt}, \quad i = 1, 2, \ldots n. \tag{19.18}$$

Then the derivatives are evaluated with the new values of c_i and another time step can be taken. The time steps have to be small, because this last equation has been obtained by truncation of the Taylor series for $c_i(t_0 + \delta t)$. (The procedure is, of course, similar to that used to integrate the nonlinear equations of motion in molecular dynamics simulations.)

19.8 CONSECUTIVE REACTIONS CAUSE DELAYS

An interesting case is a reaction that occurs in successive steps

$$A_0 \Rightarrow A_1 \Rightarrow A_2 \Rightarrow A_3 \Rightarrow A_4 \Rightarrow A_5 \Rightarrow A_6 \Rightarrow \cdots \Rightarrow A_n.$$

This gives master equations of the kind

$$\frac{dc_i}{dt} = k_{i-1}c_{i-1} - k_i c_i. \qquad (19.19)$$

If the rate constants for successive steps are the same, the $n - 1$ noninfinite relaxation times are all the same, and the simple solution does not apply. Instead one gets (Appendix F)

$$\delta c_i = \left(\sum_{j=1}^{n-1} a_{ij} t^j \right) \exp(-kt) + a_{i0}, \quad i = 1, 2, \ldots n. \qquad (19.20)$$

If at time zero all material is in the form of A_0 (concentration $= a_{00}$), then this becomes

$$\begin{aligned} c_0 &= a_{00} \exp(-kt) \\ c_i &= a_{00} \frac{k^i t^i}{i!} \exp(-kt), \quad i = 1, 2, \ldots, n-1 \\ c_n &= a_{00} \left(1 - \sum_{i=0}^{n-1} c_i \right). \end{aligned} \qquad (19.21)$$

The concentration of reactant, A_0 starts at a finite value, and decays exponentially from there to zero, while the concentration of each intermediate starts at zero, passes through a maximum and then decays to zero. The value of t where $t^j \exp(-kt)$ is a maximum, increases with j, and formation of the product is delayed more and more as intermediates are added to the system.

Figure 19.4 shows the gradual disappearance of starting reactant, successive appearance and decay of intermediates, and appearance of the product for a system with four intermediates. If the rate constants are not all the same, the concentrations grow and decay in a similar manner, but now according to Eq. 19.16.

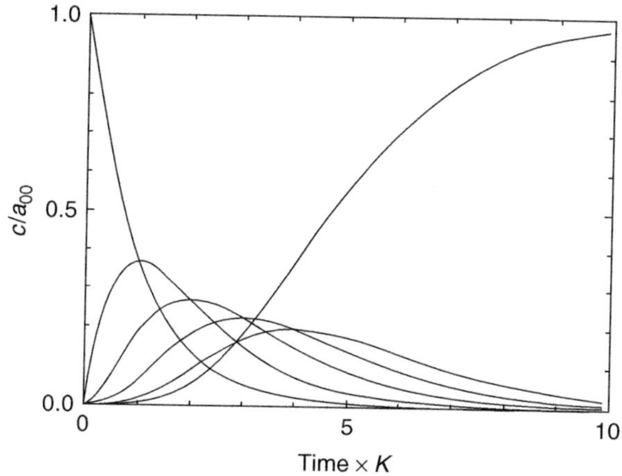

FIGURE 19.4 Progress of a series of five consecutive irreversible reactions with equal rate constant, k.

19.9 STEADY STATE ASSUMPTION: MICHAELIS–MENTEN MODEL, MICROSCOPIC REVERSIBILITY, AND CYCLIC PROCESSES

The assumption can often be made that a steady state is rapidly formed at which each intermediate's concentration is in the first approximation constant, while the concentration of the reactant(s) slowly decreases and that of the product(s) slowly increases. This is a good approximation if the reactants are in rapid equilibrium with the intermediate, which slowly converts to product.

If the concentrations of intermediates do not change significantly, one may set

$$\frac{dc_{\text{intermediate}}}{dt} = 0, \tag{19.22}$$

which then leads to equations for the rate of conversion of reactant to product.

This is the basis of the familiar *Michaelis–Menten treatment of enzyme kinetics*, where enzyme, E combines with substrate, S to form a Michaelis complex, ES, which is slowly turned over to give product, P and enzyme. (A steady-state assumption is also at the basis of the theory of reaction rates based on a model with a transition state; see Section 19.11.) The Michaelis–Menten equations are

$$\text{ES} \underset{k_-}{\overset{k_+}{\rightleftharpoons}} \text{ES} \xrightarrow{k_{\text{cat}}} \text{E} + \text{P}$$

$$\frac{d[\text{ES}]}{dt} = k_+[\text{E}][\text{S}] - (k_- + k_{\text{cat}})[\text{ES}] = 0$$

$$[\text{ES}] = \frac{k_+}{k_- + k_{\text{cat}}}[\text{E}][\text{S}]$$

$$\frac{d[\text{P}]}{dt} = k_{\text{cat}}[\text{ES}] = k_{\text{cat}}\frac{k_+}{k_- + k_{\text{cat}}}[\text{E}][\text{S}] = \frac{k_{\text{cat}}}{K_m}[\text{E}][\text{S}], \tag{19.23}$$

where K_m is the Michaelis constant,

$$K_m = \frac{k_- + k_{cat}}{k_+}. \tag{19.24}$$

As $k_{cat} \ll k_-$ (by the steady state assumption), K_m can be considered as the dissociation constant for the ES complex.

Setting the total enzyme concentration equal to E_0, gives

$$E_0 = [E] + [ES]$$

$$\frac{d[P]}{dt} = \frac{E_0 k_{cat}[S]}{K_m + [S]}. \tag{19.25}$$

Accordingly, for small [S], when few enzyme molecules are in the form of enzyme–substrate complex, the rate approaches $E_0 k_{cat}[S]/K_m$, whereas for large values of [S], when nearly all enzyme molecules are in the form of enzyme–substrate complex, the rate becomes independent of [S], approaching $E_0 k_{cat}$ (Fig. 19.5).

It appears that many metabolic enzymes have values of K_m roughly equal to the typical concentration of the substrate (when approximately half the enzyme molecules will be in the form of ES complex). Both k_{cat} and the ratio k_{cat}/K_m can be taken as measures of an enzyme's *catalytic effectiveness*; on the other hand, the quotient of k_{cat} and the rate constant of the uncatalyzed reaction, k_{non} is a good measure of an enzyme's *catalytic proficiency*; the ratio k_{cat}/k_{non} varies from 8×10^6 for carbonic anhydrase to 1.5×10^{17} for yeast orotidine 5′-phosphate decarboxylase (ODCASE). Both enzymes are effective, but because the uncatalyzed decomposition of H_2CO_3 into H_2O and CO_2 is a much faster

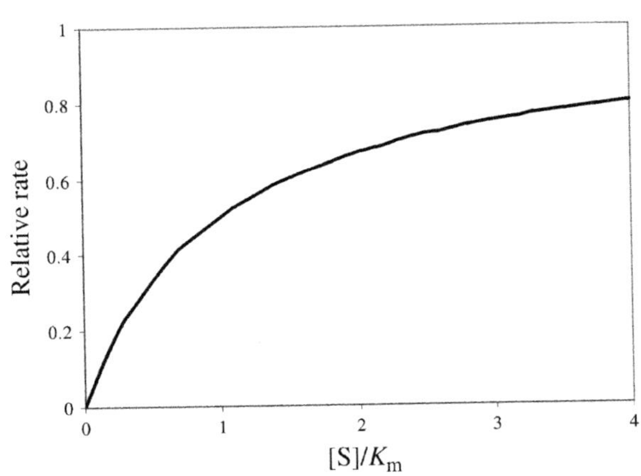

FIGURE 19.5 Catalytic rate relative to limit at high [S], according to the Michaelis–Menten kinetic scheme.

process (halftime of 0.2 s) than the decarboxylation of orotidine (halftime of 78×10^6 years), ODCase is many times more proficient than carbonic anhydrase.[7]

The principle of *microscopic reversibility* states that all chemical reactions are reversible. The Michaelis–Menten scheme does not take this principle into account and, because it does not allow for the formation of substrate from product, applies only when the reaction between substrate and product is far from equilibrium, in favor of product. The scheme applies strictly only to the initial rate in an *in vitro* study of enzyme acting on substrate, when the concentration of product is zero.

To run a reaction far from equilibrium incurs a higher cost in expenditure of free energy, ΔG. From simple thermodynamics (Section 1.7), one knows that at equilibrium the free energy for converting reactant into product, $\Delta G = 0$; otherwise,

$$\Delta G = \Delta G^{\circ} - k_B T \ln \frac{\Pi[\text{reactants}]}{\Pi[\text{products}]}, \quad \text{where}$$

$$\Delta G^{\circ} = k_B T \ln \frac{\Pi[\text{reactants}_{\text{eq}}]}{\Pi[\text{products}_{\text{eq}}]}, \tag{19.26}$$

and the subscript eq indicates equilibrium values. For the reaction to run in the forward direction, it is necessary that $\Delta G < 0$; if $\Delta G > 0$, the reaction runs backwards.

In an enzyme-catalyzed reaction, the enzyme cycles between two or more states (E and ES in Eq. 19.23), and the cycle is driven in one direction by the conversion of substrates into products. More complex coupled *cyclic processes* provide the indispensable machinery for using energy inputs (mostly, "food" or sunlight) for the purpose of biosynthesis and biomechanics, and as such are ubiquitous. One criterion to consider in biological systems is that of energy efficiency. (Another very important criterion is accuracy; thus, energy efficiency is, e.g., of little account in replication of DNA.) Total energy input is limited, and this tends to limit ΔG of cyclic reactions that must occur over and over. If the net ΔG for the cycle is not far below zero, cycling in the opposite direction is not negligible, and the reactions in the reverse sense must be included in a complete description of the cycle's kinetics.

In most cases a steady-state approximation gives an excellent description of such a repetitive cyclic process. In the steady state, the concentrations of all species are assumed constant; the concentrations of metabolite inputs and outputs are assumed to be controlled by additional biochemical processes outside the cycle in question. Given the concentrations of the metabolites, plus a knowledge of $2n$ rate constants for a cycle with n steps and n macromolecular species, and setting $dc_i/dt = 0$ for each species i, one obtains n simultaneous linear equations in the (steady-state) concentrations c_i, which are readily solved to give the fraction of each species. The difficult part remains the measurement of these $2n$ rate constants.

19.10 NUCLEATION AND GROWTH MECHANISMS IN PHASE TRANSITIONS AND BIOPOLYMER FOLDING REACTIONS

In common examples of phase transitions, the *physical properties* of the system, including thermodynamic functions, *are discontinuous*, the main exception being the chemical potentials of the components, μ_i. The phase stable at higher temperature has the higher entropy and enthalpy; the phase stable at higher pressure has the smaller volume. These are called *first-order phase transitions*; in second-order phase transitions, the discontinuity is more subtle; for example, in a second-order temperature-induced phase transition, H and S are continuous, but the specific heat, $C_P = (\partial H/\partial T)_P$ is discontinuous.

First-order phase transitions share the interesting property that a very small volume (vapor bubble or microcrystal) of the phase that is due to be formed as a result of, say, a change in temperature, is not stable until some way past the phase transition temperature. A considerable time delay in boiling or freezing may result at temperatures just above the boiling point or below the freezing point (superheated or supercooled liquids).

The kinetics of boiling or crystallization are usually described with two-step mechanisms, the first step being nucleation, which is unfavorable, and the second step being growth of the nucleus (vapor bubble or microcrystal), which is favorable (Fig. 19.6). The rate of crystal nucleation depends more strongly on the supersaturation than does the rate of crystal growth, and as a consequence crystallization from supersaturated solutions yields more, *but smaller*, crystals (in less time) as the concentration is increased.

Molecular transitions, such as folding–unfolding of globular proteins and the helix-coil transition of polypeptides share the nucleation-growth model. However, these transitions are not true phase transitions. Although unfolding of many proteins follows close to a two-state or all-or-none model, the extent of the phases is microscopic, and the transitions occur gradually over some finite range of temperature (or other variable), instead of taking place in a vanishingly small range, as is the case for true macroscopic phase transitions. (Also, as discussed in Chapter 16, helix-coil transitions of polypeptides do not follow an all-or-none model. The kinetics of protein folding are discussed in Chapter 20.)

Nucleation and growth can be an appropriate model also for the *condensation of protein subunits into long polymers*, although, again, this condensation is

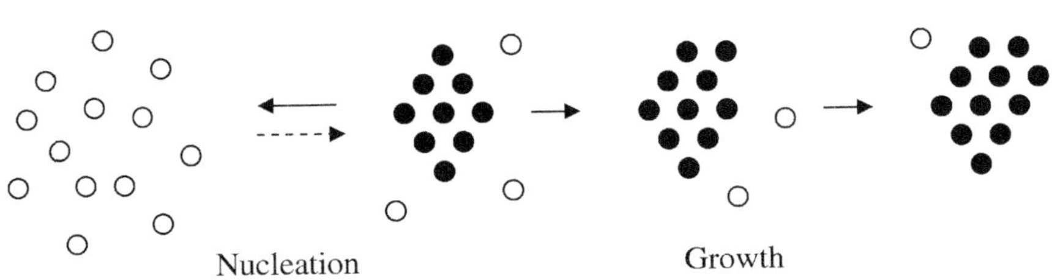

FIGURE 19.6 Nucleation and growth model of crystallization.

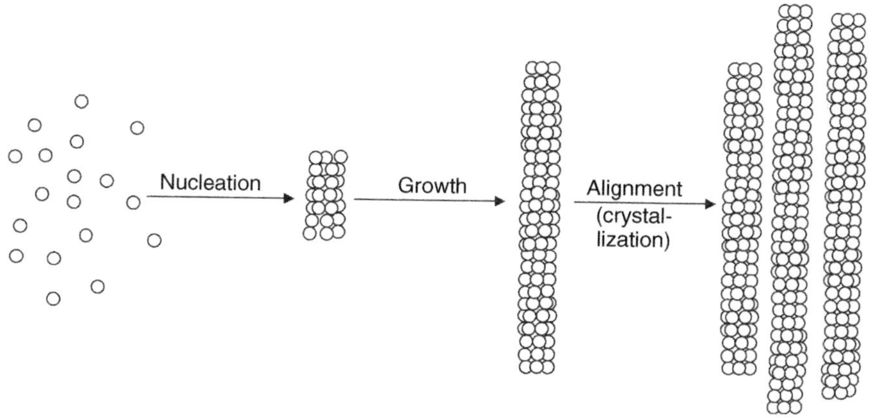

FIGURE 19.7 Nucleation and growth model of polymerization of sickle cell hemoglobin. From Hofrichter, J., Ross, P.D., Eaton, W.A. *Proc. Natl. Acad. Sci. USA* **71**: 4864–4868 (1974).

not strictly a phase transition. The sickle cell hemoglobin mutant (hemoglobin-S), which afflicts West Africans and Americans of African descent, in the deoxy form readily polymerizes to long helical rodlike assemblies that further aggregate laterally and as a result deform red blood cells and cause severe circulatory problems (Fig. 19.7). We have seen in Section 19.8 that the product of a series of consecutive reactions appears with a delay that is longer the greater the number of steps (Fig. 19.4). Polymerization of deoxyhemoglobin S also occurs with a time delay. In this case, the delay varies inversely as a very high power (30–40) of the supersaturation (the difference between concentration and solubility).[8] Analysis of equilibrium and kinetic equations based on the nucleation and growth model indicates that this power corresponds to the number of hemoglobin-S molecules required to form a stable nucleus; the number is so large because of the helical arrangement of hemoglobin-S molecules in the fiber.

19.11 KINETIC THEORY AND THE TRANSITION STATE

Many reactions are considered to pass through a *transition state*. This state has a high free energy relative to that of the reactant, and therefore a low Boltzmann probability and is rarely achieved; consequently, the rate is low. More exactly, the rate of a reaction is presumed to depend on three factors:

the rate of progress along the reaction path;
the free energy of activation;
the transmission coefficient, κ.

The reaction path is any coordinate (simple or complicated) that might be used to describe the progress of a reaction from reactant to end product.

The assumption underlying this theory is that the equilibration needed to reach the transition state from the reactant ground state is faster than the transition to product across the barrier (steady-state assumption).

19.11.1 "Ballistic" Processes

Some reactions progress rapidly once "launched." Many chemical reactions follow such a ballistic mechanism and have a similar rate of progress along the reaction path, or *transition frequency*, at 300 K equal to $k_B T/h = 6.3 \times 10^{12}/\mathrm{s} = 1/(160\ \mathrm{fs})$. The rate is then the product of this transition frequency and the fraction of molecules which is in the transition state in an equilibrium distribution or the probability, $P^{\ddagger}$ of finding the system at the transition point. The probability $P^{\ddagger}$ can be expressed in terms of the free energy of activation, or free energy barrier, $\Delta G^{\ddagger}$

$$k = \kappa \frac{k_B T}{h} P^{\ddagger} = \kappa \frac{k_B T}{h} \exp(-\Delta G^{\ddagger}/k_B T). \tag{19.27}$$

For chemical reactions the free energy of activation, $\Delta G^{\ddagger}$ contains a large contribution due to the change in electronic energy, $\Delta E^{\ddagger}$ and it is then customary to express the Boltzmann factor in the free energy difference as the product of the Boltzmann factor in the electronic energy difference, $\Delta E^{\ddagger}$ and the ratio of two partition functions that represents the difference in conformational entropy and energy of the system,

$$k = \kappa \frac{k_B T}{h} \frac{q^{*\ddagger}}{q} \exp(-\Delta E^{\ddagger}/k_B T). \tag{19.28}$$

Here the superscript $*$ indicates that the *electronic energy* factor, $\exp(-\Delta E^{\ddagger}/k_B T)$ has been taken out of the partition function, that is, $q^{\ddagger} = q^{*\ddagger} \exp(-\Delta E^{\ddagger}/k_B T)$. For chemical reactions of small molecules in vacuo, the ratio $q^{*\ddagger}/q$ is close to 1.

Activation (free) energies for chemical reactions proceeding at different rates (at room temperature) are given in Table 19.1. An application to modeling the energetics of an enzymatic reaction is discussed later in this chapter.

Applications of the transition state rate equation (19.27), especially by Eyring and Polanyi and coworkers, have established its validity for chemical reactions. The equation is sometimes associated with Eyring's name; however, the equation's derivation is apparently not associated with Eyring (for further discussion see Section 21.11).

TABLE 19.1 Activation free energies for different values of the rate constant

k (1/s)	$\Delta G^{\ddagger}/k_{\mathrm{B}}T$	$\Delta G^{\ddagger}$ at $T = 300$ (kcal/mol)
10^9	8.7	5.2
10^6	15.7	9.4
10^3	22.6	13.6
10^0	29.5	17.7

According to Eq. 19.27, when the transmission coefficient, κ is set equal to 1.

19.11.2 Diffusive Processes

Many processes can be characterized as diffusive. Here, rapid progress is impeded by collisions with solvent or parts of the same molecule, the system in fact travels back and forth in a haphazard manner, and a single transition may consist of several recrossings. In this case, the transition frequency ($k_{\mathrm{B}}T/h$ in Eq. 19.27) is replaced with a factor that represents the (average) rate at which the reaction coordinate changes in the transition region; also, we replace the Boltzmann factor $\exp(-\Delta G^{\ddagger}/k_{\mathrm{B}}T)$ with the relative probability of the system being at the transition point, $P^{\ddagger}$

$$k = \kappa \left(\frac{d\xi}{dt}\right)^{\ddagger} P^{\ddagger}. \tag{19.29}$$

The transmission coefficient, κ expresses the fraction of systems at the transition point (for which $\xi = \xi^{\ddagger}$) that are in actual fact in a barrier crossing event. We note that the value of $P^{\ddagger}$ is relative to the integrated probability distribution for the entire reactant state up until the transition point (integration over a range r). This gives

$$P^{\ddagger} = \frac{P(\xi^{\ddagger})}{\int_r P(\xi)d\xi}. \tag{19.30}$$

We shall see subsequently how the quantities κ, $(d\xi/dt)^{\ddagger}$ and $P^{\ddagger}$, and hence theoretical values of rate constants for diffusive processes can be estimated in simulations.

19.12 TRANSITION STATE IN CATALYSIS

Pauling was the first to state the principle that a catalyst, that is, an enzyme, lowers the free energy barrier and therefore lowers the difference in free energy between the (initial) ground state and the transition state. Not forgetting that binding of substrate (and sometimes also binding of product) to the enzyme lowers the free energy somewhat, one arrives at the free energy diagram depicted in Fig. 19.8.

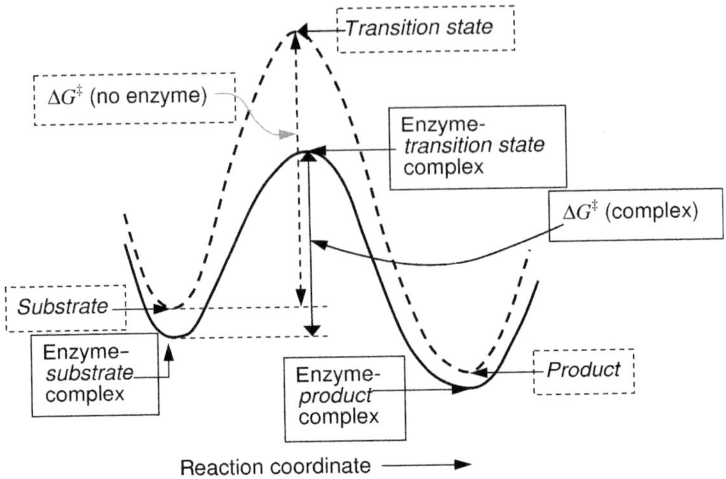

FIGURE 19.8 Free energy profile of a reaction in the absence and presence of enzyme.

Now consider the following cycle of two binding reactions and two transformations from ground state to the transition state (a "thermodynamic cycle"):

$$\begin{array}{ccc} & \Delta G^{\ddagger}_{\mathrm{E}} & \\ \mathrm{E+S} \leftrightarrow \mathrm{ES} & \rightarrow & \mathrm{ET} \\ \Delta G^{\circ}_{\mathrm{b,S}} \uparrow & & \uparrow \Delta G^{\circ}_{\mathrm{b,T}} \\ \mathrm{E+S} & \rightarrow & \mathrm{E+T} \\ & \Delta G^{\ddagger} & \end{array} \tag{19.31}$$

As the free energy change for a cyclic reaction is by definition zero, one has

$$\Delta G^{\ddagger}_{\mathrm{E}} - \Delta G^{\circ}_{\mathrm{b,T}} - \Delta G^{\ddagger} + \Delta G^{\circ}_{\mathrm{b,S}} = 0,$$

and this gives the following relation between the activation energies in the presence and absence of enzyme and the free energies of binding of transition state and substrate to the enzyme

$$\Delta G^{\ddagger}_{\mathrm{E}} - \Delta G^{\ddagger} = \Delta G^{\circ}_{\mathrm{b,T}} - \Delta G^{\circ}_{\mathrm{b,S}}. \tag{19.32}$$

Furthermore, because $\Delta G^{\ddagger}_{\mathrm{E}} \ll \Delta G^{\ddagger}$, one has

$$\Delta G^{\circ}_{\mathrm{b,T}} \ll \Delta G^{\circ}_{\mathrm{b,S}}.$$

Thus, *the transition state binds much better to the enzyme*, approximately as much (by a factor $\exp[(\Delta G^{\circ}_{\mathrm{b,T}} - \Delta G^{\circ}_{\mathrm{b,S}})/k_B T]$) as the rate is enhanced by the enzyme (by a factor $\exp[(\Delta G_E^{\ddagger} - \Delta G^{\ddagger})/k_B T])$. Many transition states are much more polar than the substrates; increased affinity of enzyme for this more polar intermediate state will result from the presence of *complementary polar groups* in the binding site. In fact, Warshel has proposed that most, if not all, enzyme-catalyzed reactions proceed via transition states that are stabilized by specific electrostatic interactions with the enzyme.[9]

An example is provided by the serine proteases and esterases, a family of enzymes with a common mechanism and similar residues in the active site. (Some well-studied examples are trypsin, chymotrypsin, and elastase.) The reaction proceeds by the attack of a serine OH group on the carbonyl carbon of the ester or amide group; the serine side chain then forms a new ester with the acid moiety of the substrate, and the amine or alcohol moiety of the substrate is the leaving group. In a second step, the attacking group is a water molecule; the serine ester is cleaved and in this step the serine is the leaving group.

Step 1:

$$\text{serine–OH} + \text{–C}^{\alpha}[\text{H, R1}]\text{–CO–NH–C}^{\alpha}[\text{H, R2}]\text{—}$$
$$\rightarrow \text{—C}^{\alpha}[\text{H, R1}]\text{–CO–O–serine} + \text{NH}_2\text{–C}^{\alpha}[\text{H, R2}]\text{—}$$

Step 2:

$$\text{H–OH} + \text{—C}^{\alpha}[\text{H, R1}]\text{–CO–O–serine}$$
$$\rightarrow \text{serine–OH} + \text{C}^{\alpha}[\text{H, R1}]\text{–COO}^{-} + \text{H}^{+}$$

In the transition state for both steps, the carbon atom has four substituents: C^{α} of residue 1, O, O – serine, and NH of residue 2 (or OH of water). The geometry of this state is tetrahedral *and* the molecule is more polar: one formally must assign a net negative charge to the now singly bonded lone O atom. The serine proteases exploit this feature of the transition state by providing a pocket, called the *oxyanion hole*, which is lined with NH groups and is situated to take optimal advantage of the change from a planar to a tetrahedral conformation at the C atom (Fig. 19.9).

19.13 INHIBITOR DESIGN: TRANSITION STATE ANALOGS

The transition state binds extremely tightly to the enzyme but itself is an unstable intermediate. This has led to the search for and discovery of "transition state analogs," stable molecules that share certain features with the transition state, and consequently bind tightly to the enzyme. These molecules would be near-perfect enzyme inhibitors, and provide a basis for development of effective pharmaceuticals.

The following is an example of substrate, putative transition state, and transition state analog inhibitors of *acid* proteases.[10] Acid proteases (which include pepsin

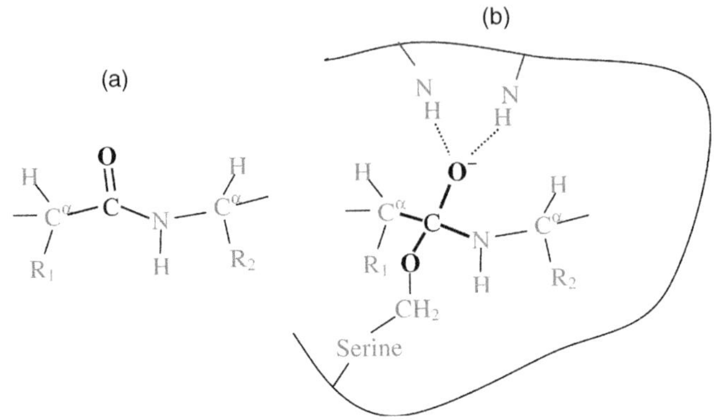

FIGURE 19.9 (a) Diagram of peptide group connecting two successive amino acid residues. In the ground state the central group is planar. (b) The transition state for substitution of the substrate's N atom with O of the enzyme's serine side chain. The center (bold face) is tetrahedral in structure, and the singly bonded oxygen atom shown with a net negative charge is stabilized by hydrogen bonds with the NH groups of the oxy-anion hole.

and the HIV protease) hydrolyze peptide bonds in a one-step mechanism in which a water molecule attaches to the amide carbon atom. The active site of acid proteases contains two carboxyl groups in close proximity to the amide group of the bound substrate (side chains of aspartic acid residues). Again, for proteases, the substrate may be symbolized as

$$\text{Peptide substrate} = \text{–C}^{\alpha}[\text{H, R1}]\text{–CO–NH–C}^{\alpha}[\text{H, R2}]\text{–}$$

where N and C are the centers of *planar* groups, R1 and R2 are the side chains of successive residues; only the C^{α} are tetrahedrally coordinated; CO and NH are moderately strong dipoles. As mentioned earlier, the reaction is presumed to pass via an intermediate,

$$\text{Tetrahedral intermediate} =$$
$$\text{–C}^{\alpha}[\text{H, R1}]\text{–C[OH, OH]–N[H, H]}^{+}\text{–C}^{\alpha}[\text{H, R2}]\text{–}$$
$$\text{or}$$
$$\text{–C}^{\alpha}[\text{H, R1}]\text{–C[OH, O}^{-}\text{]–N[H, H]}^{+}\text{–C}^{\alpha}[\text{H, R2}]\text{–}$$

in which also C and N are *tetrahedrally* coordinated; the N[H, H] group carries a net positive charge and the OH groups have a low pK. This is a clear example of a reaction in which the transition state is considerably more polar than either substrate or product. The enzyme stabilizes the intermediate via electrostatic interactions with complementary groups on the surface of the active site.

Various effective inhibitors of peptidases mimic the tetrahedral intermediate's geometry and charge distribution in different ways, as can be seen from the following list.

Hydroxy isostere = $-C^{\alpha}[H, R1]-C[H, OH]-CH_2-C^{\alpha}[H, R2]-$
Reduced peptide = $-C^{\alpha}[H, R1]-CH_2-N[H, H]^{+}-C^{\alpha}[H, R2]-$
Statine analog = $-C^{\alpha}[H, R1]-C[H, OH]-CH_2-CO-NH-C^{\alpha}[H, R2]-$
(pepstatin has $K_i = 0.05$ nM for pepsin.)
Inhibitor JG365 of HIV protease (the second residue mimics a proline in the native substrate) =
$-C^{\alpha}[H, CH_2-C_6H_5]-C[H, OH]-CH_2-N[C^{\delta}, H]^{+}-\ C^{\alpha}[H, C^{\beta}-C^{\gamma}]-$

Of course, while a pharmaceutical's action is often due to its strong inhibitory power, a powerful inhibitor *usually* is not a suitable pharmaceutical, for any of various causes, such as poor uptake, rapid degradation in the body and undesirable side effects; however, such molecules are invaluable as "lead" compounds for development of effective drugs.

19.14 THE DIFFUSION-LIMITED REACTION

Some reactions have such a low barrier in the transition state that the diffusion of the reactants to reach one another becomes rate limiting. To analyze this, it is helpful to consider a simple model developed by Smoluchowski. In this model, the enzyme is treated as a sphere, whose entire surface is reactive: whenever a substrate molecule reaches the surface (by diffusion), it is immediately destroyed. Thus, the concentration of substrate at the enzyme surface is zero, while its concentration at great distance from the surface is equal to the bulk concentration, c_0. The substrate concentration is then a function of the distance from the center of each enzyme molecule, $c(r)$, which decreases as r decreases, and becomes equal to zero when substrate reaches the enzyme surface, that is, when $r = R$.

Whenever there is a concentration gradient, $\partial c/\partial r \neq 0$, diffusion will cause a net transport, Υ of molecules in the direction along which c decreases. This transport (the product of flux and area over which the flux occurs) is given by (see Section 21.2)

$$\Upsilon = -\text{area} \times \text{flux} = 4\pi r^2 D \frac{\partial c}{\partial r}, \tag{19.33}$$

where D is the diffusion coefficient.[11] (The minus sign occurs because the transport is in the direction of decreasing r.) Because of conservation of mass, the transport is independent of r. Therefore,

$$\frac{dc}{dr} = \frac{\Upsilon}{4\pi D r^2}$$

$$c = -\frac{\Upsilon}{4\pi D r} + a, \tag{19.34}$$

where a is a constant. Because we know the boundary conditions

$$c(R) = 0$$
$$\lim_{r\to\infty} c(r) = c_o, \tag{19.35}$$

we can solve for the constant a and the transport Υ, and we get

$$c = c_o(1 - R/r)$$
$$\Upsilon = 4\pi DRc_o. \tag{19.36}$$

Thus, we find that the rate of a diffusion-limited reaction is proportional to the substrate concentration. This is the case also for the Michaelis–Menten kinetic scheme at low substrate concentration, when, as noted, the rate is $E_0 k_{cat}[S]/K_m$. The diffusion-limited rate is an upper limit that cannot be exceeded by the catalytic rate of any enzyme. Equating c_o with [S], one finds, therefore, that

$$k_{cat}/K_m \leq 4\pi DR. \tag{19.37}$$

Assuming identical radii for both enzyme and substrate,[12] and taking into account that Υ is expressed per molecule of enzyme, one finds $k_{cat}/K_m \leq 3.3 \times 10^9/(\mathrm{M} \cdot \mathrm{s})$. The small difference with the measured value for superoxide dismutase of 7×10^9 reflects the simplifying assumptions of the model and also a favorable electrostatic attraction between the negatively charged substrate (O_2^- ion) and copper and zinc (or manganese, iron, nickel) ions at the catalytic center which the model does not include. Superoxide dismutase is extraordinarily effective in clearing the cell of harmful superoxide ions. Other enzymes with activities at or near the diffusion-controlled limit include triose phosphate isomerase and acetylcholine esterase.

This simple diffusion-limited model has been refined to represent actual enzyme substrate interactions more realistically. A very simple alteration of the model is to let only part of the surface (the active site) be reactive; one can analyze the behavior of this model by random-walk simulations. A more sophisticated extension of the model uses a full atomic representation of the enzyme. The coordinates are used to compute a description of the enzyme's molecular surface, and the electrostatic potential generated by the enzyme in the solution is calculated by the continuum dielectric model of the solvent, via the Poisson or Poisson–Boltzmann equation, as described in Section 14.10. The diffusion of the substrate can then be simulated with so-called Langevin or Brownian dynamics (see Section 21.5), in essence a more sophisticated version of the random-walk model that can take into account also the electrostatic forces between enzyme and substrate, using Eq. 21.17 as the equation of motion. (See also Section 21.10.3.)

19.15 ESTIMATING REACTION RATES FROM SIMULATIONS

We have seen in Chapter 7 how to study rates and equilibria of molecular systems using molecular dynamics simulations. To summarize: if during the simulation, the system changes back and forth between different distinct states, then equilibrium and rate constants can be calculated from the fraction of the time that each state is observed, and from the number of transitions in a given simulation time (as for the example of Fig. 7.4); however, in general this will not be the case, but then the equilibrium constant between two states can still be obtained from the free energy difference between states measured as a potential of mean force (Section 7.8); however, no rate information is obtained in such simulations.

The simulation approach can be extended to also obtain estimates of rate constants in the general case by making use of transition state theory, which relates the rate constant to the activation free energy, $\Delta A^{\ddagger}$ via Eqs. 19.28 and 19.29. (Simulations give ΔA, not ΔG, unless a $\Delta(PV)$ term is explicitly included.)

19.15.1 Application to Diffusive Processes

We discuss first the applications of dynamics simulation to reactions in which no chemical bond making or breaking is involved, as macromolecular conformation change and ligand binding. In order to estimate the rate of a diffusive process, one must determine three separate quantities. These are the rate of progress along the reaction path, α, the transmission coefficient at the barrier, κ, and the activation free energy, $\Delta A^{\ddagger}$ (cf. Eq. 19.29). We illustrate this approach using as an example a conformational transition of Ace–Ala–Nme ("alanine dipeptide"), by internal rotation about the C^{α}–C bond of the alanine residue. As discussed in Section 7.7, dynamics simulations of the alanine dipeptide show the presence of two main stable regions, denoted by α and β, separated by free energy barriers at $\psi = 0$ and $\psi = -150^{\circ}$. The barrier at 0° is unusually low, so that equilibrium and rate constants could be estimated from the observed transition frequency in a 1-ns-long simulation.

Reaction Coordinate and Potential of Mean Force. For this example, the torsion angle ψ was the obvious choice of the reaction coordinate, ξ along which progress is measured (which in the general case can be a distance, a torsion angle or a function of several internal coordinates).[13] The PMF, $\Delta A(\psi)$ for changing the reaction coordinate from one stable state to the other via the transition point was then calculated as described earlier in Section 7.8 (Fig. 7.6).[14]

The kinetic behavior in the transition region was studied by simulations following removal of a restraining umbrella potential (see appendix D). An estimate of the *(angular) velocity*, $(d\psi/dt)^{\ddagger}$ was determined from the root-mean-square observed values of $\delta\psi/\delta t$ in the transition region.[15]

The *transmission efficiency* was determined from a series of paired simulations starting at the transition point. A brief period of simulation starting at the transition point can be expected to lead to one of the two stable states, that is, either α

or β. This brief simulation is then repeated from the same starting point with all conditions the same, except that *the direction of each velocity is inverted.* After the same brief simulation time, the state will most often again be one of the two stable states. Each pair of simulations that reaches opposite endpoints represents a successful barrier crossing, while a pair of simulations that reaches the same endpoint represents a failure to cross the barrier; the transmission coefficient, κ will follow from the ratio of successes over failures, and should also account for the number of recrossings observed in these simulations.[16]

With Eq. 19.29 the rate was estimated at 5×10^{10}/s, in good agreement with the earlier estimate of 4×10^{10}/s from direct observation of transitions in a 1–ns-long simulation (Section 7.7).

19.15.2 Challenge: Estimating Enzyme Catalytic Rates from "First Principles"

A related approach can be applied to reactions in which chemical bonds are made or broken and can be used successfully in studies of enzyme catalytic mechanisms; obviously, the ability to do this has much improved since the pioneering study by Warshel and Levitt.[17] We discuss here first the principle and the methods and next give a case study of one particular enzyme as an example.

Principle. Energetics of bond breaking and bond making cannot be represented adequately except with quantum mechanics (QM). The computational requirements of QM calculations greatly limit the scope and extent of simulations of chemically reacting systems, and these would not (presently) be feasible without methodology developed specifically in order to deal with this problem. An important part of the solution to this problem is to apply the transition state theory discussed in Section 19.11. In this theory, the rate is given in terms of a universal transition frequency and the Boltzmann factor of the free energy of activation (Eq. 19.27). This assumption makes it unnecessary to simulate the *dynamics* of barrier crossing in order to determine the transmission frequency.

The current approach requires detailed structural knowledge of one or more enzyme–inhibitor complexes, which nearly always will allow one to propose a structure of the (by definition unstable) enzyme–substrate complex. A structure with a bound strong inhibitor, a so-called transition state analog, will help define the transition state. *Typically, more than one mechanism, each with a different transition state, will suggest itself.* (Kinetic studies of the native enzyme and of site-directed mutants may refine the choices.) The goal of a simulation study will be to determine the best reaction path via each of the proposed transition states, and compute the free energy of activation, $\Delta G^{\ddagger}$ for each. The path with lowest value of $\Delta G^{\ddagger}$ will be the preferred path, *and*, if the theoretical rate constant calculated with Eq. 19.27 for this activation free energy agrees with the experimental value, k_{cat} then this is strong evidence that the correct mechanism has been identified. Confirmation will be obtained if the mechanism predicts changes in rate for changes in the structure of enzyme or substrate that are subsequently verified experimentally.

For a chemical reaction involving rearrangements of one or a few chemical bonds, the reaction coordinate, ξ can be defined in terms of a few interatomic distances. Taking as example a case in which a single bond is transferred from one to another nearby atom (Fig. 19.10), the reaction coordinate is chosen to be

$$\xi = r_1 - r_2; \qquad (19.38)$$

ξ then varies from negative to positive in going from reactant to product; its magnitude is on the order of 1 or a few Å. For more complex reactions, ξ can be extended with terms for breaking and making additional bonds.

Case study: 4-Oxalocrotonate Tautomerase (4-OT)[18]. On paper, the reaction catalyzed by this enzyme is very simple: the substrate is a keto acid with one C—C double bond, in which a hydrogen atom is moved to a carbon atom two C—C bonds removed; this then results also in a shift of the C—C double bond. The reaction was thought to involve a stable intermediate with a second double bond, in which the keto oxygen atom is reduced (Fig. 19.11). The catalytic rate of this enzyme is $k_{cat} = 3500/s$; according to Eq. 19.27, this corresponds to an activation free energy of 12.9 kcal/mol.

The x-ray structure shows one of the oxygens of the 1-carboxyl group and the keto oxygen hydrogen-bonded to an arginine side chain. In addition, two water molecules are not far from the keto oxygen and might temporarily donate a hydrogen to the keto oxygen, while the imino terminal NH group is in good position to serve as a base to accept a hydrogen atom from the 5-carbon and subsequently relinquish it to the 3-carbon atom.

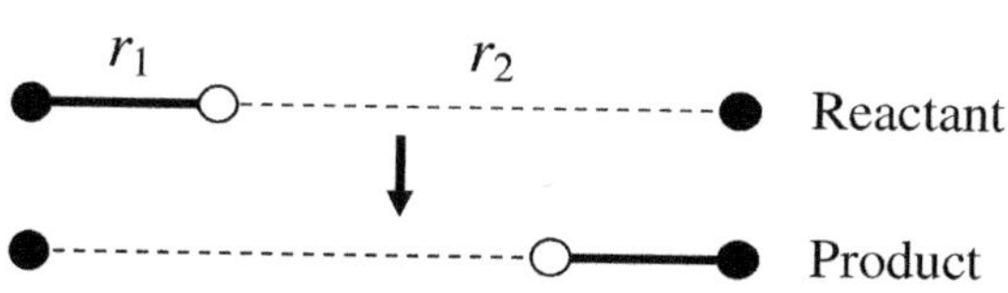

FIGURE 19.10 Definition of reaction coordinate for transfer of a bond: $\xi = r_1 - r_2$.

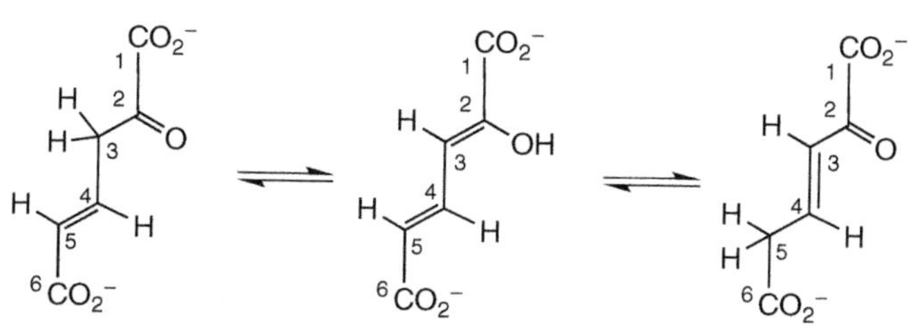

FIGURE 19.11 Isomerization catalyzed by 4-OT, showing reactant, one possible putative intermediate and product. From Cisneros, G.A., Liu, H., Zhang, Y., Yang, W. *J. Am. Chem. Soc.* **125**: 10384–10393 (2003).

For the first step, four alternative reaction paths were investigated. In each, the imino nitrogen of Pro-1 accepts a hydrogen atom from the 5-carbon. Three of the paths differ in the provenance of a hydrogen atom for the keto oxygen: from one or the other ordered water molecule, or from the arginine side chain. In the fourth path, the keto oxygen does not acquire a hydrogen atom but becomes negatively charged (Fig. 19.12). The simulations produced optimized structures of enzyme–substrate complex, and for each of the four paths a transition state complex and a complex with the intermediate. The calculated potential energy barrier for one of the four paths was 4.5 kcal/mol less than the next higher. The authors conclude in favor of the path with the lowest energy barrier; on this path, the *intermediate carries a negative charge* on the keto oxygen as in Fig. 19.12, and is *not* the putative intermediate of Fig. 19.11.

Results for the first step of the preferred reaction scheme are shown in Fig. 19.13; results for the second step have not been reproduced here, as these are very similar, except that in that step all energies start far from zero and then return to values near zero, with the PMF again passing through a slight extremum. The intermediate is separated by modest free energy barriers from both reactant and product, and thus is a true intermediate state. Interactions with the enzyme lower the free energy of the intermediate complex by over 30 kcal/mol; the enzyme–intermediate complex has 13.5 kcal/mol higher free energy than the enzyme–substrate complex. The net activation free energies are 14.5 for the first and 16.5 kcal/mol for the second step. This is slightly higher than the activation free energy of 12.9 derived from k_{cat}, but the authors feel that this is within the error of the approximations made, both in the simulations and by application of transition state theory. The enzyme–product complex was found to have free energy just 0.3 kcal/mol higher than the enzyme–substrate complex.

Details of the calculations show that the intermediate is stabilized by a hydrogen bond to one of the two ordered water molecules, and that the positive charge

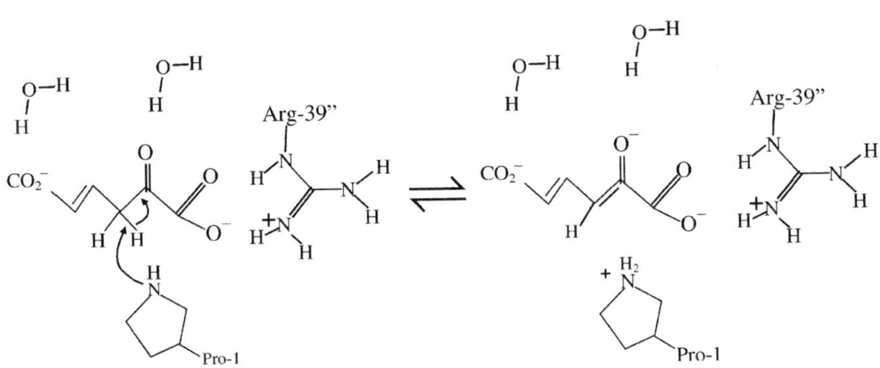

FIGURE 19.12 One of four reaction schemes for step I of isomerization catalyzed by 4-OT involving transfer of H^+ from substrate to N of Pro-1 and leading to a negatively charged intermediate. Nearby water molecules and Arg-39 do not participate in bond breaking or making. From Cisneros, G.A., Liu, H., Zhang, Y., Yang, W. *J. Am. Chem. Soc.* **125**: 10384–10393 (2003).

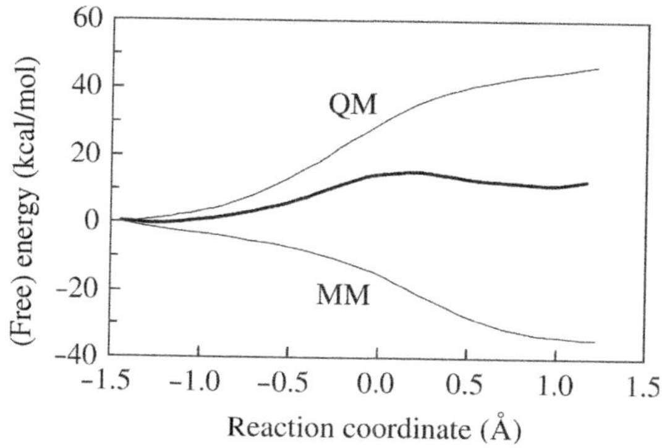

FIGURE 19.13 Potential and free energy along the reaction coordinate for the first step of isomerization by 4OT. Top thin curve: QM potential energy (excluding the QM/MM electrostatic energy); bottom thin curve: MM free energy (including the QM/MM electrostatic energy); heavy curve: net potential of mean force for conversion to the intermediate. Data from Cisneros, G.A., Liu, H., Zhang, Y., Yang, W. *J. Am. Chem. Soc.* **125**: 10384–10393 (2003).

on the arginine side chain helps stabilize the double negative charge on the intermediate. Subsequently, effects of mutations of the enzyme predicted on the basis of additional QM/molecular mechanics (MM) simulations have been confirmed experimentally.[19]

Technical Details. The details of the calculations used in this study are dominated by the need to limit number and extent of the QM calculations. This is accomplished by a number of simplifications, the most important of which is to use (high level) QM to represent the energetics of only the reactive atoms, that is, substrate and participating parts of the enzyme and perhaps a solvent molecule or two (this is termed the *QM subsystem*), but an MM energy function for the bulk of the protein and solvent (the MM subsystem). Even if it is thus reduced, dynamics simulation for generating a distribution of conformations of the QM subsystem along the reaction path is replaced by a series of energy minimizations.

We summarize the methods used in the preceding case study; for details, consult the original publication.[18] Crucial aspects of such a QM/MM implementation are to (i) provide a smooth transition between the QM and the MM subsystems wherever these are connected by chemical bonds (here accomplished with the use of the "pseudobond" method, which we do not discuss here) and (ii) to combine the energies of the subsystems into a single expression. The QM/MM electrostatic interactions are included during simulation of either subsystem. Of course, these terms contribute only once to the total energy; they have been included as part of the MM energy in Fig. 19.13.

The reaction coordinate, ξ is defined according to Eq. 19.38, with r_1 the distance between the transferred proton and substrate C-3 and r_2 the distance between this same proton and Pro-1 N.[20] A reaction path consisting of optimized coordinate sets, $\mathbf{X}_j$ of the entire system are calculated with the reaction coordinate restrained by potentials of the form

$$E_j = K(\xi - \xi_j)^2, \tag{19.39}$$

where K is a (large) force constant. The potential of mean force for changing the MM subsystem along this reaction path is then calculated by the thermodynamic perturbation method (see Appendix D).

NOTES

1. Williams, S., Causgrove, T.P., Gilmanshin, R., Fang, K.S., Callender, R.H., Woodruff, W.H., Dyer, R.B. Fast events in protein folding: helix melting and formation in a small peptide. *Biochemistry* **35**: 691–697 (1996).
2. For example, Kaplan, J.H., Ellis-Davies, G.C.R. Photolabile chelators for the rapid photorelease of divalent cations. *Proc. Natl. Acad. Sci. USA* **85**: 6571–6575 (1988).
3. Localized liberation of caged molecules by brief illumination with a focused light beam is much used also in light microscopy of cells and tissues.
4. Ansari, A., Berendzen, J., Bowne, S.F., Frauenfelder, H., Iben, I.E.T., Sauke, T.B., Shyamsunder, E., Young, R.D. Protein states and proteinquakes. *Proc. Natl. Acad. Sci. USA* **82**: 5000–5004 (1985). The abstract reads "After photo-dissociation of carbon monoxide bound to myoglobin, the protein relaxes to the deoxy equilibrium structure in a quakelike motion. Investigation of the 'proteinquake' and of related intramolecular equilibrium motions shows that states and motions have a hierarchical glasslike structure." The interested reader is strongly encouraged to read this and subsequent papers on this topic from the same laboratory.
5. Doppler line broadening of scattered light, due to molecular motion in solution, is an effective technique for determining diffusion coefficients. An incident beam of the required narrow line width is typically provided by a laser.
6. (a) Huang, G.S., Oas, T.G. Submillisecond folding of monomeric lambda repressor. *Proc. Natl. Acad. Sci. USA* **92**: 6878–6882 (1995). (b) Burton, R.E., Huang, G.S., Daugherty, M.A., Fullbright, P.W., Oas, T.G. Microsecond protein folding through a compact transition state. *J. Mol. Biol.* **263**: 311–322 (1996).
7. Miller, B.G., Wolfenden, R. Catalytic proficiency: the unusual case of OMP decarboxylase. *Annu. Rev. Biochem.* **71**: 847–885 (2002). Other than in a very few examples, we do not discuss mechanisms of individual enzymes in this book; an excellent text by Fersht is available. (See references at end of this chapter.)
8. Hofrichter, J., Ross, P.D., Eaton, W.A. Kinetics and mechanism of deoxyhemoglobin S gelation: a new approach to understanding. *Proc. Natl. Acad. Sci. USA* **71**: 4864–4868 (1974). Also *Proc. Natl. Acad. Sci. USA* **73**: 3035–3039 (1976).
9. Warshel, A. Electrostatic origin of the catalytic power of enzymes and the role of preorganized active sites. *J. Biol. Chem.* **273**: 27035–27038 (1998).
10. Sandler, M., Smith, H.J. *Design of Enzyme Inhibitors as Drugs*. Oxford University Press, Oxford, England, 1989, p. 157.
11. The diffusion coefficient in this treatment is actually the sum of the diffusion coefficients of enzyme and substrate, the second being much the larger of the two for small substrates.

12. Use Stokes' law to express D with Eqs. 21.9 and 21.14. See also Footnote 11.
13. Given a qualitative description of the mechanism, finding a reaction coordinate for a reaction involving a small number of atoms is usually feasible. However, as pointed out in Chapter 17, there is no good definition of a reaction coordinate for protein folding.
14. The equilibrium constant was found to be $K_{\alpha\beta} = 20$. The transition state occurs at $\psi = 10^\circ$; application of Eq. 19.30, with integration over all α-conformations, that is, for ψ between -150° and 0°, gives a transition point probability $P_a(y^\ddagger) = 0.0017$.
15. This gave a value of $(d\psi/dt)^\ddagger \approx \langle (d\psi/dt)^2 \rangle^{1/2} = 5 \times 10^{14}$ degree/s.
16. Of a series of 50 such simulations (each of 1-ps duration in either direction) from independent starting points, 9 produced clear transitions, 37 were clearly not transitions, of which 18 started and ended in the α conformation and 19 in the β conformation, while 4 attempts were inconclusive. Furthermore, the barrier was, on average, crossed twice. The *transmission coefficient*, k is thus estimated to be $1/2 \times 1/2 \times 9/37 = 0.06$, which has been divided by a second factor of 2 so as to count only crossings in the direction from α to β.
17. Warshel, A., Levitt, M. Theoretic studies of enzymic reactions: dielectric electrostatic and steric stabilization of the carbonium ion in the reaction of lysozyme. *J. Mol. Biol.* **103**: 227–249 (1976).
18. Cisneros, G.A., Liu, H., Zhang, Y., Yang, W. Ab initio QM/MM study shows there is no general acid in the reaction catalyzed by 4-oxalocrotonate tautomerase. *J. Am. Chem. Soc.* **125**: 10384–10393 (2003).
19. Cisneros, G.A., Wang, M., Silinski, P., Fitzgerald, M.C., Yang, W. The protein backbone makes important contributions to 4-oxalocrotonate tautomerase enzyme catalysis: understanding from theory and experiment. *Biochemistry* **43**: 6885–6892 (2004).
20. For the other three reaction schemes, ξ includes a pair of similar terms for transfer of a H^+ ion from a water molecule or Arg-39 to substrate O-2.

SUGGESTED READING

Fersht A., *Enzyme Structure and Mechanism*. New York: W. H. Freeman, 1999.

Hu, H., Yang, W. Free energies of chemical reactions in solution and in enzymes with ab initio quantum mechanics/molecular mechanics methods. *Annu. Rev. Phys. Chem.* **59**: 573–601 (2008).

20

Kinetics of Protein Folding

The protein folding problem is our problem, not the protein's.
—Ken Dill

20.1 INTRODUCTION

Given the complexities and long history of studies of protein folding equilibria discussed in Chapter 17, it should not be a surprise that a study of the kinetic mechanism of protein folding and unfolding offers new challenges. Indeed, this is as yet a not entirely finished story: the "protein folding problem" remains a subject of intense investigation.

The nature of the added complexities should be clear from previous chapters. The measurement of reaction rates requires the observation of transient signals, and more specialized equipment is needed the faster the reactions occur. Careful measurement may reveal the presence of transiently stable intermediates, whose structure may be inferred from additional measurements. However, no intermediates are observable in the transition of a protein that obeys a two-state model.

The folding rates of proteins vary enormously. At one end of the scale, some small proteins fold with halftimes on the order of microseconds, while, at the other end, one finds proteins whose folding is somehow so difficult that *in vivo* it is assisted by cellular machinery, the molecular chaperones. Unfolded polypeptide chains have many internal degrees of freedom (rotation about single bonds), and all proteins, whether fast or slow folding, must follow a path composed of *many* steps in which the conformation of each molecule successively approaches the folded state. It therefore makes sense to break down a discussion of protein folding into that of two very different questions: the first as to what specific problems limit the rate of folding of the slow folding proteins, and the second on how to understand

---

Equilibria and Kinetics of Biological Macromolecules, First Edition. Jan Hermans and Barry Lentz.

the mechanism of the most rapidly folding proteins in terms of a description of the structure and stability of the transiently unstable transition state and the nature and rates of the steps along the reaction pathway in the vicinity of the transition state.

20.2 SLOW FOLDING: MISFOLDING

Many proteins fold slowly under *in vitro* refolding conditions, and in many cases this is due to some or all of the molecules first assuming, at least temporarily, a stable, imperfectly folded state. This will slow down the formation of the native state because the activation free energy for conversion of the misfolded to the native state is increased by an amount equal to the free energy of again unfolding the intermediate, which can be several kcal/mol, although this must, of course, be less than the free energy difference between native and unfolded states (Fig. 20.1).

In vivo, misfolded proteins can be "rescued," or misfolding of not-yet-folded proteins prevented, by molecular chaperones. Much studied is the GroEL-GroES chaperone system, in which fourteen GroEL monomers are arranged in a pair of hollow cylinders, each capped by a ring of seven GroES monomers.[1] Each of these approximately cylindrical cavities can sequester a misfolded protein. Misfolded chains naturally expose more hydrophobic surface than is exposed by correctly folded proteins; the inside surface of an uncapped GroEL cavity is also hydrophobic and attracts and binds a misfolded chain, and further unfolds it to expose more hydrophobic surface with which to interact. Following capping of the cavity by GroES, the GroELES complex undergoes a conformation change in a reaction

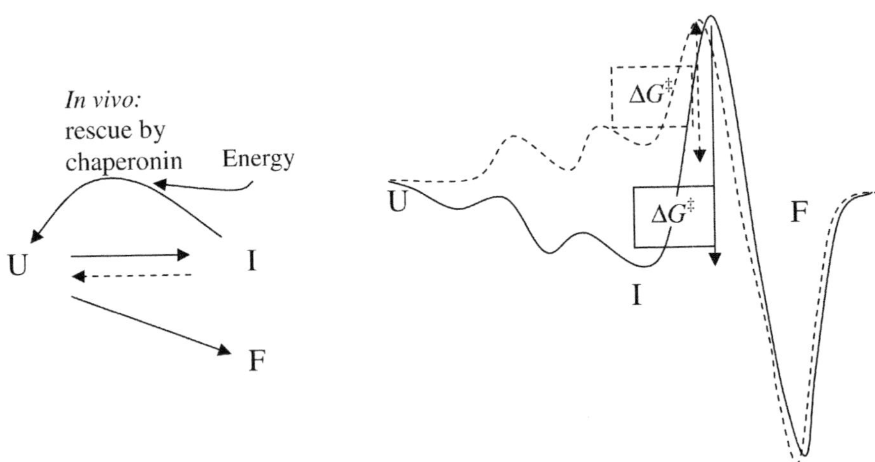

FIGURE 20.1 (Left) Unfolded protein, U predominantly forms a "misfolded" intermediate, I, rather than the native, folded state, F. The intermediate must again unfold before it can follow the folding pathway to the native state. *In vivo*, this can be accomplished with a molecular chaperone system, with input of free energy by hydrolysis of ATP. (Right) Free energy schematics in the presence (solid curve) and absence (dashed curve) of a stable intermediate.

FIGURE 20.2 Cartoon of a current model of the work cycle of one half of a GroEL chaperone complex. At far left, a misfolded or unfolded protein molecule binds to the hydrophobic inner surface of the open cavity (this state is represented as hashed), if necessary unfolding further in the process. A GroES cap and ATP molecules can now bind to this complex. Next, hydrolysis of bound ATP is coupled to a conformation change of the protein which exposes an alternate, hydrophilic, inner surface (checkerboard pattern) that releases the unfolded protein. After this folds, products and the ES cap are released and the interior again becomes hydrophobic.

coupled to ATP hydrolysis (Fig. 20.2). In this alternate conformation, the inside surface of the cavity is hydrophilic, the misfolded chain is released into the cavity, and is thereby given an opportunity to fold correctly without being able to aggregate with other not-yet-folded protein molecules.

This molecular chaperone provides an example of a protein complex in which conformation change is coupled to a chemical reaction, with hydrolysis of ATP providing the free energy necessary to drive the reaction cycle; in this case, this free energy is expended in order to strip the bound unfolded protein off the inner surface of the cavity. An interesting feature of the GroEL system is that the other half of the molecule percourses the same cycle, but that the two halves are out of phase: the other cavity is loaded with a second unfolded protein molecule and ATP, and capped before the first cavity releases its cap and contents.

Unless very dilute, unfolded proteins can easily form misfolded aggregates. This can affect *in vitro* folding experiments, and occurs for many proteins *in vivo* when these are overexpressed in engineered constructs. The aggregates form precipitates known as *inclusion bodies*; often, correctly folded protein can be recovered from these inclusion bodies by dissolving the precipitate in a denaturing solvent (such as 6 M guanidine), followed by slow dilution or dialysis.

20.3 SLOW FOLDING: CIS–TRANS ISOMERIZATION OF PROLINE

One well-recognized cause of slow folding of some proteins is that the rate limiting step is the cis–trans isomerization of proline residues, as first proposed by John Brandts.[2,3] For all residue types except proline, the equilibrium constant for the conformation change of the peptide bond preceding that residue from trans to cis is very small (10^{-3}). However, for the peptide bond preceding a proline residue, the equilibrium constant is not negligible (the trans conformation is still favored, but less strongly; Fig. 20.3). The cis-to-trans change of peptide bonds is a slow reaction;

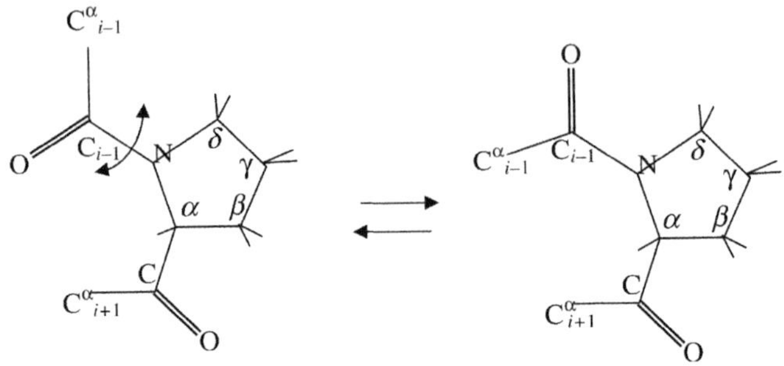

FIGURE 20.3 Equilibrium between trans and cis conformations of a proline residue (i) in a polypeptide chain. The ring carbons are indicated with letters α through δ. Positions of seven H atoms are indicated with C—H bonds. The C—N bond has a double bond character, and the energy barrier separating the trans and cis conformations is much higher than that separating rotamers of, say, butane.

thus, left in the denatured state, a protein's prolines will gradually equilibrate between these two conformations, with, as a result, a decrease in the fraction of molecules that can readily refold to the native conformation.

If all proline residues assume the trans conformation in the folded state, most of the unfolded molecules can fold without the need for a proline isomerization step (unless the protein contains many proline residues). However, proline isomerization is the main rate-limiting step if a proline peptide bond has the cis conformation in the native structure, because in that case the equilibrated denatured state will contain mostly molecules that are not ready to refold. The slow peptidyl–proline isomerization step is catalyzed by an enzyme, proline isomerase, which has turned out to be identical with the protein cyclophilin that is a binding protein for the immunosuppressive drug cyclosporin A.

20.4 SLOW FOLDING: DISULFIDE BOND FORMATION

Disulfide compounds are easily reduced to thiol (or sulfhydryl) compounds, and vice versa.[4] If the reduction of the disulfide is performed by molecules with thiol groups, the latter are themselves oxidized to form disulfides in the process. The pH dependence of this reaction indicates that the reactive reduced species is the thiolate ion, so that one must write this as

$$\begin{aligned} R_1SSR_2 + R_3S^- &\leftrightarrow R_3SSR_2 + R_1S^- \\ R_3SSR_2 + R_3S^- &\leftrightarrow R_3SSR_3 + R_2S^-. \end{aligned} \tag{20.1}$$

The rate constant is reported to be 40/M/s at high pH, when the thiol groups are predominantly in the thiolate form. (As free thiol groups have a pK_a of 9, at neutral pH only a small fraction will be in that form.)

Many proteins contain in their native, folded state, pairs of cysteine side chains linked via disulfide bonds, and if these proteins are unfolded by addition of denaturant or change of temperature, the disulfide bonds can be reduced. Such unfolded and reduced molecules are, of course, a better model of the newly synthesized polypeptide chain, of which the cysteine residues are in the thiol form. We have earlier mentioned the work of Anfinsen, who showed that slow air oxidation accompanied by gradual removal of denaturant allowed the reformation of active bovine pancreatic ribonuclease, with all four disulfide bonds correctly formed, while little or no active protein was obtained if oxidation took place first, and removal of denaturant was the second step. The eight thiol groups of a single reduced ribonuclease or hen eggwhite lysozyme molecule can be paired in $7 \times 5 \times 3 = 105$ different ways; rapid oxidation of the denatured molecule results in many or all possible arrangements, most of which are misfolded and do not correspond to an active enzyme. However, the native fold with the correct disulfide pairing is the most stable, and, under just the right conditions, any incorrectly formed disulfides links rearrange themselves, and the molecule can form the more stable native fold with the cysteines paired correctly. Folding can be optimized in a redox buffer containing reduced and oxidized forms of low molecular weight thiol compound; regeneration of active lysozyme occurs in 10–20 min.[5]

Other work led to the discovery of an enzyme, protein disulfide isomerase, that catalyzes disulfide–thiol interchange. Glutathione may serve as thiol/disulfide buffer *in vivo*. Disulfide cross-links are confined largely to extracellular proteins. The rest of the chapter will deal with folding uncomplicated by misfolding, proline isomerization, or formation of disulfide bridges.

20.5 TWO-STATE FOLDING KINETICS

We have seen in Chapter 17 that the folding equilibria of many small proteins can be described in terms of just two states, folded (F) and unfolded (U), or native and denatured. Admittedly, each state is an ensemble of conformations, a vast number for the unfolded state, and a more modest number for the folded state, but partially folded intermediates do not constitute a significant proportion of a two-state equilibrium ensemble. Partially folded intermediates may not play a role in equilibrium studies, but the role of no intermediate can safely be ignored in a study of the kinetic folding process.

If the two-state equilibrium model applies, this will carry over to the kinetics. However, we have seen that thermodynamic criteria cannot establish if the unfolding equilibrium is two-state under extreme conditions, that is, when either the folded form or the unfolded form is overwhelmingly favored, the former in the absence of denaturant, the latter at maximum denaturant concentration (both at normal temperature and pH). If the kinetic experiment follows a small perturbation from equilibrium, the relaxation will necessarily obey a two-state kinetic model. New answers are to be found by conducting measurements of kinetics of folding and unfolding over a wide range of conformational stability, in practice, over a wide range of denaturant concentration.

Addition of denaturant to the solvent will change the equilibrium constant for unfolding from $K_{f\to u}$ to $K^*_{f\to u}$, and this change is related to a change in standard free energy for unfolding according to

$$\ln K^*_{f\to u} - \ln K_{f\to u} = -\left(\frac{\Delta G^{o*}_{f\to u}}{k_B T} - \frac{\Delta G^{o}_{f\to u}}{k_B T}\right) = -\frac{\delta G^{o}_{f\to u}}{k_B T}. \tag{20.2}$$

According to transition state theory (see Section 19.11), the logarithm of the rate constant for folding will be proportional to the free energy of activation for the folding reaction, $\Delta G_f{}^{\ddagger}$, and similarly for unfolding, according to Eq. 19.29. If the rate of change along the reaction coordinate is not affected by the perturbation, then the change in folding rate constant will be related to the change in free energy of activation for the folding reaction by

$$\ln k^*_f / \ln k_f = -\frac{\delta G^{\ddagger}_{u\to\ddagger}}{k_B T}, \tag{20.3}$$

and similarly for k_u. The free energy differences are related by (see Fig. 20.4)

$$\Delta G^{\ddagger}_{f\to\ddagger} - \Delta G^{\ddagger}_{u\to\ddagger} = \Delta G^{o}_{f\to u}, \tag{20.4}$$

and hence also

$$\delta G^{\ddagger}_{f\to\ddagger} - \delta G^{\ddagger}_{u\to\ddagger} = \delta G^{o}_{f\to u}. \tag{20.5}$$

Differentiation of this equation then gives the dependence of the rate constant and equilibrium constant on the concentration of denaturant as

$$-k_B T\frac{d\ln k_u}{dc_U} + k_B T\frac{d\ln k_f}{dc_U} = \frac{d\Delta G^{o}_{f\to u}}{dc_U}. \tag{20.6}$$

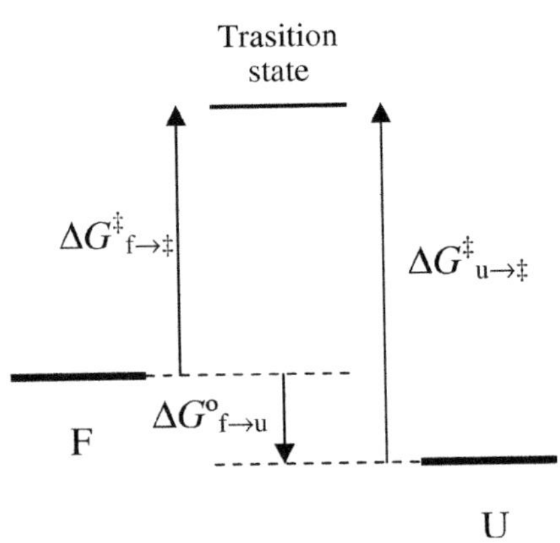

FIGURE 20.4 Free energy diagram for folded, unfolded, and transition states, showing definition of free energy differences.

Introducing the so-called kinetic m-values, m_k one obtains

$$-k_B T m_{k_u} + k_B T m_{k_f} = m_u - m_f = -m_{fu}$$
$$m_{fu} = m_f - m_u; \tag{20.7}$$

here m_u and m_f are related to the activation free energies, $\Delta G^{\ddagger}{}_{f\to\ddagger}$ and $\Delta G^{\ddagger}{}_{u\to\ddagger}$ (in that order!) in the same way as m_{fu} is related to the standard free energy for unfolding, $\Delta G^{\circ}_{f\to u}$ (as described in Section 17.7).

We have seen in Chapter 17 that, in ideal cases, the free energy of folding depends linearly on the concentration of denaturant, and that the slope m_{fu} is independent of denaturant concentration, c_U. In a perfect two-state case, this should also be true of the free energy of activation, in which case ln k_f and ln k_u both vary linearly with c_U, and the slopes m_{k_f} and m_{k_u} are independent of c_U. Accordingly, Eqs. (20.4) and (20.7) serve as a check for two-state kinetics if all three ΔGs and/or all three m-values are measured.

Linear dependence on denaturant concentration is indeed the case for the rate constants calculated from NMR line shape and T_2 relaxation measurements for λ_{6-85} protein in urea,[6] where $m_f = 0.45$, $m_u = -0.70$ ($m_u - m_f = -1.15$); the urea dependence of the free energy difference based on measurements of the equilibrium is $m_{fu} = -1.28$. (Fig. 20.5; see also Fig. 19.3 and accompanying discussion.) This may also be the case for the G46A/G48A mutant of this protein, except that no kinetic results are available below a urea concentration of 2 M. For this protein $m_f = 0.92$, $m_u = -0.18$ ($m_u - m_f = -1.10$), and $m_{fu} = -1.06$.

Exponential dependence on denaturant concentration is shown also by the rate constants for chymotrypsin inhibitor CI2.[7] In this study, the kinetics were measured from the time dependence of fluorescence intensity in rapid mixing experiments. In such a relaxation experiment, the reported rate constant is the sum of the rate constants for forward and reverse reactions (see Section 19.4)

$$k_{obs} = k_f + k_u. \tag{20.8}$$

Results are shown in Fig. 20.6; at low concentration of denaturant, $k_f \gg k_u$, and $k \approx k_f$, while at high denaturant concentration $k \approx k_u$, and under either of these conditions ln k varies linearly with denaturant concentration. The curve has a minimum near the transition midpoint, where $k_f = k_u$. For this protein, $m_f = -1.14$ and $m_u = 0.74$ ($m_u - m_f = -1.88$), while equilibrium measurements give $m_{fu} = 1.79$.

Figure 20.6 includes an inset showing similar kinetic data for a somewhat larger protein, barnase; this protein is seen to fold more slowly at low denaturant concentration than predicted by the simple two-state model (solid curve).

20.6 FOLDING RATES OF SOME PEPTIDES AND PROTEINS

α-Helical peptides. As reference system for protein folding one may use the helix-coil transition of long and short polypeptides.[8,9] Dielectric relaxation measurements

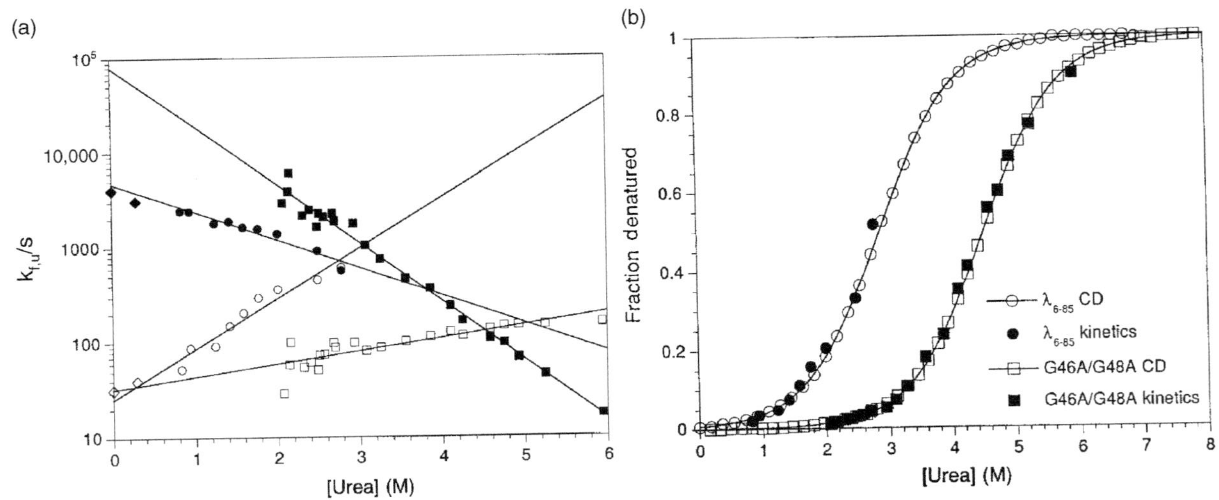

FIGURE 20.5 (a) Rate constants for folding and unfolding of λ_{6-85} protein (circles) and G46A/G48A mutant (squares) based on NMR line shapes and T_2 relaxation times. (b) Extent of unfolding of these proteins as a function of urea concentration, based on equilibrium (circular dichroism) and kinetic (NMR) measurements. From Burton, R.E., Huang, G.S., Daugherty, M.A., Fullbright, P.W., Oas, T.G. *J. Mol. Biol.* **263**: 311–322 (1996).

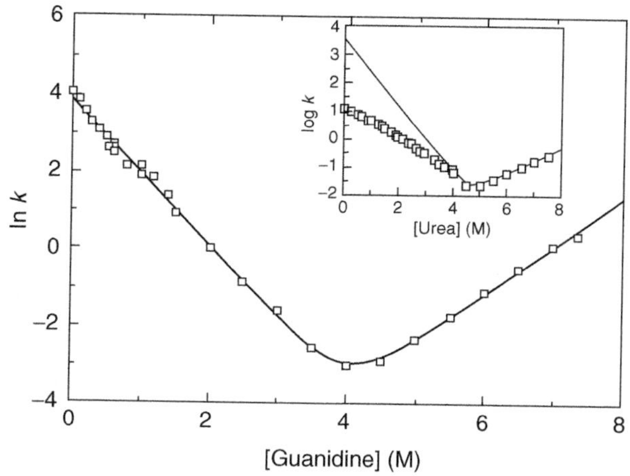

FIGURE 20.6 Rate constants for folding and refolding of CI2 in guanidine solutions, measured at 25 °C. The solid curve is the best fit of the data to a simple two-state kinetic model. (Inset) Similar results for folding and refolding of wild-type barnase in urea solutions. From Jackson, S.E., Fersht, A.R. *Biochemistry* **30**: 10428–10435 (1991).

of the helix-coil transition of poly (γ-benzyl L-glutamate) in a mixed organic solvent gave a maximum relaxation time at the transition midpoint, $\tau_{max} = 5 \times 10^{-7}$s. Temperature jump study of a 21-residue alanine-rich peptide in water gave a 25 time shorter maximum relaxation time $\tau_{max} = 2 \times 10^{-8}$ s. Ultrasonic attenuation measurements on the pH-dependent transition of poly (L-ornithine) in 85 : 15 water–methanol gave $\tau_{max} = 1.7 \times 10^{-8}$ s.

Schwarz has extended the theory of the multistate helix-coil transition to derive equations for the transition kinetics and obtained the following simple expression for mean relaxation time at the transition midpoint (which is also the maximum relaxation time).[10]

$$\tau_{max} = \frac{1}{4\sigma k_s}. \tag{20.9}$$

Here, σ is the equilibrium constant for helix initiation (see Chapter 16), and k_s the rate constant for the elementary step of helix growth (which, at the transition midpoint, is also the value for the reverse step). For poly (γ-benzyl L-glutamate) in a mixed organic solvent, an estimated value for σ of 2×10^{-4} is available based on experimental studies, and this gives a value of $k_s = 2.5 \times 10^9$/s. For the 21-residue peptide, σ was determined to be 0.0012, which gives $k_s = 1.1 \times 10^{10}$/s. No estimate is available for σ of poly (L-ornithine) in 85 : 15 water–methanol; assuming the same value of $\sigma = 0.0012$, one obtains $k_s = 1.2 \times 10^{10}$/s for poly (L-ornithine). These values for the elementary step of helix formation are in rough agreement with the rate constant for the transition of *N*-acetyl-alanyl-*N*-methyl amide, on the order of 10^{10}/s, estimated by simulation (see Section 7.7).

β-Hairpins. Another small reference system is provided by peptides capable of forming stable monomeric β-hairpin structures in aqueous solution.[11] Fluorescence

TABLE 20.1 Some kinetic parameters for folding of peptides and proteins. For references please see text in Sections 20.5 and 20.6

Molecule	Transition	τ(s)	σ	k or k_s(s^{-1})
Ace-Ala-Nme	α to β	–	–	$\sim 10^{10}$
poly (γ-benzyl L-glutamate)	helix ⇆ coil	5×10^{-7}	2×10^{-4}	2.5×10^{9}
21-Residue peptide	helix ⇆ coil	2×10^{-8}	0.0012	1.1×10^{10}
poly (L-Ornithine)	helix ⇆ coil	1.7×10^{-8}	(0.0012)	1.2×10^{10}
Peptides (n = 16, 17)[a]	β-hairpin ⇆ coil	3.5 and 15×10^{-6}	—	—
Trp cage (n = 20)	folding	4×10^{-6}	—	—
Villin domain (n = 35)	folding	4×10^{-6}	—	—
$\lambda_{6\text{-}85}$ Protein (n = 80)	folding	2×10^{-4}	—	—
CI2 (n = 60)	folding	2×10^{-2}	—	—
Barnase (n = 100)	folding	0.3	—	—

[a] n equals number of residues.

measurements of changes immediately following pulsed IR laser temperature jump gave a relaxation time $\tau = 3.5$ μs for a 16-residue peptide, and 15 μs for a 17-residue peptide.[11]

Proteins. The few examples cited so far show that folding rate constants of globular proteins vary considerably. Two small proteins, Trp cage and villin domain, have been found to fold extremely rapidly, as fast as the formation of β-hairpins[12] (Table 20.1). The extrapolated rate constant for the G46A/G48A mutant of λ_{6-85} protein at 37 °C is 8.8×10^4/s; the measured rate constant for λ_{6-85} protein, also at 37 °C, is 4.9×10^3/s, 55/s for CI2 at 25 °C, and 3/s for barnase at 25 °C. (Linear extrapolation to 0 M guanidine concentration gives a rate of 33/s for barnase.) These are all small proteins, with from 20 to 100 amino acid residues. Compared with the helix-coil reference systems, the folding of even λ_{6-85} protein is slow.

20.7 PROBING THE TRANSITION STATE: TANFORD'S β VALUE AND FERSHT'S ϕ VALUE

The nature of the transition state for folding of any particular protein can be investigated by studying the effect of changes in the system (solvent, solutes, and protein itself) on the rate constants for unfolding and refolding. The perturbation in question can be a change in denaturant concentration, presence of a small molecule such as an inhibitor or substrate, or a mutation of the protein. As a result of the perturbation, the equilibrium constant changes, and also one or both rate constants will change, as discussed in Section 20.5.

Thus, the greater the sensitivity of the unfolding rate constant to the concentration of denaturant, the more the transition state resembles the unfolded state. Accordingly, Tanford's β value, defined as

$$\beta_T = m_u/m_{fu}, \qquad (20.10)$$

is a measure of the similarity of transition state and folded state in terms of sensitivity to the presence of denaturant, which usually is taken to be the degree to which groups buried in the folded structure are exposed in the transition state, relative to the unfolded state.

The experimental results show that for λ_{6-85} protein, $\beta_T = 0.39$, while for the G46A/G48A mutant it is 0.83; for CI2 it is 0.60, and for barnase 0.65 (based on the rates near the transition midpoint between 4 and 5 M urea concentration). This indicates that the transition state for G46A/G48A mutant is considerably more solvent exposed than the transition state for the λ_{6-85} protein itself. In both cases, the transition state has solvent exposure intermediate between that of the folded and unfolded states, but whether this is achieved, as seems probable, by partial folding of the protein so that parts of it resemble the folded state, is not certain.

Information on the conformation of the transition state can be obtained by measuring the effect of individual covalent changes of the protein (i.e., mutations) on the equilibrium and rates. The sensitivity of the rates to substitution of a particular amino acid side chain indicates if the side chain does or does not contribute to the stability of the transition state. The effect is expressed in terms of the Φ value[13]

$$\Phi_F = \delta G^{\ddagger}_{f\to\ddagger}/\delta G^{o}_{f\to u} = 1 + \delta G^{\ddagger}_{u\to\ddagger}/\delta G^{o}_{f\to u}, \tag{20.11}$$

which is then a measure of the similarity of transition state and unfolded state in terms of the perturbation. If the mutation consists of replacement of an amino acid side chain, a Φ value of 0 means that the side chain plays an equivalent role in transition and folded states, while a Φ value of 1 means that the side chain does not play a structural role in the transition state. (Of course, if $\delta G^{o}_{f\to u}$ is insignificant, then the change of the side chain does not change the stability of the folded structure, and the Φ value is undefined.)

As an example, we use the cited study on barnase.[13] Of 12 single-site mutants, all less stable than wild type by at least 1 kcal/mol, Φ varies from 0.14 to 0.92. We let the authors summarize their findings in their own words: "Changing the hydrophobic residues in the core gives Φ values in the range of 0.3 ± 0.2. Thus some 70% of the interaction energy remains in the transition state. The core is weakened, but still has significant energy. In conclusion, the transition state for unfolding has significant secondary and compact structure. The major α-helix (residues 6–18) is largely intact, the hydrophobic core is weakened but still important, and the major α-helix is still docked on the central β-sheet. One loop is considerably exposed to solvent in the transition state, as are the N-termini of both helices." This conclusion agrees well with the indication from the Tanford β value that solvent exposure in the transition state is 35% of that in the unfolded state.

In the first half of this chapter we have described what can be learned about the folding process of proteins that demonstrate two-state folding, by applying kinetic theory and, in particular, the theory of a kinetic transition state. In the remainder of this chapter we explore what can be learned about the folding process when this is considered in terms of a multistep transition involving a large number of intermediates.

20.8 EARLY EVENTS IN FOLDING

We have seen (Section 19.8) that in a reaction requiring a large number of consecutive steps, the starting material disappears exponentially, intermediates successively appear and decay, and the appearance of the end product is delayed. In principle, the intermediates can be detected by measurement of transient properties. The early steps in folding have been investigated by following the time dependence of accessibility to isotopic exchange of individual hydrogens by NMR measurements.[14]

This work followed methods that had revealed the existence and approximate structure of various partly unfolded molecules, always in small proportions, under conditions where the folded state is stable. (See Section 17.8.) In the application to kinetic studies, a combination of stopped flow and hydrogen exchange is used. The unfolded protein is prepared in solution of D_2O/denaturant with all exchangeable hydrogens exchanged for deuterium (^2H). Folding can commence after rapid mixing with D_2O, in which denaturant is diluted out. After a delay, t_D another rapid mixing experiment changes the solvent to H_2O. Finally, after a short period, t_X in which accessible hydrogens exchange for ^1H, the exchange is quenched by lowering the pH, and folding is allowed to proceed to completion (Fig. 20.7).

In this experiment, only exchangeable hydrogens protected by secondary structure formed at time t_D after the return to folding conditions are labeled ^2H, all others being ^1H; the NMR spectrum is then used to identify individual labeled groups and the extent to which each is labeled.

The cited studies show the formation of structure in the early stages, before significant refolding has taken place. For cytochrome *c*, stopped flow fluorescence measurements show three relaxation processes in which the fluorescence of a single tryptophan residue is progressively quenched, with relaxation times of 18 ms, 370 ms, and 4 s, respectively. The labeling experiments show that hydrogens in two helices acquire protection from exchange in the first, most rapid stage, while hydrogen atoms in two other helices and a hydrogen atom involved in a tertiary hydrogen bond acquire partial protection from exchange only in the second stage. Obviously, these results are incompatible with a simple two-state model, and indicate the presence of partly folded transient intermediates.

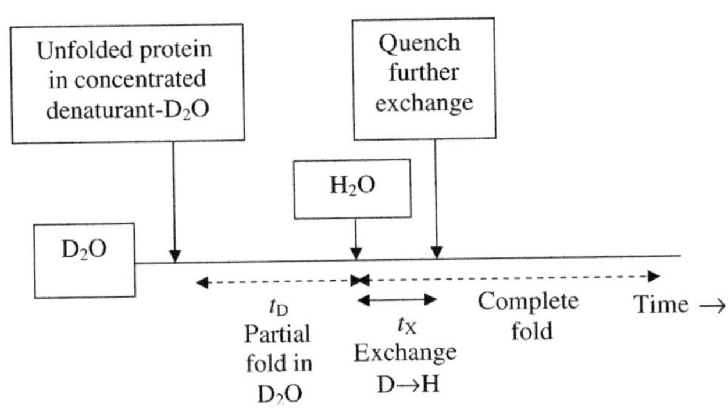

FIGURE 20.7 Scheme for quenched-flow exchange experiment.

The simplest way of modeling this multistate behavior is to introduce additional free energy minima and barriers in the free energy profile along the reaction path, with the minima successively lower in free energy and the barriers successively higher. The free energy for complete unfolding is larger in the absence than in the presence of denaturant by an amount equal to mc_{U} (see Chapter 17.7); the difference between the two free energies then changes linearly with the exposure, that is, with the reaction coordinate, ξ according to

$$(\Delta G_{\mathrm{f}\to\xi})_{c_{\mathrm{U}}=0} - (\Delta G_{\mathrm{f}\to\xi})_{c_{\mathrm{U}}} = \delta G_{\mathrm{U}} = (1-\xi)mc_{\mathrm{U}}. \tag{20.12}$$

This is illustrated in Fig. 20.8. In the presence of denaturant (solid curve), the fully unfolded conformation is the most stable form. The rate of unfolding is controlled by the first free energy barrier $(\Delta G_{\mathrm{f}\to\xi})_{c_{\mathrm{U}}=0}$, and subsequent barriers at lower values of ξ do not substantially influence the rate. When the denaturant is removed (lower curve), the conformation first shifts rapidly to partly folded stable intermediate states u′ and u″, and complete folding takes place more slowly.

There has been debate whether such an early intermediate state is characterized by the formation of partial secondary structure *or* by the formation of a collapsed partial hydrophobic core, according to the "framework" and "hydrophobic collapse" models, respectively. Either model would lead to a smaller radius of gyration, and it seems reasonable to assume that both effects combine to form the early intermediates and the transition state, although not necessarily to the same extent in different proteins. In this author's opinion, the distinction between framework and collapse models is a contrived one, as each would entail the other: formation of a hydrogen-bonded framework shrinks the molecular envelope, collapsing the hydrophobic groups, while, alternatively, a collapse to a hydrophobic entity drives the backbone peptide groups to form internal hydrogen bonds.

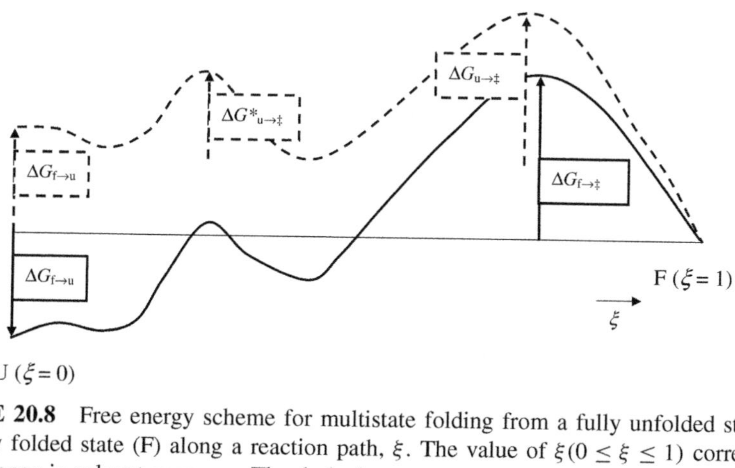

FIGURE 20.8 Free energy scheme for multistate folding from a fully unfolded state (U) to a fully folded state (F) along a reaction path, ξ. The value of $\xi (0 \le \xi \le 1)$ corresponds to the change in solvent exposure. The dashed curve, arrows, and boxes describe the profile in the absence of denaturant ($\Delta G_{\mathrm{f}\to\mathrm{u}} > 0$), and the solid curve, and so on is the profile in the presence of denaturant.

Experimental evidence has been obtained of a collapsed disordered state formed as an early intermediate when a denatured protein (say, in 6 M guanidine HCl) is suddenly brought into a new environment in which the native form is very stable (say, diluted to 0.1 M).[15] In a solution containing also a fluorescent probe (ANS, aniline naphthalene sulfonate), one observes an increase of fluorescence as an indication of formation of a hydrophobic environment for the probe, very shortly after changing the conditions to native-promoting, which then decreases over time as the protein molecules assume their native conformation. This indicates that this transiently stable state has characteristics of the equilibrium molten globule state.

20.9 (FREE) ENERGY LANDSCAPE FOR FOLDING

A complete discussion of protein folding kinetics must take into account the very large number of conformations that are possible for a polypeptide chain. A small fraction of these conformations consists of folded, biologically active molecules. Another (larger) fraction consists of molecules that are partly folded; the vast remainder collectively constitutes the unfolded state. Furthermore, folding unavoidably takes place as a sequence of steps consisting of local conformation changes tending to organize the structure into a specific final conformation.

Internal rotations about single bonds are the principal degrees of freedom that distinguish different conformations separated by free energy barriers. There are two of these in the backbone of each amino acid residue (about N—C$^\alpha$ and C$^\alpha$—C), and from 0 to 5 about single bonds in the side chain. The conformation of the unfolded polypeptide chain has been discussed in Chapter 17. A good idea of the accessibility of different backbone conformations is provided by the data of Fig. 17.13, which shows a distribution of backbone conformation of amino acid residues in high resolution crystal structures.[16]

Disregarding the typically small variations of bond lengths and bond angles, conformation space can be described in terms of a large number, n_r of torsional variables, each with a range from 0 to 2π. Because of local interactions, each internal rotation has several (average, ν_f) low energy regions separated by energy barriers; accordingly the conformational energy has on the order of $\nu_f^{n_r}$ minima, a very large number. A very large accessible volume in the multidimensional space of the n_r torsions corresponds to conformations in which the backbone of every residue is in the most populated, relatively extended conformation. A much smaller multidimensional volume corresponds to the folded conformation, in which the value of each torsion angle is defined within a narrow range.

The equilibrium distribution is determined by the relative free energies of these conformations. For only two degrees of freedom, the energy can be plotted along the vertical dimension of a three-dimensional (3D) space, akin to a relief map with mountains and valleys, that is, a free energy landscape, and this term is used also in a multidimensional context. We can assign the reference conformational free energy of zero to each of the relatively extended conformations that together occupy the largest volume of conformation space. Many conformations are not fully extended; some will have barely higher free energy, while a few will be unrealizable because

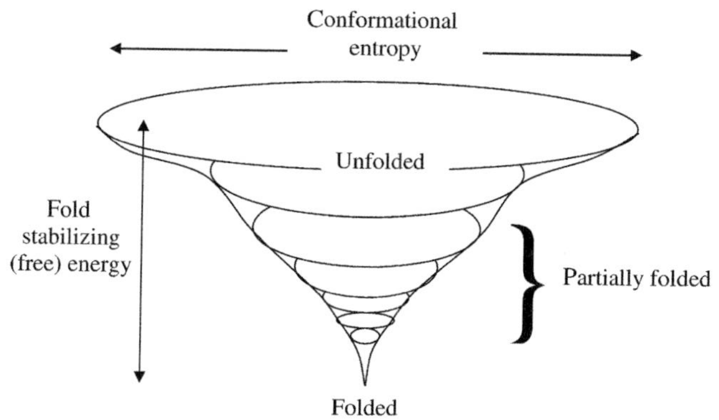

FIGURE 20.9 Schematic (smooth) free energy landscape for protein folding. Progressively narrowing circles represent increasingly folded conformations.

of steric overlap. In the very small volume of conformation space containing the folded state, the conformational free energy is large and negative. In between lie conformations that are partly folded.

Figure 20.9 represents a hypothetical free energy landscape for protein folding. When drawn at higher resolution, the landscape would show energy barriers between and also within successive circles, and would be sprinkled with high energy spikes corresponding to "impossible" conformations with steric overlap. Also at higher resolution, the folded state would be represented as a collection of relatively few conformations, differing for the most part in side chain torsions, separated by free energy barriers. The overall funnel shape expresses the idea that the free energy of partly (correctly) folded intermediates is lower as the intermediate is more extensively folded, although not necessarily in proportion to the extent of folding. We shall see next that this is a requirement needed in order to resolve the Levinthal paradox.

20.10 THE "LEVINTHAL PARADOX" AND THE FOLDING FUNNEL

In the 1960s, Cyrus Levinthal analyzed a mechanism whereby each position for each internal rotation would be tested randomly until all rotations were correct, when the molecule would be folded. As we have n_r (say 200) rotations each with on average ν_r (say 3) possible low energy states, the number of possible states is

$$N_u = \nu_f^{n_r} \approx 3^{200}, \tag{20.13}$$

and it is clear that this many states could not possibly be gone through in a random manner if the one correct answer were to come out in the time observed for folding. This seemed paradoxical at the time.

The paradox is resolved by a model in which each step in the direction of a correctly folded molecule is favored energetically. Formally, the 3^{200} states correspond to an entropy S_u, for which

$$TS_u \approx k_B T \ln N_u \approx k_B T n_r \ln \nu_f \approx 120\,\text{kcal/mol}. \tag{20.14}$$

The folded state is stabilized by a free energy (due to hydrophobic interactions, etc.) of somewhat larger magnitude. If the free energy favoring folding and opposing the entropic term $-TS_u$ is distributed evenly over all steps that lead to a folded structure, then correct steps will be favored either by an enhanced forward rate or else by a reduced reverse rate. In the direction of the folded state, the free energy surface both narrows because the number of possible conformations drops as folding progresses, and slopes downward because of stabilizing interactions; this is the folding funnel. As each of the many steps in the reaction scheme is mathematically equivalent to a step in a random walk process, the folding process can be approximated quite well in terms of a diffusive model, subject to a potential successively favoring folded states along the entire path; the effect of the potential is to enormously enhance the rate of progress in the correct direction.

In the following section, we further apply the folding funnel's free energy "landscape" to aspects of the reaction kinetics.

20.11 TRANSITION STATE(S), PATHWAY(S), REACTION COORDINATE(S)

The folding funnel can become the basis for a kinetic model with a transition state if the potential favoring folding does not change linearly with the decreasing (logarithm of the) number of intermediate states available as folding progresses. If the early steps in folding are stabilized by long-range interactions, but not as much as the last steps, the free energy will have a maximum somewhere along the folding pathway. This is of course also the model that explains the cooperativity of the helix-coil transition, except that in that case (as discussed in Chapter 16) the transition state occurs after a relatively small number of residues have formed a stretch of helix, while the transition state in protein folding occurs when the structure of a sizable fraction of the molecule has become organized.

There are two important differences between the kinetics of the helix coil transition and of protein folding. First, the uniformity of a long homopolymer molecule means that energetics and kinetics of forming a helical segment are the same at all positions (except near the chain ends), while the energetics of initiating the folding of a protein depend on the locality by virtue of the unique and nonrepetitive amino acid sequence. Second, inter-residue interactions in an α-helix are local (interactions of residue i with residues $i \pm k$ decreases rapidly as k increases); on the other hand, the stability of folded proteins depends critically on the presence of long-range contacts.[17] Rapid folding results if the potential (from long-range interactions) driving folding keeps pace with the decrease of entropy as the number of states shrinks as folding progresses; folding will be slower if the initial steps produce proportionally less stabilization than the final steps; this results in a positive free energy balance with a maximum somewhere along the reaction path, which corresponds to the transition state. The transition state requires only a few ordered residues in the case of helix formation; for this reaction the initiation equilibrium constant, σ is around 10^{-3}. Assuming a common elementary step for folding, the relaxation times can give us the equivalent for other molecules, 10^{-5} for β-hairpin formation and for folding of Trp cage and villin domain, 10^{-7} for λ_{6-85} protein, 10^{-9} for CI2, and 10^{-10} for barnase (Table 20.1). This indicates that to reach the transition state of barnase, twice as much structure must be organized, without compensating stabilization, as for the β-hairpin, a quite modest factor. Thus, the wide range of folding rates is accommodated quite well by the funnel model.

Consideration of protein folding in terms of the kinetics of a multistate system allows us to form an extremely complex picture of the process. From a single starting point, many routes lead to the folded end state at the bottom of the funnel, each with a different order in which "correct" local transitions take place. As a consequence, there are many possible successful pathways, and multiple transitions states. Because a real protein's structure is nonrepetitive, the free energy surface is irregular, and certain paths will be more favorable than other paths. Because of the irregularity, it is possible (although not guaranteed) that one of the many transition states is of sufficiently low energy that it is heavily favored over all the others. This would then be the transition state experienced by the majority of folding molecules.

One can propose a number of different measures on which to base a reaction coordinate for folding. Examples include (i) the mean square deviation of dihedral angles for single bonds from the values in the folded structure, (ii) the degree of solvent exposure, (iii) the extent to which secondary structure is formed, (iv) the radius of gyration, and (v) the sum of the squares of deviations of atomic positions from those in the folded structure. Any one of these may be useful for qualitative analysis, for a practical purpose or in a specific case, but the very diversity indicates that *no generally applicable unique reaction coordinate for folding can be specified.*

The folding funnel does not have a smooth surface, but the *surface's roughness* does not always have consequences that can be determined experimentally. Stable intermediates accumulate only if the surface is so rough that these have lower free energy than the unfolded state *and* are separated from this by a free energy barrier lower than the barrier for the reactions needed to complete folding, as illustrated in Fig. 20.1. Thus, a very rough free energy surface may correspond to the formation of transiently stable (misfolded) intermediates, and to slow folding. However, it is *not* true that slow folding necessarily implies the presence of misfolded intermediates.

20.12 COMPUTER SIMULATIONS OF PROTEIN FOLDING AND UNFOLDING

20.12.1 Background

A vast literature addresses the "protein folding problem" by computer simulation. Much of this deals not with studies of folding kinetics (or equilibria) but with the determination (prediction) of the native fold of a protein by *searching* the conformation space; a desire to make economic use of available computer facilities has led to the use of "scoring functions" made up of very simple energy functions and pseudo-energies. The latter are terms that favor structures resembling those found in proteins of known structure, and, for example, tendencies to form different types of secondary structure on the basis of the amino acid sequence. While considerable progress has been made, it is fair to say that this approach has not yet been truly successful. Collective progress in the field is reported every few years at a CASP meeting.[18]

We do not discuss structure prediction in this book; the remainder of this chapter will be devoted to studies that seek to understand protein folding in terms of detailed simulations that produce equilibrium ensembles of structures, including structures along the folding pathway, in particular at and near the transition state. The problems are the usual ones: to find an energy function for the simulations that is sufficiently accurate to adequately represent the equilibrium between folded and unfolded states, and, at the same time, is sufficiently rapidly evaluated to study a process that takes on the order of microseconds and longer in the physical system that is to be represented. One necessary factor in recent success in overcoming these problems has been the steady improvement of computer architecture; another has been the discovery of small proteins that actually fold with such short relaxation times, such as Trp cage and villin domain. Yet another very helpful factor has been the development of faster simulation methods, and finally, the application of simulations to study the unfolding process under conditions where the relaxation time is very short, within the time span available in simulations, that is, at elevated (simulated) temperatures, at which in physical experiments solutions boil and proteins decompose, which does not happen in simulations.

20.12.2 Simulated Rapid Unfolding

In a series of articles Li and Daggett analyzed the unfolding of CI2 and barnase using molecular dynamics simulations with an all-atom model and explicit solvent.[16] Simulation of the folded CI2 protein structure based on the crystal coordinates at 298 K produced insignificant change of structure (under 2 Å root-mean-square displacement of C^{α} atoms in 1.6 ns). Simulation at 498 K caused rapid unfolding; cluster analysis of the trajectory showed the presence of a transition state in the time interval of 200–230 ps after the beginning of the simulation (Fig. 20.10). Before this interval, changes in root-mean-square deviation (RMSD) were below 3 Å, while shortly afterward this reached a range of 5–7 Å; after around 1 ns the protein unfolded further to attain a final RMSD of around 12 Å. Additional simulations with starting coordinates based on three different NMR structures[19a,b] confirmed these findings. Not surprisingly, the transition state was reached at a different time, varying between 65 and 330 ps, in each simulation. Because the transition state is (as any intermediate) partly unfolded, the structures of transition states in different simulations are expected to be identical only in part. A variety of measures characterizing these structures were quite similar, including C^{α} RMSD

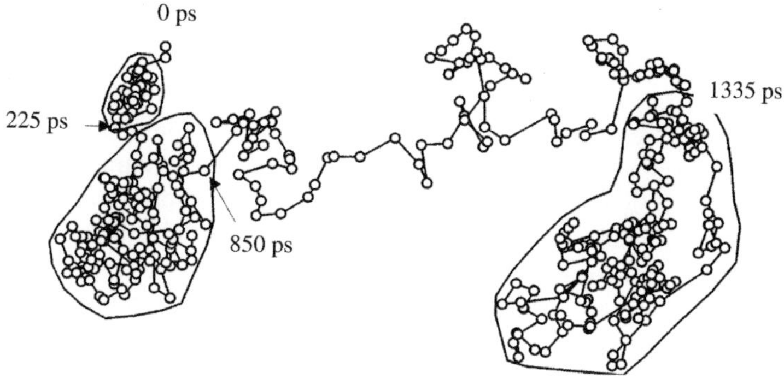

FIGURE 20.10 Progressive change of conformation during simulated unfolding of CI2. Distance between successive points represents C^{α}RMSDs between conformations 5 ps apart. Best-plane projection seeks to represent changes relative to other conformations.[24] From Li, A., Daggett, V. *J. Mol. Biol.* **257**: 412–429 (1996).

(from 3.3 to 4.3 Å), radius of gyration, solvent accessible surface and α-helix content, whereas β-structure content varied considerably. The most stringent test was a comparison of experimental Φ_F-values for various side chains[13] (see Section 20.7 and Eq. 20.11), with Φ_{MD}-value estimated for the simulated transition state structures on the basis of the number of the side-chain's van der Waals contacts, relative to those in the folded structure. The correlation coefficient between the two sets of Φ-values varied between 0.80 and 0.93.

Overall, the transition state for simulated unfolding of CI2 at a very high temperature could be described in much the same way as the transition state for physical (un)folding at room temperature discussed in Section 20.7 (Fig. 20.11). The correlation of Φ-values is convincing evidence that these simulations identify the transition state for unfolding of CI2, in spite of the difference in conditions for experiment and simulation.

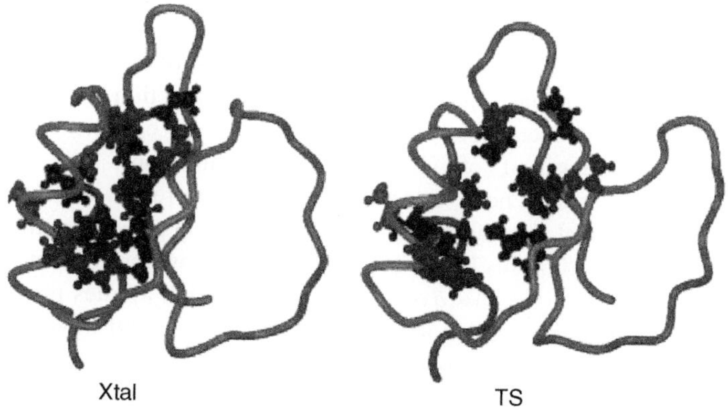

FIGURE 20.11 Comparison of transition state observed by simulations with the crystal structure. From Li, A., Daggett, V. *J. Mol. Biol.* **257**: 412–429 (1996).

20.12.3 Faster Simulation

Successful simulated folding became possible thanks to the discovery of a few very rapidly folding proteins, and, of course, improved computer resources, with many and faster processors in parallel. Villin domain was found to approximate the native fold and again unfold within a 1-μs simulation.[20] Sometime after, additional speed was attained by representing solvation not with explicit water molecules, but via a more approximate implicit solvation potential,[21] and this led to successful simulated folding of two mini-proteins, BBA5 and Trp cage.[22]

One study[22a] was carried out on a very large number of home computers (Vijay Pande's "Folding at Home" project) and allowed the accumulation of a large number (9000) of independent folding simulations, of which 16 produced a folded structure of BBA5 within 2 ns; by assuming first-order kinetics, the authors calculated a folding halftime of 6 μs. Experiments show that this protein is only marginally stable at 298 K, with a relaxation time for folding of 7.5 μs.

The two other studies[22b,c] used the same algorithm and force field, except that these used a Berendsen thermostat (Eq. 7.16) and omitted a stochastic dynamics component that combined random forces with friction coefficients intended to mimic the viscous drag of water (see Section 21.5), present in the first study. These two studies were carried out with Trp cage protein; stable folded structures appeared in tens of nanoseconds. This is many times shorter than the halftime of 4 μs for folding observed experimentally. Apparently, elimination of the effects of friction with solvent enormously speeds up the simulated folding.

The advantage of ultrafast simulated folding is offset by the fact that the time dependence of the events observed in these simulations can no longer be a useful measure of their realism. However, simulation studies can produce estimates of reaction rates by another means: on the basis of independent measurements of the potential of mean force and the rate of progress along the reaction path (see Section 19.15.) The former can be obtained with too rapid simulations, while the much less time-consuming simulations of the rate of progress can be done with a more realistic force field in which friction is accounted for by the presence of explicit solvent (or a frictional force of appropriate magnitude). Calculation of the potential of mean force is needed anyway, as it determines the free energy difference between folded and unfolded states, which is, of course, easily obtained experimentally for a comparison. Note that although a reaction coordinate for protein folding cannot be determined in general, one *can* be established for any one given protein on the basis of folding or unfolding simulations.

How much progress has been made can be seen from recent work by Ding and coworkers.[23] Additional speedup is achieved by adopting several methods that are otherwise not discussed in this book. The velocity and positions are integrated with the use of the *discrete molecular dynamics* method, which applies momentum and energy conservation laws at each collision involving protein and solvent atoms. (For this to be effective, the continuous interatomic potentials must be approximated by step functions; e.g., a pair interaction may be approximated with five steps over the range between 3.5 and 9 Å.) The latest implementation includes all protein atoms except hydrogens bonded to carbon, uses an *implicit solvation model*, and *omits Coulomb forces*, but uses an *explicit hydrogen-bond potential*. (As computers have become faster, these practical limitations have been removed, as evidenced by recent work cited at the end of this chapter.) The work also applies the *replica exchange* method, in which multiple copies of the system are simulated at different temperatures, with exchange between systems at adjacent temperatures possible on the basis of a Metropolis Monte Carlo criterion. Replica exchange can greatly speed up the exploration of the Boltzmann distribution at the lower simulated temperatures.

The results indicate that the simplifications of the potential functions have not resulted in unrealistic behavior. Trp cage and villin domain assume the experimentally determined fold within 2.5 Å RMS deviation of C^{α} positions, two other small proteins assume good secondary structure, with greater deviations in loops, and, finally, two small proteins reach near-native states during the simulations, but then again unfold. The plentiful statistics allow calculation of potentials of mean force for the first two proteins, at several temperatures, as shown for villin domain in Fig. 20.12. At 300 K, the folded state (at RMSD near 2.5 Å) is favored by circa 1 kcal/mol, while at 323 and 340 K the unfolded state (RMSD near 5.5 Å) has lower free energy.

The profile at 300 K shows a partly folded state (RMSD 4–4.5 Å) at about the same free energy as the unfolded state, which would be a transiently stable intermediate in the folding reaction, as postulated in Fig. 20.8. These intermediates are compact, with partly formed helices. The transition state lies closer along the

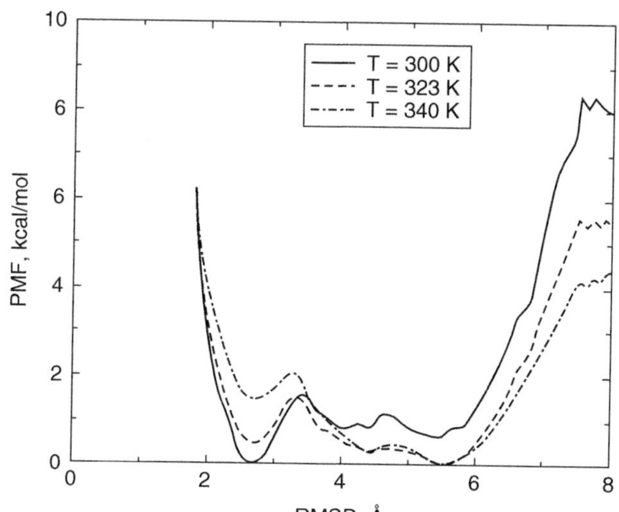

FIGURE 20.12 Conformational free energy for folding/unfolding equilibrium of villin subdomain at three temperatures, with RMS deviation from experimental C^{α} positions as reaction coordinate from a simulation study by Ding, F., Tsao, D., Nie, H., Dokholyan, N.V. *Cell* **16**: 1010–1018 (2002).

reaction coordinate to the folded state than the unfolded state, in agreement with what was concluded on the basis of Tanford's β-value of 0.39.

20.13 CONCLUSION

Between experimental studies and simulations, one thus has a good description of the folding process of very small globular proteins. A kinetic model with folded and unfolded states separated by a free energy barrier, with no or few transiently stable intermediates, gives an excellent global account of the experimental results. At higher resolution, these states are found to be ensembles of many conformations: not only is the statistical coil state of the unfolded chain inherently disordered but also the folded state includes locally disordered molecules, and the unfolded state includes molecules with ordered parts. Analysis of effects of amino-acid substitution on the kinetics in terms of Φ_F-values provides structural details of the transition state.

Simulations have become fast enough to study the process in a statistically significant manner, and confirm the above-mentioned interpretation of the experimental results. Simulations have allowed calculation of the potential of mean force for the transition between folded and unfolded states, and determination of conformational equilibrium distributions of folded and unfolded states and, most importantly, of the transition state in atomic detail. The results of Φ_F-value analysis are a key ingredient for validating the models used in the simulations.

To ask for a correct representation of all details might place too many demands on the accuracy of the potential function. This is so because the energy stabilizing the folded state is 100 kcal/mol or more, which is nearly compensated by the entropic free energy gain on unfolding, with the *net* stability being only a few kilocalories per mol. Thus, in order to reproduce the standard free energy of folding to within 1 kcal/mol, the potential function must reproduce both the stabilizing and the destabilizing terms to within 1%. It is not clear that this has been achieved with any potential function, as the fast, more approximate methods do not produce good results in all cases, and the most accurate force fields are too slow to have been exhaustively tested.

Obviously, more work remains to be done to improve simulations of protein folding with better force fields, something which will become possible as computer architecture improves further.

NOTES

1. Rye, H.S., Roseman, A.M., Chen, S., Furtak, K., Fenton, W.A., Saibil, H.R., Horwich, A.L. GroEL-GroES cycling: ATP and nonnative polypeptide direct alternation of folding-active rings. *Cell* **97**: 325–338 (1999).
2. Nall, B.T. Proline isomerization as a rate-limiting step. In: Pain, R.H., ed. *Mechanisms of Protein Folding*. Oxford: Oxford University Press, 1994, pp 80–103.

3. Brandts, J.F., Halvorson, H.R., Brennan, M. Consideration of the possibility that the slow step of protein denaturation is due to cis-trans isomerism of proline residues. *Biochemistry* **14**: 4853 (1975).
4. Gilbert, H.F. *The formation of native disulfide bonds*. In: Pain, R.H., ed. *Mechanisms of Protein Folding*. Oxford: Oxford University Press, 1994, pp 104–136.
5. Saxena, V.P., Wetlaufer, D.B. Formation of three-dimensional structure in proteins. I. Rapid nonenzymic reactivation of reduced lysozyme. *Biochemistry* **9**: 5015–5022 (1970).
6. (a) Huang, G.S., Oas, T.G. Submillisecond folding of monomeric lambda repressor. *Proc. Natl. Acad. Sci. USA* **92**: 6878–6882 (1995). (b) Burton, R.E., Huang, G.S., Daugherty, M.A., Fullbright, P.W., Oas, T.G. Microsecond protein folding through a compact transition state. *J. Mol. Biol.* **263**: 311–322 (1996).
7. Jackson, S.E., Fersht, A.R. Folding of chymotrypsin inhibitor 2. 1. Evidence for a two-state transition. *Biochemistry* **30**: 10428–10435 (1991).
8. Schwarz, G., Seelig, J. Kinetic properties and the electric field effect of the helix-coil transition of poly (γ-benzyl L-glutamate) determined from dielectric relaxation measurements. *Biopolymers* **6**: 1263–1277 (1968). Hammes, G.C., Roberts, P.R. Dynamics of the helix-coil transition in poly-L-ornithine. *J. Am. Chem. Soc.* **91**: 1812–1816 (1969).
9. Williams, S., Causgrove, T.P., Gilmanshin, R., Fang, K.S., Callender, R.H., Woodruff, W.H., Dyer, R.B. Fast events in protein folding: Helix melting and formation in a small peptide. *Biochemistry* **35**: 691–697 (1996).
10. Schwarz, G. General theoretical approach to the thermodynamic and kinetic properties of cooperative intramolecular transformations of linear biopolymers. *Biopolymers* **6**: 873–897 (1968).
11. (a) Muñoz, V., Thompson, P.A., Hofrichter, J., Eaton, W.A. Folding dynamics and mechanism of β-hairpin formation. *Nature (London)* **390**: 196–199 (1997). (b) Muñoz, V., Ghirlando, R., Blanco, F.J., Jas, G.S., Hofrichter, J., Eaton, W.A. Folding and aggregation kinetics of a β-hairpin. *Biochemistry* **45**: 7023–7035 (2006).
12. (a) Qiu, L., Pabit, S.A., Roitberg, A.E., Hagen, S.J. Smaller and faster: The 20-residue Trp-cage protein folds in 4 μs. *J. Am. Chem. Soc.* **124**: 12952–12953 (2002). (b) Kubelka, J., Eaton, W.A., Hofrichter, J. Experimental tests of villin domain folding simulations. *J. Mol. Biol.* **329**: 625–630 (2003).
13. Matouschek, A., Kellis, J.T., Jr., Serrano, L., Fersht, A.R. Mapping the transition state and pathway of protein folding by protein engineering. *Nature (London)* **340**: 122–126 (1989).
14. (a) Udgoankar, J.B., Baldwin, R.L. NMR evidence for an early framework intermediate on the folding pathway of ribonuclease A. *Nature (London)* **335**: 694–699 (1988). (b) Roder, H., Elöve, G.A., Englander, S.W. Structural characterization of folding intermediates in cytochrome *c* by H-exchange labelling and proton NMR. *Nature (London)* **335**: 700–704 (1988).
15. (a) Ptitsyn, O.B., Pain, R.H., Semisotnov, S.Y., Zerovnik, E., Razgulyaev, O.I. Evidence for a molten globule state as a general intermediate in protein folding. *FEBS Letters* **262**: 20–24 (1990). (b) Ptitsyn, O.B. Molten globule and protein folding. *Advances in Protein Chemistry*: 83–229 (1995). (c) Finkelstein, A.V., Ptitsyn, O.B, *Protein Physics*. Amsterdam: Academic Press/Elsevier Science, 2002.

16. Lovell, S.C., Davis, I.W., Arendall, W.B., de Bakker, P.I.W., Word, J.M., Prisant, M.G., Richardson, J.S., Richardson, D.C. Structure validation by C^{α} geometry: φ, ψ and C^{β} deviation. *Proteins: Struct. Funct. Genet.* **50**: 437–450 (2003).

17. The Schellman model that so successfully describes the helix-coil equilibrium is known in physics as a *one-dimensional Ising model*. A sharp distinction between the infinitely large one- and three-dimensional Ising models is that, depending on the model's parameters, the 3D model *can* show a true phase transition, while the 1D model cannot. The two-state (un)folding transition of a microscopic protein domain is a corollary of a true phase transition for a macroscopic system.

18. Most recent report: Ninth meeting on the critical assessment of techniques for protein structure prediction. *Proteins: Struct. Funct. Genet.* **79**(S10) (2011).

19. Li, A., Daggett, V. (a) Characterization of the transition state of protein unfolding by use of molecular dynamics: chymotrypsin inhibitor 2. *Proc. Natl. Acad. Sci. USA* **91**: 10430–10434 (1994). (b) Identification and characterization of the unfolding transition state of chymotrypsin inhibitor 2 using molecular dynamics simulations. *J. Mol. Biol.* **257**: 412–429 (1996). (c) The unfolding of barnase: characterization of the major intermediate. *J. Mol. Biol.* **275**: 677–694 (1998).

20. Duan, Y., Kollman, P.A. Pathways to a protein folding intermediate observed in a 1-microsecond simulation in aqueous solution. *Science* **282**: 740–744 (1998).

21. See Bashford, D., Case, D.A. Generalized Born models of macromolecular solvation effects. *Annu. Rev. Phys. Chem.* **51**: 129–152 (2000).

22. (a) Snow, C.D., Nguyen, H., Pande, V.S., Gruebele, M. Absolute comparison of simulated and experimental protein-folding dynamics. *Nature (London)* **420**: 102–106 (2002). (b) Simmerling, C., Strockbine, B., Roitberg, A.E. All-atom structure prediction and folding simulations of a stable protein. *J. Am. Chem. Soc.* **124**: 11258–11259 (2002). (c) Chowdhury, S., Lee, M.C., Xiong, G., Duan, Y. *Ab initio* folding simulation of the Trp-cage mini-protein approaches NMR resolution. *J. Mol. Biol.* **327**: 711–717 (2003).

23. Ding, F., Tsao, D., Nie, H., Dokholyan, N.V. *Ab initio* folding of proteins with all-atom discrete molecular dynamics. *Cell* **16**: 1010–1018 (2002).

24. Levitt, M., Molecular dynamics of native protein II. Analysis and nature of motion. *J. Mol. Biol.* **168**: 621–657 (1983).

SUGGESTED READING

For the interested reader we suggest the following more recent simulation studies:

Day, R., Paschek, D., Garcia, A.E. Microsecond simulations of the folding/unfolding thermodynamics of the Trp-cage miniprotein. *Proteins: Struct. Funct. Bioinformat.* **78**: 1889–1899 (2010).

Shaw, D.E., Maragakis, P., Lindorff-Larsen, K., Piana, S., Dror, R.O., Eastwood, M.P., Bank, J.A., Jumper, J.M., Salmon, J.K., Shan, Y., Wriggers, W. Atomic-level characterization of the structural dynamics of proteins. *Science* **330**: 341–346 (2010).

Lindoff-Larsen, K., Piana, S., Dror, R.O., Shaw, D.E. How fast-folding proteins fold. *Science* **334**: 517–520 (2011).

Stigler, J., Ziegler, F., Gieseke, A., Gebhardt, C.M., Rief, M. The complex folding network of calmodulin molecules. *Science* **334**: 512–516 (2011).

GENERAL REFERENCES

Maxwell, K.L., Wildes, D., Zarrine-Afsar, A., De Los Rios, M.A., Brown, A.G., Friel, C.T., Hedberg, L., Horng, J-C., Bona, D., Miller, E.J., Vallée-Bélisle, A., Main, E.R.G., Bemporad, F., Qiu, L., Teilum, K., Vu, N-D., Edwards, A.M., Ruczinski, I., Poulsen, F.M., Kragelund, B.B., Michnick, S.W., Chiti, F., Bai, Y., Hagen, S.J., Serrano, L., Oliveberg, M., Raleigh, D.P., Wittung-Stafshede, P., Radford, S.E., Jackson, S.E., Sosnick, T.R., Marqusee, S., Davidson, A.R., Plaxco, K.W. Protein folding: Defining a "standard" set of experimental conditions and a preliminary kinetic data set of two-state proteins. *Protein Sci*. **14**: 602–616 (2005).

Onuchic, J.N., Wolynes, P.G. Theory of protein folding. *Curr. Opin. Str. Biol.* **14**: 70–75 (2004).

Dill, K., Chan, H.S. From Levinthal to pathways to funnels. *Nature Str. Biol.* **4**: 10–19 (1997).

Pain, R.H., ed., *Mechanisms of Protein Folding*. Oxford: Oxford University Press; 2000.

Richards, F.M., Eisenberg, D.S., Kim, P.S., eds., *Protein Folding Mechanisms. Advances in Protein Chemistry* Vol. 53, New York: Academic Press. 2000.

Fersht, A.R. *Enzyme Structure and mechanism*, New York: W. H. Freeman, 1999.

Fersht, A.R., Daggett, V. Protein folding and unfolding at atomic resolution. *Cell* **108**: 1–20 (2002).

21

Irreversible and Stochastic Processes

21.1 INTRODUCTION

Many systems have properties that fluctuate in an apparently irregular manner. In this book, we are concerned principally with motion of molecular systems along various reaction coordinates, but traffic density on the highway and stock prices on the exchange are equally examples of systems with important fluctuating properties.

As described by quantum mechanics (e.g., by the time-dependent Schrödinger equation), the time dependence of structure and energy of molecular systems is inherently unpredictable and stochastic in nature.[1] However, even in a system in which the motion is in principle predictable (as in a dynamics simulation with Newton's equations of motion), the large number of changing molecular interactions ("collisions") render the motion so complex and irregular as to make it unpredictable in practice.

Systems containing *many* particles that individually show unpredictable behavior often can be described with simple models. Examples are the use of kinetic rate equations to describe the progress of chemical reactions, that was used extensively in previous chapters, and the classical theories of diffusion and friction in solutions containing many particles. But these equations do not describe the behavior of one or a few reactive or diffusive particles.

In order to illustrate these ideas and theories, which can, however, be applied much more widely, we shall use the motion of a small particle in solution as a familiar *example* of a system that moves along a reaction coordinate. We begin the chapter with a brief discussion of the macroscopic theory of diffusion in many

Equilibria and Kinetics of Biological Macromolecules, First Edition. Jan Hermans and Barry Lentz.

particle systems. Following this, we develop the theory of the irregular behavior of single particles with so-called stochastic[2] models, in terms of series of random events. We conclude with a discussion of ballistic and diffusive processes in terms of transition state theory and of Kramers' theory of reaction rates.

21.2 MACROSCOPIC TREATMENT OF DIFFUSION

An uneven spatial distribution of molecules in solution gradually flattens out to become smooth, and, in the absence of external forces, the distribution becomes uniform. Molecules are (apparently) transported from regions of high to low concentration, which can be described in terms of an effective *flux* (number of particles passing per unit of time through a unit of cross section). On the basis of experiments with dilute (hence ideal) solutions, the flux is found to be proportional to the concentration gradient

$$\text{Flux}_{\text{diff}} = -D\frac{\partial c}{\partial x}. \tag{21.1}$$

This empirical macroscopic relation is called *Fick's law*.

At any value of the coordinate, x where the concentration gradient is independent of x, the flux "in" is equal in magnitude to the flux "out." But wherever flux depends on the position, x, that is, when $\partial^2 c/\partial x^2 \neq 0$, conservation of mass requires that the concentration change with time according to

$$\frac{\partial c}{\partial t} = -\frac{\partial \text{Flux}_{\text{diff}}}{\partial x}. \tag{21.2}$$

If the flux is given by Eq. 21.1, this gives

$$\frac{\partial c}{\partial t} = D\,\frac{\partial^2 c}{\partial x^2}, \tag{21.3}$$

the *diffusion equation*. This differential equation can be solved for various starting conditions. Of particular interest is the case where initially all N_0 molecules are collectively at a single value of x (for which we pick $x = 0$). The concentration profile, $c(x, t)$ is a Gaussian function (Fig. 21.1):

$$c(x, t) = \frac{N_0}{\sqrt{4\pi Dt}}\exp\left(\frac{-x^2}{4Dt}\right). \tag{21.4}$$

The mean square displacement varies linearly with time according to

$$\langle x(t)^2\rangle = 2Dt, \tag{21.5}$$

while in three dimensions (x_1, x_2, and x_3)

$$\langle x_1(t)^2\rangle + \langle x_2(t)^2\rangle + \langle x_3(t)^2\rangle = 6Dt. \tag{21.6}$$

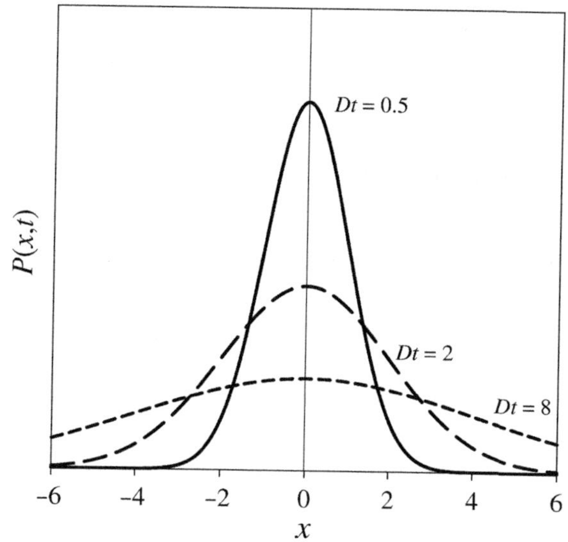

FIGURE 21.1 Distribution of solute for a diffusive process in one dimension, expressed as a probability distribution, for three different values of Dt. At time zero, all particles are at $x = 0$.

This provides a general solution for any given one-dimensional starting concentration (at time zero), as any starting concentration profile can be treated as a large number of sets of molecules, N_i in volume elements dx_i, each set spreading out independently according to a time-dependent Gaussian centered at x_i.

21.3 FRICTION FORCE OPPOSES MOTION

In the absence of an applied force, diffusion results in a uniform concentration, after which the flux becomes zero. In order to keep particles moving steadily in one direction, an external force must be applied in the same direction as the velocity, v and with magnitude proportional to the mean velocity of the particle,

$$F_{x,\text{ext}} = f v_x. \tag{21.7}$$

The proportionality constant f is called the *frictional coefficient*. In the absence of an external force, *a moving particle experiences a frictional force* of this magnitude that opposes the motion and will slow the particle to a standstill

$$F_f = -f v. \tag{21.8}$$

According to Stokes' law, the frictional coefficient of a sphere of radius r in a medium with viscosity η is given by

$$f = 6\pi\eta r. \tag{21.9}$$

An external force will provide an additional component to the flux of particles

$$\text{Flux}_{\text{ext}} = v_x c\,(x) = \frac{F_{x,\text{ext}}}{f} c(x). \tag{21.10}$$

Letting also the diffusive flux of Eq. 21.1 correspond to an effective average particle velocity, v_{diff}, one has

$$\begin{aligned}\text{Flux}_{\text{diff}} &= v_{\text{diff}} c(x),\\ v_{\text{diff}} &= \frac{-D}{c(x)} \frac{\partial c(x)}{\partial x} = -D \frac{\partial \ln c(x)}{\partial x}.\end{aligned} \tag{21.11}$$

We now attribute this motion to an internal force, equal in size to the external force that would produce this same movement

$$F_{\text{int}} = -f\,v_{\text{diff}} = -fD \frac{\partial \ln c(x)}{\partial x}. \tag{21.12}$$

The source of the force is a potential associated with the concentration gradient, clearly the chemical potential μ of the solute (always assuming ideal solution behavior)

$$F_{\text{int}} = -\frac{\partial \mu(x)}{\partial x} = -k_{\text{B}} T \frac{\partial \ln c(x)}{\partial x}. \tag{21.13}$$

Equating the two expressions for F_{int}, one finds that the frictional coefficient f is inversely proportional to the diffusion coefficient, D,[3]

$$f = \frac{k_{\text{B}} T}{D}. \tag{21.14}$$

There is obviously a *strict relation between friction and diffusion.* That this should be so follows qualitatively when one considers that diffusive motion (Brownian motion) and friction both result from collisions with solvent molecules that produce exchange of kinetic energy; a stationary particle thereby acquires a velocity, and a moving particle's velocity changes with a gradual loss of correlation with the original velocity.

21.4 RANDOM WALK AS A MODEL DIFFUSIVE PROCESS

It is well recognized that diffusion is the result of irregular molecular motion, which continues when the system has come to equilibrium and the concentrations are no longer changing. In order to describe diffusion in terms of such motion, we turn to the use of stochastic models.

Many stochastic models are based on the random-selection and *random walk models* and the related *central limit theorem.* The classical one-dimensional random walk consists of a sequence of steps, of equal magnitude, δs, to the right or to the left. The direction of each step is a random event, not influenced by prior history. The outcome of each trial of n steps is a distance traveled, and the outcome of a large number of such steps is a *distribution* of distances over a range of values ($-n\delta s$ through $+n\delta s$).

The mathematics of the random walk model and central limit theorem are discussed in Appendix B. According to this theorem, the end points, x of a random walk consisting of a large number of steps ($n \gg 1$) obey a Gaussian probability distribution:

$$P(x,n)dx = \frac{\delta s}{\sqrt{\pi n}} \exp\left[-\frac{(x/\delta s)^2}{n}\right] dx, \qquad (21.15)$$

with mean value $\langle x \rangle = 0$ and mean square $\langle x^2 \rangle = 2n(\delta s)^2$.

This result is applicable to the summation of any series of random components, as long as the positive and negative contributions of the same magnitude are likely to be equal.

If one compares the probability for the stochastic model, Eq. 21.15 with the distribution based on the diffusion equation, Eq. 21.4, and equates $c(x,t)$ with $N_0P(x,n)$, one sees that a simple random walk correctly models the macroscopic diffusion process when the parameters of the two processes are related by

$$4Dt = n(\delta s)^2. \qquad (21.16)$$

21.5 EQUATION OF MOTION FOR STOCHASTIC PROCESSES: THE LANGEVIN EQUATION

An alternative to a model that describes the motion in terms of a sequence of random steps, is one in which the particle is subject to a *randomly changing force*. This directly affects the particle's acceleration, and results also in an irregular trajectory. Application of such a force over time would increase the kinetic energy, and, in order to keep the mean kinetic energy within bounds, the motion must be *damped by a friction force*. Both of these forces represent the effects of collisions with other molecules.

Accordingly, the motion can be described in a stochastic model as the result of three forces, namely, the friction force (Eq. 21.8), a force that changes direction (and magnitude) in a random fashion, and a force equal to (minus) the gradient of an external or a chemical potential, V. For a particle of mass m we write for the force, $\mathbf{F}$ acting on the particle the following equation:

$$\mathbf{F} = -m\gamma\mathbf{v} + m\boldsymbol{\eta}(t) - \nabla V, \qquad (21.17)$$

where the frictional coefficient, f has been replaced with the product $m\gamma$

$$m\gamma = f, \qquad (21.18)$$

and $m\eta$ represents the random force. This equation of motion is the Langevin equation; in one dimension, and replacing force with the product of mass and acceleration, we have

$$\ddot{x} + \gamma\dot{x} = \frac{dv}{dt} + \gamma v = \eta(t) - \frac{1}{m}\frac{\partial V}{\partial x}, \tag{21.19}$$

where $\ddot{x} = d^2x/dt^2$ is the acceleration, $\dot{x} = dx/dt = v$ is the velocity, and the potential gradient $\partial V/\partial x$ may include both external forces and gradients in free energy. (In a highly damped system inertial effects are negligible, and the term in the acceleration, $\ddot{x}$ can be ignored.)

The random force results from collisions with solvent particles (and intramolecular collisions if the reaction coordinate is intramolecular), has a mean value of zero and is described in terms of its auto-correlation,

$$\begin{aligned}\langle \eta(t) \rangle &= 0, \\ \langle \eta(t)\eta(t+\tau) \rangle &= K(\tau),\end{aligned} \tag{21.20}$$

where the function K (the "kernel") describes how the random force changes with time. Many choices are possible for the function $K(\tau)$.

21.6 FLUCTUATION–DISSIPATION THEOREM

Diffusive motion [which is effectuated by the force $m\eta(t)$] and friction (which is effectuated by the force $-m\gamma\dot{x}$) must balance over time, so as to conserve the energy in the absence of external forces. One can obtain a general relation between random force and friction by equating the average kinetic energy increment due to the action of the random force with the average kinetic energy lost as a result of friction (in the absence of an external force)

$$\left\langle \frac{dE_\eta}{dt} \right\rangle = -\left\langle \frac{dE_\gamma}{dt} \right\rangle. \tag{21.21}$$

Both can be written as the products of force and velocity

$$\begin{aligned}\frac{dE_\eta}{dt} &= m\eta(t)v(t) = m\eta(t)\left[v\left(t_0\right) + \int_{t_0}^{t} \eta(t')dt'\right] \\ \left\langle \frac{dE_\eta}{dt} \right\rangle &= m\left\langle \int_{-\infty}^{0} \eta\,(t)\,\eta(t+\tau)d\tau \right\rangle_t.\end{aligned} \tag{21.22}$$

(as $\langle \eta v \rangle = 0$, and the integration interval, t_0 to t, is arbitrarily large), and

$$\left\langle \frac{dE_\gamma}{dt} \right\rangle = -m\gamma\langle v^2 \rangle = -\gamma k_B T, \tag{21.23}$$

where we have replaced $\langle mv^2 \rangle$ at thermal equilibrium with $k_B T$ (Eq. 7.14). If the integral in Eq. 21.22 is extended to $t = \infty$, its value doubles, and therefore

$$\left\langle \int_{-\infty}^{\infty} \eta\,(t)\,\eta(t+\tau)d\tau \right\rangle = 2\gamma\,\langle v^2 \rangle = \frac{2\gamma k_B T}{m}. \tag{21.24}$$

This last result (Eq. 21.24) expresses a relation between the system's fluctuations (due to the fluctuating force) and the energy dissipation (due to the frictional force), and this equation is a form of the fluctuation–dissipation theorem.

21.7 SPECIFIC EXAMPLES OF FLUCTUATING FORCE

As a first example, consider a special form of random force acting along the x-direction, with a constant magnitude, $m\eta_0$, and directed either in the $+x$ or in the $-x$ direction. The force is constant during each small period, δt, then its direction is again randomly chosen from among these two choices.

Integration of the auto-correlation of this force gives (omitting a factor m^2 on both sides)

$$\left\langle \int_{-\infty}^{\infty} \eta\,(t)\,\eta(t+\tau)d\tau \right\rangle = \eta_0^2 \delta t, \tag{21.25}$$

and hence with Eq. 21.24 one then has the following relation between the parameters describing random and friction forces:

$$\eta_0^2 = \frac{2\gamma k_B T}{m\delta t}. \tag{21.26}$$

A combination of a frictional force and this kind of a random force can be used effectively in dynamics simulations to provide continued thermal motion *at a given temperature*, which is set by the values of γ and η_0. (This is a possible alternative to the Berendsen thermostat mentioned in Chapter 7, section 7.)

As a second example, we consider the case in which the random force has the properties of *white noise*. For white noise the function K in Eq. 21.20 is given by

$$K(t-t') = \Gamma\delta(t-t'), \tag{21.27}$$

where Γ is a proportionality constant and δ is the Dirac delta function. (see Appendix I, Eq. AI.11) Integration of the equation of motion, Eq. 21.19 (with $\partial V/\partial x = 0$), gives[4]

$$v(t) = v(0)e^{-\gamma t} + e^{-\gamma t}\int_0^t e^{+\gamma t'}\eta(t')dt'. \tag{21.28}$$

Squaring this, and using Eqs. 21.20 and 21.27, one obtains [$\langle v(0)\rangle$ and $\langle \eta \rangle$ are zero and $v(0)$ and η are uncorrelated]

$$\begin{aligned}\langle v^2(t)\rangle &= v(0)^2 e^{-2\gamma t} + e^{-2\gamma t}\int_0^t \int_0^{t'} e^{\gamma t' + \gamma t''}\eta(t')\eta(t'')dt''dt' \\ &= v(0)^2 e^{-2\gamma t} + e^{-2\gamma t}\int_0^t e^{2\gamma t'}\Gamma dt' \\ &= v(0)^2 e^{-2\gamma t} + \frac{\Gamma}{2\gamma}(1 - e^{-2\gamma t}). \qquad (21.29)\end{aligned}$$

Letting $t \to \infty$,

$$\langle v^2 \rangle = \frac{\Gamma}{2\gamma}. \qquad (21.30)$$

Again setting $\langle mv^2 \rangle = k_B T$, we obtain

$$\frac{\Gamma}{2\gamma} = \frac{k_B T}{m}. \qquad (21.31)$$

Alternatively, this last result can be derived directly from Eqs. 21.24 and 21.20 with K defined as in Eq. 21.27.

As a third example, we analyze the actual forces experienced by water molecules in a dynamics simulation. The auto-correlation function shows oscillations in force experienced as a moving water molecule is alternatively attracted and repelled by its neighbors; these gradually decay (Fig. 21.2). Application of Eq. 21.24 gave a value of 5.5×10^{-9} M^2/s for the diffusion coefficient, which is somewhat larger than the reported value of 3.9×10^{-9} M^2/s established by other methods for the SPC model.[5]

21.8 ALTERNATIVE FORM OF THE FLUCTUATION–DISSIPATION THEOREM

To derive an alternative form of the fluctuation–dissipation theorem, we relate the work required for the simple process of dragging a molecule through a liquid by an external force, to the statistical error in this work. The particle moves under the influence of an external force, F and one measures the work done, which is equal to $\int F dx$. Because of variation in the number of collisions with solvent molecules, the measurement will return a *slightly* different result, every time it is performed: the random force averaged over the course of a measurement of finite duration is not exactly constant, and sometimes this favors the experimenter, by reducing the required work, and other times the opposite is the case.

We treat the case in which the particle moves at a constant velocity, v because of the application of a (slightly) varying external force, F. The work done, W is

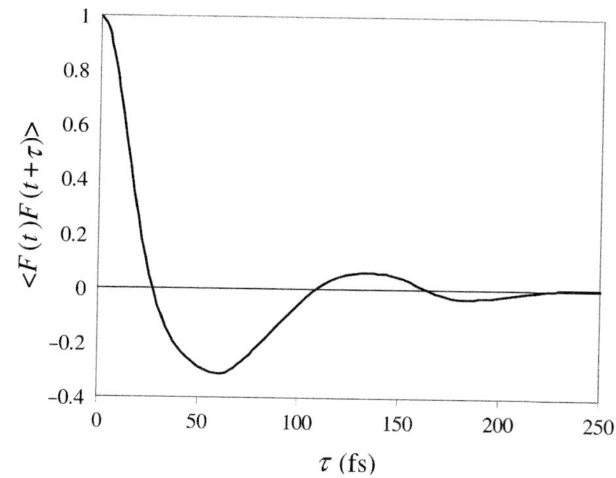

FIGURE 21.2 Auto-correlation of the force on diffusing water molecules (266 molecules; SPC model; 300 K; atmospheric pressure).

equal to the product of force and distance, s

$$W = \int_0^s F(x)dx = \int_0^t F(t')v(t')dt'. \tag{21.32}$$

Velocity and force are related via the Langevin equation (21.19), which here becomes

$$\begin{aligned} \dot{v} + \gamma v &= \eta(t) + F/m \\ F &= m\gamma v - m\eta(t). \end{aligned} \tag{21.33}$$

$\dot{v}$ being equal to zero. The work is then

$$W = \int_0^t dt\ m\gamma v^2 - \int_0^t dt\ m\eta(t)v. \tag{21.34}$$

Averaged over many experiments, mean and mean square of W are given by

$$\begin{aligned} \langle W \rangle &= \int_0^t dt\ m\gamma v^2 \\ \langle W^2 \rangle &= \langle W \rangle^2 + m^2 v^2 \left\langle \left[\int_0^t \eta(t)\, dt \right]^2 \right\rangle. \end{aligned} \tag{21.35}$$

(Linear terms in η average to zero.) Replacement of the squared integral with use of Eq. 21.24 gives[6]

$$\langle W^2 \rangle = \langle W \rangle^2 + 2m\gamma k_B T v^2 t. \tag{21.36}$$

Realizing that $m\gamma v \times vt$ = friction force × distance = $\langle W \rangle$ is the average work done in this process, we obtain

$$\mathrm{var}(W) = \langle W^2 \rangle - \langle W \rangle^2 = 2k_B T \langle W \rangle. \tag{21.37}$$

Thus we see that the variance of the frictional work in this process is equal to twice the product of $k_B T$ and the mean value of the work.

This version of the fluctuation–dissipation theorem establishes a relation between the work lost (dissipated) because of friction and the fluctuation (as the variance ≈ statistical error) in the work required for individual particles. Note that this derivation is valid only when the applied external force is small, and the processes that produce the random forces are not perturbed by the particle's being dragged along.

21.9 DIFFUSIVE MOTION AND THE LANGEVIN EQUATION

The Langevin equation can be used to describe motion in a complex set of potentials and external forces. The equation of motion can be integrated in a stepwise manner at each step, with a newly chosen random force. Many independent trajectories can be produced in this way, each representing an instance of probable behavior according to the model (useful when comparing with single molecule experiments). Averaging can be used to estimate average system properties (useful when comparing with properties of solutions containing many molecules). As an example, we use the Langevin equation to describe the displacement of a particle in Brownian motion.

The displacement is the integral of the velocity

$$x(t) = \int_0^t v(t')dt'. \tag{21.38}$$

Using Eq. 21.28 for the velocity as a function of time, one finds[7] that the mean values of x and x^2 are

$$\begin{aligned} \langle x(t) \rangle &= \int_0^t \langle v(t') \rangle dt' = 0 \\ \langle x^2(t) \rangle &= \frac{2k_B T}{m\gamma}\left[t - \frac{1}{\gamma}\left(1 - e^{-\gamma t}\right)\right]. \end{aligned} \tag{21.39}$$

In the limits of small and large t this gives

$$\begin{aligned} \lim_{t\to 0} \langle x^2(t) \rangle &= \frac{k_B T}{m} t^2 = \langle v^2 \rangle t^2 \\ \lim_{t\to \infty} \langle x^2(t) \rangle &= \frac{2k_B T}{m\gamma} t = \frac{2k_B T}{f} t = 2Dt. \end{aligned} \tag{21.40}$$

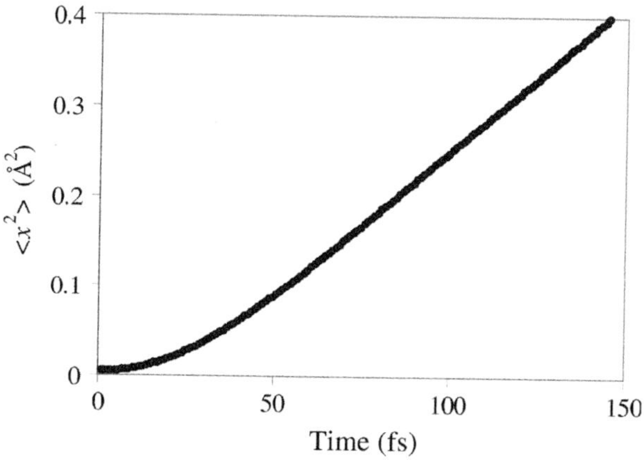

FIGURE 21.3 Mean square molecular displacement as a function of time in a dynamics simulation of liquid water (see legend of Fig. 21.2).

Here, $\langle v^2 \rangle$ is the mean square thermal velocity (see Eq. 7.14). Thus, at short times, the mean square displacement $\langle x^2 \rangle$ changes quadratically with time, while at long times it changes linearly with time. The former is called the *ballistic regime* (in which the particles still travel with their starting velocities), and the latter the *diffusive regime* (when the particles have undergone collisions and changed their direction of motion). This is demonstrated clearly by results from a simulation of liquid water (Fig. 21.3), where the quadratic ballistic behavior is evident during the first 30 fs. The diffusion equation or the random walk model can account only for diffusive behavior.

The explanation for the existence of two regimes is that the particles have some free space to travel before they collide with solvent molecules; one can think of each particle being surrounded by a cage of solvent, in which it moves rapidly with relative freedom until bounced back, while travel outside the cage is a slower process governed by random walk statistics (Fig. 21.4). One sees from the data in Fig. 21.3 that the path available for ballistic motion until collision with the cage formed by the surrounding water molecules is not very large: it is on the order of 0.2 Å.

21.10 SMOLUCHOWSKI AND FOKKER–PLANCK EQUATIONS

21.10.1 Formulation

The Langevin equation of motion describes diffusion with reference to molecular forces that can be accurately evaluated (e.g., in terms of models based on quantum mechanics or molecular mechanics). Molecular motion according to the Langevin equation implies macroscopic motion according to the diffusion equation. We can now re-examine and extend the diffusion equation on this basis.

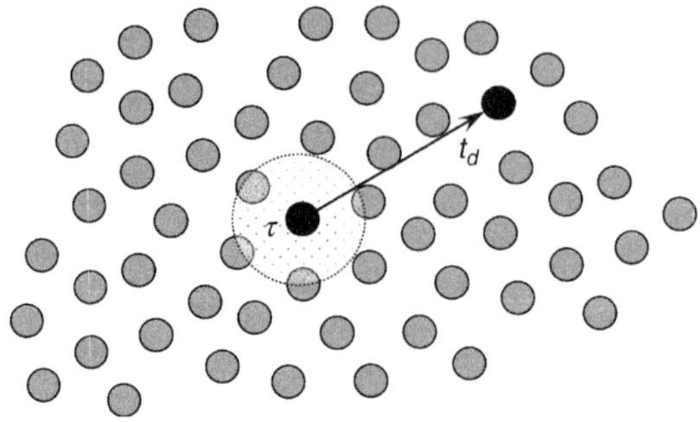

FIGURE 21.4 Ballistic motion of the central particle is restricted to the shaded volume. For a short time, τ the particle moves freely, then it collides with another particle and after that the motion is diffusive.

As diffusion results from the motion of individual solute molecules, which does not cease when the macroscopic diffusive process has reached equilibrium, the diffusion equation should be rewritten in terms of the molecular probability distribution, P as follows

$$\frac{\partial P(x,t)}{\partial t} - \frac{k_B T}{f}\frac{\partial^2 P(x,t)}{\partial x^2} = 0. \tag{21.41}$$

Furthermore, one can include in the diffusion equation the change in concentration resulting from forced motion due to a *(free) energy gradient*, $\partial V/\partial x$; this gives

$$\frac{\partial P(x,t)}{\partial t} - \frac{k_B T}{f}\frac{\partial^2 P(x,t)}{\partial x^2} - \frac{1}{f}\frac{\partial}{\partial x}\left[\frac{\partial V}{\partial x} P\,(x,t)\right] = 0. \tag{21.42}$$

This is the Smoluchowski equation; this equation reduces to the diffusion equation when $\partial V/\partial x = 0$, but accounts also for the effect of a potential, as in sedimentation or electrophoresis.

The diffusion equation and the Smoluchowski equation can be solved explicitly for only a few different initial conditions (the so-called boundary conditions), but a numerical solution is quite feasible given almost any set of boundary conditions.

Further generalization of the Smoluchowski equation leads to the *Fokker–Planck equation*,

$$\frac{\partial P(x,t)}{\partial t} - \frac{\partial^2}{\partial x^2}\left[D_1\,(x,t)\,P(x,t)\right] - \frac{\partial}{\partial x}\left[D_2\,(x,t)\,P(x,t)\right] = 0, \tag{21.43}$$

where the coefficient D_1 is a generalized diffusion coefficient, and D_2, which can be called a *drift* coefficient, represents the force due to the (free) energy gradient. Both are functions of coordinate, x and time, t, either directly, or, indirectly, through dependence on concentration of macromolecule and solvent composition.

21.10.2 Simultaneous Sedimentation and Diffusion

In a sedimentation experiment, a macromolecule with mass, m at a distance R from the axis of a centrifuge rotating at angular velocity ω, experiences a radial centrifugal force of magnitude $F_R = -\partial V_{\text{sed}}/\partial R = \omega^2 m(1-\overline{v}\rho)R$, where $\overline{v}$ is the macromolecule's partial specific volume and ρ solvent density. Sedimentation concentrates the macromolecular solute near the "bottom" (rarely the top) of the solution, while diffusion tends to undo the concentrating effect, and this can be conveniently expressed with the Smoluchowski equation. When expressed in the cylindrical polar coordinates appropriate for this experimental situation (see Appendix I, Eq. AI.7), this becomes

$$\frac{\partial P(R,t)}{\partial t} - \frac{k_{\text{B}}T}{f}\left[\frac{\partial^2 P(R,t)}{\partial R^2} + \frac{1}{R}\frac{\partial P(R,t)}{\partial R}\right] + \frac{\omega^2 m(1-\overline{v}\rho)R}{f}\left[\frac{2P(R,t)}{R} + \frac{\partial P(R,t)}{\partial R}\right] = 0. \quad (21.44)$$

This equation represents the effects of *simultaneous* diffusion and sedimentation; in the cylindrically symmetrical geometry of the sedimentation experiment, diffusion, and sedimentation are accompanied by so-called radial dilution, and this is *also* accounted for in Eq. 21.44. In a typical sedimentation experiment the initial concentration is uniform throughout the sample, that is, $P(R,0)$ is a constant, and all displacement takes place between top and bottom surfaces; these conditions represent the boundary conditions subject to which Eq. 21.44 must be solved. This expression is sufficiently complicated that traditional biophysics texts have tended to treat sedimentation and diffusion independently.

Diffusion experiments ultimately reach equilibrium, where $P(x,t)$ no longer changes. Setting $\partial P/\partial t$ to zero in the Smoluchowski equation gives

$$\frac{d^2 P_{\text{eq}}(x)}{dx^2} + \frac{1}{k_{\text{B}}T}\frac{d}{dx}\left[\frac{\partial V(x)}{\partial x}P_{\text{eq}}(x)\right] = 0, \quad (21.45)$$

and two integrations with respect to x give

$$P_{\text{eq}}(x) = \text{constant} \times \exp\left(-\frac{V(x)}{k_{\text{B}}T}\right), \quad (21.46)$$

which is just a Boltzmann distribution in the potential V. (The constant in this equation is found by imposing conservation of mass.) The corresponding equation for *sedimentation equilibrium* is

$$P_{\text{eq}}(R) = \text{constant} \times \exp\left(\frac{\omega^2 m(1-\overline{v}\rho)R^2}{2k_{\text{B}}T}\right). \quad (21.47)$$

21.10.3 Simultaneous Diffusion, Electrostatic Interaction and Catalysis

In Section 19.14 we have discussed how diffusion places an upper limit on the rate of enzyme catalysis (and of chemical reactions in general); this discussion used as a model (also due to Smoluchowski) a spherical enzyme molecule, having a reactive surface at which all substrate is immediately turned into product. The concentration of substrate on the surface of the enzyme molecule accordingly is set equal to zero, and the catalytic rate is equal to the rate at which the substrate molecules reach the surface by diffusion under steady state conditions, according to the diffusion equation (21.1).

Given an equation (21.42) that describes diffusion in a potential energy gradient, we can now develop a similar model that incorporates both diffusion and the effects of the potential energy gradient, and furthermore we can make the model more sophisticated by explicitly representing the kinetics of the reaction taking place at the enzyme surface. Although this is not strictly necessary, we again assume steady state conditions, setting $\partial P(r,t)/\partial t = 0$ in Eq. 21.42 and integrate once with respect to the coordinate r, the distance from the center of an enzyme molecule

$$-4\pi r^2\left[\frac{k_B T}{f}\frac{dc(r)}{dr} + \frac{1}{f}\frac{dV}{dr}c(r)\right] = \Upsilon, \tag{21.48}$$

where concentration c (of many particles) replaces probability distribution, P (of a single particle), and Υ is the total flow of particles *toward* the surface of the enzyme, which is independent of r according to the assumptions that were made. If enzyme and substrate both carry a net charge, it is appropriate to approximate the electrostatic potential energy V either with Coulomb's law (see Chapter 3)

$$V_C(r) = \frac{q_1 q_2}{4\pi\varepsilon_0\varepsilon r},$$

or with the Debye–Hückel equation (14.14).

The form of the solution depends on the form of the potential V; with $\beta = 1/k_B T$, and the following simple closed form is obtained for a Coulomb potential

$$\Upsilon = -4\pi D R c_0 \frac{\beta V_C(R)}{1 - e^{\beta V_C(R)}}, \tag{21.49}$$

where the value of the integration constant has been set by requiring $c(r=\infty) = c_0$, the bulk concentration of substrate.

In Section 19.14 the value of Υ was determined for the condition of extreme enzyme efficiency, under which $c = 0$ for $r = R$, R being the contact distance between enzyme and substrate. More generally, Υ can be replaced by the rate of turnover according to a kinetic equation; for example, the Michaelis–Menten rate equation (19.25) gives

$$\Upsilon = \frac{k_{cat}c(R)}{K_m + c(R)}. \tag{21.50}$$

Consider first cases where k_{cat} is large and the *rate is limited by diffusion*. In the absence of electrostatic force ($V = 0$), one recovers the rate according to Eq. 19.36,

$$\Upsilon_{V=0} = 4\pi DRc_0.$$

For the case of strong electrostatic attraction ($V_C \ll 0$) the rate is instead

$$\Upsilon_{V_C \ll 0} = -4\pi DRc_0\beta V_C(R) = 4\pi R^2 c_0 \frac{(dV_C/dr)_R}{f},$$

which, as one sees here, corresponds to substrate being dragged through the solvent toward the enzyme surface by the electrostatic force. With interaction according to Coulomb's law, the ratio of electrostatics-powered and diffusion-controlled rates is equal to $\beta q_1 q_2/(4\pi\varepsilon_0\varepsilon R)$. With both charges equal to 1 elementary charge, $\varepsilon = 80$, $T = 300$ K and $R = 10$ Å, this ratio has a value of about 0.7; that is, under these conditions the electrostatic interactions make a minor contribution to the diffusion-limited rate.

At the other extreme, which is the case for a majority of enzyme-catalyzed reactions, the *rate is strictly limited by the reaction kinetics* rather than by diffusion. In that case, the concentration of substrate will depend on distance from the enzyme according to a Boltzmann distribution in the electrostatic potential

$$c(r) = c_0 e^{-\beta V(r)}.$$

This will affect the catalytic rate, but only if the enzyme is not saturated with substrate (and $c_0 \ll K_m$ for Michaelis–Menten kinetics). Under the right circumstances, this exponential dependence on V can produce a large enhancement of the catalytic rate.

Electrostatic effects on reaction rates, and the related dependence of rates on ionic strength can thus be described with a fairly simple model. In Section 14.15 we have described numerical applications of the Langevin and Smoluchowski equations to simulate diffusive motion of charged substrate molecules in their approach to enzyme active sites, using detailed descriptions of surface and charge distribution of the protein and substrate.

21.11 TRANSITION STATE THEORY REVISITED

A previous chapter applied the concept of the transition state, and its relation to the rate of chemical reactions (Section 19.11). In this section we shall consider the transition state model a little more closely. To recapitulate, the reaction occurs between two (locally) stable states along a reaction coordinate, ξ; these states are called *reactant* and *product* states and correspond to two free energy minima along ξ. These are separated by a free energy maximum or energy barrier at the transition state. The free energy surface has a saddle point at the barrier: the free energy here is a maximum relative to ξ, but a minimum with respect to all other degrees of freedom. As a further approximation, the potentials in the free energy wells and near the barrier may each be represented by quadratic functions in ξ

$$
\begin{aligned}
V_{\mathrm{R}}(\xi) &= \frac{1}{2}K_{\mathrm{R}}(\xi-\xi_{\mathrm{R}})^2 \\
V_{\mathrm{B}}(\xi) &= -\frac{1}{2}K_{\mathrm{B}}(\xi-\xi_{\mathrm{B}})^2 + V_{\mathrm{R}}^{\ddagger} \\
V_{\mathrm{P}}(\xi) &= \frac{1}{2}K_{\mathrm{P}}(\xi-\xi_{\mathrm{P}})^2 + V(\xi_{\mathrm{P}}),
\end{aligned}
\tag{21.51}
$$

where $V_{\mathrm{R}}^{\ddagger}$ is the activation energy (or free energy) in the direction from reactant to product (Fig. 21.5).

Consider a system with all molecules in the reactant state. The transition state model divides these further into two states: one is the (reactant) ground state, which at any one time contains the great majority of molecules, and in which the energy is too low to cross the barrier ($V \ll V_{\mathrm{R}}^{\ddagger}$) and the other is the transition state, which contains a small number of molecules with energy greater than $V_{\mathrm{R}}^{\ddagger}$. The molecules in the ground state are subject to the quadratic potential $1/2K_{\mathrm{R}}(\xi-\xi_{\mathrm{R}})^2$ and are distributed over the energy levels of a harmonic oscillator (Section 2.3) according to the vibrational partition function given in Section 5.4 (Eq. 5.25):

$$q_{\mathrm{vib,R}} = \frac{\exp(-\beta h\nu_{\mathrm{R}}/2)}{1-\exp(-\beta h\nu_{\mathrm{R}})}.$$

The frequency ν_{R} is related to the force constant K_{R} and an effective mass, m_{ξ} for motion along the reaction coordinate in the usual way (Eq. 2.11).

The molecules in the transition state are *not* subject to the quadratic reactant potential, V_{R}, and are described with a "particle-in-the-box" model (Section 2.3;

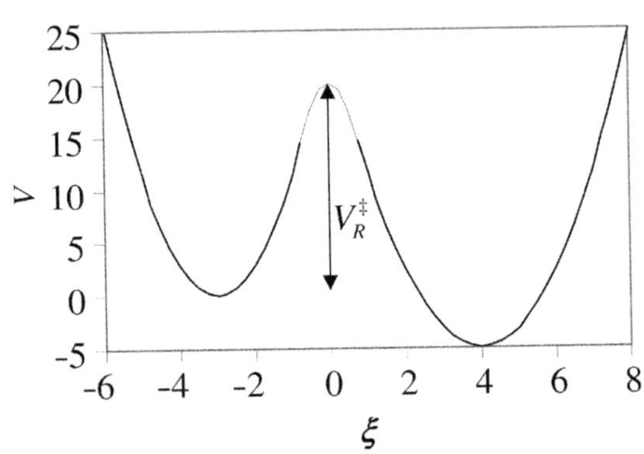

FIGURE 21.5 Potential along a reaction coordinate, consisting of three parabolic segments (arbitrary units).

Eq. 2.9) with the only new proviso that each energy level in this state is raised by the amount $V_R^\ddagger$

$$E_n = \frac{n^2h^2}{8m_\xi a^2} + V_R^\ddagger,$$

where the box size, a can remain unspecified. The velocity corresponding to this energy level is given by

$$v_n = \frac{nh}{2m_\xi a}.$$

In order to find the rate constant (= rate per molecule) at which molecules cross the box, and, therefore, cross the barrier separating reactant from product, we sum one half the velocity divided by the box size, a over the statistical weights for all energy levels of the *transition state*, and divide by the *total* partition function, q. This gives (if we replace summation with integration)

$$k = \frac{1}{2qa}\sum_n \frac{nh}{2m_\xi a}\exp\left(-\beta E_n\right) = \frac{1}{2qa}\sum_n \frac{nh}{2m_\xi a}\exp\left(-\beta\frac{n^2h^2}{8m_\xi a^2}\right)\exp\left(-\beta V_R^\ddagger\right)$$

$$k \approx \frac{1}{2q_{\text{vib,R}}a}\frac{h}{4m_\xi a}\frac{8m^\ddagger a^2}{\beta n^2h^2}\exp\left(-\beta V_R^\ddagger\right) = \frac{1}{q_{\text{vib,R}}}\frac{k_BT}{h}\exp\left(-\beta V_R^\ddagger\right). \tag{21.52}$$

Here q has been replaced with $q_{\text{vib,R}}$, as the vast majority of the molecules is in the reactant ground state. Substitution of the above expression for $q_{\text{vib,R}}$ gives

$$k_{\text{TST}} \approx [1 - \exp(-\beta h\nu_R)]\frac{k_BT}{h}\exp\left[-\beta\left(V_R^\ddagger - \frac{1}{2}h\nu_R\right)\right]. \tag{21.53}$$

Considering the limiting case of large K_R and small effective mass, for which ν_R is large, we obtain

$$k_{\text{TST}} \approx \frac{k_BT}{h}\exp\left[-\beta\left(V_R^\ddagger - \frac{1}{2}h\nu_R\right)\right], \quad \text{for } \beta h\nu_R \gg 1; \tag{21.54}$$

this expression is the basis for the discussion of transition state theory given in Section 19.11.

In the other limiting case, that is, for very small ν_R, we can approximate $\exp(-\beta h\nu_R)$ as $1 - \beta h\nu_R$, which gives

$$k'_{\text{TST}} \approx \nu_R \exp(-\beta V_R^\ddagger), \quad \text{for } \beta h\nu_R \ll 1. \tag{21.55}$$

This represents the classical limit of a wide reactant potential well and large effective mass; in this case transition state theory predicts a rate equal to the oscillator frequency of the reactant state multiplied by the fraction of molecules with sufficiently high energy to cross the barrier.[8]

One assumption in the derivation of Eq. 21.52 is that motion is "ballistic," that is, that distance traveled increases linearly with time and that a molecule, once on its way to cross the barrier, actually will do so. At the same time, transition state theory makes another assumption that transitions to higher and lower energy levels are sufficiently frequent that a Boltzmann distribution is maintained, even though reactant molecules with the highest energies cross the energy barrier to turn into product molecules and do not immediately recross the barrier to the reactant state. How the energy is transferred, whether by radiation, by anharmonic coupling of molecular vibrations, or by molecular collisions, is not specified. As each change of energy level changes the velocity, the motion is, in fact, not perfectly ballistic, and, over a sufficiently long time, velocity and position are unpredictable. Of course, it is sufficient if the motion is ballistic for the time period needed to cross the barrier; we have shown in Section 21.9 that displacement of a model with random force and friction *initially* is linear in time (ballistic regime), and only at longer time becomes proportional to the square root of time (diffusive regime; Eq. 21.40).

The duration of the ballistic regime is determined by the frictional coefficient, which is, via the fluctuation–dissipation theorem, related to characteristics of the forces that change the energy levels. The assumption is that for bond breaking–bond making reactions ("chemical reactions") the motion is ballistic over the time interval required to cross the barrier. However, when collisions are so frequent that they succeed in halting, and reversing a molecule's progress along the reaction coordinate, motion is predominantly diffusive, and in that case Eq. 19.27 does not apply. This is, for example, the case for conformational transitions of macromolecules in solution. In the following section we discuss a theoretical basis for such "diffusive" reactions.

21.12 KRAMERS' THEORY OF REACTION RATES

"A particle which is caught in a potential hole and which, through the shuttling action of Brownian motion, can escape over a potential barrier yields a suitable model for elucidating the applicability of the transition state method for calculating the rate of chemical reactions." Thus runs the abstract of Kramers' seminal 1940 article.[9] The article itself is difficult and because of that, and, perhaps, also because of Kramers' isolation in World War II and death at a young age in 1952, was not widely appreciated until much later. However, since then Kramers' theory has become fundamental to theoretical descriptions of such reactions as macromolecular conformation change in solution. We shall here summarize model and results, omitting derivations.[10]

Kramers' theory uses a one-dimensional Langevin equation (21.19) as the equation of motion along the reaction coordinate, ξ, according to

$$\ddot{\xi} + \gamma\dot{\xi} = \eta(t) - \frac{1}{m_\xi}\frac{\partial V(\xi)}{\partial \xi}, \qquad (21.56)$$

where $m_\xi \eta$ is a random force with the characteristics of white noise (see Section 21.7), $m_\xi \gamma = f$ is the corresponding frictional coefficient, m_ξ the effective mass, and $V(\xi)$ the energy function, which here assumes the form of Eq. 21.51 (Fig. 21.5). Friction and random force are, of course, related via the fluctuation–dissipation theorem (Eq. 21.24). Kramers obtained solutions of this equation in two regimes, one, for the case of small random forces and, therefore, low friction, and, the other, for the case of medium to high friction.

At *low friction*, the net flux over the barrier is limited by the rate at which the energy of the systems in the reactant well is equilibrated and reactant systems with energy above $V^\ddagger$ are generated, and at which the energy of these systems is lowered below $V^\ddagger$ once they reach the product side of the barrier. The rate constant here is found to be proportional to the magnitude of the random forces and hence to γ.

Kramers' article, in the context of the result obtained at *high friction*, generated a lot of interest among scientists. For large values of γ, the rate is, not surprisingly, found to be inversely proportional to γ, and is given by the following expression

$$k_d(\gamma) \approx \frac{\sqrt{K_R K_B}}{2\pi m_\xi \gamma} \exp\left(-\beta V_R^\ddagger\right) = \frac{\sqrt{m_\xi K_B}}{f} \nu_R \exp\left(-\beta V_R^\ddagger\right). \tag{21.57}$$

One sees that this expression for k is, except for the factor $(1/f)\sqrt{m_\xi K_B}$, equivalent to the result obtained from transition state theory (Eq. 21.55). Therefore, this "transmission factor" may be considered to represent the fraction of attempted barrier crossings that is successfully completed.

In order to derive this expression for the rate constant in the high friction regime, one may replace the Langevin equation with an equivalent Smoluchowski equation (Eq. 21.42), that is, a diffusion equation extended with a term dependent on a potential V, for which is used the energy $V(\xi)$ of Eq. 21.51. Furthermore, one can apply a steady state condition, in which the probability distribution is independent of t, with boundary conditions that prevent a return across the barrier, such as setting $P(\xi_P) = 0$. Under these conditions the reaction proceeds according to a steady state flux, which is independent of ξ and which may be equated with the rate constant, and is given by the following equation

$$\Upsilon = -\frac{k_B T}{f}\frac{dP(\xi)}{d\xi} - \frac{1}{f}\frac{dV(\xi)}{d\xi} P(\xi) = k_d. \tag{21.58}$$

Rather than reproduce a derivation of Eq. 21.57 from this starting point, consider as an approximation the following alternative to the potential of Eq. 21.51, in which V_R is, as before, quadratic in ξ, while in the barrier region $V(\xi)$ is constant, equal to $V_R^\ddagger$, and with as a boundary condition that in the product region $P(\xi) = 0$(for $\xi > \xi_{B,1}$)

$$\begin{aligned} V_B(\xi) &= V_R^\ddagger, \quad \text{for } \xi_{B,0} \ll \xi \ll \xi_{B,1} \\ P(\xi) &= 0, \quad \text{for } \xi > \xi_{B,1}. \end{aligned} \tag{21.59}$$

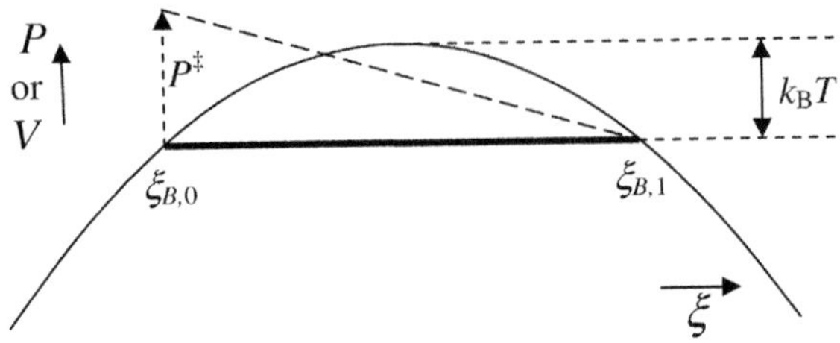

FIGURE 21.6 Potential, V in the vicinity of the barrier. In the modified potential the heavy line replaces the original quadratic potential in the interval $\xi_{B,0} < \xi < \xi_{B,1}$. Long dashes indicate the linear variation of the probability P from $P^{\ddagger}$ at $\xi = \xi_{B,0}$ to 0 at $\xi = \xi_{B,1}$ for the modified potential.

For this modified potential, the steady state probability $P(\xi)$ varies linearly with ξ in the barrier region between a high of $P^{\ddagger}$ at $\xi = \xi_{B,0}$ and zero at $\xi = \xi_{B,1}$ (Fig. 21.6), and this gives

$$k_d = -\frac{k_B T}{f}\frac{dP(\xi)}{d\xi} = -\frac{k_B T}{f}\frac{P^{\ddagger}}{\xi_{B,1} - \xi_{B,0}}. \tag{21.60}$$

If one now uses the modified potential as an approximation to the original potential of Eq. 21.51, one may relate $P^{\ddagger}$ to the (normalized) Boltzmann factor for the activation energy

$$P^{\ddagger} \approx \frac{\exp(-\beta V^{\ddagger})}{\int d\xi \exp[-\beta V_R(\xi)]} \approx \sqrt{\frac{\beta K_R}{2\pi}}\exp(-\beta V^{\ddagger}). \tag{21.61}$$

In order to make an appropriate choice of the "width" of the barrier region for the quadratic potential $V_B(\xi)$, consider that the kinetic energy along ξ is on the order of $k_B T$, and that if the potential energy is *below* the barrier energy by a significantly greater amount, that is, if $V(\xi) < V^{\ddagger} - k_B T$, then the system is a priori unlikely to pass to the product state.[10] Hence, the equivalent width of a quadratic barrier potential is on the order of

$$\xi_{B,1} - \xi_{B,0} = 2\sqrt{2k_B T/K_B}. \tag{21.62}$$

Substitution of the expressions for $P^{\ddagger}$ and $\xi_{B,1} - \xi_{B,0}$ gives a rate for the modified potential that is, except for a numerical factor, identical with the expression found by Kramers (Eq. 21.57).

One sees that the factor $\sqrt{K_R}$ is introduced into this equation as a result of the normalization of $P(\xi)$, while the factor $\sqrt{K_B}$ appears as a measure of the width of the barrier region. Frauenfelder gives an interesting alternative rationalization of the expression for the transmission factor as the ratio of the mean free path between collisions and the barrier width.[10]

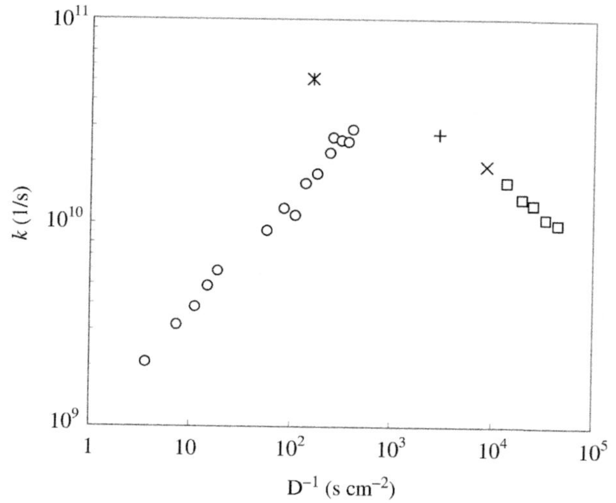

FIGURE 21.7 Rate constants of photoisomerization of stilbene in hydrocarbon media of varying density, as a function of the reciprocal of the diffusion coefficient (as a measure of friction with solvent) show increasing rate at low friction and decreasing rate at high friction, as predicted by Kramers' theory. From Fleming, G.R., Courtney, S.H., Balk, M.W. *J. Stat. Phys.* **42**: 83–104 (1986).

The results of Kramers' theory have been verified by experimental studies of the dependence of the rate of chemical reactions on viscosity[11] (Fig. 21.7 It should be remembered that the frictional coefficients for diffusive motion and for the conformational transition are not identical). The theory serves as the foundation for estimating rates of diffusive reactions such as macromolecular conformation change, by combining a knowledge of the potential of mean force along the reaction coordinate with estimates of rate of progress along the reaction coordinate and the transmission coefficient from simulations in the barrier interval, as described in Section 19.15.1.

NOTES

1. According to the Heisenberg uncertainty principle, a particle's position and momentum cannot both be known with complete accuracy. Hence, future position and velocity can only be predicted in terms of probability distributions.
2. From the Greek word for "chance."
3. According to Stokes' law, a sphere of 10 Å radius in water at 20 °C has $f = 18.8 \times 10^{-12}$ kg/s and $D = 2.20 \times 10^{-10}$ M^2/s. To move such a particle at a velocity of 1 M/s = 1 nM/ns requires a force of 18.8 pN. (The self-diffusion coefficient of water equals 2.20×10^{-9} M^2/s.)
4. This result is easily confirmed by taking the derivative and substituting the expression for $\dot{v} = \ddot{x}$ according to Eq. 21.19.

5. The integrated autocorrelation of the force vector on the water molecules was estimated to be around 3.9×10^8 kcal^2/mole^2/M^2 s. Accordingly, the self-diffusion coefficient can be estimated at $D_{\text{SPC}} \approx 3 \times 2(k_B T)^2/3.9 \times 10^8$. Because the autocorrelation function oscillates between positive and negative values, the integral is small and converges very slowly; thus, the value quoted here is not very precise.
6. This replacement is appropriate only if the interval 0 to t is large.
7. Write $x(t) = \int_0^t v(t')dt' = \int_0^t dt'\, v_0 e^{-\gamma t'} + \int_0^t dt'\, e^{-\gamma t'} \int_0^{t'} e^{+\gamma t''} \eta(t'')dt'' = \int_0^t dt'\, v_0 e^{-\gamma t'} + \int_0^t dV\, U$, solve the first integral and expand $\int_0^t dV\, U = UV|_0^t - \int_0^t V\, dU$. When squared, the result is $\gamma^2 x^2(t) = v_0^2(1 - e^{-\gamma t})^2 + e^{-2\gamma t}(\int_0^t e^{+2\gamma t'} \eta(t')dt')^2 + (\int_0^t dt' \eta(t'))^2 - 2e^{-\gamma t} \int_0^t e^{+\gamma t''} \eta(t'')dt'' \int_0^t dt' \eta(t') + O(v_0^1)$ Apply Eq. 21.27 to each of the two-integral products. On averaging, the last term gives zero. The remainder of the terms then reduces to the expression given in Eq. 21.39.
8. For transfer of a hydrogen atom from a substrate molecule to a side chain of the enzyme 4OT, which was discussed at some length in Section 19.15.2, one finds, with $K_R \approx 5$ kcal/(mol Å^2) (based on the free energy curve of Fig. 19.13) and using for $m^{\ddagger}$ the mass of a hydrogen atom, for $h\nu_R$ a value of 4.8 zJ. This is close in value to $1/\beta = k_B T = 4.142$ zJ at 300 K; hence, the factor $[1 - \exp(-\beta h \nu_R)]$ in $q_{\text{vib,R}}$ has a value of 0.7, rather than 1.0 as in the limit of *very* large $\beta h \nu_R$. Given the approximate nature of transition state theory, the difference is unimportant.
9. Kramers, H.A. Brownian motion in a field of force and the diffusion model of chemical reactions. *Physica* 7: 284–304 (1940).
10. See also Frauenfelder, H. *The Physics of Proteins: An Introduction to Biological Physics and Molecular Biophysics*. New York: Springer, 2010, Chapter 13.
11. Fleming, G.R., Courtney, S.H., Balk, M.W. Activated barrier crossing: comparison of experiment and theory. *J. Stat. Phys.* **42**: 83–104 (1986).

SUGGESTED READING

Feller, W. *An Introduction to Probability Theory and Its Applications*, New York: Wiley, 1968.

Van Kampen, N.G., *Stochastic Processes in Physics and Chemistry*. Amsterdam: Elsevier, 1992.

Appendices

Several of the appendices review specific mathematical subjects: probability theory (Appendix A), the random walk and the central limit theorem (Appendix B) and the Laplace transform (Appendix F). Three appendices contain mathematical treatments that are needed to understand the development of different theories of specific physical problems discussed in previous chapters: grand ensemble (Appendix C), potential of mean force (Appendix D), the helix-coil transition (Appendix E), and solutions of the Poisson equation (Appendix G). In Appendix H we summarize the definition of three types of molecular surface. Appendix I summarizes a collection of equations used in the text.

Equilibria and Kinetics of Biological Macromolecules, First Edition. Jan Hermans and Barry Lentz.

Appendix A

Probability

A.1 INTRODUCTION

Probability theory deals with the mathematics of random processes and random events, and with processes and events that have random components. The outcome of a random process cannot be predicted; probability theory analyzes the *distribution* of the outcome of many instances of the same random process. Probability theory makes a distinction between *probability* that has a theoretical basis and *statistics* that have an observational basis. Obviously, a theory is not correct until it agrees with the observations within statistical error.

In the context of this book, an understanding of the concepts of probability theory is, among others, needed in connection with the distribution of energy on the basis of statistical mechanics (Chapters 5 and 6), with the Monte Carlo method used to approximate the Boltzmann distribution in molecular simulations (Chapter 7), with the random walk model of the conformation of polymer molecules in solution (Chapter 15), and these concepts are basic to our discussion of stochastic processes which are the subject of Chapter 21.

In this appendix we explain basic concepts and conventions of probability theory, give a few basic equations and, finally, discuss statistical error.[1]

Equilibria and Kinetics of Biological Macromolecules, First Edition. Jan Hermans and Barry Lentz.

A.2 SAMPLE STATISTICS

Statistics of a sample (a series of instances, typically observations) of x, $\{x\}$ are expressed in terms of a distribution. The mean value of any function of x, $f(x)$ (including x itself) for this particular sample containing N instances is then defined as

$$\langle f(x)\rangle = \frac{\sum_{i=1}^{N} f(x_i)}{N}. \tag{A.1}$$

If the values of x are discrete, the number of instances of each value of x, n_x can be counted to give a distribution $\{n_x\}$. The mean value of $f(x)$ is then defined as

$$\langle f(x)\rangle = \frac{\sum_{i=1}^{m} f(x_i)n_{x_i}}{\sum_{i=1}^{m} n_{x_i}} = \frac{1}{N}\sum_{i=1}^{m} f(x_i)n_{x_i}, \tag{A.2}$$

if m is the number of different values of x in the sample and N the total number of instances in the sample.

Important characteristics of a distribution are mean, $\langle x\rangle$ and mean square deviation, $\langle(\delta x)^2\rangle$, defined with

$$\langle x\rangle = \frac{\sum_{i=1}^{N} x_i}{N}$$

$$\langle x^2\rangle = \frac{\sum_{i=1}^{N} x_i^2}{N}$$

$$\langle(\delta x)^2\rangle = \frac{\sum_{i=1}^{N} (x_i - \langle x\rangle)^2}{N} = \frac{\sum_{i=1}^{N} x_i^2}{N} - \frac{\sum_{i=1}^{N} 2x_i\langle x\rangle}{N} + \frac{\sum_{i=1}^{N} \langle x\rangle^2}{N}$$

$$= \langle x^2\rangle - \langle x\rangle^2. \tag{A.3}$$

A.3 PROBABILITY DISTRIBUTIONS

A *probability distribution* $P(x)$ is a function of x, typically derived for a model of the system. By definition, the sum of a probability over all possible instances is unity. If the values of x are discrete, then the *expectation value* of a function of

x, $f(x)$ (which can be x itself) is defined for the probability distribution, P in an analogous manner as the mean is defined for a sample distribution, with

$$\sum_{i=1}^{m} P(x_i) = 1$$

$$\overline{f(x)} = \frac{\sum_{i=1}^{m} P(x_i) f(x_i)}{\sum_{i=1}^{m} P(x_i)} = \sum_{i=1}^{m} P(x_i) f(x_i). \tag{A.4}$$

If the values of x are continuous, then the probability distribution can be expressed as a continuous function of x and the expectation value of $f(x)$ is obtained by integration over the entire allowed range of x

$$\int dxP(x) = 1$$

$$\overline{f(x)} = \frac{\int dxP(x) f(x)}{\int dxP(x)} = \int dxP(x) f(x). \tag{A.5}$$

The *variance* of x is defined with

$$\text{var}(x) = \overline{(x - \overline{x})^2} = \overline{(x^2 - 2x\overline{x} + \overline{x}^2)} = \overline{x^2} - \overline{x}^2 \tag{A.6}$$

Obviously, the variance is a counterpart of the mean square deviation.

Given the variance of a variable x, and another function y that is linearly related to x, then the variance of y is given by

$$y = ax + b$$

$$\text{var}(y) = \text{var}(ax) = a^2 \text{var}(x). \tag{A.7}$$

Given the variance of two independent variables, x and y, then the variance of the sum of these is given by

$$z = x + y$$

$$\text{var}(z) = \text{var}(x) + \text{var}(y). \tag{A.8}$$

A.4 A FEW COMMENTS

Bayesian statistics. It is very common to use a sampled distribution as a probability distribution, and thereby to bridge the gap between observation and prediction that exists when no (sufficiently accurate) predictive theory exists. An example is to correlate amino acid sequence and occurrence in specific types of secondary structure from a sample of known structures (as in the original Chou-Fasman method[2]) and use this to assign helical and other probabilities to chain segments of a protein of known sequence but unknown structure.

Gaussian distributions. Many theories end up predicting a Gaussian probability distribution of the deviation from the expectation value, as a consequence of the central limit theorem discussed in Appendix B. One then has

$$\begin{aligned}
\delta x &= x - \overline{x} \\
P(x)dx &= \frac{1}{\sigma\sqrt{2\pi}} \exp[-(\delta x)^2/(2\sigma^2)] \\
\mathrm{var}(x) &= \int (\delta x)^2 P(x)dx = \sigma^2. \qquad \text{(A.9)}
\end{aligned}$$

Combined probability. In the case of multiple kinds of chance occurrences, it is important to establish if these are independent. If they are, then the following holds: the probabilities of one *and* the other occurring in the same instance (also called *the joint probability*) is equal to the product of the individual probabilities

$$P_{x\&y}(x, y) = P_x(x)P_y(y), \qquad \text{(A.10)}$$

while the probability of one *or* the other (or both) occurring is

$$\begin{aligned}
P_{x|y}(x, y) &= P_x(x) + P_y(y)[1 - P_x(x)] \\
P_{x|y}(x, y) &= P_x(x) + P_y(y) - P_{x\&y}(x, y). \qquad \text{(A.11)}
\end{aligned}$$

A.5 FITTING THEORY TO DATA: COMPUTER-FACILITATED "LEAST SQUARES"

A least-squares fit can be used whenever a model is proposed to explain available experimental data. The process can be summarized as follows:

Collect observations: $\{F_{\mathrm{obs}}, x\}$.
Formulate a model with one or more *parameters* (constants).
Compute F_{model} for each x in the set of observations: $\{F_{\mathrm{obs}}, x\}$.
Define a mathematical criterion for better fit.
Seek the best set of parameters in the model.

If there is reason to do so, select another model, and repeat.

The fit should be optimized in terms of probability: given the observations, which set of constants has the highest probability of being correct? This can be complicated, but for fitting a model to a binding curve it is not particularly so.

Let us assume we have each value of $\ln a_L$ ($= x$, above) from a potentiometric measurement, which is very precise, and that the error is due to the uncertainty of the spectroscopic measurements. Each point is characterized by an extinction coefficient ε; in the absence of any L, $\varepsilon = \varepsilon_{min}$, and in case of saturation with L, $\varepsilon = \varepsilon_{max}$, and $\varepsilon_{min} \leq \varepsilon \leq \varepsilon_{max}$. One then computes the fraction bound with

$$f_a = \frac{\varepsilon - \varepsilon_{min}}{\varepsilon_{max} - \varepsilon_{min}}. \tag{A.12}$$

The statistical error in f_b is obviously proportional to the statistical error in ε. In a great majority of situations the best fit is found by a minimization of the sum over all observations of the squares of the difference between the observed and calculated values of f_b

$$\Phi = \sum_i (f_{calc,i} - f_{obs,i})^2. \tag{A.13}$$

The minimum is relative to the model parameters used in the expression for f_{calc}, which include the values of one or several binding constants and the values of the limits ε_{min} and ε_{max}.

The choice of minimizing a sum of squares is based on the assumption that the errors in the measurements of ε obey a Gaussian probability distribution with a mean of zero and a width that does not depend on a_L and that the errors in different measurements are independent. The use of a simple sum of squares, Eq. A.13 is appropriate if the errors are similar for all observations. If they are not, the terms must be weighted in inverse proportion to the mean square deviation, σ_i^2 of each measurement

$$\Phi = \sum_i \frac{1}{\sigma_i^2} (f_{calc,i} - f_{obs,i})^2. \tag{A.14}$$

Theorem The "least squares" formulation follows from the assumption of a Gaussian probability distribution for the errors (cf. Appendix B).

One typically has many more observations than adjustable parameters, and it is not possible to make the model agree precisely with all the observations, even if the model is correct, because each observation, i contains an error δ_i. The goal of the fitting process is to find the most probable parameters for the model (or find the maximum likelihood solution).

Assume that the errors are statistically distributed, according to a Gaussian distribution

$$P(\delta)d\delta = \frac{\sigma}{\sqrt{\pi}} e^{-\delta^2/\sigma^2} d\delta. \tag{A.15}$$

The joint probability of all the errors is the product of the separate probabilities for each (the observations are independent). The reader can easily show that maximizing the joint probability is achieved by minimization with least squares, according to

$$\Phi = \sum_i \delta_i^2 = \text{minimum}. \tag{A.16}$$

NOTES

1. In this appendix we make a rigorous distinction between statistics and probability, mean and expectation value, mean-square deviation, and variance. For mean and mean-square deviation we use the averaging symbols $< >$, for expectation value we use a superscript bar, as in $\overline{x}$, and for variance the expression var(). In the rest of the book these conventions are not rigorously adhered to.
2. Chou P. Y., Fasman G. D. Prediction of protein conformation. *Biochemistry* 13: 222–245 (1974).

Appendix B

Random Walk and Central Limit Theorem

B.1 INTRODUCTION

The random-walk model treats the statistics of processes in which two or more alternative choices are available at each of a (frequently large) number of steps. In the classical one-dimensional random walk the choices are a step to the left or a step of equal magnitude to the right. Another possible pair of alternatives is a step to the right and no step at all, and the same ideas can be applied to a stream of objects, from which some objects are selected to make a sample, and the others are rejected. The outcome of each trial of N steps is a distance traveled (for the first two examples) or a number of objects selected. The outcome of a large number of such steps is a distribution of distances over a range of values. (For the number of selected objects, this range is from 0 to N selected objects.)

B.2 RANDOM SELECTION

If every object is selected independently with probability p, then the probability of it not being selected is $1-p$. The probability that from a stream of N objects exactly m will be selected is the product of three quantities: the number of ways of selecting m from N objects, the probability of selecting m objects, and the

Equilibria and Kinetics of Biological Macromolecules, First Edition. Jan Hermans and Barry Lentz.

probability of *not* selecting $N - m$ objects, that is,

$$P_b(N, m, p) = \frac{N!}{m!(N-m)!} p^m (1-p)^{N-m}. \tag{B.1}$$

It is easy to show that the expectation value of the number selected is

$$\overline{m} = pN, \tag{B.2}$$

and the variance is given by[1]

$$\text{var}(m) = \overline{(m - \overline{m})^2} = \overline{m^2} - \overline{m}^2 = p(1-p)N. \tag{B.3}$$

For $p \ll 1$

$$\begin{aligned} \text{var}(m) &\cong pN = \overline{m} \\ \frac{\sqrt{\text{var}(m)}}{\overline{m}} &= \frac{1}{\sqrt{\overline{m}}}. \end{aligned} \tag{B.4}$$

This shows that the "uncertainty" in the sample size, as represented by the square root of the variance, is equal to the square root of the mean sample size, and the *relative* statistical error decreases in proportion to the square root of the sample size.

B.3 THE CENTRAL LIMIT THEOREM

The central limit theorem states that addition of a series of values having individually small *random* deviations results in a sum that will deviate from the mean approximately according to a Gaussian (or normal) probability distribution, with a better approximation if the number of added terms is large. One restriction is that positive and negative deviations of equal magnitude must be equally likely. As defined in the preceding section, random selection with $p = 1/2$ fits this description, as the value added is either 0 or 1.

It is easy to show that the following Gaussian distribution

$$\begin{aligned} y &= m - \frac{N}{2} \\ P_n\left(N, y, \frac{1}{2}\right) dy &= \left(\frac{2}{\pi N}\right)^{1/2} \exp\left(\frac{-2y^2}{N}\right) dy. \end{aligned} \tag{B.5}$$

has the same expectation value and variance of m as the discrete distribution, Eq. B.1. And, similarly, the following equation is a good approximation in the more general case of arbitrary value of p,

$$y = m - pN$$

$$P_n(N, y, p)\, dy = \left[\frac{1}{2\pi\, p\,(1-p)\,N}\right]^{1/2} \exp\left[\frac{-y^2}{2p\,(1-p)\,N}\right] dy. \tag{B.6}$$

This is an excellent approximation even for modest values of N. For example, with $N = 20$ and $p = 1/2$ the probabilities calculated according to Eqs. B.1 and B.5 differ by less than 0.002. The distributions P_b and P_n have identical values of $\overline{m}$ and $\overline{m^2}$. A measure of the quality of the approximation is that, for any higher power, $k < N$ the ratio of the "moment" of these two distributions differs from unity by a term that with increasing N goes to zero as $1/N$

$$\overline{m_b^k}/\overline{m_n^k} = 1 + \mathrm{O}\left(\frac{1}{N}\right). \tag{B.7}$$

A warning is in order that at values of m close to 0 or to N, the probabilities according to both equations are very close to zero, but cannot be said to be in agreement; for example, for $N = 20$ and $m = 1$, the ratio P_n/P_b exceeds 100. In order to estimate probabilities for values of m near 0, one may replace $N!$ and $(N-m)!$, but not $m!$ in Eq. B.1 with Stirling's formula

$$P_b(N, m, p) \approx \frac{N^m}{m!} p^m (1-p)^{N-m} = \frac{\overline{m}^m}{m!}(1-p)^{N-m};\ m \ll N. \tag{B.8}$$

When p is small this can be further simplified to give

$$P_b(N, m, p) \approx \frac{\overline{m}^m \exp(-\overline{m})}{m!};\ p \ll 1, m \ll N, \tag{B.9}$$

the *Poisson distribution.*

B.4 SIMPLE RANDOM WALK

In a very simple model of random progress, an object moves along a linear axis (x-axis) in steps. The magnitude of the steps is constant, δs, but the steps can be taken in either direction, so that a step can equal $-\delta s$ or $+\delta s$. If time dependence is present in the problem, the steps are taken at regular intervals, δt.

The end point of a random walk of N steps, h is the difference between the distance covered by the m steps taken in the positive x-direction, minus the distance covered by the $N - m$ steps taken in the opposite direction

$$h = m\delta s - (N - m)\delta s = (2m - N)\delta s. \tag{B.10}$$

For large N, the probability distribution of h can be found from Eq. B.6 with $p = 1/2$,

$$P_h(N, h)\,dh = \frac{\delta s}{\sqrt{\pi N}} \exp\left[-\frac{(h/\delta s)^2}{2N}\right] dh. \tag{B.11}$$

This is a Gaussian distribution with expectation value of 0, and variance of $N(\delta s)^2$. The root mean square distance between starting and end point grows as the square root of the sample size, N

$$\sqrt{\overline{h^2}} = \sqrt{N}\,\delta s. \tag{B.12}$$

The simple model can be made more sophisticated without changing the result that the distribution is Gaussian for large N. For example, the step size, δs can be different for every step, randomly chosen from a given distribution, or the direction of successive steps can be correlated. Such changes do not alter the form of the result, but only affect the value of the coefficient of h^2/N in the exponential (and the value of the factor in front of it).

NOTE

1. See Appendix A for definitions and notation.

Appendix C

The Grand Partition Function: Derivation and Relation to Other Types of Partition Functions

C.1 INTRODUCTION

In Chapter 6 we introduced the *grand ensemble* in order to describe an open system, that is, a system at constant temperature and volume, able to exchange system contents with the environment, and hence at constant chemical potential of each system component. In this appendix we first present a derivation of the partition function for this ensemble and, second, describe how it relates to other types of partition functions.

In order to simplify the presentation, we restrict this derivation to systems containing a single component. Consequently, we are able to simplify the notation: while we have in Chapter 6 used the notation $\{N_\Lambda\}_l$ to describe a particular combination of N_1, N_2, etc. molecules of different types in a multicomponent system, and summed overall values of the index l, we can describe the content of a single component simply with N and sum over all values of N.

To summarize the problem: a one-component open system can assume states characterized by composition N and quantum state N, j. The quantum state N, j has wave function $\Psi_{N,j}$ and energy $E_{N,j}$. The ensemble contains $\mathcal{R}$ systems, each

Equilibria and Kinetics of Biological Macromolecules, First Edition. Jan Hermans and Barry Lentz.

having a randomly assigned N, j, and any given quantum state can occur more than once in the ensemble, which then contains $m_{N,j}$ systems with wave function $\Psi_{N,j}$.

The distribution over different states is governed by three constraints: the total number of systems is fixed, the mean system energy is constrained to provide a constant temperature and the mean system content of material is constrained so as to provide a constant chemical potential

$$\sum_{N,j} m_{N,j} = \mathcal{R}$$

$$\sum_{N,j} m_{N,j} E_{N,j} = \mathcal{R}\langle E\rangle$$

$$\sum_{N,j} m_{N,j} N = \mathcal{R}\langle N\rangle. \tag{C.1}$$

C.2 DERIVATION

It is asked to find the most probable distribution, given a very large number of systems in the ensemble and thus a very large value of $\mathcal{R}$. The probability of a particular distribution is proportional to the combinatorial (see also Section 5.2)

$$\Omega = \frac{N!}{\prod_{N,j} m_{N,j}!}, \tag{C.2}$$

and thus we shall seek the maximum of $\ln \Omega$, but under the constraints of Eq. C.1. This constrained minimization is performed using the method of Lagrange, in which each constraint adds a term to the function that is to be optimized, multiplied by a so-called Lagrange multiplier. The three constraints of Eq. C.1 thus introduce three Lagrange multipliers α, $-\beta$, and γ, and add three expressions to $\ln \Omega$, to give the function B that is to be optimized:

$$B = \ln \Omega(\{m_{N,j}\}) + \alpha \sum_{N,j} m_{N,j} - \beta \sum_{N,j} m_{N,j} E_{N,j} + \gamma \sum_{j} m_{N,j} N. \tag{C.3}$$

The optimization with respect to all variables $m_{N,j}$ is done in the usual way

$$\frac{\partial B}{\partial m_{N,j}} = 0, N = 1, 2, \ldots, j = 1, 2, \ldots. \tag{C.4}$$

The solution is simple when each of the factorials occurring in the expression for Ω (Eq. C.2) is replaced with $n^n e^{-n}$, the leading terms of Stirling's formula for $n!$

$$
\begin{aligned}
B = {} & N \ln N - N - \sum_{N,j} m_{N,j} \ln m_{N,j} + \sum_{N,j} m_{k,j} \\
& + \alpha \sum_{k,j} m_{N,j} - \beta \sum_{N,j} m_{N,j} E_{N,j} + \gamma \sum_{N,j} m_{N,j} N \\
\frac{\partial B}{\partial m_{N,j}} = {} & - \ln m_{N,j} + \alpha - \beta E_{N,j} + \gamma N = 0 \\
m_{N,j} = {} & \exp \alpha \exp(-\beta E_{N,j}) \exp(\gamma N). \qquad \text{(C.5)}
\end{aligned}
$$

Setting the sum of all the $m_{N,j}$ equal to $\mathcal{R}$ assigns a value to α and this leads to

$$
m_{N,j} = \frac{\mathcal{R} \exp(-\beta E_{N,j}) \exp(\gamma N)}{\sum_{N,j} \exp(-\beta E_{N,j}) \exp(\gamma N)}, \qquad \text{(C.6)}
$$

$$
P_{N,j} = \frac{m_{N,j}}{\mathcal{R}} = \frac{\exp(-\beta E_{N,j}) \exp(\gamma N)}{\sum_{N,j} \exp(-\beta E_{N,j}) \exp(\gamma N)},
$$

where $P_{N,j}$ is the probability of a molecule in the system containing N molecules and being in the jth quantum state with energy $E_{N,j}$.

We define the *grand canonical or grand partition function* as

$$
\Xi = \sum_{N,j} \exp(-\beta E_{N,j}) \exp(\gamma N), \qquad \text{(C.7)}
$$

and thus

$$
P_{N,j} = \frac{\exp(-\beta E_{N,j}) \exp(\gamma N)}{\Xi}, \qquad \text{(C.8)}
$$

and

$$
E_{N,j} = -\frac{1}{\beta}(\ln P_{N,j} + \gamma N - \ln \Xi).
$$

C.3 CONNECTION WITH THERMODYNAMIC FUNCTIONS

In this section we obtain expressions that relate the Lagrange multipliers β and γ to thermodynamic functions, namely, temperature and chemical potential. The following derivation follows Hill's Statistical Mechanics, which the reader should consult for a deeper understanding.

We first note that the mean system energy, which is equated with the thermodynamic energy, is given by

$$E = \langle E_{N,j} \rangle = \sum_{N,j} E_{N,j} P_{N,j}. \tag{C.9}$$

From this expression for E we obtain

$$dE = \sum_{N,j} E_{N,j} dP_{N,j}. \tag{C.10}$$

Substituting for $E_{N,j}$, we obtain

$$dE = -\frac{1}{\beta} \sum_{N,j} (\ln P_{N,j} + \gamma N - \ln \Xi) dP_{N,j}. \tag{C.11}$$

Writing

$$d \left\{ \sum_{N,j} (P_{N,j} \ln P_{N,j}) \right\} = \sum_{N,j} (dP_{N,j} \ln P_{N,j}) + \sum_{N,j} dP_{N,j}, \tag{C.12}$$

where the second term on the right is equal to zero, as $\sum_{N,j} P_{N,j} = 1$ by definition, we get

$$\begin{aligned} dE &= -\frac{1}{\beta} \sum_{N,j} (\ln P_{N,j} + \gamma N - \ln \Xi) dP_{N,j} \\ &= -\frac{1}{\beta} d \left[\sum_{N,j} P_{N,j} \ln P_{N,j} \right] - \frac{1}{\beta} \sum_{N,j} \gamma N dP_{N,j} + \frac{1}{\beta} \ln \Xi \sum_{N,j} dP_{N,j} \\ &= -\frac{1}{\beta} d \left[\sum_{N,j} P_{N,j} \ln P_{N,j} \right] - \frac{\gamma}{\beta} d\langle N \rangle. \end{aligned} \tag{C.13}$$

We now compare this with the corresponding thermodynamic relation for a system at constant volume (see Eq. 1.2, with $dV = 0$)

$$dE = TdS + \mu d\langle N \rangle, \tag{C.14}$$

and conclude that the Lagrange multipliers are related to the *chemical potential* by

$$\mu = -\frac{\gamma}{\beta}, \tag{C.15}$$

and to the *temperature* by

$$\frac{1}{k_B T} = \beta. \tag{C.16}$$

Furthermore, the *entropy* is equated with

$$S = -k_B \sum_{N,j} P_{N,j} \ln P_{N,j}. \tag{C.17}$$

Finally, we rewrite our expression for the grand partition function as follows:

$$\Xi = \sum_{N,j} \exp(-\beta E_{N,j}) \exp(\beta \mu N)$$

$$\Xi = \sum_{N,j} \exp\left(-\frac{1}{k_B T} E_{N,j}\right) \exp\left(\frac{1}{k_B T} \mu N\right). \tag{C.18}$$

C.4 RELATION TO OTHER TYPES OF PARTITION FUNCTIONS

We have discussed in Chapter 6 that this last expression can be rewritten in terms of canonical partition functions Q for each composition (in this simple example with a single component, one term for each value of N) multiplied by the Boltzmann factor of the free energy $G = N\mu$ (see discussion in Section 6.6):

$$\Xi = \sum_{N} Q(N, V, T) \exp(\beta N \mu)$$

$$Q(N, V, T) = \sum_{j} \exp(\beta E_{N,j}). \tag{C.19}$$

Remembering that the terms in the partition functions Ξ and Q both describe the partitioning (or equilibrium distribution) of a large number of systems over allowed states, we reach the (not unexpected) conclusion that the partitioning of a closed system at constant composition (constant N) is found by selecting only terms with that particular composition from the partition function for the open system (the grand partition function). (The common factor $\exp(\beta N \mu)$ in the terms selected from Ξ does not, of course, alter the partitioning when N is fixed at a single value.)

The derivation of an expression for the grand partition function given in the first two sections of this appendix is readily extended from a system with a single component, to a system with multiple components indexed by $\Lambda = 1, 2$, etc. up through the number of different components in the system, for all possible compositions (indexed with k). The resulting expression for Ξ (which has been given in Eq. 6.30) is reproduced here

$$\Xi = \sum_{\{N_\Lambda\}_k} \left[Q\left(\{N_\Lambda\}_k, V, T\right) \prod_{\Lambda} e^{\beta N_\Lambda \mu_\Lambda} \right].$$

Applying what we learned above for a one-component system, we find first that the partitioning of a closed system (a system having a fixed composition) is obtained by selecting from this sum only the term for that particular composition, and that this is the distribution corresponding to the canonical partition function. Second, if we consider a system that is closed for only a single component (component $\Lambda = 1$) and open for the other components ($\Lambda = 2$ and higher), we can describe its partitioning by a new type of partition function

$$\Gamma = \sum_{\{N_{\Lambda>1}\}_k} \left[Q\left(N_1, \{N_{\Lambda>1}\}_k, V, T\right) \prod_{\Lambda>1} e^{\beta N_\Lambda \mu_\Lambda} \right], \tag{C.20}$$

obtained simply by retaining only the terms in Ξ for a given value of N_1, but omitting the common factor $\exp(\beta N_1 \mu_1)$.

This modified grand partition function or semi-grand partition function is used to describe situations in which the concentration of just one component is fixed and one is interested in its preferential interactions with other components; an example of such an experiment is, of course, the classic equilibrium dialysis procedure that is used to study interactions between different molecular species as described in Section 8.3.1. This partition function was applied in Chapter 10 to develop a statistical mechanical treatment of binding of ligand (variable component) to macromolecules (fixed component).

In the first part of Chapter 6 we saw that the canonical partition function Q defines the thermodynamic state function A (Helmholtz free energy) according to Eq. 6.3,

$$k_B T \ln Q = -A.$$

In the second part of Chapter 6, we found (Eq. 6.30) that for a *macroscopic* system (in which the ensemble is completely dominated by systems whose thermodynamic properties are indistinguishable) the grand partition function defines the thermodynamic state function PV

$$k_B T \ln \Xi = -k_B T \ln Q + \sum_{\Lambda} N_\Lambda \mu_\Lambda = -A + G = PV,$$

G being the Gibbs free energy. We now extend this to the semi-grand partition function and obtain the following important relation which shows that the semi-grand partition function defines the chemical potential of the fixed component,

$$k_B T \ln \Gamma = -k_B T \ln Q + \sum_{\Lambda>1} N_\Lambda \mu_\Lambda = -A + G - N_1\mu_1 = PV - N_1\mu_1. \tag{C.21}$$

The vast majority of experimental studies of solutions in biological macromolecules is carried out under conditions of constant pressure, rather than constant volume, and in solution studies at atmospheric pressure the PV term is insignificant. Only when analyzing studies of the effect of high pressure, is it necessary to take into account the differences in volume between systems with different composition. One may modify the canonical partition function in order to take into account the PV energy of each system configuration. This is achieved by adding PV to the internal energy of each quantum state, therefore multiplying the Boltzmann factor of the internal energy of each term in the canonical partition function by $\exp(-\beta PV)$; in addition an integration over all possible values of V is required to take into account that the volume can fluctuate, and to find the value of V that minimizes the system's free energy. For a macroscopic system, where a single configuration dominates, we can write the equivalent of the canonical partition function for the isobaric ensemble as

$$\begin{aligned} Q_{\text{isobaric}} &= Q\exp(-\beta PV) \\ k_{\mathrm{B}}T\ln Q_{\text{isobaric}} &= k_{\mathrm{B}}T\ln Q - PV = -A - PV = -G. \end{aligned} \tag{C.22}$$

This shows that the thermodynamic state function defined by the isobaric partition function is the Gibbs free energy G. The semi-grand partition function Γ for the isobaric ensemble simplifies Eq. C.21 to

$$k_{\mathrm{B}}T\ln\Gamma_{\text{isobaric}} = k_{\mathrm{B}}T\ln Q_{\text{isobaric}} + \sum_{\Lambda>1} N_\Lambda\mu_\Lambda = -N_1\mu_1. \tag{C.23}$$

SUGGESTED READING

Hill, T.L., *Statistical Mechanics. Principles and Selected Applications*, New York: McGraw-Hill, 1964.

Appendix D

Methods to Compute a Potential of Mean Force

D.1 INTRODUCTION

In this book we have discussed examples of potentials of mean force (PMF) computed from molecular simulations (Chapters 7 and 19). In this appendix we discuss four different methods for doing this that have been used in published studies: thermodynamic integration, slow growth, thermodynamic perturbation, and umbrella sampling.[1]

We remind here that PMF is simply another term for a free energy profile along a reaction coordinate. The free energy change is equal to the work done on the system in a reversible isothermal process and therefore equal to the integral of an (externally applied) force over the coordinate. Because a reversible process takes place infinitely slowly, we can express this in terms of the expectation (mean) value of the force, hence the term *PMF*.

Only two of the above-mentioned methods can be derived from this principle of thermodynamics, and we use instead a common approach based on statistical mechanics starting with the canonical partition function for a system with a classical energy, Eq. 6.17. Writing the potential energy V as a function of what may be either a reaction coordinate or a more general coupling parameter, λ, this gives the free energy as

$$A = -\frac{1}{\beta} \ln \left\{ \int \exp[-\beta V(\lambda)] d\mathbf{X} \right\}. \tag{D.1}$$

Equilibria and Kinetics of Biological Macromolecules, First Edition. Jan Hermans and Barry Lentz.

D.2 THERMODYNAMIC INTEGRATION

The thermodynamic integration method estimates A by approximating the integration of the mean force over the path. Differentiation of the free energy with respect to λ gives the expectation value of the force (see also Eq. 7.1)

$$\frac{\partial A}{\partial \lambda} = -\frac{1}{\beta}\frac{\displaystyle\int (-\beta \partial V/\partial \lambda)\exp(-\beta V)d\mathbf{X}}{\displaystyle\int \exp(-\beta V)d\mathbf{X}} = \overline{\partial V/\partial \lambda} = -\overline{F(\lambda)}, \tag{D.2}$$

in agreement with the thermodynamic definition of A. Hence one also has

$$A = \int d\lambda\, \overline{\partial V/\partial \lambda}. \tag{D.3}$$

A series of simulations at equally spaced values of λ produces mean values of $\partial V/\partial \lambda$ and summation gives an approximate value of the free energy difference

$$\Delta A_{\lambda_0 \to \lambda_1} \approx \sum_{i=1}^{n} \Delta\lambda \left\langle \frac{\partial V(\lambda)}{d\lambda} \right\rangle_{\lambda=\lambda_0+i\Delta\lambda}$$

$$\Delta\lambda = \frac{(\lambda_1 - \lambda_0)}{n}. \tag{D.4}$$

It should be noted that, if the free energy varies rapidly and irregularly with the reaction coordinate, then the values of λ will have to be very closely spaced in order to attain good precision.

D.3 SLOW GROWTH

In the slow growth method the PMF is approximated as the work, W done in a single simulation in which the value of the coupling parameter, λ is changed by $\delta\lambda$ with each integration time step

$$W_{\lambda_0 \to \lambda_1} = \sum_{i=1}^{N_s} \delta\lambda \frac{-\partial V(\lambda)}{d\lambda}_{\lambda=\lambda_0+i\delta\lambda}$$

$$\lambda_1 = \lambda_0 + N_s \delta\lambda, \tag{D.5}$$

where i is the time step and N_s is the total number of time steps.

The work equals the change of free energy plus a contribution from energy dissipation ("friction"), W_f, plus a statistical error component that averages to zero

$$\overline{W_{\lambda_0 \to \lambda_1}} = A_{\lambda_1} - A_{\lambda_0} + \overline{W_f}. \tag{D.6}$$

If the simulation is run in the opposite sense, the free energy difference will have the opposite sign, but friction will be the same

$$\overline{W_{\lambda_1 \to \lambda_0}} = -\left(A_{\lambda_1} - A_{\lambda_0}\right) + \overline{W_f}. \tag{D.7}$$

Hence, by using a pair of simulations, in which λ changes in opposite sense, subtracting the computed values of W and dividing by two, one eliminates the friction term

$$\overline{W_{\lambda_0 \to \lambda_1}} - \overline{W_{\lambda_1 \to \lambda_0}} = 2\left(A_{\lambda_1} - A_{\lambda_0}\right). \tag{D.8}$$

Jarzynski has provided an excellent theoretical foundation of the slow growth method.[2] The precision of the slow growth method increases with N_s, or by averaging over more than one pair of simulations.

D.4 THERMODYNAMIC PERTURBATION

One obtains the basis for the thermodynamic perturbation method by writing the difference of the free energies of the system in two different states, that is, at two different values of λ

$$\begin{aligned} A_{\lambda_0+\Delta\lambda} - A_{\lambda_0} &= -\frac{1}{\beta}\ln\left\{\int \exp\left[-\beta V(\lambda_0 + \Delta\lambda)\right] d\mathbf{X}\right\} \\ &\quad + \frac{1}{\beta}\ln\left\{\int \exp\left[-\beta V(\lambda_0)\right] d\mathbf{X}\right\} \\ \Delta A &= -\frac{1}{\beta}\ln\frac{\int \exp\left[-\beta V(\lambda_0)\right]\exp\left(-\beta\Delta V\right) d\mathbf{X}}{\int \exp\left[-\beta V(\lambda_0)\right] d\mathbf{X}} \\ \Delta V &= V(\lambda_0 + \Delta\lambda) - V(\lambda_0). \end{aligned} \tag{D.9}$$

One sees that the free energy difference for the perturbation is equal to the expectation value of the Boltzmann factor for the energy difference, $\exp(-\beta\Delta V)$ and can be approximated as an average for a sample obtained in a simulation at constant $\lambda = \lambda_0$ according to

$$\Delta A_{\lambda_0 \to \lambda_1} \approx -\frac{1}{\beta}\ln\left\langle \exp\left\{-\beta\left[V(\lambda_1) - V(\lambda_0)\right]\right\}\right\rangle_{\lambda_0}. \tag{D.10}$$

Further analysis shows that this expression converges poorly unless the conformational ensembles at the two values of λ have a good overlap. Whether or not convergence is adequate can be determined in practice for any given choice of λ_0 and λ_1 by repeating the simulation at fixed value of $\lambda = \lambda_1$ and computing an estimate of $\Delta A_{\lambda_1 \to \lambda_0}$. This should give the same answer, but with the opposite sign. The precision of the method is increased by doing a series of such simulations over many successive, small intervals to bridge the range from λ_0 to λ_1.

D.5 UMBRELLA SAMPLING

One may in principle approximate a PMF on the basis of a histogram of a distribution over a range of λ, produced in a simulation in which no constraint is placed on the value of λ

$$A(\lambda) = -\frac{1}{\beta}\ln\left[P_0(\lambda)\right]$$
$$A(\lambda_i) \approx -\frac{1}{\beta}\ln n_i, \tag{D.11}$$

where n_i is the number of instances in which the system is found with the value of λ in a fixed small interval in the neighborhood of λ_i. A problem with this simple approach is that the free energy must not vary greatly over the range of interest, because else some values of λ will rarely or never occur in a sample of finite size, which results in a large statistical error in A at those values of λ. However, the system can be stabilized at any value of λ by addition of an artificial term to the potential energy, usually a quadratic "umbrella" potential

$$V_j(\lambda) = \frac{1}{2}K\left(\lambda - \lambda_j\right)^2. \tag{D.12}$$

As a result, the probability distribution is perturbed

$$P(\lambda) = P_0(\lambda)\exp\left[-\beta V_j(\lambda)\right], \tag{D.13}$$

while the free energy profile is now found by weighting each instance by multiplication with $\exp[\beta V_j(\lambda)]$,

$$A(\lambda_i) \approx -\frac{1}{\beta}\ln n_i$$
$$n_i = \sum \exp\left[\beta V_j(\lambda)\right], \tag{D.14}$$

the summation taking place over only instances in the neighborhood of λ_i.

Typically, a series of simulations with different umbrella potentials is done, each with a different value of λ_j in order to obtain significant statistics at all values of λ. A final complication results from the fact that the individual corrected distributions obtained with different umbrella potentials are not normalized and do not share one normalization factor. A common basis for all histograms can then be obtained by choosing each normalization factor so as to maximize the agreement in all regions where two or more histograms have significant overlap ("weighted histogram analysis method" or WHAM).[3]

D.6 CONCLUSION

Each of these four methods has its advantages and disadvantages, in terms of ease of use and precision achievable for a given computational effort, and each has its adherents. Our recommendation is that any of these four methods will produce good results if used with care. Of particular importance is to check the results for convergence, for example, by increasing the simulation times or by repeating the same calculations with different starting conditions (which can be as simple as a different set of initial velocities).

NOTES

1. For a more extensive discussion of these methods, see van Gunsteren, W.F., Weiner, P.K., eds *Computer Simulation of Biomolecular Systems*. Leiden: Escom, 1989.
2. Jarzynski, C. Nonequilibrium equality for free energy differences. *Phys. Rev. Lett.* **78**: 2690–2693 (1997).
3. Kumar, S., Bouzida, D., Swendsen, R.H., Kollman, P.A., Rosenberg, J.M. The weighted histogram analysis method for free-energy calculations on biomolecules. I. The method. *J. Comput. Chem.* **13**: 1011–1021 (1992).

Appendix E

Theory of the Helix-Coil Transition

E.1 INTRODUCTION

Here, we present two solutions to the problem of reducing the partition function for Schellman's model of the helix-coil equilibrium from a sum with very many terms (infinitely many for an infinitely long polymer) to a closed expression. This model has been introduced in Chapter 16, in which also the partition function for a polymer with n residues, q_{hc}, has been given in Eq. 16.6, which is reproduced here

$$\frac{q_{\mathrm{hc}}}{q_{\mathrm{coil}}} = \sum_{i=0}^{n} \sum_{j} m_{ij} \sigma^{j} s^{i}. \tag{E.1}$$

where q_{coil} is the partition function of an all-coil molecule. Here s is the equilibrium constant for lengthening an existing helix by one residue, σs is the equilibrium constant for starting a helix, i is the number of residues in helical segments, j is the number of helical segments, and m_{ij} is a combinatorial term that describes in how many ways these i residues and j segments can be distributed over the n residues. We describe two different approaches to the problem.

Equilibria and Kinetics of Biological Macromolecules, First Edition. Jan Hermans and Barry Lentz.

E.2 MAXIMUM TERM SOLUTION

A common approach in statistical mechanics is to replace a partition function by its maximum term. For this to be correct, the system must be very large, preferably macroscopic. In this case, that means that the polymer molecules should be infinitely long, and thus the solution applies only in the limit of high molecular weight. (This limitation of the treatment is absent in the alternative treatment by matrix algebra, given below.) The maximum term solution is presented here as an example of this approach, which we have also seen in the derivation of the Boltzmann distribution in Chapter 5.

Gibbs and DiMarzio proposed that in order to find the solution, we must express m_{ij} in terms of i, j, and n, and then maximize $m_{ij}\sigma^j s^i$ with respect to i and j.[1] Obviously, for the extreme case, $i = j = 0$ (no helix) $m_{ij} = 1$. Next consider the case $j = 1$. The i helical residues can be located starting at residue 1, 2, ..., or $n - i + 1$, so that $m_{i1} = n - i + 1$.

There are obviously more possibilities in a polymer with $j = 2$ helical segments and i helical residues. We consider the case in which the residue at one end ("the beginning") of the molecule is helix and the residue at the other end, coil. There are then two coil segments and two helix segments. One can think of such a conformation as being "assembled" from alternating segments of helix and coil, and obtain these segments by "slicing" two "loaves"; one loaf contains all the helical residues and the other contains all the coil residues (Fig. E.1). In this case, each loaf needs to be sliced once; the helix loaf can be cut at $i - 1$ locations, and the coil loaf at $n - i - 1$ places. Consequently,

$$m'_{i2} = (i - 1)(n - i - 1). \tag{E.2}$$

(The prime indicates the restriction we placed on the molecule's ends.) For $j = 3$, we cut the loaves twice. The first slice in the coil loaf can be made at $i - 1$ locations, and the second slice in $i - 2$ places, but the total number of possibilities is $(i - 1)(i - 2)/2$, in order to avoid double counting. Similarly, the two cuts in the coil loaf can be made at $n - i - 1$ and $n - i - 2$ places, respectively, and this gives

$$m'_{i3} = \frac{(i - 1)(i - 2)}{2}\frac{(n - i - 1)(n - i - 2)}{2}. \tag{E.3}$$

which can be written as

$$m'_{i3} = \frac{(i - 1)!}{(i - 3)!2!}\frac{(n - i - 1)!}{(n - i - 3)!2!}. \tag{E.4}$$

Further consideration leads to an expression for any value of i,

$$m'_{ij} = \frac{(i - 1)!}{(i - j)!(j - 1)!}\frac{(n - i - 1)!}{(n - i - j)!(j - 1)!}. \tag{E.5}$$

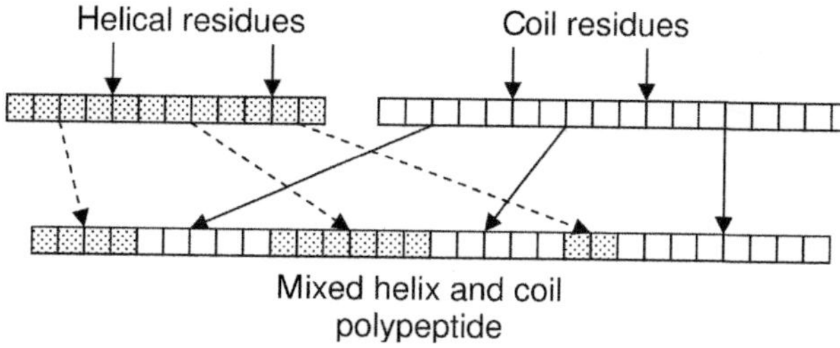

FIGURE E.1 A *particular* state of the polymer (bottom) is assembled by alternating stretches of helical residues with stretches of coil residues. One can imagine the stretches being obtained by "slicing" (top row of arrows) a "loaf" of i helical residues and another "loaf" of $n-i$ coil residues (top) and "moving" these (lower row of arrows) to form the n-residue polymer (bottom).

Something similar can be done to find expressions for molecules having coil or helix at both chain ends, but it turns out that for all cases the maximum term occurs at the same values of i and j.

In order to find the maximum term in the sum of Eq. E.1 we replace each factorial with the leading term of Stirling's formula

$$m'_{ij}\sigma^j s^i \approx \frac{(i-1)^{i-1}}{(i-j)^{i-j}(j-1)^{j-1}} \frac{(n-i-1)^{n-i-1}}{(n-i-j)^{n-i-j}(j-1)^{j-1}} \sigma^j s^i. \quad \text{(E.6)}$$

The maximum term can now be found by setting the partial derivatives of the natural logarithm of this expression with respect to i and j to zero. Approximating further for large n, i, etc., we get

$$\begin{aligned} 0 &= \ln s + \ln i - \ln(n-i) - \ln(i-j) + \ln(n-i-j) \\ 0 &= \ln\sigma + \ln(i-j) - 2\ln j + \ln(n-i-j), \end{aligned} \quad \text{(E.7)}$$

or

$$\begin{aligned} f &= i/n, \\ g &= j/n, \\ \frac{1-f}{f}\frac{f-g}{1-f-g} &= s, \\ \frac{g^2}{(f-g)(1-f-g)} &= \sigma. \end{aligned} \quad \text{(E.8)}$$

The midpoint of the helix-coil transition occurs when $f = 1/2$. One sees that for $f = 1/2$, $s = 1$,

$$f(s=1) = \frac{1}{2}, \quad \text{(E.9)}$$

and substitution of $f = 1/2$ in the equation for σ gives

$$g^2 = \sigma(f - g)(1 - f - g). \tag{E.10}$$

For $f = 1/2$, this becomes

$$g = \frac{j}{n} = \frac{\sqrt{\sigma}}{2(1 + \sqrt{\sigma})}. \tag{E.11}$$

In general, Eq. E.8 relates a pair of values of f and g to any pair of values of s and σ, and vice versa.

E.3 SOLUTION VIA MATRIX ALGEBRA

Zimm–Bragg theory. An alternative solution of Eq. E.1 was used by Zimm and Bragg.[2] Their result gives the partition function for the helix-coil equilibrium of polypeptides of any length. The process which they followed uses the repeated multiplication of a small matrix to generate terms containing the statistical weight of all possible molecular states, and then uses simple matrix algebra to accomplish the summation.

Given the sequence of helix (h) and coil (c) state residues in a polypeptide of n residues, we can write the statistical weight of this sequence as a product, if every c is replaced with "1", every h with the symbol s and, in addition, the symbol σ is inserted once for every ch (and if the sequence begins with h). For example, the molecular state ccchhhhcchhh has statistical weight $111\sigma ssss11\sigma sss = \sigma^2 s^7$. If it were not for the need to include the factors σ, all possible sequences could be generated by writing out the 2^{12} terms of the expression $(1 + s)^{12}$, each term corresponding to a different sequence of 1's and s's (12 total). Obviously, a more sophisticated operation is required in order to also introduce the factors σ, and this is done with the aid of matrix algebra.

The required statistical weights can be generated by repeated multiplication by the following 2×2 matrix

$$\mathbf{M} = \begin{bmatrix} 1 & 1 \\ \sigma s & s \end{bmatrix}. \tag{E.12}$$

To see how this works, we square this matrix to get

$$\mathbf{M}^2 = \begin{bmatrix} 1 + \sigma s & 1 + s \\ \sigma s + \sigma s^2 & \sigma + s^2 \end{bmatrix}. \tag{E.13}$$

Note that the sum of the elements in the first column of $\mathbf{M}^2$ equals the sum of the statistical weights of the four molecular states, cc (1), ch (σs), hc (σs), and hh (σs^2), and represents the ratio of the partition functions q_{hc} and q_{coil} Similarly, the sum of the terms in the first column of $\mathbf{M}^3$ gives this ratio for the trimer, and so forth. The following equation gives the multiplications that extract the sum of the elements of the first column from the matrix $\mathbf{M}^n$,

$$\frac{q_{hc}}{q_{coil}} = \begin{pmatrix} 1 & 1 \end{pmatrix} \mathbf{M}^n \begin{pmatrix} 1 \\ 0 \end{pmatrix}. \tag{E.14}$$

This can be reduced by converting the matrix $\mathbf{M}$ to the diagonal form

$$\mathbf{M} = \mathbf{T}\Lambda\mathbf{T}^{-1} = \mathbf{T}\begin{pmatrix} \lambda_1 & 0 \\ 0 & \lambda_2 \end{pmatrix}\mathbf{T}^{-1}, \tag{E.15}$$

where the matrices $\mathbf{T}$ and $\mathbf{T}^{-1}$ are inverse matrices

$$\mathbf{T}\mathbf{T}^{-1} = \mathbf{I}. \tag{E.16}$$

Consequently, the product $\mathbf{M}^n$ becomes

$$\mathbf{M}^n = \mathbf{T}\Lambda\mathbf{T}^{-1}\mathbf{T}\Lambda\mathbf{T}^{-1} \ldots \mathbf{T}\Lambda\mathbf{T}^{-1} = \mathbf{T}\Lambda^n\mathbf{T}^{-1} = \mathbf{T}\begin{pmatrix} \lambda_1^n & 0 \\ 0 & \lambda_2^n \end{pmatrix}\mathbf{T}^{-1}. \tag{E.17}$$

The eigenvalues, λ_1 and λ_2 are the roots of the quadratic secular equation

$$\begin{vmatrix} 1-\lambda & 1 \\ \sigma s & s-\lambda \end{vmatrix} = (1-\lambda)(s-\lambda) - \sigma s = \lambda^2 - (s+1)\lambda + s - \sigma s = 0$$

$$\lambda_{1,2} = \frac{s + 1 \pm \sqrt{(s-1)^2 + 4\sigma s}}{2}. \tag{E.18}$$

The partition function, q_{hc} contains a term in $\lambda_1{}^n$ and one in $\lambda_2{}^n$. For large n, the smaller term (by convention, the latter) becomes negligible. This then gives (only terms in n are retained)

$$\ln(q_{hc}/q_{coil}) \approx n \ln(s + 1 + \sqrt{(s-1)^2 + 4\sigma s}) - n \ln 2. \tag{E.19}$$

However, for finite n, the following complete expression for q_{hc} should be used

$$q_{hc}/q_{coil} = [\lambda_1^{n+1}(1-\lambda_2) - \lambda_2^{n+1}(1-\lambda_1)](\lambda_1 - \lambda_2). \tag{E.20}$$

The chapter on the helix-coil transition (Chapter 16), shows how one obtains the fraction helix and other parameters from the partition function, as a function of s, σ, and n. The reader may check that (for large n) this produces the same answers for midpoint and helical segments at the midpoint, as predicted by the maximum-term solution Eqs. E.9 and E.11.

NOTES

1. Gibbs, J.H., DiMarzio, E.A. Statistical mechanics of helix-coil transitions in biological macromolecules. *J. Chem. Phys.* **30**: 271–282 (1959).
2. Zimm, B.H., Bragg, J.K. Theory of the phase transition between helix and random coil in polypeptide chains. *J. Chem. Phys.* **31**: 526–535 (1959).

Appendix F

Laplace Transform

F.1 SOLVING LINEAR DIFFERENTIAL EQUATIONS WITH THE LAPLACE TRANSFORM

The Laplace transform is a mathematical tool with which to solve differential equations (such as occur in kinetic problems). This is a process in four steps:

All terms in the equation(s) are transformed.

The Laplace transform converts all differential quotients to algebraic expressions.

The new algebraic equation(s) is (are) solved.

The solution is back-transformed to obtain the solution of the differential equation(s).

F.2 THE LAPLACE TRANSFORM

The Laplace transform is performed by the following integration:

$$\widehat{f}(s) = \int_0^\infty e^{-st} f(t)dt. \tag{F.1}$$

Notice that the variable t of the original function disappears as a result of the integration (definite integral), and a new variable s appears.

Equilibria and Kinetics of Biological Macromolecules, First Edition. Jan Hermans and Barry Lentz.

Each function $f(t)$ has a Laplace transform $\widehat{f}(s)$ associated with it (provided that the integral of Eq. F.1 exists)

Many functions $\widehat{f}(s)$ have associated with them an inverse Laplace transform $f(t)$.

It is frequently easier to find the Laplace transform of a given function (that occurs in the differential equation) than to find the inverse Laplace transform of a given function (that occurs in the solution obtained in third step). For this reason one usually consults a Table of inverse Laplace transforms when working on the solution of the fourth step. An online (www) reference to a table with 129 entries is given in Section F.7.

F.3 TWO KEY PROPERTIES OF THE LAPLACE TRANSFORM

1. The Laplace transform of a sum of terms is the sum of the Laplace transforms, that is, if

$$f(t) = f_1(t) + f_2(t), \tag{F.2}$$

then

$$\widehat{f}(s) = \widehat{f_1}(s) + \widehat{f_2}(s). \tag{F.3}$$

2. The Laplace transform converts a derivative of a function into a simple algebraic expression, as follows:

$$\widehat{f'}(s) = \int_0^\infty e^{-st} f'(t)\, dt; \tag{F.4}$$

first apply the partial integration

$$\int_0^\infty u\, dv = uv|_0^\infty - \int_0^\infty v\, du, \tag{F.5}$$

with

$$\begin{aligned} u &= e^{-st};\ du = -s\, e^{-st}\, dt \\ dv &= f'(t)\, dt;\ v = f(t), \end{aligned} \tag{F.6}$$

so that

$$\begin{aligned} \widehat{f'}(s) &= \int_0^\infty e^{-st} f'(t)\, dt = e^{-st} f(t)|_0^\infty + s\int_0^\infty e^{-st} f(t)dt \\ \widehat{f'}(s) &= -f(0) + s\widehat{f}(s). \end{aligned} \tag{F.7}$$

For the second derivative

$$\widehat{f'}(s) = -f'(0) - sf(0) + s^2\widehat{f}(s), \tag{F.8}$$

and for higher derivatives

$$\widehat{f^{(n)}}(s) = s^n\widehat{f}(s) - \sum_{i=1}^{n} f^{(i-1)}(0)s^{n-i}. \tag{F.9}$$

F.4 EXAMPLE 1: THE POISSON PROCESS (OR CONSECUTIVE REACTIONS)

This system has been introduced in Chapter 19. The differential equations (master equations) for this process are as follows:

$$\dot{p}_n = p_{n-1} - p_n, \tag{F.10}$$

with boundary conditions

$$p_n(t = 0) = \delta_{n,0}, \tag{F.11}$$

or, in other words,

$$p_0(t = 0) = 1;\ p_{i\neq 0}(t = 0) = 0. \tag{F.12}$$

When the Laplace transformation is applied to Eq. F.10, one gets

$$\begin{aligned} \hat{\dot{p}}_n(s) &= \widehat{p}_{n-1}(s) - \widehat{p}_n(s) \\ s\widehat{p}_n(s) - p_n(0) &= \widehat{p}_{n-1}(s) - \widehat{p}_n(s). \end{aligned} \tag{F.13}$$

For $n = 0$ this gives

$$\begin{aligned} s\widehat{p}_0(s) - p_0(0) &= \cancel{\widehat{p}_{-1}(s)} - \widehat{p}_0(s) \\ s\widehat{p}_0(s) - 1 &= -\widehat{p}_0(s) \\ \widehat{p}_0(s) &= 1/(s+1). \end{aligned} \tag{F.14}$$

To find the corresponding function $p_0(t)$, we consult the *Table of inverse Laplace transforms*, and find

$$\widehat{f}(s) = \frac{1}{s+a} \quad \ldots \quad f(t) = e^{-at}. \tag{F.15}$$

Consequently,

$$p_0(t) = e^{-t}. \tag{F.16}$$

For $p_1(t)$, we get

$$s\widehat{p}_1(s) - p_1(0) = \widehat{p}_0(s) - \widehat{p}_1(s), \tag{F.17}$$

and substitution gives

$$\begin{aligned} s\widehat{p}_1(s) - 0 &= \frac{1}{s+1} - \widehat{p}_1(s) \\ \widehat{p}_1(s) &= \frac{1}{(s+1)^2}. \end{aligned} \tag{F.18}$$

On continuing this we find that

$$\widehat{p}_n(s) = \frac{1}{(s+1)^{n+1}}. \tag{F.19}$$

Returning to the *Table of inverse Laplace transforms*, we find

$$\widehat{f}(s) = \frac{1}{(s+a)^n} \quad \ldots \quad f(t) = \frac{t^{n-1}e^{-at}}{(n-1)!}, \tag{F.20}$$

and therefore

$$p_n(t) = t^n e^{-t}/n!. \tag{F.21}$$

This is the general solution; for a graph, see Fig. 19.4.

F.5 EXAMPLE 2: GENERAL CASE OF LINEAR KINETIC EQUATIONS

This system was introduced in Chapter 19. The master equation for such a system is

$$\dot{p}_i(t) = \sum_{j=1}^{n} a_{ij}\, p_j(t). \tag{F.22}$$

The Laplace transformation produces a system of simultaneous linear equations

$$\begin{aligned} \hat{\dot{p}}_i(s) &= \sum_{j=1}^{n} a_{ij}\, \widehat{p}_j(s) \\ s\widehat{p}_i(s) - p_i(0) &= \sum_{j=1}^{n} a_{ij}\, \widehat{p}_j(s), \end{aligned} \tag{F.23}$$

which can be rearranged and written in the standard matrix/vector form

$$\begin{pmatrix} a_{11}-s & a_{12} & \dots & a_{1n} \\ a_{21} & a_{22}-s & \dots & \dots \\ \dots & \dots & \dots & \dots \\ a_{n1} & a_{n2} & \dots & a_{nn}-s \end{pmatrix} \begin{pmatrix} \widehat{p_1} \\ \widehat{p_2} \\ \dots \\ \widehat{p_n} \end{pmatrix} = \begin{pmatrix} p_1(0) \\ p_2(0) \\ \dots \\ p_n(0) \end{pmatrix} \tag{F.24}$$

$$\mathbf{A}\widehat{\mathbf{p}} = \mathbf{p}(0)$$

The solution for $\widehat{p_i}$ can be written as the determinant of the matrix **A** in which the i-th column has been replaced with the vector $\mathbf{p}(0)$, divided by the determinant of the matrix **A** (Cramer's rule). For example, $i = 2$,

$$\widehat{p_2} = \frac{N_2}{D} = \begin{vmatrix} a_{11}-s & p_1(0) & \dots & a_{1n} \\ a_{21} & p_2(0) & \dots & \dots \\ \dots & \dots & \dots & \dots \\ a_{n1} & p_n(0) & \dots & a_{nn}-s \end{vmatrix} \Bigg/ \begin{vmatrix} a_{11}-s & a_{12} & \dots & a_{1n} \\ a_{21} & a_{22}-s & \dots & \dots \\ \dots & \dots & \dots & \dots \\ a_{n1} & a_{n2} & \dots & a_{nn}-s \end{vmatrix}. \tag{F.25}$$

The denominator D is a polynomial of degree n with n roots, $-\lambda_1$ through $-\lambda_n$, and therefore

$$\widehat{p_i} = \frac{N_i}{D} = \frac{N_i}{(s+\lambda_1)(s+\lambda_2)\dots(s+\lambda_n)}, \tag{F.26}$$

and this can be written as

$$\widehat{p_i} = \frac{b_{i,1}}{(s+\lambda_1)} + \frac{b_{i,2}}{(s+\lambda_2)} + \dots + \frac{b_{i,n}}{(s+\lambda_n)}, \tag{F.27}$$

where b_{ij} are constants (i.e., independent of s). The values of b_{ij} and λ_i depend on the rate constants, and the values of b_{ij} also depend on the initial concentrations, $p_i(0)$. λ_i are the *eigenvalues* of matrix **A**. *Conservation of mass* causes relations between the elements of matrix **A**, as a consequence of which the eigenvalues λ_i are all real and positive, with an exception of one eigenvalue, λ_1, which is zero. Application of the inverse Laplace transformation gives

$$p_i(t) = b_{i,1} + b_{i,2}e^{-\lambda_2 t} + \dots + b_{i,n}e^{-\lambda_n t}. \tag{F.28}$$

Special case: if *k roots have the same value*, λ_x, then each expansion for $\widehat{p_i}$ contains a term in $1/(\mathrm{s}-\lambda_x)^k$, and application of the inverse Laplace transform to this function shows that the solution, that is, every $p_i(t)$ will in this case contain terms of the form $b_{i,x}t^k e^{-\lambda_x t}$.

F.6 EXAMPLE 3: COUPLED HARMONIC OSCILLATORS—NORMAL MODES

Single oscillator. The harmonic oscillator is governed by the equation

$$\ddot{x}(t) = \frac{d^2}{dt^2}[x(t)] = -a^2 x(t). \tag{F.29}$$

where x represents the excursion from the equilibrium position. For a simple mechanical harmonic oscillator consisting of a mass and a spring, the coefficient a^2 equals the spring's force constant divided by mass. Application of the Laplace transform gives

$$\hat{\ddot{x}}(s) = -\dot{x}(0) - sx(0) + s^2\widehat{x}(s) = -a^2\widehat{x}(s). \tag{F.30}$$

One may choose the initial conditions as a finite deviation from equilibrium and zero velocity,

$$\begin{aligned} \dot{x}(0) &= 0, \\ x(0) &= b_x, \end{aligned} \tag{F.31}$$

in which case

$$\begin{aligned} \widehat{x}(s) &= \frac{sb_x}{(s^2 + a^2)} \\ x(t) &= b_x \cos(at), \end{aligned} \tag{F.32}$$

cf. the Table of inverse Laplace transforms.

If one instead chooses the initial conditions as zero deviation from equilibrium and a finite velocity,

$$\begin{aligned} \dot{x}(0) &= b_v \\ x(0) &= 0, \end{aligned} \tag{F.33}$$

in which case

$$\widehat{x}(s) = \frac{b_v}{(s^2 + a^2)}$$

$$x(t) = \frac{b_v}{a}\sin(at) = \frac{b_v}{a}\cos\left(at + \frac{\pi}{2}\right). \tag{F.34}$$

The general case can be described with an amplitude and phase (both determined by the initial conditions), according to

$$x(t) = b\cos(at + \varphi). \tag{F.35}$$

Coupled oscillators. For a set of coupled oscillators the equations are

$$\ddot{x}_i(t) = \sum_j a_{ij} x_j(t)$$
$$\ddot{\mathbf{X}} = \mathbf{AX}. \tag{F.36}$$

Matrix **A** is related to the second derivative or Hessian matrix, which describes the curvature of the potential energy surface near the energy minimum,

$$U(\mathbf{X}) = U(\mathbf{X}_{\min}) + \sum_{i=1}^{n}\sum_{j=1}^{n} \frac{\partial^2 U}{\partial x_i \partial x_j} x_i x_j = U_{\min} + \sum_{i=1}^{n}\sum_{j=i}^{n} h_{ij} x_i x_j$$
$$a_{ij} = \frac{h_{ij}}{\sqrt{m_i m_j}}, \tag{F.37}$$

A is symmetric, and we can write

$$\ddot{\mathbf{X}} = \mathbf{C}^{-1} \Lambda \mathbf{CX}$$
$$\mathbf{C}\ddot{\mathbf{X}} = \Lambda \mathbf{CX}, \tag{F.38}$$

where Λ is a diagonal matrix with elements $\lambda_1{}^2, \lambda_2{}^2, \ldots \lambda_n{}^2$. This corresponds to a set of n equations,

$$\sum_j c_{ij} \ddot{x}_j(t) = \frac{d^2}{dt^2}\left[\sum_j c_{ij} x_j(t)\right] = \lambda_i^2 \sum_j c_{ij} x_j(t), \tag{F.39}$$

each of which has the form of Eq. F.29, and has as solution

$$c_{ij} x_j(t) = \frac{b_i}{\lambda_i} \cos(\lambda_i t) \quad j = 1, 2, \ldots n. \tag{F.40}$$

The motion consists of a superposition of oscillations with amplitudes of (atomic) motion proportional to the elements of each eigenvector of matrix **A** (= the elements of a row of matrix **C**) with frequencies determined by the eigenvalues of matrix $\mathbf{A}(\lambda_i)$, and the overall amplitude (i.e., the coefficients b_i) and phase (not indicated in these equations) of each of these "*normal mode*" oscillations, as determined by the starting conditions.

The three translational and three rotational modes have eigenvalues equal to zero, and a molecule consisting of m atoms has $3m - 6$ vibrational modes.

Finally, it is clear that vibration according to the i-th eigenvector contributes terms of relative magnitude $1/(c_{ij}c_{ik})$ to the covariance $< x_j x_k >$. In fact, the eigenvectors and eigenvalues can be calculated from a covariance matrix containing contributions from all n normal mode vibrations. This calculation is (mathematically) possible even when the motion does not consist purely of a superposition of n harmonic vibrations, but in that case the resulting eigenvectors are called *the principal components*. The so-called principal component decomposition can be applied to an arbitrary covariance matrix, and is an important tool in statistical analysis.

F.7 TABLE OF INVERSE LAPLACE TRANSFORMS

The following is a short list of Laplace transforms and corresponding functions:

$\widehat{f}(s)$	$f(t)$
$1/s$	1
$1/s^n$	$t^{n-1}/(n-1)!$
$1/(s+a)$	$\exp(-at)$
$1/(s+a)^n$	$t^{n-1}\exp(-at)/(n-1)!$
$1/(s^2+a^2)$	$\sin(at)/a$
$s/(s^2+a^2)$	$\cos(at)$
...	...
$\exp(-ks)$	$\delta(t-k)$ (Dirac delta function)
...	...

Additional formulas are given in Section 29.3 of *Handbook of Mathematical Functions*, edited by M. Abramowitz and I.A. Segun, Dover publications, NY, for 129 cases. An online copy of this handbook is available at: http://www.convertit.com/Go/ConvertIt/Reference/AMS55.ASP, Chapter 29. Laplace transforms begins on page 1019; Section 29.3 Table of Laplace Transforms is given on pages 1021–1029.

Appendix G

Poisson Equation

G.1 FORMULATION

The electrostatic potential Φ of charges distributed in a medium of uniform dielectric constant ε is given by *Coulomb's equation*. (See also Section 3.3, "Noncovalent forces between atoms and molecules".) This can be expressed in terms of integrals over a continuous charge distribution, or, alternatively, in terms of sums over a set of point charges q located at positions $\mathbf{r}_i$, as[1]

$$\Phi(\mathbf{r}') = \int d\mathbf{r} \frac{\rho(\mathbf{r})}{4\pi\varepsilon_0\varepsilon|\mathbf{r} - \mathbf{r}'|} = \sum_i \frac{q(\mathbf{r}_i)}{4\pi\varepsilon_0\varepsilon|\mathbf{r}_i - \mathbf{r}'|}. \tag{G.1}$$

The electrostatic energy is then given by

$$U_{\text{el}} = \frac{1}{2}\int d\mathbf{r}\rho(\mathbf{r})\Phi(\mathbf{r}) = \frac{1}{2}\sum_i q(\mathbf{r}_i)\Phi(\mathbf{r}_i). \tag{G.2}$$

Here, ε_0 is the permittivity of vacuum and ε the dielectric constant of the medium (or relative permittivity). If the dielectric constant is not uniform, Coulomb's equation no longer applies and one must replace Eq. 3.11 with a more complicated expression for the electrostatic potential Φ in terms of the charge density ρ

Equilibria and Kinetics of Biological Macromolecules, First Edition. Jan Hermans and Barry Lentz.

and the dielectric constant ε, *Poisson's equation*. This is a second-order differential equation,

$$\nabla \cdot \varepsilon \nabla \Phi = -\rho/\varepsilon_0. \tag{G.3}$$

The gradient operator ∇ operates on a scalar and the result is a vector; in Cartesian coordinates

$$\nabla \Phi = \left(\frac{\partial \Phi}{\partial x}, \frac{\partial \Phi}{\partial y}, \frac{\partial \Phi}{\partial z} \right) = \frac{\partial \Phi}{\partial x} \mathbf{u}_x + \frac{\partial \Phi}{\partial y} \mathbf{u}_y + \frac{\partial \Phi}{\partial z} \mathbf{u}_z. \tag{G.4}$$

where $\mathbf{u}_x$, etc., are the unit vectors along the coordinate axes. The Poisson equation is

$$\nabla \varepsilon \nabla \Phi + \varepsilon \nabla^2 \Phi \equiv \frac{\partial \varepsilon}{\partial x}\frac{\partial \Phi}{\partial x} + \frac{\partial \varepsilon}{\partial y}\frac{\partial \Phi}{\partial y} + \frac{\partial \varepsilon}{\partial z}\frac{\partial \Phi}{\partial z} + \varepsilon \left(\frac{\partial^2 \Phi}{\partial x^2} + \frac{\partial^2 \Phi}{\partial y^2} + \frac{\partial^2 \Phi}{\partial z^2} \right) = -\frac{\rho}{\varepsilon_0}. \tag{G.5}$$

Wherever the dielectric coefficient ε is constant, this becomes

$$\frac{\partial^2 \Phi}{\partial x^2} + \frac{\partial^2 \Phi}{\partial y^2} + \frac{\partial^2 \Phi}{\partial z^2} = \nabla^2 \Phi = -\frac{\rho}{\varepsilon_0 \varepsilon}, \tag{G.6}$$

which, as we shall see, is equivalent to Coulomb's law, Eq. 3.11.

The following boundary conditions apply to the solution of the Poisson equation for a finite solute in a bath of solvent

$$\Phi(\mathbf{r}) \to 0 \quad \text{and} \quad \nabla \Phi(\mathbf{r}) \to 0 \quad \text{as} \quad |\mathbf{r}| \to \infty. \tag{G.7}$$

In case the volume is divided into parts with different, but in each part uniform, dielectric constant, Eq. G.6 applies within each part, and the following boundary conditions apply at the interfaces between parts

$$\begin{aligned} &\Phi(\mathbf{r}) \text{ is continuous} \\ &\mathbf{n} \cdot (\varepsilon(\mathbf{r}) \nabla \Phi(\mathbf{r})) \text{ is continuous} \end{aligned} \tag{G.8}$$

where $\mathbf{n}$ is the normal to the surface at which ε is discontinuous.

G.2 EXACT SOLUTION FOR A SIMPLE CASE: THE BORN MODEL

As an important example, we consider the case of a spherical particle of radius R having uniform charge density ρ_0. Inside, the particle has a uniform dielectric constant $\varepsilon_b = 1$, and outside the dielectric constant is ε_a. Poisson's equation, Eq. G.6, applies throughout, but the solution has a different form in the two parts.

The system is spherically symmetrical, all properties being a function of only the radial coordinate, r. With the expression of the Laplace operator for a spherical coordinate system, Poisson's equation becomes (cf. Eq. I.7)

$$\frac{1}{r^2}\frac{d}{dr}\left(r^2\frac{d\Phi(r)}{dr}\right) \equiv \frac{1}{r}\frac{d^2(r\Phi)}{dr^2} = \frac{-\rho}{\varepsilon_0\varepsilon}. \tag{G.9}$$

We are interested in the solutions of this differential equation outside the sphere, where $\rho(\mathbf{r}) = 0$, and inside the sphere, where $\rho(\mathbf{r}) = \rho_0$. Integrating twice, one finds the following solutions for these two cases; outside the charged sphere it is

$$\Phi(r) = \frac{c_1}{r} + c_2 \quad \text{for} \quad \rho(r) = 0, \tag{G.10}$$

and inside it is

$$\Phi(r) = -\frac{\rho_0}{6\varepsilon_0\varepsilon_\mathrm{b}} + \frac{c_3}{r} + c_4 \quad \text{for} \quad \rho(r) = \rho_0. \tag{G.11}$$

The values of the integration constants c_1 through c_4 are then determined as follows. The potential outside the particle $\Phi_\mathrm{a}(r)$ obeys Eq. G.10 with $c_2 = 0$ (because Φ must tend to zero for large r), and that inside $\Phi_\mathrm{b}(r)$ obeys Eq. G.11, with $c_3 = 0$ (because Φ must remain finite for $r = 0$), that is,

$$\begin{aligned}\Phi_\mathrm{a}(r) &= \frac{c_1}{r}\\ \Phi_\mathrm{b}(r) &= -\frac{\rho_0}{6\varepsilon_0\varepsilon_\mathrm{b}}r^2 + c_4.\end{aligned} \tag{G.12}$$

Applying the two boundary conditions given in Eq. G.8 at the surface of the sphere, we obtain

$$\begin{aligned}-\frac{\rho_0}{6\varepsilon_0\varepsilon_\mathrm{b}}R^2 + c_4 &= \frac{c_1}{R}\\ -\frac{\varepsilon_\mathrm{b}\rho_0}{3\varepsilon_0\varepsilon_\mathrm{b}}R &= -\frac{\varepsilon_\mathrm{a}c_1}{R^2}.\end{aligned} \tag{G.13}$$

Solving for c_1 and c_4, and substituting the total charge $q = \int \rho dV = 4\pi\rho_0 R^3/3$, we get

$$\begin{aligned}\Phi_\mathrm{a}(r) &= \frac{1}{4\pi\varepsilon_0}\frac{q}{\varepsilon_\mathrm{a}r}\\ \Phi_\mathrm{b}(r) &= \frac{1}{4\pi\varepsilon_0}\left(-\frac{qr^2}{2\varepsilon_\mathrm{b}R^3} + \frac{q}{2\varepsilon_\mathrm{b}R} + \frac{q}{\varepsilon_\mathrm{a}R}\right).\end{aligned} \tag{G.14}$$

The potential outside the particle is simply Coulomb's law for a particle of charge q immersed in a medium with dielectric constant ε_a.

With known electrostatic potential and charge density, the electrostatic energy can be easily computed by integrating the product of the two over the entire volume. For this particular case, only the inside of the particle contributes to the integral

$$U_{\text{el}} = \frac{1}{2}\int_{r<R} \rho\Phi_b(r)dV = \int_0^R 4\pi r^2 \rho\Phi_b(r)dr. \tag{G.15}$$

Of interest is the electrostatic contribution to the *solvation energy* (actually, a free energy), which equals the electrostatic energy of the particle in the solution, less the energy for the same particle in a vacuum, which corresponds to a change in the medium's dielectric constant ε. Only the third term in the sum in Eq. G.14 depends on ε; consequently,

$$\Delta G_{\text{el,sol}} = \frac{1}{2}\frac{1}{4\pi\varepsilon_0}\int_{r<R} \rho dV \left(\frac{q}{\varepsilon_{\text{a}} R} - \frac{q}{R}\right)$$
$$\Delta G_{\text{el,sol}} = \frac{1}{2}\frac{-q^2}{4\pi\varepsilon_0 R}\left(1 - \frac{1}{\varepsilon_{\text{a}}}\right). \tag{G.16}$$

Analytical solutions to the Poisson equation can be found for a few simple problems, while numerical solutions can be computed for more complex situations.

G.3 ACCOUNTING FOR IONIC STRENGTH: POISSON–BOLTZMANN EQUATION AND DEBYE–HÜCKEL THEORY

It is possible to account for the presence of (other) charged particles (ions) in the solution in the context of our description of the solution as a polarizable dielectric by attributing to the solution a nonzero charge density due to the ions. In the absence of an electrostatic potential due to a neighboring solute, positive and negative charges cancel and the charge density is zero. But if the electrostatic potential is positive, this will result in an excess concentration of negative ions and an excess negative charge density, and the other way around. The solution's charge density is proportional to the *excess* concentration of negative or positive ions.

If each type of ion obeys a Boltzmann distribution in the electrostatic energy, Φze (where ze is the ion's charge) one has

$$c_z = c_{z,0}\exp(-\beta\Phi ez). \tag{G.17}$$

and writing the charge density as the sum over all ion species

$$\rho = \sum_i ez_i c_i = \sum_i ez_i c_{i,0}\exp(-\beta\Phi ez_i). \tag{G.18}$$

Substitution of this expression for the charge density of the solution into the Poisson Eq. G.3 gives the *Poisson–Boltzmann equation*. This equation can be solved numerically by the same methods that can be used to solve the Poisson equation itself for arbitrarily complex charge distributions (see Chapter 14.)

An important simplification gives the *linearized Poisson–Boltzmann equation*; this requires expanding the exponential in powers of $\beta\Phi e z_i$ and retaining only terms in the first power of Φ

$$\rho = \sum_i e z_i c_i (1 - \beta\Phi e z_i \ldots) = -\Phi \sum_i \beta (e z_i)^2 c_i$$

$$\nabla^2 \Phi = -\frac{\rho}{\varepsilon_0 \varepsilon} = \Phi \frac{\sum_i \beta (e z_i)^2 c_i}{\varepsilon_0 \varepsilon} = \Phi \kappa^2. \tag{G.19}$$

This equation is the basis for the Debye–Hückel theory of electrolyte solutions. The constant κ has units of inverse length and $1/\kappa$ is called the *Debye–Hückel screening length*; κ depends only on the composition of the solvent, and its square is proportional to the ionic strength $((1/2)\sum z_i{}^2 c_i)$.

Integration of the linearized Poisson–Boltzmann equation for the spherical particle that was earlier used for the Born model gives the potential in the solvent as

$$\Phi_\mathrm{a}(r) = c_5 \frac{e^{-\kappa r}}{r} + c_6 \frac{e^{\kappa r}}{r} = c_5 \frac{e^{-\kappa r}}{r}. \tag{G.20}$$

(The boundary conditions require $c_6 = 0$.) This expression replaces Φ_a of Eq. G.14; as the concentration of ions is zero inside the particle, the expression for the potential inside the sphere Φ_b is the same as before. Applying the boundary conditions at the interface ($r = R$) and solving for c_5, one obtains

$$\Phi_\mathrm{a}(r) = \frac{q e^{-\kappa r}}{4\pi \varepsilon_0 \varepsilon r} \frac{e^{\kappa R}}{1 + \kappa R}, \tag{G.21}$$

and, at the surface of the sphere,

$$\Phi_\mathrm{a}(R) = \frac{q}{4\pi \varepsilon_0 \varepsilon r} \frac{e^{\kappa R}}{1 + \kappa R}. \tag{G.22}$$

One sees that the potential around the macromolecule is reduced due to the presence of the dissolved electrolyte, although the gradient $d\Phi/dr$ at the surface of the sphere is the same for Coulomb and Debye–Hückel potentials (Fig. 14.4).

For a 1–1 electrolyte at 0.01 mol/l, $1/\kappa$ is circa 30 Å.

NOTE

1. Older texts may use a different definition of the unit of charge (as well as cgs units), in which case the factor $4\pi\varepsilon_0$ does not appear in Coulomb's equation and a factor 4π instead of $1/\varepsilon_0$ appears in Poisson's equation.

Appendix H

Defining Molecular Boundaries

Our current understanding of what constitutes a macromolecular surface was developed by Lee and Richards[1] and by Connolly,[2] who realized the utility of defining three different surfaces of any given molecule.

To understand the significance of these surfaces, we start with the previously well-known *van der Waals surface*, which is determined as the surface of a set of intersecting spheres, every atom having its typical radius that determines molecular packing, r_v.[3] As bond lengths are typically much shorter than contact distances between nonbonded atoms, the spheres of two atoms connected by a chemical bond intersect. In a folded macromolecule, the van der Waals spheres of two hydrogen-bonded atoms also intersect, and, as a result, the van der Waals surface of a folded macromolecule is topologically complex, enclosing a volume with many narrow channels. If the macromolecule is well packed (as is the case for a typical globular protein), then these channels will be too narrow to admit even a water molecule. (This does not exclude the presence of wider channels filled with water molecules and allowing passage of substrate or product molecules.) Effectively, only the outside parts of a given van der Waals surface are encountered by a water molecule.

This realization led to the introduction of a new surface, called *the solvent-accessible surface*; this is the surface traced by the *center* of a *solvent-sized probe* ball that is rolled all over the van der Waals surface, always in contact with one, two, and sometimes three, atomic van der Waals spheres, without ever intersecting any of these. The solvent-accessible surface consists entirely of adjoining convex segments of spherical surface.

Equilibria and Kinetics of Biological Macromolecules, First Edition. Jan Hermans and Barry Lentz.

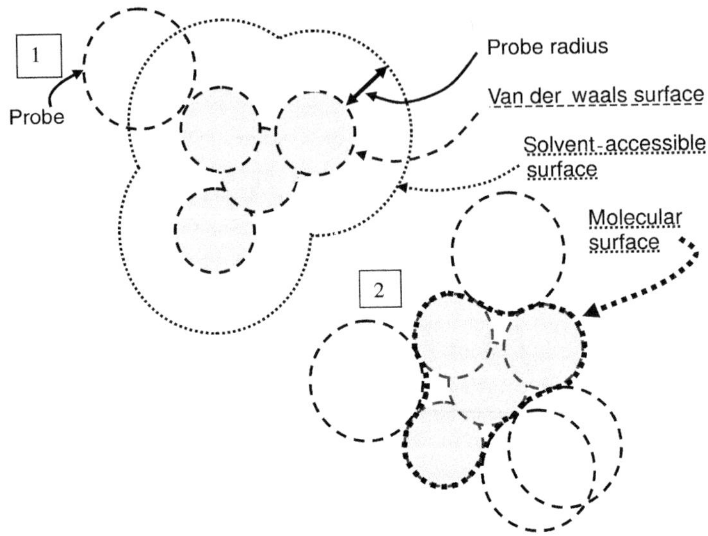

FIGURE H.1 Definition of various molecular surfaces.

This same operation of rolling a probe sphere over the van der Waals surface is used to define a third surface, the molecular surface, defined by the inner surface traced by the collection of probe spheres. This surface includes any parts of the van der Waals surface that are at any time touched by the probe sphere, and these parts form a disconnected set of convex surface elements. These convex elements are connected by concave elements (the reentrant part of the molecular surface) that result whenever the probe sphere "rolls along" touching two van der Waals spheres, or ends up in a site where it touches three van der Waals spheres.

This is illustrated in two dimensions in Fig. H.1. (In two dimensions the surfaces are circles, and the surface elements are arcs of circles, the molecular surface has convex and concave elements; the latter occur whenever the probe circle touches two van der Waals circles.)

NOTES

1. (a) Lee, B., Richards, F.M. The interpretation of protein structures: Estimation of static accessibility. *J. Mol. Biol.* **55**: 379–400 (1971). (b) Richards, F.M. Areas, volumes, packing and protein structure. *Ann. Rev. Biophys. Bioeng.* **6**: 151–176 (1977).
2. (a) Connolly, M.L. Solvent-accessible surfaces of proteins and nucleic-acids. *Science* **221**: 709–713 (1983). (b) Connolly, M.L. Analytical molecular surface calculation. *J. Appl. Crystallogr.* **16**: 548–558 (1983).
3. The contact distance $r_{v,i} + r_{v,j}$ corresponds to the energy minimum.

Appendix I

Equations

I.1 STIRLING'S FORMULA AND COMBINATORIALS

Stirling's formula for $n!$

$$n! \approx n^n e^{-n} \sqrt{2\pi n}$$

$$\ln n! \approx n \ln n - n + \frac{1}{2} \ln n, \tag{I.1}$$

is an excellent approximation. For example, for $n = 10$, the value of $n!$ is 3,628,800 and Stirling's approximation gives 3,598,696. The *relative* error in $n!$ decreases rapidly as n increases: it is 8% for $n = 1$, 4% for $n = 2$, 1% for $n = 10$, 0.08% for $n = 100$. For large n, the first term in $\ln n!$ ($n \ln n$) dominates and is itself an excellent approximation to $\ln n!$

Combinatorials represent the number of possible selections of a limited number of items from a larger set. For the simplest case, in which one selects i items from a total of n, one has

$$\binom{n}{i} = C_i^n = \frac{n!}{[i!(n-i)!]}, \tag{I.2}$$

Equilibria and Kinetics of Biological Macromolecules, First Edition. Jan Hermans and Barry Lentz.

where we have used two alternate common notations. If multiple subsets containing, respectively, i_1, i_2, etc. items are selected, the combinatorial becomes

$$C^n_{\{i\}} = \frac{n!}{\left[(n-m)!\prod_j i_j!\right]}$$

$$m = \sum_j i_j, \tag{I.3}$$

where $\{i\}$ refers to the set i_1, i_2, etc.

Combinatorials are generated as coefficients of the series expansion of powers of polynomials. For binomials this gives

$$(a+b)^n = \sum_{i=0}^{n} C^n_i a^i b^{n-i}. \tag{I.4}$$

The coefficients of powers of longer polynomials have the combinatorial form given in Eq. I.3.

I.2 INTEGRALS OF GAUSSIAN DISTRIBUTIONS

$$\int_{-\infty}^{\infty} \exp(-ax^2)dx = 2\int_0^{\infty} \exp(-ax^2)dx = \sqrt{\frac{\pi}{a}}$$

$$\int_{-\infty}^{\infty} x^2 \exp(-ax^2)dx = 2\int_0^{\infty} x^2 \exp(-ax^2)dx = \frac{1}{2a}\sqrt{\frac{\pi}{a}}. \tag{I.5}$$

I.3 CARTESIAN AND SPHERICAL POLAR COORDINATES

If the coordinate systems are aligned as shown in Fig. I.1, the Cartesian coordinates are related to the spherical coordinate set by

$$x = r\sin\theta\cos\varphi$$

$$y = r\sin\theta\sin\varphi$$

$$z = r\cos\theta, \tag{I.6a}$$

and to the cylindrical polar coordinate set by

$$x = R\cos\varphi$$

$$y = R\sin\varphi. \tag{I.6b}$$

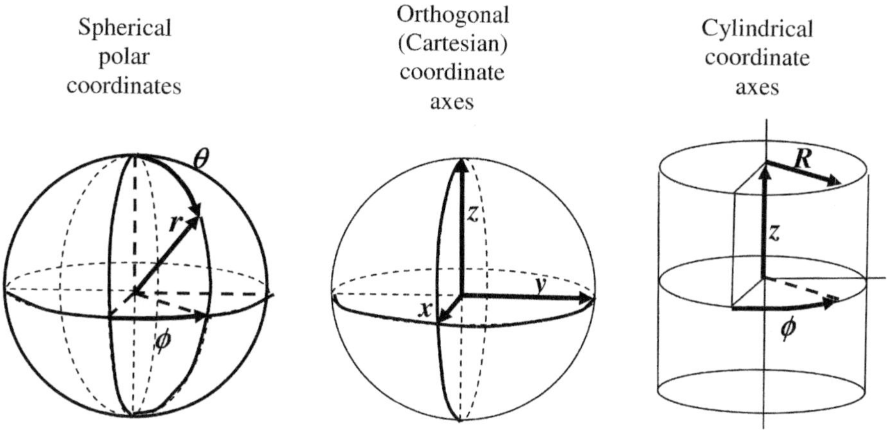

FIGURE I.1 Polar and orthogonal coordinate systems.

I.4 LAPLACE OPERATOR IN THREE-DIMENSIONAL CARTESIAN, POLAR, AND CYLINDRICAL COORDINATES

$$\nabla^2\psi = \frac{\partial^2\psi}{\partial x^2} + \frac{\partial^2\psi}{\partial y^2} + \frac{\partial^2\psi}{\partial z^2}$$

$$= \frac{1}{r^2}\frac{\partial}{\partial r}\left(r^2\frac{\partial\psi}{\partial r}\right) + \frac{1}{r^2\sin\theta}\frac{\partial}{\partial\theta}\left(\sin\theta\frac{\partial\psi}{\partial\theta}\right) + \frac{1}{r^2\sin^2\theta}\frac{\partial^2\psi}{\partial\varphi^2}$$

$$= \frac{1}{R}\frac{\partial}{\partial R}\left(R\frac{\partial\psi}{\partial R}\right) + \frac{1}{R^2}\frac{\partial^2\psi}{\partial\varphi^2} + \frac{\partial^2\psi}{\partial z^2}. \tag{I.7}$$

(For two-dimensional, circular polar coordinates use the expression for cylindrical polar coordinates, with constant z.)

I.5 SUMS OF GEOMETRIC SERIES (VALID FOR $s^2 < 1$)

$$A_0 = \sum_{i=0}^{\infty} s^i = \frac{1}{1-s} \qquad A_1 = \sum_{i=0}^{\infty} i s^i = \frac{s}{(1-s)^2}$$

$$A_2 = \sum_{i=0}^{\infty} i^2 s^i = \frac{s(1+s)}{(1-s)^3} \qquad A_{m+1} = s\frac{dA_m}{ds} \tag{I.8}$$

and

$$\sum_{i=0}^{\infty} i^m s^i = \sum_{i=0}^{N} i^m s^i - \sum_{i=N+1}^{\infty} i^m s^i$$

$$\sum_{i=0}^{\infty} i^m s^i = A_m - s^{N+1} \sum_{j=0}^{\infty} (j + N + 1)^m s^i$$

$$\sum_{i=0}^{\infty} i^2 s^i = A_2 - s^{N+1} \left\{ A_2 + s\,(N+1)\,A_1 + (N+1)^s A_0 \right\}. \tag{I.9}$$

I.6 THE KRONECKER AND DIRAC DELTA FUNCTIONS

The Kronecker delta function is defined by the following two equations:

$$\delta(0) = 1$$
$$\delta(i) = 0, \quad \text{for } i \neq 0. \tag{I.10}$$

The Dirac delta function is a related function that is defined by the following integrals:

$$\int_{-\infty}^{\infty} \delta(x) dx = 1$$
$$\int_{-\infty}^{\infty} f(x)\delta(x - x_0) dx = f(x_0). \tag{I.11}$$

Actually, it is sufficient if the limits of integration bracket the value of 0 in the first and of x_0 in the second integral. (If they do not, the value of the integral is zero.)

I.7 USEFUL RELATIONS BETWEEN DIFFERENTIAL QUOTIENTS

Inversion

$$\left(\frac{\partial x}{\partial y}\right)_z = \frac{1}{\left(\frac{\partial y}{\partial x}\right)_z} \tag{I.12}$$

The *Euler chain rule* (z not an explicit function of x)

$$\frac{dz(y)}{dx} = \frac{dz(y)}{dy}\frac{dy(x)}{dx}, \tag{I.13}$$

and the related

$$\left(\frac{\partial z}{\partial y}\right)_x \left(\frac{\partial x}{\partial z}\right)_y \left(\frac{\partial y}{\partial x}\right)_z = -1. \tag{I.14}$$

For $f(x, y, z)$ when both $x(z)$ and $y(z)$

$$\left(\frac{\partial f}{\partial x}\right)_z = \left(\frac{\partial f}{\partial x}\right)_y + \left(\frac{\partial f}{\partial y}\right)_x \left(\frac{\partial y}{\partial x}\right)_z. \tag{I.15}$$

Finally, if df in the following expression is an exact or total differential

$$df = g(x, y)dx + h(x, y)dy, \tag{I.16}$$

then

$$g(x, y) = \left(\frac{\partial f}{\partial x}\right)_y$$
$$h(x, y) = \left(\frac{\partial f}{\partial y}\right)_x, \tag{I.17}$$

and also

$$\left(\frac{\partial g}{\partial y}\right)_x = \frac{\partial^2 f}{\partial g \partial h} = \left(\frac{\partial h}{\partial x}\right)_y. \tag{I.18}$$

I.8 RANDOM NUMBERS

Random number generators play a vital role in simulation methods. For example, they are an integral part of the Metropolis Monte Carlo method discussed in Chapter 7 and are used at the start of molecular dynamics simulations in order to assign a random set of atomic velocities according to a Boltzmann distribution.

In principle, a true random number generator can be based only on a random physical phenomenon, such as radioactive decay events in a set time interval in a sample. Those used in simulations are programmable arithmetic constructs; the best of these produce long sequences of uncorrelated values. For example, in a (once) widely used method, multiplication of two k-bit binary numbers m and s produces a $2k$-bit product. (Typically, $k = 32$.) Of this product, the low order k bits are used to represent a new positive k-bit integer; although the low order k bits are perfectly predictable given the values of m and s, its magnitude as new integer is not correlated with the magnitudes of m and s. The result, when normalized by division by 2^k, becomes a uniformly distributed (pseudo) random number in the range 0–1. Also, this becomes the new value of s for generating the next number in the sequence. The entire sequence is of course perfectly reproducible given the values of the multiplier, m, and of the "seed," s, used to start the sequence;

nevertheless, extensive tests have established the validity of using it as a sequence of random numbers.

More sophisticated methods have since been developed that produce more perfect sequences of numbers (with longer sequences before the same sequence repeats, and absence of correlations between successive values), and for producing numbers obeying nonuniform distributions, for example, a Gaussian distribution.

Index

Equilibria and Kinetics of Biological Macromolecules, First Edition. Jan Hermans and Barry Lentz.

CONSTANTS

Elementary charge	$e = 1.602189 \times 10^{-19}$ C
Atomic mass unit	amu $= 1.660566 \times 10^{-27}$ kg
Proton rest mass	$m_p = 1.672648 \times 10^{-27}$ kg $= 1.0073$ amu
Electron rest mass	$m_e = 9.109534 \times 10^{-31}$ kg
Boltzmann's constant	$k_B = 1.380662 \times 10^{-23}$ J K^{-1}
	$= 8.31441$ J K^{-1} mol^{-1} $= 1.987$ cal K^{-1} mol^{-1}
Planck's constant	$h = 6.626176 \times 10^{-34}$ J s
Avogadro's number	$N_A = 6.022 \times 10^{23}$ mol^{-1}
k_BT at 300 K	$k_BT = 4.142$ zJ $= 2.494$ kJ mol^{-1} $= 0.596$ kcal mol^{-1}
Freezing point of water	273.15 K
Viscosity of water at 20 °C	1.003×10^{-3} Pa s
Gravity	$g = 9.8066$ m s^{-2}
Permittivity of vacuum	$\varepsilon_0 = 8.854 \times 10^{-12}$ C^2 N^{-1} m^{-2}

PREFIXES OF THE METRIC SYSTEM

Name	Prefix	Value	Name	Prefix	Value
yotta	Y	10^{24}	yocto	y	10^{-24}
zotta	Z	10^{21}	zepto	z	10^{-21}
exa	E	10^{18}	atto	a	10^{-18}
peta	P	10^{15}	femto	f	10^{-15}
tera	T	10^{12}	pico	p	10^{-12}
giga	G	10^{9}	nano	n	10^{-9}
mega	M	10^{6}	micro	μ	10^{-6}
kilo	k	10^{3}	milli	m	10^{-3}
- - -	- - -	10^{0}	- - -	- - -	10^{0}

Also deka (da, 10^1), hekto (h, 10^2), deci (d, 10^{-1}) and centi (c, 10^{-2}).

GREEK ALPHABET

α	A	alpha	η	H	eta	ν	N	nu	τ	T	tau
β	B	beta	θ, ϑ	Θ	theta	ξ	Ξ	xi	υ	Y, Υ	upsilon
γ	Γ	gamma	ι	I	iota	o	O	omicron	ϕ	Φ	phi
δ	Δ	delta	κ	K	kappa	π	Π	pi	χ	X	chi
ε	E	epsilon	λ	Λ	lambda	ρ	P	rho	ψ	Ψ	psi
ζ	Z	zeta	μ	M	mu	σ, ς	Σ	sigma	ω	Ω	omega

Suggested pronunciation:
like "pie": π (pie), χ (kie);
like "bee": ξ (ksee), ϕ (fee), ψ (psee)

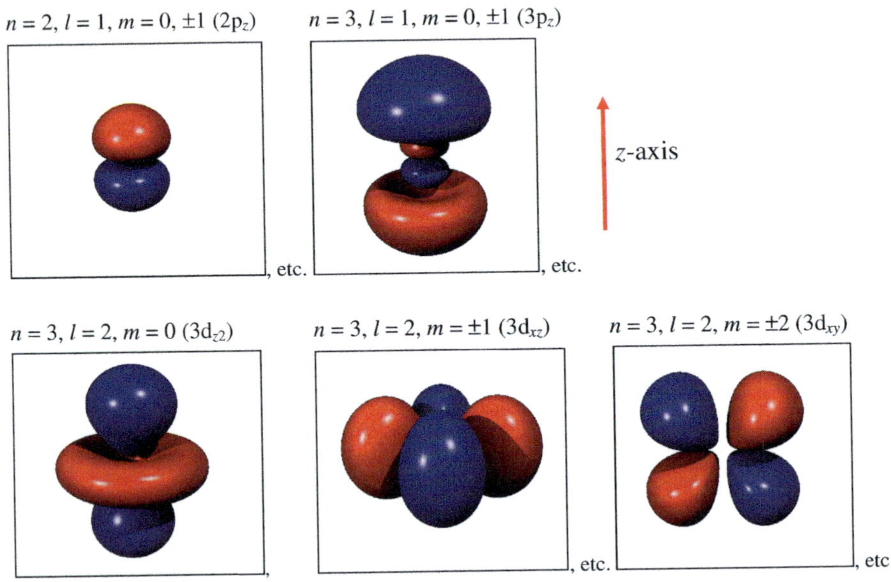

FIGURE 2.7 Asymmetry of some hydrogen atom wave functions. The direction of the z-axis is "up" tilted toward the viewer. Not shown are wave functions that are obtained from these by exchange of x, y, and z axes, indicated with "etc." Colors indicate positive (red) and negative (blue) values of ψ. From M. Hanrath, University of Cologne, http://www.tc.uni-koeln.de/people/hanrath/.

Equilibria and Kinetics of Biological Macromolecules, First Edition. Jan Hermans and Barry Lentz.

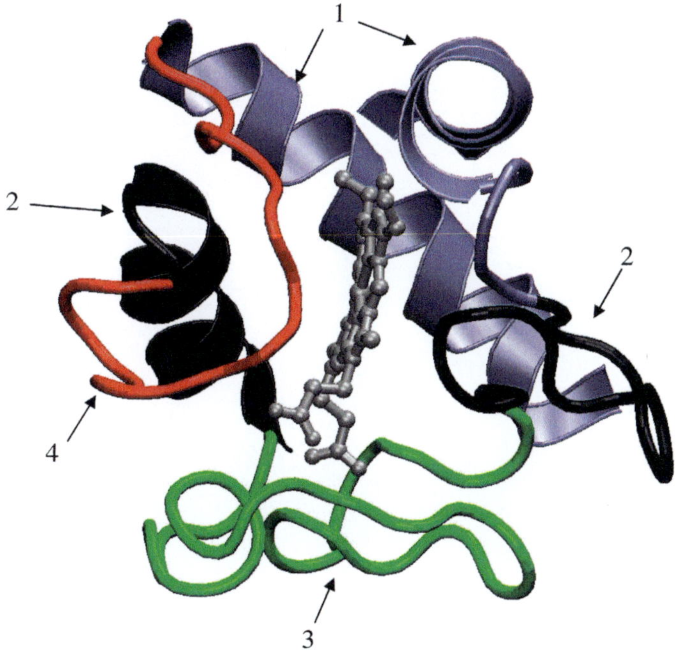

FIGURE 17.12 Four clusters in cytochrome c having different unfolding isotherms (in order of decreasing stability: colored blue, black, green, and red). The protein segments defined by the four clusters account for the whole cytochrome c protein. From Bai, Y., Milne, J.S., Mayne, L., Englander, S.W. *Proteins: Struct. Funct. Genet.* **17**: 75–86 (1993); vmd graphic.

Printed and bound by CPI Group (UK) Ltd, Croydon, CR0 4YY
16/09/2026
14963421-0006